THE

MATHEMATICAL CAREER

OF PIERRE DE

FERMAT

1601–1665

Michael Sean Mahoney

Second Edition

Princeton University Press
Princeton, New Jersey

Published by Princeton University Press, Princeton, New Jersey
In the United Kingdom, Princeton University Press, Chichester, West
Sussex

Library of Congress Cataloging-in-Publication Data

Mahoney, Michael S. (Michael Sean)
   The mathematical career of Pierre de Fermat, 1601–1665 / Michael
Sean Mahoney)
      p. cm.
   Includes bibliographical references and index.
   ISBN 0-691-03666-7
      1. Fermat, Pierre de, 1601–1665. 2. Mathematicians—France—
Biography. I. Title.
QA29.F45M33 1994
510′.92—dc20
   [B]   94-1553

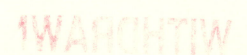

Princeton University Press books are printed on acid-free paper and
meet the guidelines for permanence and durability of the Committee
on Production Guidelines for Book Longevity of the Council on
Library Resources

First printing of the second edition, in paperback format, 1994

Printed in the United States of America

10   9   8   7   6   5   4   3   2   1

FOR JEAN

CONTENTS

PREFACE (1994)

Revising a work written more than twenty-five years ago at the start of my academic career has posed the problem of how far to go in reshaping it. Since its appearance, history of science, and with it history of mathematics, has taken an ever more critical and sophisticated approach to its subject, insisting on the inherently social nature of science and mathematics and seeking interests where we once perceived progress. At the time I wrote *Fermat*, I consciously sought to replace an older, positivist historiography with a new, contextually sensitive approach, inspired by my masters and then senior colleagues, Thomas S. Kuhn and Charles C. Gillispie, I prided myself in having kept the work of Fermat firmly rooted in the seventeenth century, hewing closely to his notation, taking his problems and solutions in the terms in which he cast them and in which his contemporaries understood them, and in general working his material with his tools as found in the books he read and the letters of people with whom he corresponded.

Nonetheless, as I reread the text, I found a younger self still taking much of symbolic algebra as an inherently progressive idea whose time had come and looking for it to assume familiar form, rather than inquiring why it took whatever shape it did at the time and since. It has been tempting to grab the pen from the hand of that neophyte and set him straight. I have refrained from doing so. It goes without saying that with thirty years' experience I would now write a very different book from the one I wrote after five years in the field. But I am not writing that book, and I shall not take it from the person who did. The revisions consist of corrections of errors indicated by reviewers and friends (sometimes the same people, sometimes not) and inclusion of results from more recent literature and my own subsequent research where they seemed to bear directly on the story my younger self was trying to tell.

INTRODUCTION

The chapters that follow strive toward two goals, one definite and the other open-ended. Their definite goal is to give as full an account as possible of the life and mathematical career of Pierre de Fermat. That goal requires little apology or explanation. In the history of mathematics Fermat enjoys a reputation for genius shared only by such figures as Archimedes, Newton, Euler, and Gauss. One of the inventors of analytic geometry, he laid the technical foundations of differential and integral calculus, established (together with Blaise Pascal) the conceptual guidelines of the theory of probability, and created modern number theory. Today his name graces several mathematical theorems and equations, one of the famous unsolved problems of mathematics, and a physical principle still employed in optics. And yet there exists nothing akin to a full biography of the man. This study aims to fill that lacuna in the history of the human mind.

The salient feature of Fermat's career in mathematics dictates the nature of such a biography. Fermat was a lone wolf. His original contributions to mathematics, many of them without precedent and without peer at the time, were the product of his creative imagination alone. He had no collaborators, but only correspondents to whom he reported his achievements and with whom he exchanged problems. He set his own directions of research. In some areas, they coincided with those of his contemporaries; in others, they did not. With few exceptions, the coincidence or divergence had no effect on the development of his ideas (though failure to interest others in his number theory distressed him personally). Fermat marched to his own drummer.

On the one hand, then, his biography neither can nor should be a "life and times." Rather, it must trace the odyssey of a single mind in quest of self-determined goals. Ultimately, Fermat is the only source of information about Fermat. More often than not, he is an uncooperative source. Secretive and taciturn, he did not like to talk about himself and was loath to reveal too much about his thinking. Hence, his biography perforce involves an intensive analysis of his mathematical writings with the aim of stripping away specific procedures and solutions to uncover fundamental underlying concepts and habits of thought. In a quasi-Platonic sense, mathematics is the bond that links Fermat and the modern biographer. Knowledge of and sensitivity to mathematics constitutes the biographer's most important tool for understanding Fermat and presenting his thought. For that reason, much of this text is mathematical in style and content; it often becomes technical, and it frequently deals with unfamiliar problems and ways of solving them. This approach may place something of a burden on the reader (as it has on the author), but it is unavoidable and indispensable to any biography of Fermat, especially given the relative isolation in which he worked.

On the other hand, time and place define the two-dimensional matrix of history. Fermat was not any mathematician anywhere at any time. He was a French mathematician of the first two-thirds of the seventeenth century. His thought, however original or novel, operated within a range of possibilities limited by that time and that place. His odyssey had its boundaries; his drummer beat to a tune of the times. Hence, any analysis of the structure of his thought requires an historical context to guide its direction and suggest its categories. This biography also strives, then, to fix the background against which Fermat worked and to locate the inspirational sources of his goals and methods. That background, which ties Fermat's work together and makes it a unitary phenomenon in time, is complex and to a surprising extent not yet part of the secondary literature in the history of mathematics. Many of its fine details would overburden an already long and involved biography, and therefore they have been left in Part I of the author's dissertation, "The Royal Road" (Princeton, 1967), which in some ways forms the *prolegomena* to what follows. The details essential to an historical understanding of Fermat do, however, form an integral part of the biography below; they are as indispensable to it as the mathematical analysis.

If the nature of Fermat's career and his extant works, combined with the historical axiom just discussed, prescribe the path toward the first of the goals of this biography, they also suggest how the first may serve the second. For the axiom has a converse. Just as it requires that Fermat's career be set in an historical context, it also implies that Fermat's career sheds light on that context. A better understanding of Fermat means a better understanding of seventeenth-century mathematics. The converse is far from trivial. If the following biography has succeeded, through the special nature of its subject, in uncovering the concepts and habits of thought fundamental to Fermat's mathematics, it should have also revealed some fundamental aspects of mathematical thought at a time and place in which the discipline underwent truly revolutionary changes. To the extent that Fermat's contemporaries and successors accepted or rejected his work, their reaction says something about them. The second goal of the following chapters, then, is to make a contribution to a deeper understanding of mathematics in the seventeenth century, both directly by revealing the thought of one of its foremost practitioners and indirectly by analyzing contemporary reaction to it. The goal is open-ended because no one man could embody the variety of mathematical thought at the time. If, however, this biography of Fermat suggests an approach to that goal, it will have accomplished its aim.

Fermat's peculiar career not only dictates the general nature of his biography, it also suggests its more specific form. First of all, not enough is known about his personal life and professional career to enable the biographer to link them to the progress of Fermat's mathematical thought in any but the grossest manner. What little is known, however, reveals something of Fermat's personality and so provides hints about his motivation to pursue mathematics and the nature of his commitment to the subject. By lending insight into his mathematical behavior *vis à vis* his contemporaries, the material is best understood in reference to those contemporaries. Hence, Chapter I includes a brief discussion of mathematics and its practitioners in the late sixteenth and early seventeenth centuries.

As Chapter I will show, not all of Fermat's contemporaries had operative influence on him. Not a professional discipline at the time, mathematics was divided into different "schools." Fermat belonged to the school of François Viète, whose works constituted his main source of inspiration and whose goals directed his research. Chapter II provides a summary of the main features of Viète's approach to mathematics and then surveys Fermat's mathematical career in that context.

Although the influence of Viète and his school explains the primary characteristics of Fermat's behavior as a mathematician, ultimately one must turn to intensive analysis of Fermat's works themselves to discover the bases of his mathematical thought. That is the purpose of Chapters III through VI. They are simultaneously chronological and topical in arrangement. That they can be so reflects another strange aspect of Fermat's career. Except for number theory, on which he worked steadily throughout his life, different subjects received his attention at different times. His restoration of Apollonius' *Plane Loci* and the system of analytic geometry that grew out of it (Chapter III) occupied him during the period 1629–1636; thereafter he never returned to the subject. The method of maxima and minima, together with the method of tangents derived from it, first emerged from his studies in 1629, but then really occupied the focus of his attention during the period 1637–1643 (Chapter IV). Again, though Fermat established some basic results in quadrature early in his career, especially around 1636, his most brilliant achievements stem from concentrated research carried out between 1643 and 1657 (Chapter V). Chapter VI breaks the chronological pattern. It deals with Fermat's number theory, which spans his career and unites many of the conceptual themes that emerge separately in other areas. It thereby provides an opportunity to sum up Fermat's career.

The major problem encountered in setting out Fermat's mathematics has been that of notation. Any historian of mathematics conscious of the perils and pitfalls of Whig history quickly discovers that the translation of past mathematics into modern symbolism and terminology represents the greatest danger of all. The symbols and terms of modern mathematics are the bearers of its concepts and methods. Their application to historical material always involves the risk of imposing on that material a content it does not in fact possess. By the same token, the purpose of mathematical symbolism—and its foremost virtue—is to lay bare the basic structure of concepts and methods. Hence, judiciously applied, it should serve historical analysis by enabling one to cut through to the core of past mathematics without introducing anachronisms. Moreover, its use where feasible lessens the demands on the reader that he simultaneously accustom himself to unfamiliar symbolism and terminology and to unfamiliar concepts and techniques. The presentation below reflects a compromise between these two sides of the question of notation. Where modern symbolism would block historical insight, it turns to the language of the original sources. At certain points, one must simply be ready to read mathematics as men such as Fermat and Viète wrote it. Where translation into modern symbolism promises to foster insight into mathematical structure, or where it can ease matters for the reader without distorting the information it is meant to convey, it has been freely used. Where the choice involves a borderline decision, the issue is usually discussed in a footnote. Clarity and consistency have been the goals; in most cases it has been possible to achieve them.

Although this biography strives for a total picture of Fermat's mathematics, it does not attempt encyclopedic coverage of his writings. It delves, often quite deeply, into those works that offer insight into the essence of Fermat's mathematical thought and into his behavior as a practicing mathematician. Some few works provide no insight more profound than that Fermat, at some point in his career, tried his hand at the subject. That is, they round out the picture of his overall career. A few other works simply repeat what is contained in his major treatises or elucidate it through further concrete examples. Such works are mentioned only in passing and in the conspectus of Fermat's mathematics in Appendix II, but they do not receive detailed exposition. In addition, Fermat's occasional efforts in the realm of mechanics and optics do not really afford insight into the development of his mathematics. The mechanics is, frankly, pedestrian even for its times, and the optics has already undergone a thorough investigation at the hands of A. I. Sabra in his *Theories of Light from Descartes to Newton* (London, 1967). Finally, Fermat's exchange with Pascal on the subject of probability does not, for all its inherent interest, directly involve Fermat's mathematics. Although any full biography must include some discussion of this material, the present biography does so as an appendix (Appendix I).

ACKNOWLEDGMENTS

This book has been long a-borning and, consequently, has had many intellectual and spiritual midwives to whom it owes a debt of gratitude. Foremost among them are Professors Thomas S. Kuhn and Charles C. Gillispie, each my threefold teacher, friend, and colleague in the Program in History and Philosophy of Science at Princeton. Having guided the project to first fruition in my doctoral dissertation, Professor Kuhn then read, with the conscientiousness and critical acumen peculiar to him, the extensive additions and revisions produced by subsequent research. His continuing interest in my work made him not only a superb critic of written drafts but also an understanding and responsive audience for ideas only half-articulated. In a project fraught with ramifications, I could always count on him to remind me of the main branches. Professor Gillispie first emboldened me to undertake a biography of Fermat and has been a constant source of encouragement during the years it has taken to complete it. He has lent valuable criticism to several of the chapters. In addition, as successive Directors of the Program, both men did much to afford me opportunities and resources to carry out my research.

The research and writing for three of the chapters of the book took place during the academic year 1969–1970 at the Institut für Geschichte der Naturwissenschaften der Universität München, where I spent my tenure as an NSF–NATO Postdoctoral Fellow. The Director of the Institut, Prof. Dr. Helmuth Gericke, and his colleagues not only provided me with all the material needs for my research, but, more importantly, offered me the sense of belonging among them. Several sessions of their staff colloquium were devoted to hearing my work in progress, and their patience, interest, and constructive comments did much to solidify my thinking on many issues. Besides Prof. Gericke, three others of my German colleagues deserve special thanks. Dr. Ivo Schneider became my constant friend and companion during the year; perhaps more than any other person, he watched and heard this biography progress from crude beginnings to a finished product. His sincere interest, his considerable expertise in the history of mathematics, and his steady willingness to listen and, when needs be, argue were of great value to me, and I am deeply indebted. Prof. Dr. Joseph E. Hofmann, who, as the footnotes will show, has done extensive work on Fermat, has graciously and unselfishly supported my research from its very beginnings. He is a giant on whose shoulders I have frequently had to stand. Finally, it is fitting that this book should have reached near-completion in Munich in the presence of the man under whom I began the study of the history of mathematics over a decade ago. Prof. Dr. Kurt Vogel, Director–Emeritus of the Munich Institute, befriended me as a young graduate student on a DAAD Fellowship to Munich during the years 1960–1962. Not only for all he taught me, but also for his kindness, his interest in a young career, and his unfailing good humor, I shall remain forever grateful.

Back in Princeton, I have benefited immensely merely from being in contact with all my colleagues in the Department of History and the Program in History and Philosophy of Science. Among them in particular, Professor David D. Bien (now at the University of Michigan) gave me considerable help in probing aspects of Fermat's legal career, especially in regard to the nature of French *parlements* in the seventeenth century, and Professors Theodore K. Rabb and Jerrold E. Seigel frequently helped me to orient my thinking on the cultural and intellectual history of early modern Europe. Several ideas resulted from provocative discussions with Professor Albert W. Tucker of the Princeton Department of Mathematics with whom it was my good fortune to teach a course a few years ago.

Among my students, I owe particular thanks to Mr. Philip S. Kitcher, who read the penultimate draft of the book in its entirety and offered many helpful comments; to Miss Toby A. Appel, whose investigation of the Wallis–Fermat controversy was of considerable help to me; and to Mr. Jed Z. Buchwald and Miss Patricia West, who as my research assistants contributed to this book in innumerable ways.

Several institutions have helped to make this work possible. The National Science Foundation granted me a Graduate Fellowship in 1964–1965 and a NATO Postdoctoral Fellowship in 1969–1970 and thus provided crucial financial support at the very beginning and toward the very end of the project. Princeton University in turn allowed me sabbatical and leave of absence during 1969–1970 in order to take up the second NSF fellowship. The Department of History at Princeton generously supported two summers of research on this project from funds at its disposal. I have already mentioned my debt to the Institut für Geschichte der Naturwissenschaften in Munich.

The preparation of the final version of the book was considerably eased by the willingness of the Princeton University Press to read an incomplete draft. To Mr. John Hannon, the Science Editor, and to Mrs. Gail Filion, the copy editor for the book, I wish to signal my appreciation for their help. At the final stage, Professors Dirk J. Struik and B. L. van der Waerden read the manuscript in its entirety and offered both encouragement and helpful comments.

As every married author knows, the ultimate debt of gratitude goes to that one person without whose support and devotion the whole endeavor would lose much of its meaning. I shall not attempt here to articulate what I owe to her; it is too personal, and she will understand.

Michael S. Mahoney

Princeton, New Jersey, 17 April 1972

BIBLIOGRAPHICAL
ABBREVIATIONS

FO   *Oeuvres de Fermat*, ed. Charles Henry and Paul Tannery, 4 vols., Paris, 1891–1912. *Supplément*, ed. Cornelis de Waard, Paris, 1922. Reference to this edition takes the form FO. (volume no. or *Suppl.*). (page no.), as e.g. FO.II.256 for vol.II, p. 256.

AT   *Oeuvres de Descartes*, ed. Charles Adam and Paul Tannery, 12 vols., Paris, 1897–1913. Citations follow the model of FO.

CM   *Correspondance de Mersenne*, ed. Cornelis de Waard, 11 vols. thus far, Paris, 1945 –. Citations follow the model of FO.

Alquié   *Descartes, Oeuvres philosophiques*, ed. Ferdinand Alquié, 2 vols. thus far, Paris, 1963 –. Citations follow the model of FO.

THE MATHEMATICAL CAREER OF PIERRE DE FERMAT

CHAPTER I

The Personal Touch

Many will pass by, and knowledge will increase.

Francis Bacon[1]

I. MATHEMATICS IN 1620

In Eric Temple Bell's royal house of mathematics, Pierre de Fermat stands as the "Prince of Amateurs," just behind the Crown Prince, Carl Friedrich Gauss.[2] Yet, Julian Lowell Coolidge purposely omitted Fermat from his *Mathematics of Great Amateurs*, "because he was so really great that he should count as a professional."[3] Both men, of course, meant only to honor Fermat's mathematical genius: Bell considered him an "amateur" because he achieved his stunning results in the spare time left him from his professional career as lawyer and jurist; Coolidge thought those same results so important and sophisticated that their author merited the title "professional." In one sense, then, both men concur, despite their apparent disagreement. If, however, one seeks to probe beyond Fermat's genius to the patterns of mathematical behavior through which it was expressed, then their difference of terminology focuses on an important point. Fermat's profession, in the strict sense of the word, was law. He pursued mathematics as a hobby. But he also pursued mathematics at a time when it was not a profession, at a time when "amateur" and "professional" could not be meaningfully applied to mathematics.

[1] This form of the passage that Bacon took from the Book of Daniel (XII, 4: Plurimi pertransibunt, et multiplex erit scientia) appeared as motto on the frontispiece of the first edition of the *Novum Organum* (London, 1620). Fermat quoted it several times in his works and correspondence throughout his life and seems to have treated it as a *leitmotiv* for his own career in science. Its original form is perhaps an even better *leitmotiv* for the present chapter.

[2] *Men of Mathematics* (New York, 1937), Chap. IV.

[3] *The Mathematics of Great Amateurs* (Oxford, 1949; repr., New York, 1963), p. vi.

Perhaps only at the time of Fermat's death in 1665, if indeed then, can one find the elements of an emerging profession of mathematics. During his early life and career, at the time that he received his education and formed his basic habits as a mathematician, one is hard pressed to find even a single, unified discipline of mathematics. Throughout the sixteenth and early seventeenth centuries, mathematics meant many different things to many different people. Various sixteenth-century treatises discussing mathematics as a whole show that there was widespread difference of opinion on what mathematics is or should be, on the end to which it should be pursued, on the problems to be investigated, on the methods to be used in solving them, and on standards of achievement.[4] In view of such basic disagreement, it is not surprising that practitioners of the time also disagreed on what constituted "correct" behavior on the part of the mathematician. In short, not only were there no professional mathematicians in Fermat's day but there were also no professional standards to which an "amateur" might strive to conform. Fermat's mathematical behavior stemmed in part from this variegated state of mathematics. To understand him, one must have some picture of the varied context in which he operated.[5]

For all the variety among its practitioners, mathematics in this period was no Tower of Babel. With due allowance for exceptions, one may sort those practitioners into six broad categories: the classical geometers, the cossist algebraists, the applied mathematicians, the mystics, the artists and artisans, and the analysts. Although only one or two of these categories constitutes what one might call a "school" of mathematics, and although the work of many individuals falls into several categories, nonetheless each category distinguishes itself from the others by characteristic attitudes toward the nature and purpose of mathematics, its problems and methods of solution. Each category has a distinctive style, and the different styles often conflict in essential ways.

The classical geometrical tradition was an offshoot of Renaissance humanism and largely shared its Italian and French locale. Its foremost sixteenth-century representative was Federigo Commandino (1509-1575), who single-handedly prepared Latin translations of Euclid, Apollonius, Archimedes,

---

[4]Contrast, for example, Federigo Commandino's *De disciplinis mathematicis brevis dissertatio* (Leiden, 1647), John Dee's "Mathematicall Preface" to the 1571 Billingsley edition of Euclid's *Elements*, and Petrus Ramus' *Scholarum mathematicarum libri unus et triginta* (Basel, 1569; often republished).

[5]In gaining an overview of the subject, one is still best served by H. G. Zeuthen's *Geschichte der Mathematik im 16. und 17. Jahrhundert* (Leipzig, 1903; repr., New York/Stuttgart, 1966); though some of Zeuthen's material has been emended and expanded in the intervening secondary literature, his book as a whole has yet to be replaced. J. E. Hofmann's *Geschichte der Mathematik* (3 vols., Berlin, 1957; vol. 1, 2nd ed., 1963) provides a useful outline of developments together with the most complete published guide to the primary literature.

Aristarchus, Autolycus, Hero, Pappus, Ptolemy, and Serenos.[6] His translations enjoyed widespread use not only because of his superb command of Greek and sense of Latin style but also because his own talents as a mathematician enabled him to unravel passages in the original Greek that centuries of copyists had blurred. Each translation was accompanied by extensive commentaries elucidating vague passages, filling in mathematical details, and drawing comparisons and contrasts within the text itself and with reference to other treatises. Commandino's work as a translator set the style for others who participated in the recovery and circulation of classical Greek mathematics, for example, Claude Bachet de Meziriac's (1581–1638) Latin edition of Diophantus of Alexandria's *Arithmetic*.[7] By the early seventeenth century, their combined efforts had made almost all of the extant corpus of Greek mathematics easily available to those interested in the subject. The result, evident from the middle of the sixteenth century on, was a rapid and marked rise in the level of sophistication of European mathematics.

That increased sophistication becomes immediately apparent when men such as Commandino turn from the translation of extant works to the restoration of lost material or to the extension of Greek results and methods. The Italian translator's own treatise *On the Center of Gravity of Solids* (Bologna, 1565), which derives results assumed in Archimedes' *On Floating Bodies* but nowhere explicitly determined in the Greek's works, amply attests to his command of the methods of quadrature employed by Archimedes in the *Quadrature of the Parabola*. It set the standard for a spate of similar work by such writers as Gianalfonso Borelli (1608–1679) and Vincenzo Viviani (1622–1703). Even Galileo participated in this task of reconstruction by devoting an appendix of the *Two New Sciences* to an extension of Commandino's results on centers of gravity of solids.

The classical tradition that emerged from these translations and reconstructions shared some of the fundamental limits of the humanist movement of which it was a part. Like most humanists, the classical geometers tended to accept their Greek forebears as an exclusive model for conduct and hence to ignore developments that could not claim Greek paternity. The works of Commandino, or even those of Galileo, give no hint of the algebraic research that flourished simultaneously in Italy. Schooled in classical Greek philosophy, the geometers accepted the canons of Plato's and Aristotle's philosophies of mathematics: e.g. the strict separation of arithmetic and geometry, the demonstrative superiority of geometry, the need for absolute syllogistic

[6]There is yet no modern, scholarly account of the life and work of this amazing Renaissance figure, and one must continue to rely on Bernardino Baldi's 1587 biography published in the *Giornale de' letterati d'Italia*, vol. 19 (Venice, 1714).

[7]Paris, 1621. The Latin text of this Greek–Latin edition was based in part on the translation by Wilhelm Holtzmann (Xylander), which appeared in Basel in 1575. The Bachet edition served as Fermat's source for Diophantus and was republished with Fermat's commentaries by his son, Clément-Samuel, in Toulouse in 1670.

rigor. For Galileo, the book of the world was written in geometrical figures, and hence it was to geometry in the Greek style that he reduced the science of motion. It was Archimedes the geometer who bore the epithets "divine" and "superhuman." Bonaventura Cavalieri's *Geometry by Indivisibles* (Bologna, 1635) displays the hold of the Greek model over the classical geometer even when he cautiously ventured beyond the pale. Within the new classical tradition, research activity was directed toward the eminently humanistic goal of restoration of classical Greek texts in their original form; of translation into a refined, classical Latin; of exegesis and commentary; and ultimately of emulation. The geometers placed greater emphasis on style of presentation than on novelty of results, and purity of style excluded the adoption of much that was new and valuable. With these limitations, the new classical tradition shared the locus of its parent movement; it was largely pursued in humanist academies and circles well outside the scholastic university, and its pursuit depended on the support of wealthy patrons. By and large, it drew its practitioners from the same ranks of society as did the parent movement.

The predominantly German and Italian tradition of cossist algebra, like the classical tradition, operated outside the precincts of the university but, unlike the classical tradition, placed much less emphasis on style than on innovation in the solution of problems. The body of mathematical knowledge with which its practitioners concerned themselves had its roots in a lively and developing Arabic algebraic enterprise that in turn stretched back to the mathematics of the Babylonians.[8] It had followed several paths into Latin European culture. Leonardo Fibonacci of Pisa, for example, gathered the material for his *Liber abacci, Liber quadratorum,* and *Flos* during his merchant travels throughout the Mediterranean littoral.[9] Fibonacci's works in turn constituted an important source for the eminent cossist Fra Luca Pacioli (1445–1514). The basic source and prototype for Latin algebra, however, like that of Latin science in general, entered the culture through Spain. Muhammad b. Mûsâ al-Hwârizmî's *Kitâb al-jabr w'al-muqabâla (Book of Opposition and Restoration),* translated into Latin by Robert of Chester in the twelfth century,[10] provided not only a model for the cossist enterprise but also the names of the enterprise. "Algebra" is simply the Latinized form of *al-jabr,* and the Latin and Italian translations of al-Hwârizmî's technical term for the first power of the unknown quantity (*shai'* = Lat. *res* = Ital. *cosa* = Engl. "thing") lent the art the subtitles *ars rei et census*[11] and *l'arte della cosa,*

---

[8] See Roshdi Rashed, *Entre arithmétique et algèbre: Recherches sur l'histoire des mathématique arabes* (Paris: Les Belles Lettres, 1984).

[9] See the account by Kurt Vogel, "Fibonacci," *Dictionary of Scientific Biography,* Vol. IV (New York, 1971), pp. 604–613.

[10] Ed. L. C. Karpinski, New York, 1915. An Arabic-English edition of the original was published by F. Rosen (London, 1831).

[11] *census* = Arab. *mâl* = "wealth"; al-Hwârizmî used the term to denote the square of the unknown.

respectively. From the Italian form came the name applied to practitioners of the art, the *cossists*.

Cossist algebra constituted essentially a sophisticated form of arithmetical problem-solving. Often it is difficult when reading a cossist text to discern where computational arithmetic leaves off and algebra begins. The vagueness of the boundary derives not only from the mathematical nature of the art but also from the fact that both Fibonacci and al-Ḥwârizmî introduced Latin Europe to the Hindu–Arabic system of numeration and computation.[12] Since their algebras employed the new system, they often circulated in company with a treatise describing it; indeed, Leonardo's *Liber abacci* begins with an exposition of Hindu–Arabic reckoning.

Although cossist algebra employed an abbreviatory symbolism, its characteristic format of presentation consisted of a large collection of representative problems, the specific solutions of which were meant to serve as paradigms for similar problems. Despite increased sophistication in the pedagogical arrangement and systematization of the art in such works as Rudolff's *Coss* (Strassburg, 1525),[13] Cardano's *Ars magna* (Nuremberg, 1545), or Clavius' *Algebra* (Rome, 1608), this basic format changed little. Recurrent paradigms might be translated into operational recipes, but they seldom became formulas in the true sense of the word. Techniques of solution might, in keeping with al-Ḥwârizmî's prototype, receive theoretical justification through appeal to theorems from Books II and VI of Euclid's *Elements*, but little research was directed at amplifying the theoretical foundations of the art. Cossist algebra remained from beginning to end a problem-solving approach to mathematics.

By its very nature, cossist algebra had no need of a scholarly ambience to guarantee its continued life. As the mundane context of the problems found in the texts amply attests, it served a recognized and appreciated role in the mercantile community, a role enhanced by the art's intimate ties to an efficient new system of numeration and computation. Together with the art of Hindu–Arabic reckoning, it provided the means to carry out the increasingly complex trade and exchange problems of a burgeoning European (especially Italian) commercial class. Though many texts contain problems obviously without real application, cossist writers of the sixteenth century continued to proclaim the practical importance of their subject. The sensitivity of these writers to changes in prices and exchange rates often makes their texts valuable sources of information on contemporary economic conditions.[14] More-

---

[12] *Algorismus* (as in Johannes de Sacroboso's *Algorismus vulgaris*) is the Latinized form of al-Khwârizmî. Cf. Kurt Vogel, *Mohammed ibn Musa Alchwarizmi's Algorismus, Das Früheste Lehrbuch zum Rechnen mit indischen Ziffern* (Aalen, 1963). On medieval mathematics in general, see M. S. Mahoney, "Mathematics," in *Science in the Middle Ages*, ed. D. C. Lindberg (Chicago, 1978), Chap. 5.

[13] The full title is revealing: *Behend und Hubsch Rechnung durch die kunstreichen regeln Algebre so gemeinicklich die Coss genennt werden.*

[14] Appendix 2 of Herbert Hunger and Kurt Vogel, "Ein byzantinisches Rechenbuch des

over, the non-applicable problems quite likely represent a body of mathematical lore with which the practical problem-solvers entertained themselves in a form of algebraic "busman's holiday."

Taken at face value, the appeals to practical utility and the prevalence of cossist texts—they constitute the majority of mathematical works published in the fifteenth and sixteenth centuries—may mask more than they reveal about the behavior of the cossists. Despite the claims found on many title pages, cossist algebras constituted a body of literature for specialists. They did not fill an educational role for large numbers of businessmen, nor in all probability were they really designed to do so. As the German title of *Rechenmeister* suggests, the more complex and sophisticated problems of the mercantile community required the services of men skilled in the art of solving them. Niccolò Tartaglia (*ca.* 1500-1557), for example, gained at least part of his income from performing such services, and he was only one of many. As men deriving a livelihood from their mathematics, the algebraists operated and published under a strange combination of pressures. On the one hand, publication represented a form of advertising. The cossist strove to display his possession and command of rapid and effective means of problem-solving by setting out his procedures in as straightforward and easily understandable a manner as possible. He often did so at the expense of style. Few, if any, cossist works even approach the canons of classical Greek mathematical exposition, and no other field of mathematics at the time can boast of as heavy a concentration of works written in the vernacular. Moreover, success in competition with other cossists demanded an element of novelty, i.e. new and more efficient means of solving problems. On the other hand, only other cossists could really appreciate the strengths and subtleties of a colleague's work. To that extent, the cossists wrote not so much for a broad public as for each other. Yet, that public conditioned the manner in which they could write for each other. A cossist's ability to solve problems his competitors could not solve gave him an advantage he was loath to surrender through publication. The famous story of Tartaglia and Cardano illustrates the effect of competition on publication and is worth a short digression.[15]

Having arrived at a means of solving cubic equations (the solution of quadratic equations had been known since antiquity), the Bolognese algebraist Scipione del Ferro (*ca.* 1456-1526) took care to keep his technique a secret, revealing it only to close friends and pupils. Why he did so is clear in light of the common practice among cossists of the day to compete against one another in a form of mathematical joust. The object was to solve the challenge problems posed by one's opponent while in turn posing problems that he could not solve. The winner enjoyed as a reward not only the immediate

---

15. Jahrhunderts," *Denkschriften der österr. Akad. d. Wiss., Phil.-hist. Kl.*, Vol. 78, No. 2 (Vienna, 1963), provides an excellent example taken from another setting.

[15]Short account, with bibliographical references, in Hofmann, *Gesch. d. Math.*, I, pp. 133-134.

monetary prize (plus side bet) but also fame with its indirect but no less tangible benefits. Therefore, someone able to solve cubic equations would have a distinct advantage over any competitor. It was in the hope of using that advantage that one of Ferro's students, A. M. Fior, challenged Tartaglia in 1535. Shortly before the actual day of competition, Tartaglia worked out his own solution of the cubic equation, which, unbeknown to him, was essentially the same as that of Ferro. Hence, he was able to make short work of Fior's challenges. Unfortunately for Fior, cubic equations were about the only thing he could solve, and Tartaglia quickly disposed of him. Hearing of Tartaglia's victory, Cardano pressed him to reveal his solution, and Tartaglia did so, but only after swearing Cardano to absolute secrecy. When, however, Cardano learned through independent means of the details of Ferro's solution, he immediately assumed it to be the source of Tartaglia's and considered himself freed of any obligation to remain silent. He broke his promise to Tartaglia by publishing the solution in his *Ars magna* in 1545. Moreover, Cardano's own student, Luigi Ferrari, needed only the hint of Tartaglia's solution of the cubic equation to extend it to a solution of the quartic, and the latter solution was also included in the *Ars magna*. Even though Cardano gave Tartaglia full credit for his invention, Tartaglia was furious, broke off all relations with Cardano, and never forgave his breach of trust. For a time, the exchange of angry charges and countercharges was a *cause célèbre* in Italian intellectual circles.[16]

To the modern scholar imbued with the notion that, as R. K. Merton has put it, "an idea is not really yours until you give it away" (i.e. through publication),[17] Tartaglia's attitude seems strange. Yet, in sixteenth-century Italy, where mathematics was a competitive business among the cossists, the attitude is more than reasonable. And it becomes even more understandable when one considers that the very notion of individual ownership of ideas was itself novel and strange. Reflected in Galileo's suit of Baldassare Capra for plagiarism, for example, is not only the real diminution in financial return that Capra's theft of Galileo's proportional compasses would involve, but also the confusion wrought by the new Renaissance sense of intellectual property rights.[18] In sum, the competitive nature of the cossist enterprise lent it both a style of exposition and an attitude toward publication all its own, one quite different from what eventually emerged at the end of the seventeenth century.

Although the members of the third group, the applied mathematicians, shared the practical, problem-solving orientation of the cossists, they seem to

---

[16] As Mario Gliozzi points out ("Cardano," *Dictionary of Scientific Biography*, III, p. 65), Tartaglia's subsequent vilification of Cardano, which circulated widely, probably contributed to the latter's poor reputation in succeeding times and proved a perhaps overly effective form of retribution.

[17] In a lecture delivered at Princeton University, Fall 1970.

[18] For details of the suit, which occurred in 1607, see L. Geymonat, *Galileo Galilei* (Engl. trans. S. Drake, New York, 1965), p. 26ff.

have taken a less competitive and hence less secretive approach to their enterprise. The applied mathematicians, who can be found throughout Europe in the sixteenth and seventeenth centuries but who were concentrated primarily in England and the Low Countries,[19] relied perhaps on a sense that possession of the appropriate mathematical tools brought the practitioner only halfway toward the solution of his problem. The invention of efficient navigational techniques, the establishment of accurate maps and charts, the design of harbors and fortifications, the planning of effective artillery placement, all required ingenuity that transcended mathematics. Even though, as in the case of the Belgian engineer Gemma Frisius (1508-1555), this attitude led to no less jealous protection of the practitioner's ideas and methods (through the imperial or royal *privilegium*, an early form of copyright and patent),[20] the applied mathematicians showed a greater willingness to publish their latest results and methods.

Their written works reveal a curious amalgam of the attitudes of the classicists and the cossists.[21] Much of their mathematical material, including both problems and methods of solution, derived from classical Greek sources, but those sources had lain in antiquity on the periphery of the Greek geometrical tradition. It was to Archimedes' *On Floating Bodies* and *Equilibrium of Planes* that they turned, rather than to his more straightforwardly mathematical treatises. Ptolemy's *Planisphere, Analemma*, and *Cosmography* took precedence over Apollonius' *Conics*, Menelaus' trigonometry and Theodosius' spherical geometry over Pappus' *Mathematical Collection*. Because, in the eyes of the applied mathematicians, the ancient material was so obviously incomplete and open to emendation and extension, it did not restrict the range of research problems, the means of their solution, or the style in which they were presented in their works. Even when a man such as Simon Stevin (1548-1620) strove to model his works on Euclid and Archimedes, he did so with a full sense of the originality and innovation contained in these works. Stevin did not hesitate to imitate the style of Greek geometry at the same time that he argued the superior virtues of Dutch over Latin and Greek as a scientific language.

In particular, the applied mathematicians did not let their admiration of Greek achievements blind them on the one hand to other, more recent forms of mathematics or on the other hand to the practical applications of whatever

[19] Italians such as Tartaglia, Giambattista Benedetti, and Guid' Ubaldo dal Monte constitute an outstanding exception, as does the Arsenal of Venice as a locus for such activities. Moreover, the widespread warfare of the period assured a fairly uniform distribution of military engineers throughout Europe.

[20] See Ivo Schneider, "Urheberrechtliche Sicherung im naturwissenschaftlichen Scrifttum des 16. Jarhunderts: Buchprivilegien bei Gemma Frisius (1508–1555)," *Aus dem Antiquariat* 30 (1974), A145–A152 (supplement to *Börsenblatt für den Deutschen Buchhandel* 43 [1974]).

[21] For England, see E.G.R. Taylor, *The Mathematical Practitioners of Tudor and Stuart England* (Cambridge, 1954) and J. A. Bennett, *The Divided Circle: A History of Instruments for Astronomy, Navigation and Surveying* (Oxford, 1987).

mathematics they pursued. The works of Ptolemy, Menelaus, and Theodosius served as only one source for the complete system of plane and spherical trigonometry developed by the applied mathematicians. With the Greek rudiments they freely mixed material taken from Arabic texts and applied algebra where feasible. And the system that emerged was essentially their own creation. Moreover, since the purpose and goal of mathematical research lay precisely in its practical applications, the applied mathematicians were more than willing to violate the canons of Greek mathematical philosophy in order to realize those applications. Trigonometrical theorems, for example, meant little to the navigator until translated into computational techniques for determining position and laying a course in degrees, minutes, and seconds with the aid of extensive numerical tables. And the preparation of such tables in turn required that one surrender geometrical exactitude for accurate arithmetical approximation and interpolation. Hence, in contrast to the canons of Greek mathematics, the navigator, the map-maker, the engineer of the sixteenth and seventeenth centuries, all treated arithmetic and geometry as realms with a common, open border. Like merchants, they measured the success of their enterprise by how often they fruitfully wandered back and forth over that border. If an essentially numerical enterprise such as algebra offered a passport into the realm of geometrical triangles, they accepted it and used it. The same Stevin who honored Archimedes' geometry also devised a system of decimal place-value fractions to speed and ease computation and did research in algebra.[22] John Napier and his followers may have conceived of logarithms in terms of the motions of points along lines, but Euclid was ignored as the ratios of those motions were computed in numbers and rounded off at the fifth or sixth decimal place. The goal was not theoretical rigor, but a sufficiently accurate technique for manipulating large numbers and trigonometrical functions.

The same practical spirit overrode the classical Greek distinction between the mathematical and the mechanical and fostered increasing research on mathematical instruments. Galileo's geometrical compass was only one of a myriad of similar proportional compasses, designed not only for the immediate task of reduction and increase of scale in technical drawings but also for an ever growing number of computational operations restricted only by the number of different scales that could be etched on the instrument's legs.[23] Within six years of the publication of Napier's *Description of the Wonderful Canon of Logarithms* (Edinburgh, 1614), and within three years of Briggs' *First One Thousand Logarithms* (London, 1617), Richard Gunter and Edmund Wingate had reduced logarithmic computation to a small instrument,

---

[22]Cf. *De Thiende* (*The Tenth*) (Leiden, 1585); *Appendice algebraice*, Leiden, 1594. Stevin's *Principal Works* have been appearing in a modern edition since 1955.

[23]See the recent study by Ivo Schneider, "Der Proportionalzirkel—Ein universelles Analogrecheninstrument der Vergangenheit," *Abhandlungen und Berichte des Deutschen Museums*, 38, Heft 2, 1970.

the first slide rule.[24] Dürer's "Melancholia," with its panoply of mathematical instruments, is but an early witness to the proliferation of such mechanical devices during the sixteenth and early seventeenth centuries; Jacques Besson's *Theater of Mathematical and Mechanical Instruments* (Lyon, 1579) attests further to their popularity and development. Rarely illustrated are the often even more sophisticated machines required to produce the instruments, and the mathematics on which the machines' design is based.[25] Taken as a whole, the instrumental aspect of applied mathematics represents a definite break with traditional methods and standards of solution, a break one finds in a different form in the work of Fermat and Descartes. Computation took the place of proof, and, within the applied mathematical tradition, at least, an operational approach to mathematics predominated.

The applied mathematicians shared this operational attitude not only with the cossists, with whom they also shared a sense of the practical utility of mathematics, but in addition with the mathematical magicians, of whom John Dee was perhaps the foremost sixteenth-century exponent. One has only to read Dee's "Mathematical Preface" to Henry Billingsley's 1571 edition of Euclid's *Elements* to discern his obvious ties to the applied mathematical tradition. The major portion of the preface is devoted to extensive discussion and praise of geodesy, geography, cosmography, perspective, mechanics, etc. Indeed, Dee's commitment to an operational view of mathematics leads him to present the earliest known operational concept of number, in which he takes the positive integers as given and sees the negative integers, fractions, and irrational surds as the products of the demand that the basic operations of arithmetic each produce numerical results.[26]

Yet, it is equally clear from the opening sections of the preface that mathematics had for Dee a meaning far more profound than it had for the applied mathematicians, the cossists, or even the classicists. None of them would have written that:[27]

All things (which from the very first originall being of thinges, have bene framed and made) do appeare to be Formed by the reason of numbers. For this was the principall example or patterne in the minde of the Creator. O comfortable allurement, O ravishing perswasion, to deale with a Science, whose Subiect, is so Auncient, so pure, so excellent, so surmounting all creatures, so used of the Almighty and incomprehensible wisdome of the Creator, in the distinct creation of all creatures: in all their distinct partes, properties, natures, and vertues, by order, and most absolute number,

---

[24] First published in their *Construction, description et usage de la règle de proportion* (Paris, 1624). F. Cajori's *History of the Slide Rule* (New York, 1909) remains the standard modern account of the subject.

[25] The Deutsches Museum in Munich holds a fine collection of both instruments and the machines used to make them.

[26] "Preface," unnumbered p. 5.

[27] *Ibid.*, unnumbered pp. 3–4.

brought, from Nothing, to the Formalitie of their being and state. By Numbers propertie therefore, of us, by all possible meanes, (to the perfection of the Science) learned, we may both winde and draw our selves into the inward and deep search and vew, of all creatures distinct vertues, natures, properties, and Formes: And also, farder, arise, clime, ascend, and mount up (with Speculative winges) in spirit, to behold in the Glas of Creation, the Forme of Formes, the Exemplar Number of all thinges Numerable: both visible and invisible: mortall and immortall, Corporall and Spirituall.

It requires the special talents of a Frances Yates to unravel the complex content of this passage. One need not unravel it, however, to see that Dee operated within a very special mathematical tradition.[28] Hermeticism, neo-Pythagoreanism and neo-Platonism, magical astrology, cabalism, and the Lullian search for the symbolic key to universal knowledge all formed an intricately interwoven backdrop against which the magician viewed mathematics. Ancient number theory flourished again, probing the secrets of the integers. The symbolic characteristics of cossist algebra took on new meaning, and the name "algebra" itself was thought to be derived from the Middle Ages' most famous alchemist, Jabir b. Hayyan.[29] The mysterious qualities of the five perfect solids assumed again deep metaphysical and cosmological significance. But, within the tradition itself, mathematical research *per se* surrendered priority to the mystical interpretation of results already achieved. Mathematics served as a tool for unlocking the secrets of ancient texts or of the universe surrounding man, but it did not serve as an end in itself. Although the magical tradition in sixteenth-century mathematics could and did offer inspiration to men such as Descartes and Kepler, it was not a tradition within which one operated primarily as a mathematician, nor was it one's mathematical work that formed the basis of one's reputation. For all his brilliant insights, John Dee could produce nothing of lasting value in mathematics.

The artists and artisans of the sixteenth century offered as a by-product of their endeavors the mathematical field of perspective and projective geometry. The subject was an outgrowth on the one hand of the optician's practical concerns with the design of lenses and theoretical inquiry into the nature of

---

[28] On Dee, see Nicholas H. Clulee, *John Dee's Natural Philosophy: Between Science and Religion* (London, 1988) and John L. Heilbron's introduction to *John Dee on Astronomy*, ed. Wayne Schumaker (Berkeley, 1978).

[29] I.e. the famous Geber; cf. Petrus Ramus, *Algebra*, ed. Schoner (Frankfurt, 1627): "Nomen Algebrae Syriacum putatur, significans artem et doctrinam hominis excellentis. Nam Geber Syris significat virum, idque nomen interdum est honoris, ut apud nos Magister aut Doctor. Etenim insignis mathematicus quidam fuisse fertur, qui suam Algebram Syriaca lingua perscriptam ad Alexandrum Magnum miserit, eamque nominaverit *Almucabalam*, hoc est, librum de rebus occultis, cuius doctrinam Algebram alii dicere maluerunt." The tie between the mathematical procedure al-H̲wârizmî called *al-muqabâla* and the Hebrew cabala existed, of course, only in Ramus' imagination and that of many others. Dee, by contrast, was too much of a scholar and Arabist not to know the precise etymology both of *algebra* and of *cabala*.

vision and on the other hand of the Renaissance artist's striving to present his audience with as faithful a reproduction of reality as possible.[30] Though its practitioners could look to a research tradition that stretched back to antiquity (Euclid and Ptolemy), they pursued their goals rather in the spirit of the applied mathematicians, feeling no compulsion to conform to ancient models.[31] Most important for the present discussion, however, they did not consider themselves mathematicians. Those who did were the exception rather than the rule. Nor, for that matter, were they, again with some exceptions, considered to be mathematicians by their contemporaries. One must look hard among mathematical texts of the sixteenth and seventeenth centuries to find even mention, much less discussion, of the work of, say, Alberti or Dürer. For all its brilliance, and for all its author's reputation as a mathematician in other areas, Girard Desargues' projective geometry provoked little reaction among contemporary mathematicians.[32] For the most part, and especially in the sixteenth century, the mathematical behavior of the artists and artisans was dictated by the non-mathematical groups to which they felt primary allegiance.

Finally, the analysts, latecomers to the mathematical scene, will form the subject of a more detailed examination in Chapter II, since it was to that group that Fermat belonged. At first predominantly French in makeup, the group shared with the classicists their deep regard for the ancient Greek sources and their desire to restore them at the same time that it shared with the cossists their commitment to algebra as a powerful problem-solving technique. Yet, it also shared with the applied mathematicians their sense that the Greek models represented only a beginning on which it was their task to build further, even if doing so required a revision of Greek mathematical canons. And it shared with the mystics their desire to devise a symbolic art of reasoning that would unite and open all of mathematics to the inquirer.

One explanation for the continued coexistence of the six different groups discussed above may lie in the fact that the university as the traditional seat of learning formed the locus for none of them. Despite the several reforms of the university curriculum in the sixteenth century, in particular the various *Rationes studiorum* of the Jesuits, mathematics retained its minor, basically propaedeutic place in the arts faculty.[33] As had been the case in the high

---

[30] See, for example, Samuel Y. Edgerton, Jr. *The Renaissance Rediscovery of Linear Perspective* (New York: Basic Books, 1975; Harper & Row, 1976). The elements of projective geometry growing out of the optical perspective tradition is best illustrated by the optical theorems of Kepler and Newton.

[31] Indeed, in the realm of cartography, the technical interests of the perspectivists and the applied mathematicians overlapped.

[32] Cf. René Taton, *L'Oeuvre mathématique de G. Desargues* (Paris, 1951).

[33] On the nature and development of the mathematical curriculum in the sixteenth and seventeenth centuries, see for Germany Sigmund Günther's magisterial *Geschichte des mathematischen Unterrichts in Deutschland bis zum Jahre 1525* (Berlin, 1887). For France, see François de Dainville, "L'Enseignement des mathématiques dans les Collèges Jésuites de France du XVIᵉ au XVIIIᵉ siècle," *Rev. hist. sci.* 7 (1954), pp. 6–21, 109–123; and Jean-Claude Margolin,

Middle Ages, university students of the sixteenth and early seventeenth centuries learned little more than the first six books of Euclid's *Elements* as preparation for reading the introductory sections of Ptolemy's *Almagest*. Only at a few German universities in the late fifteenth century could students hear lectures on the art of algebra or gain some training in advanced geometry and trigonometry.[34] The function of the arts faculty in readying students for one of the three higher faculties—law, medicine, or theology—did little to reinforce any attempt to strengthen the mathematical curriculum. Despite the importance of mathematics in the philosophy of science pursued at Padua, for example, little if any mathematics *per se* was produced there, nor did it have any more extensive part in the curriculum than elsewhere. Galileo's experience on the arts faculty there from 1592 to 1610 attests to mathematics' lack of status even at a university supposedly sensitive to its importance. And Galileo, one should recall, learned mathematics not at the University of Pisa, but from a private tutor, Ostilio Ricci.[35] When, moreover, Petrus Ramus sought to foster the study of mathematics, he could do so only by transforming his own chair of rhetoric at the Collège Royal into a chair of mathematics;[36] no French university followed his lead.

Only with the establishment of the Savile Chair at Oxford in 1619 and the Lucasian Chair at Cambridge in 1664 did the status of mathematics in the university change at all in the seventeenth century. As a student at Cambridge in the 1640s, Seth Ward could find no one among his masters who could explicate some elementary theorems of trigonometry.[37] His experience is fully representative. Although they were almost all university graduates, the major figures of seventeenth-century mathematics nonetheless received their mathematical training outside the walls of their *alma mater*. Descartes learned mathematics from Isaac Beeckman and Johann Faulhaber, van Schooten learned from Descartes, Christiaan Huygens from van Schooten. John Wallis was largely self-taught, as was Newton for all the influence of Barrow.[38]

"L'enseignement des mathématiques en France (1540–1570): Charles de Bovelles, Finé, Peletier, Ramus," in *French Renaissance Studies*, ed. Peter Sharratt (Edinburgh, 1976), 109–55.

[34] For example, MSS Dresden C 80, Munich Clm 4387, and Clm 26639 contain copies of the algebraic lectures delivered in 1467 and 1468 at the University of Erfurt by Gottfried Wolack de Bercka; the text was published by E. Wappler, "Zur Geschichte der Mathematik," *Zeitschrift f. Math. u. Phys.* 45(1900), Hist.-litt. Abtlg., pp. 47–56.

[35] Cf. Geymonat, *Galileo*, Chaps. I–II.

[36] Gilles Personne de Roberval held the Ramus Chair throughout the period of Fermat's mathematical career. By the terms of Ramus' will, the occupant had to defend the chair in open mathematical competition every three years. Roberval did so successfully from 1634 until his death in 1675. Cf. Léon Auger, *Gilles Personne de Roberval (1602-1675)* (Paris, 1962), p. 10. As Auger notes, the need to beat off competitors often led Roberval to withhold publication of his results.

[37] W. T. Costello, *The Scholastic Curriculum at Early Seventeenth-Century Cambridge* (Cambridge, Mass., 1958), pp. 102–103.

[38] M. S. Mahoney, "Barrow's Mathematics: Between Ancients and Moderns," in *Before Newton*, ed. Mordechai Feingold (Cambridge, 1990), Chap. 3.

A graduate of the University of Leipzig, one of the few institutions that had taught algebra in the late fifteenth century, Leibniz first began the serious study of mathematics with Huygens in Paris in 1672. Valerio, Torricelli, and Cavalieri took their lead from Galileo. So too Fermat first undertook the pursuit of mathematics after leaving the university, from the followers of François Viète in Bordeaux.

At the turn of the seventeenth century, then, mathematics comprised a hodge-podge of varied activities by various groups of people. In their interaction, these people could not always depend on much common ground. They could honestly disagree over what mathematics was or should be, over the purposes and goals toward which it should strive, over the problems to be investigated, over the methods to be applied in solving those problems, and over the standards by which achievement and distinction should be measured. The absence of any academically communicated, unifying discipline fostered continued fragmentation. Mathematicians even lacked a common name. The humanists called themselves *geometrae*, "geometers," and the continued dominance of the title *géomètre* in French today attests to the popularity of the designation. But Germans proudly bore the title of *Rechenmeister*; Stevin preferred his self-concocted title *wisconstler*; and "cossist" derives from contemporary usage, as does "algebraist." Fermat did use *geometrae* when referring to mathematicians at large, but preferred to be called an "analyst" (*analysta*) himself. Yet, on one thing these differently named men could agree: they were not *mathematici*. *Mathematicus* retained in the sixteenth and seventeenth centuries the meaning it had for the Middle Ages; it meant "astrologer" or "astronomer." That is the sense in which Copernicus spoke of mathematics being written for mathematicians. It was as astrologer and astronomer, and not as mathematician in the modern sense, that Kepler bore the title of "court mathematician" to the Emperor of the Holy Roman Empire. It was as astrologer and astronomer, and not as mathematician and mechanician, that Galileo occupied the same post in the Duchy of Tuscany. Indeed, it is only with that special meaning in mind that one can speak of "professional mathematicians" in the sixteenth and seventeenth centuries. Astrologers and astronomers formed the only body of practitioners of a mathematically related enterprise that could conceivably claim status as a profession.

For reasons that will become clear in Section III below, the fragmentation of the mathematical enterprise and the corollary lack of any mathematical profession meant that the idiosyncrasies of mathematicians played a particularly important role in their practice. Hence, in dealing with mathematicians of the sixteenth and seventeenth centuries, the historian must be especially sensitive to their personalities. Before attempting to elucidate this point, it would help to have an overview of the private life and professional career of the particular mathematician under study here.

14

II. FERMAT'S LIFE AND CAREER IN PARLEMENT

Fermat est Gascon, moi non.
René Descartes [39]

Descartes called him a "braggart," Pascal termed him "the greatest mathematician in all of Europe," Mersenne referred to him as "the learned councillor from Toulouse," and Wallis thought of him as "that damned Frenchman." His parents christened him Pierre Fermat, after his uncle. He was baptized (and most probably born) on 20 August 1601 to Dominique Fermat, a prosperous leather merchant and bourgeois second consul of Beaumont-de-Lomagne (now in Tarn et Garonne), and his wife Claire, *née* de Long, the daughter of a prominent *famille de robe*. According to Charles Taupiac, who as mayor of Beaumont in the late 1870's first unearthed in the town archives evidence of Fermat's birth and early life,[40] Fermat had a brother, Clément, and two sisters, Louise and Marie. The lack of any evidence to the contrary led Taupiac to conjecture that Fermat's family remained resident in Beaumont throughout his childhood and that he therefore probably received his primary and secondary education at the monastery of Grandselve, operated by the Cordeliers (Franciscans), before matriculating at the University of Toulouse. Various documents uncovered and published by Taupiac, and later by Blaquière and Caillet, suggest that the Fermat family was gaining steadily in wealth and status during Pierre's early life. In view of his father's mercantile wealth and his mother's family background, Fermat's choice of a legal career seems both natural and typical of his time; it represented the most common path of upward social mobility and of the translation of financial into political power.

[39] Franz van Schooten to Christiaan Huygens, 19.IX.1658, FO.IV.122.

[40] Following Libri's discovery of the Fermat manuscripts and the subsequent call for a nationally financed edition of Fermat's works (for details, see below, App. II), the question of his birthplace, long assumed to have been Toulouse, and the exact date of his birth, estimated from 1590 to 1608, took on new emphasis. As Taupiac freely admitted, it was local pride more than scholarly interest that led him to search his town archives for evidence of Fermat's life. The discovery that Fermat had been a native of Beaumont translated that local pride into a series of articles, in which Taupiac showed conclusively that the Pierre Fermat of his records was indeed the famous mathematician and that Beaumont always retained Fermat's first loyalty. The particular historical value of Taupiac's articles derives from his liberal publication of the actual documents on which his arguments were based. See his: "Pierre Fermat," *Biographie de Tarn-et-Garonne* (Montauban, 1860), pp. 468–514; "Fermat, notice biographique," *Bull. archéol. et hist. de la Soc. archéol. de Tarn-et-Garonne*, VII(1879), pp. 177–213 (republished as "Fermat, sa vie privée avec pièces justificatives," Montauban, 1880).

Some of Taupiac's source material, together with some new documentation, was again published in 1957 by H. Blaquière and M. Caillett in a pamphlet issued on the occasion of the dedication of the Lycée Pierre de Fermat in Toulouse in 1957 and entitled *Un mathématicien de génie, Pierre de Fermat 1601–1665* (Toulouse, 1957). In that pamphlet, Blaquière reported that the Fermat family home at St. Antoine (Gers) mysteriously burned to the ground on the night of the death of the last member of the family (1880) and that the remaining family papers presumably burned with it.

Sometime prior to 1631, Fermat received the degree of Bachelor of Civil Law at the University of Orléans; his *aggrégation* into the law faculty at the University of Toulouse bears the date 1 May 1631.[41] By then, he had already purchased the offices of *conseiller au Parlement de Toulouse and commissaire aux requétes du Palais* some five months earlier. The price of 43,500 *livres* paid to the widow of the former holder illustrates the comfortable wealth of the Fermat family.[42] Fermat acquired the provisions of the offices on 22 January 1631. Four months later, he acquired a wife, his cousin fourth removed, Louise de Long. Again, the dowry of 12,000 *livres* illustrates that money would not be a major problem for the young lawyer. The marriage contract reflects his family's rise in status: his uncle, Samuel de Long, was the *président au présidial et juge mage en la Sénéchaussée d'Armagnac*; another uncle, Dominique Guy, a *conseiller* of the same *Sénéchaussée*; Fermat's brother, Clément, an *avocat au Parlement de Toulouse*; and his new father-in-law a fellow *conseiller*.[43] All parties, then, with the exception of Clément Fermat were members of the high *noblesse de robe*. Fermat's offices made him a member of that social class also and entitled him to add the "de" to his name, which he did from 1631 on.

Sparse as are the details of his early years, even less is known about Fermat's subsequent private life. Five children are known to have issued from the marriage: Clément-Samuel, Jean, Claire, Cathérine, and Louise.[44] Clément-Samuel, the eldest, was apparently closest to Fermat and shared his mathematical interests (though not his talent); it was he who edited the posthumous editions of Fermat's *Observations on Diophantus* (1670) and *Various Mathematical Works* (1679). As eldest son and executor of Fermat's will, Clément-Samuel also inherited his father's office of *conseiller*, which he in turn passed on to his own son, Jean-François. According to the custom of the time, the family's *noblesse de robe* thereby became permanent. The direct male line in office appears to have ended, however, with Jean-François, as he left his provisions to his nephew, Claire's grandson. Fermat's other son, Jean, became archdeacon of Fimarens. Of the daughters, only Claire married; her two younger sisters took holy orders.[45] Though one would, of course, like to know much more about Fermat and his family, these few details do suffice to indicate that it was a typical family of the period.

Fermat was associated with three towns during his lifetime. His primary residence after 1631 was Toulouse, where he functioned in his main capacity as *conseiller* of the Parlement. Beaumont continued, however, to claim both

---

[41] Record of reception reproduced in Blaquière-Caillet, item #9.

[42] Contract reproduced in *ibid.*, item #21.

[43] Cf. marriage contract, *ibid.*, item #5.

[44] Although Taupiac was able to establish the order of birth of Fermat's children, he could find actual baptismal dates for the two youngest only. Cathérine was baptized (and most probably born) on 20 August 1641, her father's fortieth birthday. Louise was born in Castres in 1645 but was baptized in Beaumont in 1655 (28 June), a fact which Taupiac took to indicate Fermat's profound native loyalty to his birthplace.

[45] Tannery, *Notes bio-bibliographiques*, FO.IV.238.

his affection and his attention, and he frequently presided over its *conseil général.*[46] In addition, Fermat often traveled to Castres, a nearby Huguenot stronghold, where he first pleaded before and then sat as a member of the *Commission de la Chambre de l'Édit* (of Nantes) from 1632 on. He was engaged in this last capacity when he died in Castres on 12 January 1665.

Except for a possible illness in 1643,[47] Fermat enjoyed good health until 1652 or 1653, when the plague that struck Toulouse on its way north almost claimed him as a victim. In May 1653, a friend, the philosopher Bernard Medon, reported Fermat's death to the prominent Dutch man of letters, Nicholas Heinsius. Shortly thereafter, he corrected himself: "I informed you earlier of the death of Fermat. He is still alive, and we no longer fear for his health, even though we had counted him among the dead a short time ago. The plague no longer rages among us; would that the war were not a greater evil!"[48]

Fermat recovered, but by 1660 he was again ailing. Unable to withstand the rigors of a long journey to Clermont to meet the equally frail Blaise Pascal, Fermat suggested that the two meet at some halfway point.[49] Pascal's health, however, would not permit even that compromise, and the meeting never took place. Fermat himself might not have really been up to the trip; on 4 March he had seen fit to make his last will and testament, in which he named Clément-Samuel his sole heir and executor.[50] Somewhat more than a year later, on 13 December 1661, he took steps to assure his son's succession to his parliamentary offices by asking Pierre Séguier, the chancellor of France, to waive the requirement that all successors present themselves to him personally.[51] By the summer of 1662 Fermat had written his last letter of scientific content. On 9 January 1665, he signed what was to be his last *arrêt* for the *Chambre de l'Édit* at Castres; three days later he was dead. His body was buried in the Church of St. Dominique in Castres.[52] His wife, Louise, survived him by more than twenty-five years; she was still alive but very frail in October 1690.[53]

---

[46] Cf. Blaquière-Caillet, item #26, and Taupiac (1879).

[47] Fermat to Carcavi, Summer 1643, FO.II.247: "Pour la Géométrie, je n'ose pas encore m'y attacher fortement depuis mon incommodité."

[48] Bernard Medon to Nicholas Heinsius, kal.Maii 1653, in Pieter Burmann, *Sylloges epistolarum . . . a viris illustribus scriptarum libri quinque* (Leiden, Luchtmans; 1727), vol. V, letter #543: "Priori monueram te de morte Fermatii, vivit adhuc, nec desperatur de ejus salute, quamvis paulo ante conclamata. Pestis non amplius furit apud nos, utinam non esset majus malum bellum." The announcement of Fermat's death is added as a P.S. to the previous letter: "Fato functus est maximus Fermatius."

[49] Letter in FO.II.450.

[50] Blaquière-Caillet, item #16.

[51] Fermat to Séguier, 13.XII.1661, FO.II.455–456.

[52] Blaquière-Caillet, item #17: "Le douzième du mois de janvier 1665 décéda, ayan reçu tous les sacrements, M. Pierre de Fermat, conseiller du Roi en son Parlement de Toulouse et commissaire en la Chambre de l'Édit séant à Castres, et fut enseveli le treizième dans l'église des Révérends Pères de St. Dominique, où les messieurs du Vénérable Chapitre ont fait l'office."

[53] Taupiac (1879), pp. 194, 210–211.

The details of Fermat's activities as a lawyer and *parlementaire* are almost as obscure as those concerning his private life. During his lifetime and at his death, he was lauded as a great jurist, but there is little evidence that would enable one to separate sincere esteem from mere *politesse*. Fermat did help to organize the customary laws of his birthplace, Beaumont, and did frequently act as legal counsel to the town, but there is nothing to document Dubédat's attribution to him of a commentary on the *Digest*.[54]

Fermat began his parliamentary career on 14 May 1631, when he was received into the offices of *conseiller au Parlement* and *commissaire aux requêtes* to replace Pierre de Carrière, from whose widow he had purchased the positions. He signed his first known *arrêt* on 6 December 1632.[55] Three weeks later he reported a case to the *Chambre de l'Édit* and signed the resulting *arrêt*.[56] Whether or not Fermat was a regular member of the Chambre at this time is not entirely clear; his later description of the organization of the Parlement of Toulouse would seem to rule out that possibility.[57] Fermat remained a member of the *Chambre des Requêtes*, the lowest of the chambers of the Parlement, until 16 January 1638, when he was received into the *Chambre des Enquêtes* as *conseiller lay (aux enquêtes)* to replace a Pierre de Renaldy.[58] Fourteen years later, he reached the highest councils by moving to the *Tornelle*, or criminal court. The plague that almost swept him away wreaked havoc on the other members of the Parlement, and, in a body where promotion depended most heavily on seniority, Fermat's survival meant rapid advancement. By March 1654 he could report to President d'Augeard of the *Chambre de l'Édit de Guienne* in Paris that he was third in order of seniority in the *Tornelle* and expected to move to the *Grand'Chambre* the following St. Martin's Day.[59]

Despite his advance through the ranks, due mostly to accumulated seniority, Fermat does not seem to have been too successful as a *parlementaire*. When, in 1642, he petitioned Chancellor Séguier for appointment to some unnamed office, perhaps President of *Enquêtes*, perhaps a permanent place in the *Chambre de l'Édit*, his nomination was delayed. In his letters to Mersenne, Fermat expressed the belief that Séguier's physician and intellectual confidant, Marin Cureau de la Chambre, through whom Fermat had placed his request, might be working against him, and he hinted that failure in office might be one of the reasons:[60] "I don't know what my standing

---

[54] J.-B. Dubédat, *Histoire du Parlement de Toulouse* (Paris, 1885), vol. II, p. 208ff. On Fermat's legal activities in Beaumont, see Taupiac (1879).

[55] Blaquière-Caillet, item #24.

[56] *Ibid.*, item #25.

[57] See below, n.64.

[58] FO.IV.22, n.1.

[59] Fermat to d'Augeard, 14.III.1654, FO.IV.16: "Il y a près de deux ans que je suis hors des Enquêtes et par la révolution que les maladies ont causé, je me trouve présentement le troisième de la Tornelle et en état d'être de la Grand'Chambre à la Saint Martin prochaine."

[60] Fermat to Mersenne, 10.XI.1642, FO.II.244.

will be in the mind of M. de la Chambre, since the commission at Castres failed so badly." According to a later letter to Mersenne, Fermat eventually obtained the desired appointment, whatever it was, and the matter disappeared from his correspondence after February 1643.[61] Whether or not nomination to the *Chambre de l'Édit* at Castres was the office in question, Fermat was directed by Séguier in 1648 to preside over a meeting of that body.[62]

Perhaps due to his scholarly reputation in Paris, Fermat became official spokesman for the Parlement of Toulouse in its relations with Chancellor Séguier. In 1648 he presented to Séguier, through de la Chambre, a detailed explanation and justification of the actions of the Parlement in suspending the enforced collection of the *taille* in Aquitaine.[63] In his report, Fermat argued that the continued use of force to collect taxes from an area which had been promised relief from this burden would only foment civil disturbances. Fermat suggested that the King instead allow the communities to borrow the requisite amounts and guarantee the debts by declaring them privileged over all others. Such a move, he felt, would bring back into circulation money that was being hoarded for fear of bankruptcy. Nothing more was heard of this proposal, nor can one find an explanation for the interference of the Parlement of Toulouse in an area nominally under the jurisdiction of the Parlement of Bordeaux.

Fermat again acted as spokesman for his Parlement in 1654 in a series of letters to President d'Augeard of the *Chambre de l'Édit de Guienne* in Paris. In these letters, Fermat explained in some detail the organization and history of the Parlement of Toulouse; his description provides welcome information to the historian of French institutions in the seventeenth century.[64]

---

[61]Fermat to Mersenne, 16.II.1643, FO.II.251: "Je vous remercie de vos soins à l'endroit de M. de la Chambre, et à lui-même de ceux qu'il a prit à Lyon pour moi. M. de Marmiesse, notre avocat-général, m'ayant confirmé ce que vous venez de m'écrire, lorsqu'il sera temps, je ne doute point qu'il n'ait assez de crédit pour faire tenir cette vieille promesse que M. le Chancelier a faite depuis si longtemps en ma faveur."

[62]Cf. FO.II.278, n.2.

[63]Fermat to de la Chambre, with accompanying report, 18.VIII.1648, FO.II.#65, 66.

[64]Fermat's three letters to d'Augeard (FO.IV.15–22) yield, in brief, the following picture. The Parlement is divided into five chambers: the *Grand'Chambre* (G.C.) with 19 members, the *Tornelle* (T.) with 13 members, the two *Chambres des Enquêtes* (E.) with approximately 28 members each, and a *Chambre des Requêtes* (R.) with 11 members. The officers include a First President (G.C.) with six presidents *à mortier* (3 in G.C., 3 in T.), 2 presidents each for E. and 2 for R. Unlike other parlements, Toulouse considers G.C. and T. as two parts of the same body. No one may belong simultaneously to both; rather, the members of the two bodies exchange two men a year on St. Martin's Day (the opening of Parlement?) according to seniority, with the exception that the two oldest members of the G.C. remain in that body. Also, no clerical members of the Parlement, of which there were 12 in 1654, may serve in T. (Fermat gives no reason, but it clearly derives from the fact that T. was a criminal court with the right to impose the death penalty. Clerics were forbidden by Church and excused by custom from having anything to do with the shedding of blood.) G.C. and T. contain the 32 oldest members of the Parlement, and younger men moving up from E. must go first to T. Members of R. are considered full-fledged members of the Parlement with seniority rights. Should one of them have served longer than anyone in T., he may move directly to G.C. Parlement sits

The most candid appraisal of Fermat's abilities as a jurist, and one that runs counter to the usual adulation, comes from a secret report of Claude Bezin de Bésons, *intendant* of Languedoc, to Minister Colbert in 1663. Speaking of the *conseillers* and their relations to the suspect First President, Gaspar de Fieubet, Bezin said of Fermat: "Fermat, a man of great erudition, has contact with men of learning everywhere. But he is rather preoccupied; he does not report cases well and is confused. He is not among the friends of the First President."[65] Indirect evidence would seem to justify Bezin's remark. By 1664 Fermat had managed to maintain his position in a body that Paris found most dangerous, while at the same time remaining on good terms with the central authorities. It would conform to the sense of Fermat's naïveté in personal relations that his correspondence conveys, that people such as Séguier simply felt that Fermat was not a good enough lawyer and jurist (or politician) to be dangerous.

III. MOTIVATION TO MATHEMATICS

I will share all [my results] with
you whenever you wish and do so
without any ambition, from which
I am more exempt and more distant
than any man in the world.

   Pierre de Fermat[66]

Underlying the diversity of mathematics in the sixteenth and seventeenth centuries was the diversity of its practitioners. They came from and followed all walks of life. Napier was a Scots baron; Tartaglia, the impoverished son of a post rider. Clavius was a high-ranking Jesuit; Cavalieri, a simple monk; Oughtred and Wallis, Protestant divines. Stevin was a military engineer; Hariot, an adventurer; Viète, a lawyer; Roberval, a university professor. Fermat and Descartes never had to worry about money; Galileo never had enough. The diversity transcended merely socio-economic differences. Men

---

4 days a week—Monday, Wednesday, Thursday, and Saturday—and on the eves of all festivals. Only one trial a day per member is held for which a major president receives 6 livres 10 sou, a president of E. $4\frac{1}{2}$ 1. $7\frac{1}{2}$ sou, and a *conseiller* 3 1. 5 sou. When in working session, the various chambers are broken up into bureaus, which in G.C. are called *guets*, each headed by a president, or the oldest member in his absence. The *Chambre de l'Édit* of Castres, one of several bireligious groups set up under the provisions of the Edict of Nantes (1598) to guarantee minority justice, is under the aegis of the Parlement of Toulouse. The latter body sends a deputation of one president *à mortier* (by turns), who serves as president of the commission, 3 members of G.C., 3 of T. and 2 each from E. The Toulousains are all Catholic (like Fermat); Castres, one of the Hugenot strongholds secured under the Edict of Nantes, supplies the 10 Protestant members of the commission. Since Fermat is mentioned as president of the *Chambre de l'Édit* at one point, he must have risen to the rank of president *à mortier* of the Parlement of Toulouse.

[65] Blaquière-Caillet, item #33.
[66] Fermat to Mersenne, 26.IV.1636, FO.II.5.

did mathematics differently from one another because they sought different goals in mathematics. Its study meant one thing to the philosophically oriented humanist, another to the practically oriented cossist or applied mathematician, and still another to the mystically oriented hermetic magician.

There was no professional discipline of mathematics to override the diversity and impose uniformity. With few exceptions, mathematicians of the period were free from the pressures to conform that a profession exercises on its members. Few depended on mathematics as a livelihood. There were no positions to be gained or held. There was no ladder of advancement leading into a hierarchical elite. Within very broad limits, each practitioner could choose his own style of mathematics. Since there was no single set of standards nor particular group of judges, each man's choice could claim validity. In cases of conflict, there was no clear-cut arbiter. Nor, except for personal gratification, was there reward or sanction.

In this regard, Fermat is typical of the mathematicians of his day. Pursuing mathematics as a sideline, he operated in the social context of the professional lawyer and jurist. Answerable to no one for his mathematical behavior, he could afford even fundamental disagreement over the nature and goals of mathematics. His success or failure in mathematics could affect little more than his self-esteem or his reputation among those whose judgment he valued. Not all of them were mathematicians. Some were quite likely to mix their appraisal of his mathematics with their evaluation of him as a lawyer or as a classicist. The mathematicians to whom he turned for approval, some of them also professional lawyers, were on the whole men who shared his particular mathematical tastes, or were at least sympathetic to them. But Fermat did not need their approval. He could and did proceed without it. He was to that extent a free agent.

So too were most, if not all, of his contemporaries. That is what gives the famous mathematical controversies of the seventeenth century—Fermat vs. Descartes, Fermat vs. Wallis, Newton vs. Leibniz—their peculiar flavor. So little that was tangible hung on the outcome. No one's professional career was at stake, whatever the retrospective importance of the issues disputed. At heart, they were issues of style and method, and the fragmentation of the enterprise and independence of its practitioners meant that the controversies could only end in a draw. Insofar as the participants felt their reputations to be at stake, they appealed to different groups within which they held those reputations. Each, that is, was playing to his own audience. Since the decision rested on methodological and stylistic taste, none really could lose. In the end, Cartesians preferred Descartes' approach to maxima and minima; the Roberval group, Fermat's. In the end, Wallis and his supporters still found number theory useless; Fermat and his supporters continued to extol its virtues. In the end, the English would defend Newton's priority; the Continental mathematicians, Leibniz's independence. Each man retained his reputation among his supporters.

The setbacks, if any, were personal. Descartes prided himself on the novelty of his *Geometry*. He thought absolute preeminence as a mathematician vital to his long-range philosophical program. He was bitterly disappointed to find in his enemies' camp a mathematician of equal caliber, whom he could not vanquish, whose style and methods were not demonstrably inferior to his own. By the same token, Fermat was deeply hurt when his methods, which seemed to him so simple and clear, evoked from others the charge that he was just "lucky." Later, his failure to convince Wallis and others of the beauty and challenge of number theory was a source of anguish and frustration to Fermat. But, it did not alter his own opinion, nor turn the course of his own research. Win, lose, or draw, he, like his opponents, remained a free agent.

Mathematics in the seventeenth century, then, depended heavily on individual personalities and their idiosyncrasies. How men such as Fermat did mathematics was largely a matter of personal choice. Their particular motivation to pursue mathematics determined which of the many, equally valid approaches they would take to the subject. If a particular school gave their work its direction, it did so by their choice. Two things, therefore, become important for understanding a mathematician of the seventeenth century. First, one must try to determine what sort of man he was and why he pursued mathematics. Second, one must locate the category or school of mathematics within which he chose to operate and the extent to which he employed its methods and pursued its goals.

Chapter II will explore in some depth the school of mathematics to which Fermat felt primary loyalty and the ways in which his mathematical career reflects the tenets of that school. Fermat was an analyst and a follower of Viète. At the moment, however, it is worth attempting to cull from the sparse sources some insight into Fermat's personality and its reflection in his pursuit of mathematics. The very scarcity of the sources reveals one important trait: Fermat was a man who kept to himself. That trait, perhaps more than any other, marks his career as a mathematician.

There are, however, hints in his mathematical correspondence that combine with the scant details of his private life and professional career to reveal other traits and to form, at least in rough outline, a picture of the man. It is a picture of apparent paradoxes. Merely to have chosen a legal career, to have pursued it throughout his life, and to have survived in a corporate group that represented one of the main barriers to Bourbon dreams of absolute monarchy, Fermat must have possessed that toughness and ambition so characteristic of the seventeenth-century French bourgeoisie.[67] The letters to d'Augeard reflect a pride of accomplishment in having reached the highest ranks of the Parlement at a time when it was coming to dominate Toulouse's traditionally aristocratic government and set the pace of its social and political

---

[67]As a colleague, Theodore K. Rabb, pointed out to me in conversation, one could on the contrary argue that Fermat's longevity in the Parlement bespeaks nothing more than a high tolerance for boredom and a deep sense of family and social status.

life.[68] Fermat expressed full devotion to the procedures and policies of that body and participated actively in its legislative and judicial functions, including criminal trials. Yet, from all indications, he was a gentle, retiring, even shy man. None of his writings displays the egoistic bluster and bellicosity of a Descartes, a Roberval, or a Wallis. Indeed, when Fermat does rarely talk of himself, it is most frequently in self-deprecatory terms that transcend the norms of *politesse*; he is, for example, the "laziest of men." For a man whose very career embroiled him in controversy and even intrigue, his correspondence is remarkably devoid of reference to such burning issues as Jansenism or the Fronde, issues that came so naturally to the pens of his colleagues. Perhaps the delicacy of his position is sufficient explanation, but there is much to suggest that he simply did not like controversy and that he shied away from it whenever possible. His legal activities in Beaumont and Castres, at least, emphasized his qualities as a mediator; and he apparently felt most comfortable in that role. When Catholic representation in the *Chambre de l'Edit* required a stronger, more partisan stance, as perhaps in 1642, Fermat was less effective.

If these characteristics are true, they suggest a motivation to pursue mathematics that goes a long way in explaining Fermat's behavior as a mathematician. Mathematics, that is, may well have represented for Fermat a refuge from controversy, a field of inquiry in which questions had unique answers, in which right and wrong were clear, in which disputation yielded to ineluctable proof. But how paradoxical, then, that Fermat should figure in two of the most vehement and most famous mathematical disputes of his day! And yet, when one examines his role in those controversies, one is struck by his obvious reticence, by his shock at their vehemence, and by the fact that he actually took very little part in them. Indeed, as Chapter VI will show, whatever mild belligerence marks Fermat's dispute with Wallis in 1657–1658, it must be taken as a sign of the extreme frustration he felt at his inability to interest others in his new number theory. Fermat liked mathematical controversy no more than any other kind.

Fermat would not be the first man to turn from the rough-and-tumble world of politics to the solace of abstract contemplation; Cicero offered him an ancient example, Bacon a more modern one. If this was his motivation, it would explain perhaps the most curious aspect of his mathematical behavior, to wit, his steadfast refusal to edit and publish his results or to allow them to appear in print under his name. To publish would be to enter the mathematical ring, to turn a refuge into another arena, to be forced to put the finishing touches on results that have lost their interest and challenge, to vie with others before a large audience. He was, he assured Mersenne, a man "without ambition." The letters to Séguier and d'Augeard alone suffice to show that this self-characterization applied only to his involvement with mathematics.

---

[68] Philippe Wolff et al., *Les Toulousains dans l'histoire* (Toulouse: Privat, 1984), 109.

Fermat's refusal to publish under his own name is a constantly recurring theme of his correspondence, and his adamancy often annoyed and frustrated his friends. As early as the spring of 1637, Roberval and E. Pascal offered to edit and publish some of Fermat's work on plane and solid loci. "Whatever of my work is judged worthy of publication," Fermat replied,[69] "I do not want my name to appear there." A year later, Descartes insisted that the letters exchanged on the subject of his *Dioptrics* and Fermat's method of maxima and minima be published in order to submit the issues in dispute to a wider audience; Fermat refused absolutely, since anonymity would have been impossible. By 1651, Bernard Medon, the friend who would later report Fermat's near-death in the plaque, grasped at straws in his frustration; he wrote to Heinsius:[70]

> You are also greeted by the great Fermat, of whose mathematical knowledge, which is greater than any mortal possesses, nothing can be extorted unless the most outstanding of queens, Christina [of Sweden], at some time adds to the voices of all the intellectuals of this age, including that of the Council of France, her own commands, to which, I think, he would not be deaf. If your efforts could bring this about, you would do all of Europe a great favor.

Even had Heinsius succeeded in getting Christina to issue such a plea,[71] it is doubtful her urgings would be any more effective than those of Pascal in 1654 or those of Huygens in 1659. Fermat knew what the preparation of his work for publication would entail. It would turn a fascinating, yet restful hobby into a drudging bore.

And yet, Fermat's professional career bespeaks an *amour-propre* that his obvious genius for mathematics could only have reinforced. Fermat could not have done the mathematics he did for himself alone. He had to share his results and receive from others the approbation and admiration those results deserved. Hence, the same man who shunned publication nonetheless circulated his work and placed stock in others' reaction to it. It was a compromise measure. Circulation of results by letter brought the rewards of approval and fame at the same time that it avoided the full commitment and responsibilities that would rob mathematics of its recreational role. He could scatter

[69] Fermat to Roberval, 20.IV.1637, FO.II.106.

[70] Burmann, *Sylloges epistolarum*, vol. V, p. 626.

[71] Fermat's treatise on porisms (*Porismatum Euclideorum renovata doctrina et sub forma isagoges recentioribus geometris exhibita*), written in 1655 or 1656, contains a passage that suggests, though inconclusively, that Heinsius may indeed have succeeded (FO.I.77): "Postquam enim Suevicum sidus omnibus disciplinis illuxit, frustra scientiarum arcana tanquam mysteria quaedam abscondemus: nihil quippe impervium perspicacissimo incomparabilis Reginae ingenio, nec fas censemus occultare doctrinam quam vel unico duntaxat aut inspirantis aut mandantis nutu, quandocumque libuerit, detectam iri vix possumus dubitare." If Fermat's reference to Christina here is a result of Heinsius' efforts, it is a poor response to a request for publication of his works.

results unsystematically, he could skirt over difficulties, he could excuse himself from the drudgery of chasing out final details, he could show work in progress, omit proofs or merely sketch them, drop subjects for a while and pick them up again later. Throughout his correspondence Fermat promised to fill in gaps "when leisure permits." But the sensitive reader knows better; doing so would have required a far greater degree of self-discipline than Fermat was willing to devote to mathematics.

In this light, his desire for anonymity begins to make sense. He took pride in his work and enjoyed the reputation that work brought him. He was deeply hurt when Franz van Schooten published a restoration of Apollonius' *Plane Loci* without even mentioning Fermat's earlier restoration, which had circulated in Paris from 1636 on. Johann Hudde's attempts to claim priority for the method of maxima and minima also upset him, just as Huygens' vehement defense of his priority pleased him. Fermat did not wish to remain unknown. Rather, his often repeated desire for anonymity must be understood with respect to the issue of publication. On the one hand, he was not willing to bring his own work to the state of perfection that he felt publication would demand, i.e. a state of perfection that would preclude criticism and controversy. On the other hand, he was ready to allow others to publish the material, making in it whatever improvements they felt necessary or desirable; that was the offer he made to Pascal and Carcavi in 1654. But, should others undertake the task of publication, Fermat would not accept the full responsibilities of authorship; he would not enter the arena with weapons partly forged by others, especially since he did not particularly want to enter the arena at all.

Fermat's own personality, then, coupled with the exigencies and pressures of his professional career, provides an explanation of some of his most fundamental patterns of behavior as a mathematician, in particular his response to criticism and controversy and his attitude toward publication. It also goes a long way in explaining the form his mathematics took: the emphasis on problem-solving, the lack of attention to details, the absence of proofs, the failure to follow up promising leads, and the overly laconic style of presentation. Personality factors, however, do not alone explain these formal aspects, nor do they at all lend insight into Fermat's more specific behavior, e.g. the commitment to an algebraic, analytic approach to mathematics, the emphasis on derivation as a form of proof, an ambivalent attitude toward classical Greek mathematics. What problems Fermat chose to investigate, how he investigated them, and what he took to be the proper goals of mathematical research, in short the shape of his mathematical career, depend, as the first section of this chapter has pointed out, on the source of his mathematical training and the particular school of mathematics to which he felt primary commitment. That will be the subject of Chapter II.

CHAPTER II

Nullum Non Problema Solvere: Viete's Analytic Program And Its Influence On Fermat

I do not think you would want to neglect such an important man, who represents for us a second Viète. J. Chapelain to Chr. Huygens[1]

François Viète (1540–1603) died when Fermat was two years old.[2] Yet, he shaped Fermat's mathematical career as decisively as if the two men had stood in a living master-student relationship. Chapelain was not the only contemporary of Fermat to see in him a successor to Viète; in both style of exposition and patterns of thought, everything that Fermat did in mathematics, however original, bears Viète's imprint. Through his works, Viète provided Fermat with the algebraic symbolism and tools of analysis that Fermat employed throughout his career. More important, Viète imbued Fermat with a then new mathematical tradition and a program of mathematical research that determined the sorts of problems Fermat chose to investigate and the manner in which he treated them.

One must be aware of Viète's influence to understand even the grossest outlines of Fermat's career. Fermat appears to have been selective in his reading, perhaps through ignorance, perhaps by choice, perhaps for both reasons. If, for example, he ever saw or read Clavius' great *Algebra* or any of the cossist treatises on algebra, he never mentioned them in his works nor reacted to them. Despite his location in Toulouse and his firsthand knowledge of Italian literary classics,[3] he knew little or nothing of the Italian mathematical school until after opening contact with Mersenne. If, indeed, he read all of Archimedes, he cited only two works in his writings, the *Quadrature of the Parabola* and *On*

[1] 12.VI.1664, FO.IV.135.
[2] J. Grisard's *Francois Viète, mathématicien de la fin du seizième siècle: Essai bio-bibliographique* (Doctorat de troisième cycle, École Pratique des Hautes Études, Centre de Re-

Spirals, though he was clearly responding at one point to the *Sphere and Cylinder* and Eutocius' commentary on it. Despite Fermat's consuming interest in number theory, the name of Nicomachus appears nowhere in his letters on the subject or in the *Observations on Diophantus*. He did no work in trigonometry or perspective and, except for some pedestrian efforts in the area of magic squares, apparently had no interest in the mystical aspects of mathematics. By contrast, Apollonius, Pappus, and Diophantus dominated Fermat's attention, though not in the sense of the neoclassical tradition. Clearly, only a portion of the overall mathematical activity of Fermat's time had operative influence on him. His ties to Viète explain why that was so and, in addition, delineate the operative portion.

Viète told Fermat what to read. He did so in part by telling Fermat how to read it. And, as the following chapters will show, how Fermat read the sources on which his work was based was just as important as which of the available sources he did read. In Fermat we have one of the most important links in the transition from ancient to modern mathematics.[4] The transition resulted in part from his reading ancient works in a manner which their authors never intended, and for information that, strictly speaking, they did not contain. Not all of Fermat's contemporaries approached the ancient sources in the same way; not all of them took part in the transition. Fermat's approach was peculiar to the analytical school of Viète. Fermat became its foremost student.

For all its retrospective importance, the school hardly dominated the mathematics of its day. Viète's direct disciples still suffer today the obscurity that the master did a century ago; Pierre and Jacques Aleaume, Alexander Anderson, James Hume, Jean de Beaugrand, and others appear today only in the indexes to scholarly works. If Thomas Hariot's *Practice of the Analytic Art* (*Artis analyticae praxis*, London, 1621) and William Oughtred's *Key to Mathematics* (*Clavis mathematicae*, London, 1631) derive their content from Viète,[5] their authors certainly enjoyed a better press. For all Isaac Beeckman's mathematical erudition, he does not seem to have known about Viète. Beeckman's most famous protégé, Descartes, claimed he first read Viète's work only after the appearance of his own *Geometry*, and the recorded

---

cherche d'Histoire des Sciences et des Techniques, n.d.) updates Frédéric Ritter, "François Viète (1540–1603), inventeur de l'algèbre moderne. Essai sur sa vie et son oeuvre," *Revue occidentale* (1895), pp. 234–274 and 354–415. Viète's *Opera mathematica*, edited by Frans van Schooten in 1646 and published by Elzevier in Leiden, appeared in a reprinted edition with an extensive introduction by J. E. Hofmann (Hildesheim, 1969). See also Warren van Egmond, "A Catalog of François Viète's Printed and Manuscript Works," in *Mathemata: Festschrift für Helmuth Gericke*, ed. M. Folkerts and U. Lindgren (Stuttgart, 1985), 359–96. On Viète's mathematics, see M. S. Mahoney, "The Royal Road" (Ph.D. diss., Princeton, 1967), Chaps. I and II.

[3] In his correspondence, Fermat occasionally quotes Ariosto's *Orlando furioso* and Tasso's *Gerusalemme liberata*.

[4] For an analysis of that transition, see M. S. Mahoney, "Die Anfänge der algebraischen Denkweise im 17. Jahrhundert," *Rete* 1(1972), pp. 15–31.

[5] Cf. Chap. XVI of the *Clavis mathematicae*, which is taken directly from Chap. V of Viète's *Introduction to the Analytic Art*. Remarkably, Viète is not cited as the source.

genesis of the latter treatise in the *Rules for the Direction of the Mind* (*ca.* 1628) acts to support that claim.[6] Mersenne and the Parisian circle of mathematicians about him knew about Viète but did not follow his mathematical lead.[7] Viète's far superior algebraic works had been in print for more than a decade when Clavius published his *Algebra*, so different in tone and style, in retrospect so backward. The miscopying by Italian mathematicians of Viètan technical terms used in Fermat's writings offers further evidence that Viète was unknown to the Italians.[8]

One wonders how Fermat first heard about Viète and his school. The sources provide no hint, and not enough is known at the moment about Viète's immediate followers and the spread of his work to warrant conjecture. One thing seems clear. Sometime in the late 1620s, Fermat went to Bordeaux, where Viète had worked and died. There he made contact with students of Viète and immersed himself in Viète's mathematics. With regard to both time and mathematical content, Fermat's career began in Bordeaux. Hence, it is important to know what he found there.

I. ANALYSIS, ALGEBRA, AND THE ANALYTIC ART

In Tours in 1591, Viète published his *Introduction to the Analytic Art (In artem analyticen isagoge)*, the first of a series of treatises containing a new system of symbolic algebra under a new name, the analytic art. The work opens with a discussion of analysis highly reminiscent of the discussion that opens Book VII of Pappus of Alexandria's *Mathematical Collection*:[9]

There is in mathematics a way of inquiring after truth which is said to have first been invented by Plato, and which was called "analysis" by Theon and defined by him as "the assumption of the thing sought as if admitted [and the arrival] by consequences at something truly admitted." Opposed to this is synthesis, the assumption of something admitted and the arrival by consequences at the goal and comprehension of the thing sought. And although the Ancients proposed only two forms of analysis, *zetetic* and *poristic*, to which Theon's definition most aptly applies, I have set up in addition a third species, which is called *rhetic* or *exegetic*. As commonly understood,

[6]Cf. Descartes to Mersenne, 20.II.1639, Alquié.II.126: "Je n'ai aucune connaissance de ce géomètre dont vous m'écrivez [Beaugrand?], et je m'étonne de ce qu'il dit, que nous avons étudié ensemble Viète à Paris; car c'est un livre dont je ne me souviens pas avoir seulement jamais vu la couverture, pendant que j'ai été en France."

[7]The section on algebra in Pierre Hérigone's *Cursus mathematicus*, for instance, took its wording directly from Viète, but adopted neither the symbolism nor the spirit of Viète's analytic art.

[8]The important technical term *syncrisis*, central to Fermat's original method of maxima and minima, was regularly misread as *syneresis*. Cf. Ricci to Torricelli, 4.II.1645, *FO.Suppl.*132.

[9]Viète, *Introduction to the Analytic Art* (ed. Leiden, 1646), p. 1; cf. *Pappi Alexandrini Collectionis quae supersunt*, ed. F. Hultsch (Berlin, 1877), Vol. II, pp. 634, 1–636, 14 (trans. in M. S. Mahoney, "Another Look at Greek Geometrical Analysis," *Archive for History of Exact Sciences*, 5[1968], p. 322).

zetetic is that by which is found the equality or ratio of the magnitude which is sought with those [magnitudes] which are given. Poristic is that by which the truth of the theorem set up concerning the equality or ratio is examined. Exegetic is that by which the magnitude which is sought is derived (*exhibetur*) from the equality or ratio set up. And therefore the whole threefold art of analysis fulfilling this task should be defined as the "doctrine of finding well in mathematics" (*doctrina bene inveniendi in mathematicis*).

What is striking about this passage is not so much the similarity it bears to Pappus' classical presentation of analysis, but that it should appear at all in a treatise on algebra. It betrays both its novelty and its unaccustomed surroundings by the manner in which it alters the classical meaning of the terms "zetetic" and "poristic" and adds a third, non-classical division of analysis, "exegetic." Viète is clearly bending classical Greek analysis to a new purpose. He is the first to identify algebra with analysis, and the identification becomes the hallmark of his school of mathematics.

From its emergence as a recognizably distinct and generally defined procedure in mathematics, classical Greek analysis constituted a form of geometrical heuristic.[10] It provided a means of attacking two sorts of general problems: to test the validity and find the proof of a theorem, and to solve a problem, particularly a problem in geometrical construction. As described by Pappus,[11] one proceeded analytically by assuming for the moment that the theorem in question is valid or the problem solved and by then following out the logical implications of the theorem or the solution until one arrived at a conclusion known to be true or false. In the case of a theorem, one could argue directly from a false conclusion the invalidity of the theorem, and the analysis itself constituted a valid *reductio ad absurdum* disproving the theorem.

Should the conclusion arrived at through analysis be true, however, one faced the logical problem inherent in analysis. By its nature, analysis yielded a chain of inferences leading from a premise of unknown truth value to a conclusion of known truth value, that is, an inference of the general form $P \longrightarrow K$. Logically, the falsity of K implies that of P, but the truth of K says nothing of that of P unless one can turn the inference around, that is, unless it also holds that $P \longleftarrow K$, or $P \longleftrightarrow K$. Mathematically, two related phenomena strengthened the effectiveness of analysis. First, as Aristotle pointed out, most theorems of geometry have valid converses, so that one may move directly from $P \longrightarrow K$ to $K \longrightarrow P$. Secondly, when the converse does not immediately hold, it normally can be made to do so by the addition of

[10] For a full discussion of the subject, see Mahoney, "Another Look at Greek Geometrical Analysis," pp. 318–348.

[11] The late date of Pappus (he lived during the third century A.D.) need not disturb us. He was working from texts of the fourth and third centuries B.C. (Euclid, Appollonius, Archimedes, *et al*.) and his discussion reflects the state of the art at that earlier time.

supplementary conditions, called by the Greeks *diorismoi* (sing., *diorismos*). For example, the theorem that any three lines can be combined to form a triangle requires for its validity the *diorismos* that the sum of any two of the lines exceeds the third. Once the necessary *diorismoi* for reversing the inference $P \longrightarrow K$ have been established (and often analysis aids in establishing them), the reversed inference constitutes a synthesis, or logically rigorous deduction of the theorem. It was the awareness of the pitfalls inherent in the reversing of inferences that led all Greek mathematicians to insist on formal, synthetic demonstration of all results achieved by analysis. In the presence of that synthetic demonstration, however, any analysis became superfluous and, with rare exceptions, was deleted from finished treatises.

The logical drawbacks of analysis played less of a role in problem-solving than in theorem-proving. Here, analysis offered the research mathematician a path of attack on the problem facing him. Assuming the construction to have been carried out, he used analysis to uncover the implications of that construction. What else, for example, was constructed thereby? What further constructions did the original imply? The goal of problem-solving analysis was a solution to the problem. In every case, the validity of the solution, posed in the form of a theorem, could be tested, and a proof of the theorem found, by application of theorem-proving analysis. Hence, problem-solving analysis connoted to the Greek geometer not a form of reasoning, but a body of mathematical literature particularly useful in attacking new problems. For example, Euclid's *Data* is essentially his *Elements* reformulated in a manner more directly applicable to mathematical research. It tells the mathematician what is implicitly given by his data and whether his data suffice for various constructions. Pappus codified this body of problem-solving analytical treatises in Book VII of the *Mathematical Collection* under the heading *topos analyomenos*, or "field of analysis"; it included: Euclid's *Data*, *Porisms*, and *Surface Loci*; Apollonius' *Conics*, *Section of Ratio*, *Section of Area*, *Determinate Section*, *Contacts*, *Inclinations*, and *Plane Loci*; Aristaeus' *Solid Loci*; and Eratosthenes' *Means*. All of these treatises aimed at facilitating the reduction of a given problem to an equivalent problem, for which either a solution was already known or a particular line of attack had proved useful. This "reduction analysis" offered the added benefit of revealing the intimate relationship between apparently diverse problems. For example, Hippocrates of Chios reduced the problem of doubling a cube to that of constructing two mean proportionals between two given magnitudes. Archimedes later showed that the problem of dividing a sphere into segments bearing a given ratio reduced to the same construction. Moreover, he showed that the construction of two mean proportionals itself reduced to the construction of two intersecting conic sections.[12]

[12]On Hippocrates' reduction, see T. L. Heath, *History of Greek Mathematics* (Oxford, 1921), Vol. I, pp. 244–246. On Archimedes' reduction (*Sphere and Cylinder* II, 4), see Mahoney, "Another Look," pp. 337–340.

Except for a few examples in Euclid and Archimedes, Greek geometrical analysis shared the fate of Pappus' *Mathematical Collection* and most of the works cited in Book VII.[13] Largely ignored or unknown by the Arabic writers, it lay hidden in Greek manuscript until the sixteenth century, when it shared in the general revival of Greek mathematics. The recovery of Pappus and Apollonius, especially following their translation into Latin by Commandino, brought geometric analysis into renewed focus. Intrigued, and often annoyed, by the way in which the synthetic style of Greek geometrical exposition deprived the reader of any sense of the manner in which the author first discovered his theorems, mathematicians of the sixteenth and early seventeenth centuries eagerly turned their attention to Pappus's promise of insight into the matter. They began a concerted effort to recover the ancient method of analysis.

Descartes best expresses the combination of curiosity and frustration that motivated the search for the details of the method of analysis; in Rule IV of the *Rules for the Direction of the Mind*, he complained:[14]

But when I afterwards bethought myself how it could be that the earliest pioneers of Philosophy in bygone ages refused to admit to the study of wisdom any one who was not versed in Mathematics, evidently believing that this was the easiest and most indispensable mental exercise and preparation for laying hold of other more important sciences, I was confirmed in my suspicion that they had knowledge of a species of Mathematics very different from that which passes current in our time. . . . Indeed I seem to recognize certain traces of this true Mathematics in Pappus and Diophantus, who though not belonging to the earliest age, yet lived many centuries before our times. But my opinion is that these writers then with a sort of low cunning, deplorable indeed, suppressed this knowledge.

Though not everyone shared his distrust in the motives of the ancient writers, Descartes did speak for most mathematicians of his time in decrying the manner in which the Greek texts obscured what interested them most. Only a few, however, agreed with his further diagnosis: "Finally, there have been some most ingenious men who have tried in this century to revive the same [true mathematics]; for it seems to be nothing other than that art which they call by the barbarous name of "algebra," if only it could be so disentangled from the multiple numbers and inexplicable figures that overwhelm it that it

[13]The broader epistemological concept of analysis and synthesis was not lost in the Middle Ages, but medieval discussions were based on Aristotle and Galen and aimed at elucidating the respective roles of analysis and synthesis (now called "resolution" and "composition") in natural philosophical research. See A. C. Crombie, *Robert Grosseteste and the Origins of Experimental Science, 1100-1700* (Oxford, 1953), Chaps. II–IV, and J. H. Randall, Jr., "The Development of Scientific Method in the School of Padua," *Renaissance Essays* (ed. P. O. Kristeller and P. P. Wiener, New York, 1968), pp. 217–251.
[14]Trans. R. M. Eaton, *Descartes Selections* (New York, 1927), pp. 52–53.

no longer would lack the clarity and simplicity that we suppose should obtain in a true mathematics."[15]

Few "men of talent" indeed discerned algebra at the root of Greek geometry; few indeed saw an intimate tie between the treatises of Apollonius, Archimedes, or Pappus, on the one hand, and the "art of the coss" on the other. One of the first who did, and the very first to try to give algebra the "clearness and simplicity" it should possess, was Viète. Whether or not Descartes was conscious of the fact in 1628, he was following in Viète's footsteps. The symbolic algebra that underlies the "genuine mathematics" of the *Geometry*, in particular the theory of equations in Book III, is in essence nothing other than Viète's "art of analysis."

Viète may have taken the idea of identifying algebra with analysis from Petrus Ramus (1505-1572), the famous French pedagogue, who hinted at it in several of his works. In his *Twenty Seven Books of Geometry* (1569), Ramus pointed to the algebraic substance of Book II of Euclid's *Elements*.[16] Elsewhere, he discerned in the *Arithmetic* of Diophantus "an art of admirable subtlety which is called by the vulgar [NB: Descartes said "barbarous"] Arabic name of 'Algebra'; the antiquity of the art, however, is apparent from this ancient author (for he is cited by Theon)."[17] Finally, in his *Algebra*, he explicitly stated that "the Greeks called algebra analytic."[18] From here, given Ramus' demonstrable influence on Viète, it was not all too great a step to Viète's citation of "the algebra which Theon, Apollonius, Pappus, and other ancient analysts taught."[19] The step to the "analytic art," however, was a giant step, and it required a mathematical talent that Ramus did not possess.[20]

---

[15] *Regulae ad directionem ingenii*, AT.X.376-377.

[16] Ramus, *Arithmeticae libri duo et Geometriae septem et viginti*, ed. Lazarus Schoner (Frankfurt, 1627); *Geometria*, cap. xii, p. 86. Ramus notes that Euclid II, 4, the geometric translation of the algebraic identity $(a + b)^2 = a^2 + 2ab + b^2$ can be illustrated by the technique of the "analysis of the square side" (taking of the square root) and adds: "Et hic geometricae analyseos usus superest, ut postea in cubo, cum alius in totis elementis nullus sit."

[17] *Petri Rami . . . Scholarum mathematicarum libri unus et triginta. Dudum quidem a Lazaro Schonero recogniti & aucti, nunc verò in postrema hac editione innumeris locis emendati & locupletati* (Frankfurt, 1627; 1st ed. Paris, 1569), cap. i, p. 35.

[18] Ramus, *Algebrae libri duo*, ed. Schoner (Frankfurt, 1627), p. 190: "A qua resolutione Algebra Graecis dicta fuit analytica, quibus absoluta Arithmetica dicebatur synthetica."

[19] Viète, *Apollonius Gallus*, in *Opera*, ed. van Schooten (Leiden, 1646), Appendicula I, p. 339. Whether or not, as has been suspected, Viète met Ramus in Paris in 1571, Ramus' influence on him emerges clearly from his *Introduction to the Analytic Art*. The definition of analysis as "doctrina bene inveniendi in mathematicis" and the reference to the three laws of method in Chapter VI (and see below, n.49) are but two of the Ramist characteristics of the *Introduction to the Analytic Art*. For a description of Ramism, see W. J. Ong, *Ramus: Method and the Decay of Dialogue* (Cambridge, Mass., 1958).

[20] Ong, in the work just cited, consistently underestimates Ramus' knowledge of mathematics, if not his talent in the subject. For a more balanced view, see J. J. Verdonk, *Petrus Ramus en de wiskunde* (Assen, 1966).

At first glance, this identification of algebra and analysis bespeaks acute mathematical insight. In the late nineteenth century, Zeuthen and Tannery confirmed Ramus' judgment of the fundamentally algebraic content of the Pythagorean doctrine of the application of areas (Euclid, Books II and VI), and historical research of the past half-century has located the source of that algebra in the mathematics of the Babylonians.[21] Moreover, insofar as Apollonius reduced the *symptômata*, or essential defining properties, of the conic sections to problems in the application of areas (whence the names of the conic sections), one might well want to speak of an algebraic input into the *Conics*.[22] Also, an examination of Archimedes' use of analysis in his *Sphere and Cylinder* reveals the presence of an elementary algebra of line segments and their ratios.[23]

A closer look, however, at the identification of algebra and analysis reveals an underlying lack of comprehension of the nature of Greek geometry. The analytic geometries of Fermat and Descartes alone suffice to show by contrast that Greek geometry was not algebraic in any essential way. Nor does Greek geometrical analysis rest on algebraic foundations. At heart, algebra and classical Greek geometry represent two substantially different approaches to mathematics and reflect different demands on mathematical knowledge.[24] Nothing better illustrates the differences than the Greeks' felt need to translate inherited Babylonian algebraic techniques into an essentially alien geometrical form. If analysis included that inherited material, it nonetheless was the product of purely Greek geometrical thought. No one before the Greeks had ever thought to prove theorems, and theorem-proving analysis constitutes a logical and epistemological procedure that transcends the particular form of mathematics to which it is applied. Similarly, although algebra may enhance the heuristic capabilities of problem-solving analysis, it does not thereby make analysis dependent on it; problem-solving analysis subsumes algebra.

Reflection on the matter makes it clear, on the one hand, why men such as Ramus, Viète, and Descartes considered algebra analytic in nature and, on the other, why in taking the further step of equating algebra and analysis they necessarily did violence to the classical doctrine of analysis. Algebra links one or more unknown quantities with known quantities in an equation and operates on that equation as if all the quantities were equally known. It aims to express the unknown quantities in terms of combinatory products of the

---

[21] First set forth in Otto Neugebauer's *Vorgriechische Mathematik* (Berlin, 1934; 2nd unrevised ed., 1969), the ties between Babylonian mathematics and early Greek mathematics were emphasized more explicitly by B. L. van der Waerden, *Ontwakende Wetenschap* (Groningen, 1950; Engl. trans., *Science Awakening*, Groningen, 1954), Chaps. IV-V. Cf. M. S. Mahoney, "Babylonian Algebra: Form vs. Content," *Studies in History and Philosophy of Science*, 1(1970), pp. 369–380.

[22] See E. J. Dijksterhuis, *Archimedes* (Copenhagen, 1956), pp. 59–63.

[23] Mahoney, "Another Look," pp. 337–338.

[24] On the differences between algebra and geometry, see M. S. Mahoney, "Die Anfänge der algebraischen Denkweise im 17. Jahrhundert;" *Rete*, 1(1972), pp. 16–18.

known and thereby to make the unknown quantities known. What procedure could better fit the classical description of analysis? In the passage cited above, however, Viète goes beyond treating algebra as an example of analysis. He replaces the classical definition with a new one based on the procedures of algebra itself. That is, he not only views algebra as analytic, he also makes analysis algebraic. To do so, he must alter the constituents of the classical definition. Where zetetic had denoted for Pappus theorem-proving analysis and poristic, problem-solving analysis, Viète now makes zetetic analysis the procedure by which a problem is transformed into an algebraic equation linking known and unknown quantities and poristic analysis, the procedure by which the validity of the equation is confirmed. More important, in equating algebra with analysis, Viète is forced to introduce a third category of analysis completely alien to Pappus's discussion but absolutely essential to algebra, to wit, the procedures by which the equation set up by zetetic is solved. Whereas for Pappus zetetic and poristic played roughly equal roles in analysis, for Viète the third category, exegetic, bears the major burden; it encompasses nothing less than the theory of equations addressed to their solution. Despite the similarities between the passages that open Book VII of the *Mathematical Collection* and the *Introduction to the Analytic Art*, the two passages are talking about very different subjects. In the latter, analysis has become algebra.

More accurately stated, analysis has become the analytic art, for the algebra with which Viète equates analysis is a form of mathematics very different from what the cossists and their predecessors had called "algebra." The same sorts of motives that led him to alter the nature of analysis also brought about a change in the nature of algebra so radical that Viète and his successors found it appropriate to drop the name "algebra" altogether.

From its beginnings in Babylonia to its culmination in the Renaissance cossist tradition, algebra constituted a sophisticated form of arithmetical problem-solving.[25] It presupposed the six basic combinatory operations of arithmetic—addition, subtraction, multiplication, division, raising to a power, and extraction of a root—and treated the unknowns as numbers.[26] It was addressed solely to the solution of problems, that is, to the determination of the numerical value of the unknown. Although, beginning with Diophantus of Alexandria, various systems of abbreviatory symbolization were employed, the symbols always denoted numbers and played nothing but an abbreviatory role. Indeed, up to the sixteenth century, the symbols used derived from standard paleographic abbreviations for the words "unknown (number)," "square," "cube," etc.; for example, Diophantus' ς (derived most probably

---

[25]On cossist algebra, see P. Treutlein "Die deutsche Coss," *Abhandlungen zur Geschichte der Mathematik*, 2(1879), pp. 1–124; J. Tropfke, *Geschichte der Elementar-Mathematik*, Vol. II (3rd. ed., Berlin, 1933), Chap. A.

[26]Indeed, as noted in Chap. I above, cossist texts often opened with an exposition of arithmetical computation by means of the new Hindu-Arabic system.

from a ligature abbreviation for *arithmos*), Δ^Y (*dynamis* = "square"), K^Y (*kybos* = "cube"), or the cossists' ʉ (*res*), ʒ (*census*), or ɣ (*cubus*). Hence, the totally verbal algebra of al-Hwârizmî differed in no essential way from that of Diophantus or the cossists. Each proceeded by example, producing a paradigmatic solution of a problem. Although "recipes" or algorithms emerged for the solution of problems of various sorts, no formulas *per se* were derived. Beyond such explanations as, for example, that ʉ times ʉ equals ʒ, or ʉ times ʒ equals ɣ, medieval and Renaissance algebra did not really compute with symbols.

Aiming in his *Introduction to the Analytic Art* to establish an entirely new form of algebra with essentially different goals and procedures, Viète began with a radically new form of symbolization: "In order that [the setting up of equations] be aided by some art, it is necessary that the given magnitudes be distinguished from the unknown ones being sought by a constant, perpetual, and highly conspicuous convention (*symbolo*), such as by designating the magnitudes being sought by the letter A or some other vowel *E, I, O, U, Y*, and the given [magnitudes] by the letters *B, G, D*, or other consonants."[27]

First of all, the new notation made it possible to divorce algebra from a style of exposition rooted in examples and verbal algorithms. Viète's literal symbolism enabled the mathematician to treat the data of a problem as parameters and hence to treat the problem itself as a general type.[28] Where, for example, cossist algebra could present the solution of the quadratic equation only by carrying out the direct solution of a specific example, say, 3 ʒ + 5 ʉ = 20, Viète's analytic art could deal directly with the general quadratic equation $B \cdot A \ quad + G \ planum \cdot A = Z \ solido$ and derive a solution as a formula.[29]

This liberation of algebra from the necessity of dealing with particular examples involving specific numerical coefficients had two far-reaching corollaries. First, by eliminating the possibility of actually carrying out combinatory operations involving the parameters, it focused attention on those combinatory operations and hence on the procedures of solution rather than on the solution itself. In doing so, it also emphasized the way in which those

[27]Viète, *Introduction to the Analytic Art* (1646), p. 8.
[28]Cf. C. B. Boyer, *History of Analytic Geometry* (New York, 1956), pp. 59–60: "Using vowels to designate unknown quantities and consonants to represent quantities assumed to be known, Viète made it possible to distinguish not two, but three, types of magnitudes in algebra—specifically given numbers, parameters, and variables. . . . It became possible to build up a general theory of equations—to study, not cubic equations, but *the* cubic equation." In fact, one step separates Viète's algebra from fully general equations. Viète followed Greek mathematics in restricting the concept of number to positive numbers. Hence, his parameters always carry a positive sign, and the equation $x^2 + ax + b = 0$ is unimaginable. His symbolic algebra broke the path, however, to a fully general concept of number ranging over the complete real number system and even allowing for complex numbers. One sees the results clearly in Descartes' *Geometry*, Book III. See J. Klein, *Greek Mathematical Thought and the Origins of Algebra* (Cambridge, Mass., 1968).
[29]*Planum* and *solido* are homogeneity designations; see below, p. 41ff.

procedures of solution might be applied to quantities other than numbers, provided that the combinatory operations were suitably defined. That is, the wholly symbolic solution of general equations readily suggested that the solution could be applied to other than numerical problems; to problems involving, for example, geometrical line segments or angles. Indeed, through its literal symbolism alone, Viète's analytic art suggested its applicability to problems involving any sorts of objects for which one could define the sum, difference, product, quotient, power, and root. Unlike numerical logistic, Viète pointed out in Chapter IV of the *Introduction*, the analytic art constituted a logistic of species, an arithmetic "set forth in terms of the species or forms of things, such as the letters of the alphabet." The species Viète had in mind was the species of quantity, and the application of the analytic art depended, as he went on to say in Chapter IV, only on the possibility of setting forth what it meant to add two things together, to subtract one from the other, to multiply them together, and so on. In the *Introduction to the Analytic Art*, as in the whole of the *Analytic Art* itself, algebra was transformed from a sophisticated sort of arithmetical problem-solving into the art of mathematical reasoning itself, insofar as that reasoning was based on combinatory operations. Where cossist algebra had represented a subdivision of arithmetic, the analytic art rose to a position subsuming all combinatory mathematics, whether arithmetic, geometry, or trigonometry.

Second, and again because symbolically denoted combinatory operations were not replaced by their products (as 2 + 3 is replaced by 5 in numerical logistic), the symbolic solution of equations drew attention to the structure of the solution and hence to the structure of the original equations. It rendered clear, for example, the relationships between the roots of an equation and its coefficient parameters. Hence, Viète's analytic art not only emphasized procedures of solution over solution itself, but it also fostered investigation of the structure (Viète's word is *constitutio*) of the equations being solved. In short, Viète's analytic art led directly to his major contribution to the development of algebra, the first consciously articulated theory of equations. The shift in focus is reflected in the different subtitles algebra bears in the sixteenth and seventeenth centuries; as a result of Viète's reformulation of its goals and procedures, the "art of the coss" becomes the "doctrine of equations."

The subjects of the remaining works that comprise the *Analytic Art* best illustrate that reformulation. Following the *Introduction* and beginning from the four basic combinatory operations set forth symbolically there, the *Prior Notes on Specious Logistic*[30] further articulates the combinatory manipulation of symbolic quantities. Here Viète derives the many algebraic identities fundamental to the transformation of equations into the canonical forms

---

[30]First published by Jean de Beaugrand (Paris, 1631); cf. Ritter, "François Viète," p. 380ff.

treated by his theory of equations: for example, $(A + B)^2 = A^2 + B^2 + 2AB$, or $(A + B)^2 + (A - B)^2 = 2A^2 + 2B^2$. The identities derived, which constitute practically an exhaustive catalogue of the identities known at the time, increase steadily in complexity, culminating in such expressions as

$$(B^2 + D^2)(F^2 + G^2) = \begin{cases} (BG + DF)^2 + (BF - DG)^2 \\ (BG - DF)^2 + (BF + DG)^2, \end{cases}$$

an identity on which Fermat would later build an imposing theory of decomposition of primes and their powers into squares.

On the basis of this extended system of symbolic calculation, Viète is then able in the *Five Books of Zetetics* [31] to attack a broad range of traditional algebraic problems drawn from varied sources and to translate them into equations treated by the analytic art. The *Zetetics* thus serves several purposes. First, it reinforces by example the claims made in the *Introduction* of the wide applicability of the art and of the new status of algebra as the superordinate vehicle of analysis. Second, it further explicates and illustrates the symbolic manipulation of quantities set forth in the *Prior Notes*. Most important, however, it constitutes the main guide for the manner in which one in fact applies algebra to mathematical problems, that is, how one goes about translating verbal statements of arithmetical, geometrical, or trigonometrical relationships into equations. Insofar as the equations that emerge fall immediately into combinatorial categories already treated in the *Prior Notes*, their solution following immediately from their structure, the *Zetetics* also provides an introduction to the more detailed analysis of the structure of equations and of the solution of equations that follows in Viète's next works. Its main goal, however, is to show how the algebraist—or, as Viète now says, the *analyst*—sets up the equations he will study.

The *Two Treatises on the Recognition and Emendation of Equations* [32] represents the central element of Viète's theory of equations. Its assigned role in the *Analytic Art* is to set forth and justify the procedures by which the infinite variety of equations established by the application of zetetic analysis to problems is transformed into a small number of canonical forms for which the general solution is known. The *Recognition and Emendation of Equations* constitutes, therefore, Viète's translation into the analytic art of the most prominent and heuristically powerful aspect of traditional analysis, the analysis that reduced an unknown problem to a known one. To take just one example that will appear below in the discussion of Fermat's work, Viète knew—most probably from Cardano's *Ars magna*—of Tartaglia's general solution of the cubic equation $x^3 + ax = b$. He could readily translate that solution, which itself constituted a reduction of a cubic equation to a quadratic, into the symbolism of the analytic art. The *Recognition and Emendation of*

[31] On the date, see Grisard, 60.
[32] Published by Alexander Anderson (Paris, 1615); Ritter, p. 389.

Equations vastly extends the importance of that solution by showing that any cubic equation can be reduced to the form $x^3 + ax = b$ (which thereby becomes the canonical form of the cubic equation), by establishing and justifying standard procedures for this reduction, and by determining the relationship between the solution of the reduced equation and that of the original. In doing so, it also presents an ancillary apparatus of results concerning, for example, the relationship between the roots of equations that contain the same terms but in which those terms bear different signs, the relationship between the roots of an equation and the coefficients of its terms (i.e. the elementary symmetric functions), the procedures for elimination of terms through adjustment of the roots, and the possibility of reduction of the degree of an equation.

Although the *Recognition and Emendation of Equations* includes the general solutions of various canonical equations, it places emphasis less on those solutions themselves than on the manner in which they constitute the reduced solutions of whole families of equations. In doing so, it subtly alters the nature of the algebraic enterprise in ways that prove important for understanding the work of Fermat. The shift from direct problem-solving to the theoretical investigation of the structure of equations brought new attention to questions of solvability, of classification of problems, and of the relationship among problems. It reflected, if indeed it did not inaugurate, the striking reorientation of mathematics in the seventeenth century away from specific problems toward *methods* of solution.

Whatever the long-range implications of Viète's theory of equations, the *Analytic Art* could not ignore the final stage of problem-solving analysis, the translation of general solutions into specific results. *On the Numerical Resolution of Affected Powers*[33] takes up, then, the problem of interpreting the symbolic quantities and operations of the analytic art in concrete numerical terms. Similarly, the *Canonical Recension of Geometrical Constructions*[34] illustrates in concrete terms how those symbolic quantities may be interpreted as line segments and the symbolic operations as geometrical construction procedures. Since the *Geometrical Recension* restricts itself to Euclidean construction procedures based ultimately on the circle and straightedge alone, the *Supplement to Geometry* and the *Pseudo–Mesolabe*[35] complete the translation of the analytic art into geometry by establishing the correspondence between certain solution procedures on the one hand and constructions involving so-called "mechanical" curves on the other. Finally, the *General Theorems on Angular Sections*[36] (stated by Viète but proved by his student, Alexander Anderson) provides the concrete foundations on which to rest the application of the analytic art to trigonometry.

[33] Published by Marino Ghetaldi (Paris, 1600); Ritter, p. 383.
[34] Tours, 1593; Ritter, p. 402.
[35] *Supplementum* (Tours, 1593); Ritter, p. 403; *Pseudo-Mesolabe* (Paris, 1595); Hofmann, Intro. to Viète's *Opera*, p. xxii.
[36] Paris, 1615; Ritter, p. 406.

The number and variety of the works just cited illustrate the central feature of the analytic art, the feature that makes it as much a program of research as a body of mathematical results. The elevation of algebra from a subdiscipline of arithmetic to the art of analysis deprived it of its content at the same time that it extended its applicability. Viète's *specious logistic*, the system of symbolic expression set forth in the *Introduction*, is, to use modern terms, a language of uninterpreted symbols. Just as the letters denote nothing more than quantities or magnitudes and thus require in any application of the art interpretation into numbers or line segments, so too the symbolic operations of addition, subtraction, multiplication, and so on, demand concrete interpretation in terms of arithmetical or geometrical combinatory procedures. As a formal language, specious logistic can itself generate problems of syntax only. That, indeed, is what both the *Prior Notes* and the *Recognition and Emendation of Equations* do. The former builds on the basis of a few primitive syntactical elements procedures for formulating increasingly complex statements in the language of specious logistic. The latter, by contrast, analyzes complex combinations into their primitive elements. Although elaboration of syntax, i.e. the theory of equations, provides its own problems for research, it cannot of itself determine which problems will lead to successful application of specious logistic to concrete problems. It cannot, that is, specify which syntactical problems will prove particularly enlightening when interpreted arithmetically or geometrically. Hence, the analytic art must look beyond itself for the problems it will investigate.

Viète had no lack of stimulus. Cossist algebra and the *Arithmetic* of Diophantus alone provided him with an extensive amount of interpreted material to be analyzed syntactically. Insofar, however, as those sources represented the traditional, numerical use of algebra, their translation into specious logistic, though it made possible a hitherto unachieved systematization of material, did nothing to extend the realm of applicability of the analytic art. For that extension, Viète looked to geometry and to the primary source of geometrical analysis, the works cited by Pappus in Book VII of the *Mathematical Collection* as comprising the "field of analysis."

Incorporation of this material into the analytic art posed two major problems: recovering it in its original geometrical form and translating it into the symbolic language of the art. The first problem derived from the fact that very few of the works cited had survived the holocaust of late antiquity. Euclid's *Data* was alone entirely extant.[37] Only the first four books of Apollonius' *Conics* were available in Europe in 1600; Books V–VII were translated from the Arabic in 1661, and Book VIII seems to be irrevocably lost. None of the other treatises of analysis were even partially extant. Yet, the situation was not entirely hopeless. Pappus not only discussed each work in more or

---

[37]The *Data*, together with the commentary of Marinus of Sichem, appeared in a Latin translation by Zamberti (Venice, 1505) and a Greek-Latin edition by Claude Hardy (Paris, 1625).

less general terms, he also set down in Book VII over 400 lemmas pertinent to them. The first task facing the new school of analysts, therefore, was clear: reconstruct as closely as possible the content of the lost works of classical geometrical analysis cited by Pappus.

Viète himself began this project of restoration with a reconstruction of Apollonius' *On Contacts*.[38] Marino Ghetaldi, a Ragusan patrician who had studied with Clavius in Rome before meeting Viète in Paris and becoming his devoted student, followed the master's lead by supplementing the restoration of *On Contacts* and adding his own reconstruction of Apollonius' *Inclinations*.[39] Wilebrord Snell countered Viète's "Apollonius Gallus" with an "Apollonius Batavus," a restoration of the *Determinate Section*, to which he later added those of the *Section of Ratio* and the *Section of Area*.[40] It was in this tradition of preliminary restoration prior to translation into the analytic art that Fermat began the restoration of Apollonius' *Plane Loci* while still in Bordeaux. Through the system of analytic geometry suggested by that restoration, he thought he saw the path to a reconstruction of Aristaeus' *Solid Loci* and Euclid's *Surface Loci*. Indeed, long after his own research had rendered much of this restorative activity superfluous, Fermat was still trying to reconstruct Euclid's *Porisms*.[41] Since these efforts at recovery of the analytic texts were only preparatory to their translation into the analytic art, the restorations themselves were carried out in the style and context of classical Greek geometry and hence resembled closely the work of the classical school of mathematics discussed in Chapter I. The underlying motivation, however, was quite different, and the two sorts of restorational activity should not be confused.

Once the constituent works of Pappus' field of analysis were restored, the analysts faced their second task: to translate the geometrical content of the "field of analysis" into the language of the analytic art, i.e. into specious logistic, or symbolic algebra. The difficulty of the task varied with the nature of the treatise. The *Section of Ratio*, for example, contained only metric problems demanding a point solution; e.g., divide a given line segment into two parts bearing a given ratio. For such problems, a model for translation already lay at hand in the sections of Euclid in which Ramus and Viète had first discerned traces of algebra applied to geometry and received the original impetus toward developing algebra as the art of analysis. At first, then, only

---

[38]*Francisci Vietae Apollonius Gallus seu exsuscita Apollonii Pergaei* περὶ ἐπαφῶν *Geometria* (Paris, 1600; *Opera*, pp. 325–346).

[39]*Marini Ghetaldi, Patritii Ragusensis: Apollonius redivivus seu restituta Apollonii Pergaei inclinationum Geometria. Supplementum Apollonii Galli seu exsuscitae Apollonii Pergaei tactionum Geometriae pars reliqua* (Venice, 1607).

[40]*Wilebrordi Snellii R. F.* περὶ λόγου ἀποτομῆς καὶ περὶ χωρίου ἀποτομῆς *resuscita Geometria* (Leiden, 1607); *Apollonius Batavus seu exsuscita Apollonii Pergaei περὶ διωρισμένης τομῆς Geometria* (Leiden, 1608).

[41]FO.I.76ff. The preface to this work testifies to Fermat's knowledge of all previous attempts at the restoration of the lost works cited by Pappus.

those treatises, of which the problems could be expressed as determinate equations in one unknown, lay at the analysts' algebraic command. Indeed, the use of algebra may at times have aided in their restoration. The analytic treatises involving loci, however, foremost among them Apollonius' *Conics*, posed translational problems for which there existed no model. How does one express loci in the form of equations? It remained for Fermat and Descartes to devise the necessary technique.

It was, then, the overriding goal of applying an algebra with historical roots in arithmetical problem-solving to the realm of geometrical analysis that set up the main program of research connected with Viète's analytic art. Before the equation-theoretical content, i.e. the syntactical content, of the geometrical "field of analysis" could itself be analyzed and compared with that of numerical algebra, the problem of translation itself had to be solved. Beyond any specific technical difficulties of translation, one major hurdle stood in the way, the problem of homogeneity. In adopting the classical Greek interpretation of the four basic combinatory operations in the realm of geometry, Viète confronted an element alien to numerical algebra: dimension. His algebraic treatment of dimension led his analytic art into a fundamental conceptual difficulty, from which Fermat and Descartes had to extricate it before they could proceed in realizing its goals.

Numbers have no dimension. Hence, the product of any combinatory operation on numbers is itself a number. Greek geometry, however, operated with elements that do have differing dimension; a point has none, a line is one-dimensional, a rectangle two-dimensional, etc. The classical Greek translation of the combinatory operations of addition and multiplication (with their inverses) into geometrical constructions took account of this aspect of dimension. First of all, one could add together only elements of the same dimension, i.e. a line to a line, a rectangle to a rectangle. Second, and more important, as the problem of multiplying two elements was presented to them—e.g. given two lines, to construct their product—the Greeks could see only one solution: to construct a figure of higher dimension. Hence, the product of two lines became the rectangle constructed from the lines as adjacent sides. The product of a rectangle and a line was a parallelepiped. In modern terms, the algebraic operations did not remain closed when applied to sets of geometrical elements.

Viète accepted the Greek translations of the operations. He could hardly do otherwise, given the fact that those translations, as found in Euclid II and VI, had in part inspired his whole project. He recognized the dimensional aspect of the operations when interpreted geometrically and he canonized them in what he called "the first and perpetual law of equations or proportions," the Law of Homogeneity:[42] one can add only homogeneous magnitudes, and their sum is likewise homogeneous with them; the product of any two magni-

---

[42] Viète, *Introduction to the Analytic Art* (1646), Chap. III, pp. 2–4.

tudes is heterogeneous with them. As a result, every equation of Viète's analytic art had dimension, and the dimension was directly linked to the degree of the equation. For example, the solution of a cubic equation corresponded to a solid construction in three-dimensional space (as in Archytas' construction of two mean proportionals via the intersection of a cone, a cylinder, and a torus) or, less satisfying to traditional canons of geometry, to the use of mechanically constructed "solid curves" (as in the determination of two mean proportionals via intersecting parabola and hyperbola; cf. Eutocius' commentary on *Sphere and Cylinder* II,4). Viète did not, however, accept what would seem to be an unavoidable implication of his canon, to wit, that equations of the fourth degree and higher had no meaning in the realm of geometry. Their interpretation was possible, though it required the introduction of "mechanical" procedures like the *Pseudo-Mesolabe* for generating fourth, fifth, sixth, etc., proportionals to two given line segments. Indeed, in Viète's eyes at least, one beauty of the analytic art lay in its ability to describe such mechanisms abstractly and to reveal their link to higher-degree problems. Moreover, Pappus provided those very "mechanical" construction elsewhere in the *Mathematical Collection*.

For the problems Viète was treating, the link between dimensionality and the degree of an equation offered no hindrance to handling the problems algebraically. All his problems, as noted above, aimed at point constructions; e.g. find the center of a circle under given metric conditions, or find the point that divides a line into given segments. The solution sought, a determinate point, was always dimensionless. Whether it resulted from a two-dimensional or a three-dimensional construction was a matter of esthetic concern only. But both Fermat and Descartes saw clearly the difficulties that Viète's translation of geometry into algebra caused if one tried to apply it to a realm of problems which themselves involved dimensionality, e.g. locus problems. If one wishes to determine the nature of a locus, which itself may be dimensionless or have a dimension depending on the conditions, one cannot use a vehicle of analysis that is already burdened by dimensionality. Both men independently asserted that the dimension of a problem lay not in its degree but in the number of unknowns necessary to analyze it. A point solution corresponds to a determinate equation in one unknown or a system of n equations in n unknowns; a two-dimensional solution, which contains the one-dimensional as a degenerate case, corresponds to an indeterminate equation in two unknowns, etc.

Although Fermat paid lip service to Viète's Law of Homogeneity, his use of algebra in geometric situations clearly shows that he no longer attached dimension to the degree of an expression. Fermat often operated as if all combinatory operations on geometrical magnitudes yielded homogeneous products. At no time, however, does he justify his behavior. He seems to have assumed *sotto voce* a line of reasoning made explicit by Descartes in the opening pages of the *Geometry*. Descartes too had begun at the same point as

Viète, that is, in attaching dimensionality to the combinatory operations of geometry. But, in thinking about the ideal nature of the "true mathematics," he came to the conclusion that the dimensional connotations attached to the operations robbed that mathematics of its full abstractness. "I myself," he recorded in the *Rules for the Direction of the Mind*,[43]

> was long deceived by these names ["square," "cube," etc.]; nothing seemed to be more clearly proposed to my imagination, after the line and the square, than the cube and other figures conceived in a similar way to these. Moreover, I resolved not a few problems with their aid. But finally I observed after many experiments that I had found nothing by this way of conceiving that I could not have recognized far more simply and distinctly without it, and that such names are to be altogether rejected, lest they confuse thought. . .

To reject such dimensional ideas from algebra, Descartes applied a basic principle of Euclid's theory of ratio and proportion to an idea suggested by the algebraic notation he himself was devising. Like Viète's notation, Descartes' clearly displayed the powers of the unknowns as powers, that is, as products of the repeated multiplication of the unknown by itself. Where Viète had retained the one last vestige of verbal algebra (and its intuitively geometrical overtones) by writing *A quadratus, A cubus*, and so on, Descartes introduced numerical superscripts, the one essential difference between the two systems of notation. Both men, moreover, observed that the successive powers of the unknown constituted a continued proportion, i.e. $x : x^2 = x^2 : x^3 = x^3 : x^4 = \ldots$.[44] Descartes went beyond Viète, however, in noting first that the continued proportion began with a unit magnitude, i.e. $1 : x = x : x^2 = x^2 : x^3 = \ldots$,[45] and second that, according to Euclid, ratios obtain only between homogenous quantities. Hence, he reasoned, the powers of the unknown must all be homogeneous with one another and with the unit that serves as basis for the continued proportion. To maintain homogeneity in multiplication, one needed only a unit magnitude to which all other magnitudes could be re-

---

[43]*Regulae ad directionem ingenii* (AT.X), Regula XVI, pp. 456–457. On the development of Descartes' algebra and of his *Geometry*, see M. S. Mahoney, "Descartes, René." *Dictionary of Scientific Biography*, Vol. IV (1971), pp. 55–58.

[44]Viète, *Introduction to the Analytic Art* (1646), p. 1, speaking of the powers of the unknown: ". . . & inde constituta, ut fit, solemni magnitudinum ex genere ad genus vi suâ proportionaliter adscendentium vel descendentium serie seu scalâ, quâ gradus earundem & genera in comparationibus designentur ac distinguantur."

[45]Descartes, *Regulae*, Rule XVI: "It is therefore most strongly to be noted that the root, the square, the cube, etc., are nothing other than continuously proportional magnitudes to which one always supposes prefixed that assumed unity of which we have already spoken above; to which unity the first proportional is referred immediately and by a unique relation, the second by way of the first and thus by two relations, the third by way of the first and second, and by three relations, etc. Hence, we will hereafter call the first proportional that magnitude which is called in algebra the "root," the second proportional that which is called the "square," and so on for the others."

referred. Then to multiply two given line segments, for example, a and b, one needed only to devise a construction which corresponds to the proportion $1 : a = b : ab$.[46] That is, construct a triangle of sides 1 and a and then construct a triangle similar to it in which the side corresponding to 1 is b; the side corresponding to a will then be ab.

The mark of Descartes' genius lies in his bold assertion that, if the unit is not given by the problem, it may be chosen arbitrarily, provided only that all other magnitudes are then referred to it. His approach to the problem of homogeneity resulted from his pursuing in a more consequential way the idea underlying the analytic art. In studying the structure of a geometrical problem, one could treat its particular metric as irrelevant. As Descartes strongly insisted in the *Geometry*, the assumption of a unit did not make arithmetic of geometry; rather, it rendered geometry susceptible to algebraic expression and algebraic analysis.

Fermat's works contain no such explicit solution to the problem of the dimensionality of combinatory operations in geometry. They do, however, show Fermat treating his problems as if he had come to the same conclusion. Both Fermat and Descartes had to overcome the problem in order to introduce into the theory of equations the analytical content of ancient treatises on loci, content that extended that theory to the analysis of the structure of indeterminate equations and provided the foundations for a gradually emerging concept of the function.

In sum, Fermat inherited from Viète a highly sophisticated system of algebra, which included a theory of equations and a definite program of research aimed at extending its applicability over a widening range of mathematical problems. With respect to its technical content, Viète's system offered few problems of internal articulation and explication. Although Fermat made slight improvements here and there (discussed below in Chapters III and IV), he largely accepted Viète's *Analytic Art* as his model and extended it to new sorts of problems.

Yet, Fermat was sensitive to the drawbacks of Viète's system. Although he quite rightly considered the choice between Viète's *A,E,I;B,G,D* notation and Descartes' *x,y,z; a,b,c* notation a matter of taste and remained faithful to the former throughout his career, he could and did appreciate the great simplicity and clarity that characterized Descartes' theory of equations. Viète's humanist flair for elegant Latin and for the frequent use of Greek neologisms, combined with what seems to have been a lawyer's penchant for casuistry, makes the *Analytic Art* an extremely difficult book to read and understand. Not only did a limited concept of number, which led Viète to restrict the range of his parameters to the domain of positive integers, force him to divide equations into a frustratingly large number of particular cases, but he also built that division into cases into his system by giving each treatment of a

---

[46] *Geometry* (Leiden, 1637), pp. 297–299.

case a special name. Many of his techniques that bear different names are not in fact different, but merely variant applications of the same basic technique to different cases of equations. As Chapters III and IV below will show, Fermat recognized the essential unity of many of Viète's equation-theoretical methods. Nonetheless, he neither altered Viète's system nor abandoned the bewildering terminology. His devotion to Viète frequently made his own works difficult to read, especially in circles relatively unfamiliar with Viète, e.g. the circle of Italian mathematicians.

Even in circles that knew Viète's works, Fermat's steadfast commitment to the model of the *Analytic Art* ultimately placed him at a disadvantage. It meant that his own work would share to some extent the fate of Viète's upon the appearance of Descartes' *Geometry* with its brilliant new theory of equations in Book III.[47] Though perhaps not as complete as Viète's system, Descartes' took advantage of a more broadly conceived notion of number to strip away treatment by cases and lay bare the essential procedures for the transformation and reduction of equations to canonical forms. As a result, many aspects of the structure of equations, though they had been present in Viète's theory, emerged with far greater clarity in Descartes'. For example, although Viète's method of *syncrisis* discussed in detail below in Chapter IV provides a means of generating the elementary symmetric functions of any equation, the cumbersome manner in which it is presented tends to block the reader's view of the importance of those functions (the importance is not retrospective; both Viète and Fermat were fully conscious of it). By contrast, Descartes' approach to equations brought out their importance with brilliant clarity. Any comparison between the works of the two men makes it evident why Descartes' carried the day to the rapid obscurity of Viète's. Even though Fermat's command of the theory of equations equalled, or even excelled, that of Descartes, his continued use of the notation and terminology of the *Analytic Art* increasingly prevented others, now schooled in the Cartesian model, from appreciating the fact and diminished the effect of Fermat's work.

Despite differences in notation and terminology, Fermat and Descartes shared with Viète a basic style of exposition that marked them as members of the analytic school of mathematics and frequently led to dispute with members of other schools. All three men were convinced that the translation of classical Greek analysis into the new algebra of the analytic art removed the logical difficulties inherent in analysis and so obviated the need for synthetic proof of analytically derived results. Fermat offered no explicit justification for his belief, though his conviction shows through at every point in his

---

[47] As Descartes himself recognized; cf. Descartes to Mersenne, 29.VI.1638, Alquié.II:70: "Pour M. Fermat, son procédé me confirme entièrement en l'opinion que j'aie eue dès le commencement, que lui et ceux de Paris avaient conspiré ensemble pour tâcher à décréditer mes écrits le plus qu'ils pourraient: peut-être à cause qu'ils ont eu peur que, si ma Géométrie était en vogue, ce peu qu'ils savent de l'Analyse de Viète ne fût méprisé."

works. If, as suggested in Chapter I, his personal motives for pursuing mathematics explain to some extent his tendency to slight the rigorously demonstrative aspects of the subject, his commitment to Viète's style of mathematics completes the explanation.

Chapter VI of Viète's *Introduction* provided Fermat with sufficient justification. In its entirety, it reads:[48]

> Zetetic having been carried out, the analyst moves from hypothesis to thesis and sets forth the theorems that he has found in the order of the art, subject to the laws *kata pantos, kat' auto [sic]*, and *kath' holou proton*. While these [theorems] derive their soundness and demonstration from zetetic, nevertheless they are to be subjected to the law of synthesis, which is thought to be a more logical way of demonstrating. And, when it is necessary, they are proved in this manner by a great miracle of the inventive art; for to do so one repeats the traces of the analysis. But this is also analytical and, due to the use of specious logistic, it is not difficult. For if a strange theorem, or one found by accident, is proposed, then the way of poristic should be tried first, from which the return to synthesis is then easily made. Examples of this are laid out by Theon in the *Elements*, XIII,1-5, and Apollonius of Perga in the *Conics*, and by Archimedes himself in various books.

Viète's citation of Ramus' meaningless three "laws of method" should not detract from the sense of this passage.[49] For Viète, the foundation of the analytic art on the biconditional inferences of equality and substitution meant that analysis alone sufficed to guarantee the correctness of a derived result. For those who nevertheless insisted that the result be confirmed by rigorous synthetic demonstration, the analytic art offered further aid. It preserved symbolically all the inferences comprising the analysis. Hence, since all those inferences were biconditional, one had only to reverse the sequence in order to arrive at a synthesis. Moreover, to the extent that any algebraic statement could rapidly and systematically be examined for its validity (ultimately by reduction to an identity), the analytic art provided a sure means for finding the proof of any theorem. That was the "great miracle" of the analytic art. Firm in his conviction that synthesis was always possible, Viète himself never carried it out; for him, synthesis became a matter of esthetics.

It was not, then, merely Fermat's personal lack of concern for putting the finishing touches on results that prompted him to remark with a frequency

---

[48]Viète, *Introduction to the Analytic Art* (1646), p. 10.

[49]On Ramus' three laws of method, which he took from a complex discussion in Aristotle's *Posterior Analytics* and elevated to the rank of guiding principles of his pedagogy, see Ong, *Ramus*, pp. 258-262. Their appearance in the passage just cited may be taken as a mark of Viète's Ramist tendencies. An analysis of what purpose Viète meant them to serve here is given in Mahoney, "Royal Road," p. 198ff.

that dismayed many of his correspondents: *facilis est ad synthesin regressus*, "the reversion to synthesis is easy" (and so, by implication, he will omit it). Rather, the source on which he was modeling his mathematics provided ample justification for taking that attitude. Another source in the same tradition spelled out the justification in detail, though it is not clear Fermat in fact knew about it. In converting to Viète's style of algebra, Marino Ghetaldi[50] not only participated, as noted above, in the effort to restore the lost works of analysis, but also undertook to support Viète's claims regarding analysis and synthesis in an extensive work entitled *On Mathematical Resolution and Composition*, which appeared posthumously in Rome in 1630. There Ghetaldi paid systematic attention to two of the aspects of the analytic program discussed above. First, he carried out in step-by-step, problem-by-problem fashion the translation of geometrical analysis into the language of the analytic art, taking his material largely from Euclid and from the section treatises he and Viète had restored. Second, he spelled out precisely how, having arrived by algebraic analysis at a solution of a problem, one reversed the sequence of reasoning and picked up the "traces" (*vestigia*) of that analysis to produce a rigorous synthesis of the problem in the ancient mode. For those who knew it and used it, Ghetaldi's work provided both a theoretical justification of the analysts' neglect of synthetic proofs and a useful collation of analytical results already achieved. Fermat's analyses are often incomplete, stopping at a problem to which he assumes the reader knows the solution. In most cases, the reader could find the solution in Ghetaldi's work. To that extent, Ghetaldi's *On Mathematical Resolution and Composition* represented an algebraic version of Pappus' "field of analysis."

The patterns of mathematical behavior deriving from the analytic tradition provide insight into what the following chapters will reveal as a fundamental paradox of Fermat's mathematics. Fermat constantly maintained that he was doing mathematics in keeping with classical Greek analysis. In topic after topic he began from classical sources. Yet, in each instance, he eventually took steps that carried him across the frontiers and limitations of those classical sources and into a realm of mathematical concepts unknown to the ancient Greeks. Especially where, in his mature method of quadrature, Fermat began to operate with infinitesimals and limit procedures, the reversion to synthesis became, as mathematicians from Newton to Cauchy were to find, far from easy. In freeing Fermat from the obligation to provide syntheses, the

---

[50] On Ghetaldi's life and work, see M. Saltykow, "Souvenirs concernant le Géomètre Yougoslave Marinus Ghetaldi, conservés à Doubrovnik, en Dalmatie," *Isis*, 29(1938), pp. 20–23; E. Gelcich, "Eine Studie über die Entdeckung der analytischen Geometrie mit Berücksichtigung eines Werkes des Marino Ghetaldi Patrizier Ragusaer," *Abhandlungen zur Geschichte der Mathematik*, 4(1882), pp. 191–231; and the collection of essays published from an international symposium on Ghetaldi, *Geometrija i algebra početkom XVII stoljeća, povodom 400-godišnjice rođenjy Marina Getaldića* (Zagreb, 1969), esp. the article by Karin Reich, "Quelques remarques sur Marinus Ghetaldus et François Viète," pp. 171–174.

analytic tradition conditioned him to pass by unawares the point at which those syntheses were no longer merely dispensible, but at the time impossible to provide.

Before turning, however, to the precise details of Fermat's mathematics, it would help to gain an overview of his career and of the general, thematic ways in which that career exemplifies the analytic program of François Viète.

II. FOLLOWING THE "PRECEPTS OF THE ART"

A. *The Early Years (to 1636)*

In light of the above, Itard's complaint that "we are totally ignorant of [Fermat's] first teachers and of the scientific education he received" is not wholly accurate.[51] Nonetheless, the precise path Fermat took to acquire the mathematical apparatus with which he was fully equipped when he appeared on the mathematical scene in 1636 remains obscure. The few reminiscences that he allowed himself in his correspondence indicate that, sometime during the late 1620s, he spent time in Bordeaux before moving on to Orléans, where he received his law degree sometime before 1631.[52] His reasons for going to Bordeaux are far from clear. Perhaps he did so to further his legal studies; his acquaintances there were members of the legal profession. Yet, looking back on the experience, he spoke only of his mathematical activities. Indirect evidence strengthens a conjecture that Fermat took a respite from study of the law during the sojourn in Bordeaux and devoted himself to his side-interest, mathematics. Certainly his family's wealth would have made such a respite possible, and a break in his legal studies would go at least part of the way in explaining why he completed them some six or seven years later than was usual at the time (Descartes, for example, received his law degree from the University of Poitiers at the age of 20).

Without information about Fermat's mathematical education before going to Bordeaux, it is impossible to determine if he went there with any specific mathematical goal in mind. He almost certainly had covered the usual amount of arts-faculty mathematics which, at Toulouse even in the 1620s, probably did not extend much beyond the first six books of Euclid's *Elements* and some arithmetic.[53] But there were no mathematicians or scientists of note in Toulouse at the time to spur his interest or guide his early training, and, from all appearances, his mathematical education substantially began in Bordeaux, where he encountered the works and influence of François Viète.

Fermat's correspondence mentions four people in connection with Bordeaux. In his first letter to Mersenne in 1636, Fermat said that he had given the autograph copy of his restoration of Apollonius' *Plane loci*, albeit lacking

---

[51] J. Itard, *Pierre Fermat* (*Kurze Mathematiker-Biographien*, Nr.10), Basel, 1950, p. 2.
[52] Information on Fermat's activities in Bordeaux is all gained from his letters to Roberval and Mersenne in 1636 and 1637.
[53] The precise content of the scientific curriculum in French schools and universities during the seventeenth century is a critical lacuna in historical scholarship.

the "question la plus difficile et la plus belle," to a M. Prades "some six years ago" and that he had shown some of propositions to M. de Beaugrand. Writing to Roberval later that year, Fermat reported that:[54]

> On the matter of the method of maxima and minima, you should know that, since you have seen the one that M. d'Espagnet has given you, you have seen mine, which I gave him about seven years ago when I was in Bordeaux. And I recall that at that time M. Philon, having received one of your letters, in which you proposed to him the problem of finding the greatest cone having a conical surface equal to a given circle, sent it to me, and I gave its solution to M. Prades to send on to you. If you search your memory, you may perhaps recall the incident, and also that you proposed this problem as difficult and hitherto unsolved. If I can find your letter (which I saved at the time) among my papers, I will send it to you.

A year later, Fermat informed Roberval that the second book of *Plane loci* had been written eight years earlier, at which time he gave out two copies, one to "M. Despagnet, *conseiller* au Parlement de Bordeaux," and the other to someone whose name is apparently illegible; the earlier letter to Mersenne suggests it was Prades.

Little is known about three of these people and not much more about the fourth. Pierre Prades competed at least once for the Ramus chair of mathematics and corresponded with the man who later held that chair for forty-one years (1634–1675), Roberval.[55] François Philon was more a philologist than a philosopher. His ties with Fermat may have depended less on mathematics than on Fermat's own concern for classical letters, illustrated at the end of Volume I of the *Oeuvres*.[56] Étienne d'Espagnet was more a friend than a mentor. Son of Jean d'Espagnet, a Bordeaux *parlementaire* who had some reputation as an alchemist and himself a *conseiller*, Étienne was some ten years' Fermat's senior and by the latter's own description an *ami intime*.[57] By 1636, d'Espagnet held at least some of Fermat's mathematical papers and hence must have shared a mathematical interest with him. Perhaps more important for the present discussion is the fact that Jean d'Espagnet held some

[54] Fermat to Roberval 22.IX.1636, FO.II.71–72.

[55] Cf. Prades' *Celeberrimus Platonis locus de nuptiis, pro laurea professionis Rameae publice explicandus . . . Una cum exercitatione ad 64. et 65. proposit. Lib. I Arithmeticorum Francisci Maurolici* (Paris, 1625). Other than the title of this one work, I have been unable to find any further information on Prades. The identification of the man whom Fermat refers to only as "M. Prades" is tentative.

[56] *Oeuvres de Maistre François Philon contenant la traduction des douze livres de l'Aenéide de Virgile et autres pièces.* (Agen, 1640). Again, the identification is tentative.

[57] Jean d'Espagnet's works include: *Enchiridion physicae restitutus* (Paris, 1623); *Arcanum Hermeticae Philosophiae Opus* (Geneva, 1653); an edition of Pierre Choinet's *Le Rozier des guerres composé par le feu roi Louis XI . . . pour M. le dauphin Charles son fils . . . et ensuite un traité de l'institution d'un jeune prince, fait par ledit sieur . . . d'Espagnet* (Paris, 1616). Étienne had been serving as *conseiller* in Bordeaux since 1617. For Fermat's account of their friendship, cf. Fermat to Mersenne. 20.IV.1638, FO.II.136–137.

manuscripts of Viète's work, which by 1638 had been deposited, either directly or through Étienne, with Fermat.[58]

Jean Beaugrand first appears in the historical record in 1630, when he became mathematician to Gaston d'Orléans.[59] By then, his standing among Viète's followers stood high enough to earn J. L. Vaulezard's dedication of his French translation of the *Five Books of Zetetic* (Paris, 1630). A year later, Beaugrand published an edition of Viète's *Introduction to the Analytic Art* with commentaries, some of which van Schooten included in the 1646 edition of Viète's works. In 1635, after having served on the official commission to examine Morin's method for determining longitude, Beaugrand was named *secrétaire du Roi* under Chancellor Pierre de Séguier. He served in that capacity until his death in 1640.

Beaugrand and Fermat apparently had close ties, both intellectually and personally. If one must guess that Beaugrand introduced Fermat to Viète's analytic art, one can be certain that Fermat's early interest in geostatics and the mechanics of free fall stemmed directly from Beaugrand. In early correspondence with Mersenne, Roberval, and Pascal, Fermat stoutly defended Beaugrand's position that the weight of a body varied as its distance from the center of the earth, a position that was drawing heavy fire from scientists in Paris and that sparked a generally negative reaction to Beaugrand's *Geostatics*.[60] It was Beaugrand who sent Fermat a pre-publication copy of Descartes' *Dioptrics*, thus touching off the famous Fermat-Descartes controversy of 1637–1638.[61]

Their personal attachment seems clear. From all reports, Beaugrand behaved rather badly in Paris and managed to alienate most of his scientific colleagues. He carried on a nasty polemic with Desargues and attacked Descartes several times in anonymous pamphlets.[62] Even the unusually tolerant and neutral Mersenne was angered by Beaugrand's misuse of the pre-publication copies of Descartes' *Essays*. Through it all, and despite some personal discomfort at being embroiled in a controversy with Descartes, the overtones of which he did not fully understand, Fermat always spoke kindly

[58] On d'Espagnet's possession of Viète's manuscripts, see the preface to the 1646 Elzevier edition of Viète's *Opera*. Responding to an inquiry from Mersenne, who was gathering material for the edition, Fermat wrote in February 1638 that what he possessed had already been published, except for "des exemples plus étendus et quelques propositions de nombres multangulaires, qui se trouvent en d'autres livres, de sorte que l'impression de ses oeuvres n'en profiterait guère. Outre que je les ai reçus de M. Despagnet, à la charge de ne les bailler à personne que par son aveu" (FO.II.133).

[59] See Henry Nathan, "Beaugrand, Jean," *Dictionary of Scientific Biography*, Vol. I (1970), pp. 541–542, from which some of the information below is taken. The record is sparse, Nathan notes: ". . . what little is known or surmised about Beaugrand has had to be pieced together from sources dealing with his friends and enemies, and only rarely with him directly. There are few manuscripts or letters, and no records (*Ibid.*, 541).

[60] *Geostatice, seu de vario pondere gravium secundum varia a terrae (centro) intervalla dissertatio mathematica* (Paris, 1636). The treatise drew especially heavy fire from Descartes.

[61] See below, Chap. IV, §IV.

[62] It was, apparently, Beaugrand who first accused Descartes of plagiarizing Viète's algebra. He may well be "ce géomètre" to whom Descartes was referring in the passage cited above, n. 6.

of Beaugrand and referred to him as a friend. In turn, Beaugrand tended to treat Fermat as a protégé. On a trip to Italy in 1635, he brought his friend's work to the attention of Cavalieri, so that Fermat's later achievements did not come as a complete surprise to Italian mathematicians.[63] Beaugrand was not, however, above passing off Fermat's work as his own, as a letter of his to Thomas Hobbes clearly shows. Fermat does not seem to have known about this, nor does he seem to have appreciated the extent to which his relations with Beaugrand placed him at an initial disadvantage upon entering the Parisian mathematical scene in 1636.

Almost all of Fermat's fundamental innovations in mathematics have their roots in the research of the Bordeaux period. The first version of his method of maxima and minima, from which an algebraic method of tangents quickly followed, resulted from his investigation of Viète's theory of equations and his own attempts to explicate algebraically a passage in Pappus' discussion of a lemma to Apollonius' *Determinate Section*.[64] The letter to Roberval quoted above shows that by 1629 in Bordeaux the method of maxima and minima had been sufficiently articulated to enable the solution of a rather difficult problem.

Although Fermat's analytic geometry reached final form only sometime around 1635, its essential roots and its motivation also stretch back to Bordeaux. There, in keeping with the programmatic aspects of Viète's *Analytic Art*, Fermat began a restoration of Apollonius' *Plane Loci*, one of the treatises cited by Pappus as belonging to the "field of analysis" but no longer extant. Though able to complete much of the restoration, Fermat left Bordeaux before he could finish it all. As Chapter III below shows, one theorem in particular proved recalcitrant, and Fermat continued to work on it after his return to Toulouse in 1631. Resolving the difficulties in that theorem through a new algebraic approach to the problem, Fermat quickly arrived at the foundations of his analytic geometry. By 1635, he was able to apply the new theory to the classical problems of the three- and four-line locus, a demonstration of which he sent to Beaugrand.[65] He set down the theory itself in a work entitled *Introduction to Plane and Solid Loci*,[66] a copy of which he dispatched to Paris sometime late in 1636 or early in 1637, after his

---

[63] Cf. Cavalieri to Gianantonia Rocca, 11.XI.1635, CM.V.466–467: "[Beaugrand] mi disse che da un tal Senatore di Tolosa gli era stato proposto questo problema, cioè: Descrivere una parabola che passi per quattro dati punti (vogliono però esser talmente posti che se ne possi formare un quadrilatero, due de' lati del quale almeno non sieno paralleli) e che l'aveva sciolto . . ." For Fermat's solution of the problem see FO.1.84–87 and *Varia*, pp. 144–145; DeWaard notes in a commentary on this passage in CM, that Cavalieri's remarks date that solution far earlier in Fermat's career than its place in FO would suggest. Beaugrand visited Cavalieri on 23 October 1635. Beaugrand later tried to pass off Fermat's method of tangents as his own in a letter to Thomas Hobbes entitled "De la manière de trouver les tangentes des lignes courbes par l'algèbre et des imperfections de celle de S[ieur] des C[artes]" (FO.*Supple.* 102–113).

[64] See below, Chap. IV.

[65] *Loci ad tres lineas demonstratio*, FO.I.87–90.

[66] *Ad locos planos et solidos isagoge*, FO.I.91–110.

correspondence with mathematicians there had begun. Like Descartes' *Geometry*, of which Fermat was totally unaware at the time, the *Introduction to Plane and Solid Loci* effectively solved the problem of the translation of locus problems into the language of the analytic art by establishing a correspondence between loci and indeterminate equations in two unknowns.

Following his return to Toulouse, then, Fermat continued research on topics immediately growing out of his Bordeaux experience. During this early period up to 1636 he began the research into number theory that would claim his constant attention throughout his career. At the time, the research seems to have centered on a classical complex of problems involving the sums of the aliquot parts (proper divisors) of integers. Several subsidiary results discussed in his correspondence of the late 1630s, perhaps including the famous "little theorem" (for mutually prime a and p, and p prime, $a^{p-1} \equiv 1$ (mod p)), probably date from this period.

Fermat's research first branched out into new paths when, sometime in 1634 or 1635, he read Galileo's *Two World Systems*. His own research in geostatics led him to focus on the Italian mathematician's discussion of free fall. His dissatisfaction with that discussion centered, however, on the mathematical rather than the physical aspects of Galileo's treatment. Objecting to the statement that a freely falling cannon ball would follow a semicircular path toward the center of the earth, Fermat determined (as Galileo later agreed) that, to accord with the principles of the motion of bodies, the path would have to be a spiral, and he provided a proof of the result in a short paper that ultimately found its way into Galileo's hands.[67] More important, an investigation of Galileo's spiral ($\rho = \alpha^2$, generalized to spirals of the form $\rho = \alpha^n$) in terms of Archimedes' treatise *On Spirals* produced not only a method of quadrature for the new spirals[68] but also insight into the manner in which that method could be reformulated for application to the quadrature of curves of the form $y = x^n$, including the ordinary parabola.

As noted above, Fermat's Bordeaux friends, especially Beaugrand, remained the prime audience for these pre-1636 endeavors. At the same time, however, that he maintained his ties to his former colleagues in Bordeaux, Fermat found new friends in Toulouse, foremost among them another young *conseiller*, Pierre de Carcavi.[69] In addition to their common profession, Fermat

---

[67]Cf. FO.V.1-19. Cf. also Fermat's own paper on the mechanics of free fall, FO.V.20-43. The correspondence which resulted from these papers between Galileo and Fermat via Carcavi is contained *ibid.*, pp. 46-71.

[68]Cf. Fermat to Mersenne, 3.VI.1636, FO.II.#3, and below, Chap. V, §II.

[69]*Conseiller* of the Parlement of Toulouse until 1636, Pierre de Carcavi (1600?-1684) then became a member of the *Grand Conseil* in Paris. Made *bibliothécaire du Roi* in 1663, he participated in the founding of the *Académie des Sciences*. Himself no original mathematician, Carcavi understood and appreciated the work of others more talented than he. See Charles Henry, "Pierre de Carcavy, intermédiaire de Fermat, de Pascal, et de Huygens," *Bollettino di bibliografia e storia delle scienze matematiche e fisiche*, 17(1884), 317-391, 879; and H. H. Busard, "Carcavi, Pierre de," *Dictionary of Scientific Biography*, III, 63-64.

and Carcavi shared an interest in mathematics and science, and Fermat con-fided in his friend the results of his work. When Carcavi went to Paris as *bibliothécaire du Roi* in 1636, he spoke of his talented colleague to France's walking scientific journal, the Reverend Père Marin Mersenne. Mersenne at that time was a member of a group of mathematicians and scientists which included Roberval; Etienne Pascal, leader of the group; Claude Hardy, the mathematician; and a number of less well known thinkers.[70] Mersenne acted as recording and corresponding secretary for the group, soliciting in their name solutions to outstanding problems from those whom he felt best suited to handle them. Carcavi's report of Fermat's work in geostatics and mathe-matics interested Mersenne, and he addressed an invitation (now lost) to Fermat to share his findings with Mersenne's circle.

B. Joining the Scientific Community (1636-1643)

On 26 April 1636 Fermat replied to Mersenne in the first of a long series of letters that formed his only link with mathematicians outside Toulouse. The opening paragraph of the letter reveals a young man somewhat awed by the attention being paid him by his illustrious correspondent: "I remain much obliged to you for the favor you have done me in giving me hope of con-ferring with you by letter, and it is not the least of the debts I owe M. Carcavi, who procured me this favor. It saddens me that doubtlessly my reply to points in your letter will not satisfy you, but I would rather appear ignorant in answering you badly than seem indiscreet in not responding at all."[71]

If Fermat was surprised at having been addressed in the first place, what followed the opening paragraph of his reply put Mersenne on the same foot-ing. In addition to accepting the invitation to discuss the mechanics of free fall, Fermat enclosed the results of his research on spirals, announced the complete restoration of Apollonius' *Plane Loci*, and remarked, almost in passing: "I have also found many sorts of analyses for diverse problems, numerical as well as geometrical, for the solution of which Viète's analysis could not have sufficed. I will share all of this with you whenever you wish and do so without any ambition, from which I am more exempt and more distant than any man in the world."[72] Fermat's request that Mersenne ap-prise him of what books in mathematics had appeared over the past five or six years could only have added to Mersenne's astonishment. Nothing had ap-

---

[70] On the makeup of this group, one of the nuclei joined to form the Académie des Sciences, see Léon Auger, *Un savant méconnu, Gilles Personne de Roberval* (Paris, 1962), pp. 154ff. Auger's study serves as an introduction to the life of Roberval (Roberval [Oise], 1602-Paris, 1675) and to his work. Etienne Pascal (1588-1651), though resident in Paris, served at this time as *Président en la Cour des Aides de Clermont-Ferrand*.

[71] FO.II.3.

[72] *Ibid.*, p. 5.

peared which could match the achievements of this young *parlementaire* working alone in Toulouse.

The letter to Mersenne put an end to Fermat's isolation. The material contained in it and in the letters that followed amply impressed Mersenne's colleagues. Since Fermat's propositions on geostatics ran counter to the theories of Pascal and Roberval, these two men soon took up direct correspondence with Fermat. As the disagreement over geostatics was found to rest on irreconcilable first principles, the subject of the correspondence shifted to the realm of mathematics.

To his first letter to Mersenne Fermat had appended two problems involving the determination of maxima, which he asked Mersenne to pass on to the Parisian mathematicians. Later letters to both Mersenne and Roberval contained more problems ranging over loci, aliquot parts, figurate numbers, sums of series, and the determination of areas and volumes of figures. To all these problems Fermat had solutions, which he forwarded, however, only after letting those in Paris try their hand at them. He spoke of a method by which he found his results. It was contained, he said, in a memoir on the determination of maxima and minima written while he was in Bordeaux. He had asked d'Espagnet, who had the only copy, to send it to Roberval, but, he added in a letter of September, 1636:[73]

If M. d'Espagnet has set forth my method to you only as I sent it to him at the time, you have not seen its most beautiful applications. For, with slight alterations, I make it serve

1. for finding propositions similar to those on the conoid which I sent you in my last letter;
2. for finding tangents to curved lines, on the subject of which I propose this problem to you: to find the tangent to a given point on the conchoid of Nicomedes;
3. for finding the centers of gravity of all sorts of figures, even figures different from the ordinary, like my conoid and infinitely many others, examples of which I will show you whenever you wish;
4. for numerical problems which deal with aliquot parts and which are all very difficult. . . .

I have left out the main application of my method, which is for the determination of plane and solid loci.

Roberval and his friends found all of the problems Fermat was sending to Paris "very difficult." Their mathematics, which represented the highest state of the traditional style in 1636, was not always capable of solving Fermat's problems, which were arriving with alarmingly increasing frequency. Roberval and Mersenne, therefore, avidly solicited Fermat's solutions and the methods

---

[73] Fermat to Roberval, 22.IX.1636, FO.II.72,74.

employed in gaining them. During 1637, their solicitations brought Fermat's *Method For Determining Maxima and Minima and Tangents to Curved Lines*[74] to Paris from Bordeaux. Fermat himself next sent the completed restoration of Apollonius' *Plane Loci* and the memoir outlining his new method of algebraic geometry, the *Introduction to Plane and Solid Loci*. As the discussion of geostatics turned to the important corollary question of centers of gravity, Fermat filled letters to Roberval with analytic demonstrations of the quadrature and centers of gravity of various solids of revolution. His work in this area was crowned by a 1638 paper, *The Center of Gravity of the Parabolic Conoid, Determined by the Method of Maxima and Minima.*[75] During this period also Fermat began his efforts to extend to three-dimensional figures the algebraic methods of the *Introduction to Plane and Solid Loci*. These efforts took final shape in the *Introduction to Surface Loci*,[76] sent to Carcavi (who by then had succeeded Beaugrand and d'Espagnet as the *dépositaire* of Fermat's papers) in January 1643.

The Parisian mathematicians to whom all this work was directed took action on their assurances of admiration by coaxing Fermat to publish the details of his methods and by offering their assistance in such a venture. As early as the spring of 1637, Roberval wrote to Fermat:[77]

> Even though I received your demonstration of the plane locus last Monday, nevertheless my business, both public and personal, did not permit me to examine it until Thursday, when I presented it in your behalf to a group of our mathematicians, who met that day at the home of M. de Montholon, *conseiller*. There it was received, examined, admired with astonishment, and your name was exalted to the heavens. I was charged personally with thanking you in the name of the company and with asking you to send me the complete construction of the solid locus with a brief demonstration, so that we may publish the two pieces either with or without your name, as you wish. In this regard we will take the trouble to supplement whatever seems too concise for the public. . . .
>
> As for me, I can promise no leisure until the next three months have passed and I am delivered of my public lectures. Even when I have that leisure, I am not assured of finding the solid locus, which I foresee as very difficult. That is why I shall, for the moment, if you want, offer you full declaration of my impotence, in order that, without teasing me any longer and with regard for the request of such a company as I have told you about, you will share your invention with us, which is the one which the great

[74]*Methodus ad disquirendam maximam et minimam et de tangentibus linearum curvarum*, FO.I.133–136; see below, Chap. IV.
[75]*Centrum gravitatis parabolici conoidis, ex eadem methodo*, FO.I.136–139; see below, Chap. V, § III.
[76]*Isagoge ad locos ad superficiem*, FO.I.111–117; see below, Chap. III. § V.
[77]Roberval to Fermat, 4.IV.1637, FO.II.102–103.

geometer of centuries past took particular glory in having perfected on the basis of the inventions of those who preceded him. Think of how you have an opportunity to bring glory to yourself for having found it at a time in which it was in the same state as if it had never been known.

As indicated in Chapter I, such fulgent praise evoked a dual response from Fermat. Though it clearly pleased him, he intransigently rejected all responsibility for publishing the results in finished form. While he certainly appreciated not only the praise but that for which he was praised—the restoration of ancient analysis—his desire to keep his "amateur" status blocked this first attempt to put his work in print as it did all succeeding efforts.

The brilliance of Fermat's achievements nonetheless sufficed, without direct publication, to establish his name among European scientific circles. Roberval's and Mersenne's group publicized Fermat's talents at every opportunity. Mersenne edited some of his material on geostatics and number theory for the 1637 edition of the *Universal Harmony* and the 1644 *Physicomathematical Thoughts*. [78] Pierre Hérigone added a version of the method of maxima and minima, edited most probably by Mersenne, to the 1644 supplement to his *Course in Mathematics*. The manuscripts and letters that Fermat sent to Paris were passed from hand to hand and dispatched all over the Continent. Through Beaugrand's and Mersenne's trips and letter packets to Italy, Fermat's results, beginning with his correction of Galileo, became well known to Italian mathematicians such as Ricci, Torricelli, Cavalieri, and even Galileo himself. Later, Dutch mathematicians sojourning in Paris, in particular Franz van Schooten, took home with them copies of Fermat's papers.

If, however, Roberval and his immediate circle thought highly of Fermat's efforts, not everyone in Paris shared their enthusiasm. Fermat's habit of communicating his results piecemeal and in the form of challenges soon began to annoy mathematicians such as Bernard Frenicle de Bessy and Pierre Brûlart de St.-Martin, with whom Fermat had begun to exchange problems in number theory in 1638. [79] It frequently seemed to them that Fermat was outrightly ignoring their pleas for more complete details. At one point, they angrily accused Fermat of posing impossible problems and threatened to break off correspondence. Fermat's reply, in which he provided solutions but continued to withhold their derivations, did little to assuage the situation. In

[78] Cf. FO.II.nos.2_a,3_a,3_b,4_a,4_b.

[79] Brûlart was a fellow member of Carcavi's of the *Grand Conseil* in Paris. His only major work is an unpublished treatise, *Les causes et les admirables effects des météores ou diverses impressions de l'air*, written sometime before 1664. Cf. FO.V.120, n.2 Frenicle (1605–1675) was *Conseiller à la Cour des Monnaies* in Paris. His most important work is *Solutio duorum problematum circa numeros cubos et quadratos, quae tanquam insolubilia universis Europae mathematicis a clarissimo viro D. Fermat sunt proposita . . .* (Paris, 1657). Frenicle's works were edited and published as Vol. V of the *Mémoires de l'Académie des Sciences* by Philippe de la Hire (Paris, 1693; 2nd ed., den Haag, 1729). For details of the contretemps concerning the number-theoretical problems Fermat was sending to Paris, see below, Chap. VI.

part, as Chapter VI will show, such negative reaction to Fermat derived from a fundamental misunderstanding among the parties about the subject they were discussing. Fermat was asking new sorts of questions and looking for new sorts of answers, and with little help from him his correspondents failed to grasp the novelty of his results.[80] In part also, however, Frenicle's and Brûlart's attitude reflected the ironic aftermath of Fermat's famous dispute with Descartes over the method of maxima and minima.

That involved controversy, which will be discussed in detail in Chapter IV, took place during late 1637 and early 1638 and at one point engaged the attention of practically the entire mathematical community. It began with Fermat's somewhat off-hand criticism of Descartes' *Dioptrics*, criticism to which Descartes reacted all the more strongly for the fact that Fermat should not have had access to the galleys of the work. When, in addition, just prior to the actual publication of the *Discourse on the Method* with its three essays, Descartes saw for the first time copies of Fermat's analytic geometry and method of maxima and minima, his worst fears about the possible failure of the *Discourse* to achieve its aim seemed confirmed. Through the total novelty and unique brilliance of the essays, Descartes hoped to shock French intellectual circles into asking to see more of his philosophy. Fermat's criticism of the *Dioptrics* cast doubt on Descartes' brilliance, and his independent development of much of the new material of the *Geometry* robbed that work of its novelty. Descartes' subsequent attack on the method of maxima and minima was as vehement as it was ill-considered. He clearly had not taken the trouble to understand Fermat's method, as Fermat finally succeeded in showing him after a long series of exchanges between The Hague and Paris on the one hand and Paris and Toulouse on the other. Indeed, it became clear to both men that their respective methods of determining extreme values rested on the same foundations in the algebraic theory of equations. Having discerned the theoretical validity of Fermat's method, Descartes increasingly focused his attack on Fermat's style as a mathematician. He accused Fermat of generating non-general methods to fit specific results and of bringing to mathematics nothing more than some talent in problem-solving; Fermat's mathematics lacked method, lacked system, lacked generality.

The personal aspects of the ensuing dispute reveal that naïveté in Fermat, of which Chapter I spoke. He seems to have had little sense of the part personalities might play in a scientific debate. He could not understand why Descartes would be so upset by his criticism of the *Dioptrics* nor why he would so vehemently attack the method of maxima and minima when he clearly had not made much effort to grasp its principles. Fermat welcomed Roberval's support in the matter, even though the mutual hatred that obtained between Descartes and Roberval (stemming mostly from a nasty incident in Paris in 1626) was well known in mathematical circles and beyond that clearly evi-

---

[80] Again, for details, see below, Chap. VI.

dent in the exchange of letters.[81] To him, this dispute, and those yet to come, were matters for detached scientific debate. He does not even seem to have discerned how his own rather elusive behavior as a mathematician might aggravate others. Fermat was not "confused," as Bezin put it; he just naïvely assumed that everyone was as sincere and detached as he himself.

Shown the errors of his criticism of the method of maxima and minima, Descartes appeared to accept his defeat gracefully in a letter to Fermat dated 11 November 1638: "I well know that my approval is not necessary for you to judge what opinion you should have of yourself, but if it can contribute anything, as you have done me the honor of writing to me, I feel obliged to assure you openly here that I have never known anyone who has shown me that he knows as much as you in mathematics."[82]

In fact, however, the defeat stung Descartes, not only because he did not like to lose arguments but also because Fermat's work had dimmed the splendor of the *Geometry*. Hence, Descartes' famous remark made to Franz van Schooten in a Dutch garden more accurately reflects his attitude toward Fermat: "Fermat is a Gascon [braggart], I am not. It is true that he has found many pretty, special things, and that he is a man of great mind. But, as for me, I have always endeavored to examine things quite generally, in order to be able to deduce rules that also have application elsewhere."[83] It was that more sincere reaction that shaped the attitude of Descartes' supporters in Paris toward Fermat. Having won the specific point in question during the controversy, Fermat emerged from it with his mathematical reputation badly damaged in some quarters. During the early 1640s there seems to have been widespread feeling that he was not so much talented as "lucky" in mathematics and that he operated more by trial and error than by systematic analysis.

The rumors reached Fermat's ears in Toulouse and they clearly hurt. They had, however, at least for posterity's sake, the virtue of drawing Fermat out of his shell. Forced to defend himself, he issued a series of memoirs explicating the precise origins and foundations both of his method of maxima and minima and of the method of tangents deriving from it. His *Method of Tangents Explained for M. Descartes*,[84] sent to Descartes in June 1638, is the document that ostensibly clarified issues for Descartes and prompted his admission of defeat. Continued murmurings concerning the fortuitous ele-

---

[81] On the mutual feelings of Descartes and Roberval, see Auger *Roberval*, pp. 161ff.
[82] Descartes to Fermat, 11.X.1638, FO.II.167.
[83] Reported by van Schooten in a letter to Huygens, 19.IX.1658, FO.IV.122. Cf. Descartes to Mersenne, 4.III.1641, FO.IV.113: "Je voudrais bien que vous n'eussiez point envoyé de copie de ma Métaphysique à M. Fermat; . . . entre nous, je tiens M. Fermat pour l'un des moins capables d'y faire de bonnes objections; je crois qu'il sait des mathématiques, mais en philosophie j'ai toujours remarqué qu'il raisonnait mal."
[84] "Méthode *de maximis et minimis* expliquée et envoyée par M. Fermat à M. Descartes," FO.II.154–162.

ments of the method of maxima and minima were the target of the *Analytic Investigation of the Method of Maxima and Minima* [85] which, though written sometime around 1640, nonetheless recounted research done in Bordeaux. To counter charges that the method of tangents lacked generality, Fermat then composed the *Doctrine of Tangents*, [86] in which he applied his method to the cissoid, conchoid, cycloid, and quadratrix. Finally, he reconsidered his method of maxima and minima in light of the new theory of equations set forth in Book III of Descartes' *Geometry* and presented a new version of it in a letter to Brûlart in 1643. [87]

That new version of the method was but one result of a spate of research in the theory of equations, in which Fermat was engaged at the end of the 1630s and beginning of the 1640s. A more careful and detailed investigation into the relationships between algebraic reduction procedures and methods of graphical solution of equations transformed a brief appendix to the *Introduction to Plane and Solid Loci* into a full-scale treatise, the *Tripartite Dissertation on the Solution of Geometrical Problems by the Most Simple Curves Properly Belonging to Each Class of Problems.* [88] Here, Fermat took issue with another topic in Descartes' *Geometry*, the classification of curves and equations. In addition, a clearer understanding of the structure of polynomial equations resulting from the substitution of a binomial expression, say, $x + a$, for the unknown x led Fermat to a fully systematic treatment of the "single" and "double equations" employed by Diophantus in the *Arithmetic*. If he continued to withhold his insights from correspondents such as Frenicle and Brûlart, he did reveal them to his friend at home, Père Jacques de Billy, who brought them together in a treatise entitled *A New Invention of the Analytic Doctrine*, first published as an appendix to the 1670 edition of the *Arithmetic* that also included Fermat's famous *Observations on Diophantus.* [89]

Despite the rumors that prompted works such as the *Tripartite Dissertation*, and despite the fervence of the criticism of Descartes' mathematics contained in that work, Fermat retained a deep respect and admiration for his erstwhile adversary. He may not, it is true, have realized that Descartes himself was the original source of the rumors. Their direct correspondence, begun toward the end of the controversy, was short-lived due to the subsequent separation of their interests; by 1638 Descartes was largely through with mathematics. But, even in the *Tripartite Dissertation*, Fermat could still write, "So great is my admiration of this, if I may use strange words, most portentous mind, that I hold in greater esteem an erring Descartes than many correct pedants." None-

---

[85] *Analytica eiusdem methodi investigatio*, FO.I.147–153.

[86] The title derives from the opening words of the treatise, FO.I.158–167.

[87] Fermat to Brûlart, 31.III(?).1643, FO.*Suppl*.120–125.

[88] *De solutione problematum geometricorum per curvas simplicissimas et unicuique problematum generi proprie convenientes dissertatio tripartita*, FO.I.118–131; see below, Chap. III, § V.

[89] French trans. in FO.III.325–398; see below, Chap. VI, § III,B.

theless, the controversy with Descartes capped Fermat's brilliant introduction to the French mathematical community. By 1643 he had established his reputation as one of the best (if also more eccentric) mathematicians in Europe and received both the adulation and enmity that seem inevitably to accompany such a reputation.

C. His Career Interrupted (1643-1654)

Fermat did not have much opportunity to exploit his fame in the next decade. His mathematical activities were continually interrupted by his parliamentary duties in Toulouse and Castres. He took these duties seriously and repeatedly dropped mathematics under the press of local business.[90] Ill health and troubles with the authorities in Paris during the early 1640s, the *Fronde* and its dire consequences for the Parlement of Toulouse in 1648, the Spanish raids on Languedoc in 1649, the plague in 1651, all kept Fermat from pursuing his mathematical studies. The work did not, however, stop altogether.

Fermat's correspondence with Digby in 1657 and 1658 indicates that he had, around 1644, fundamentally revised his approach to the problem of quadrature and developed a technique that allowed the direct quadrature of curves of the forms $y^q = ax^p$ and $x^p y^q = a$ (with the exception, of course, of the ordinary hyperbola, $xy = a$). The technique became the foundation for a novel theory of algebraic quadrature that aimed at the reduction of the quadrature of any algebraic curve to that of one of the two basic families of curves just cited and that included a procedure very close to what ultimately became integration by parts. The title Fermat gave to his exposition of the new theory, which he did not set down until 1657, clearly reveals its inspiration in the theory of equations of Viète: *On the Transformation and Emendation of Local Equations for the Manifold Comparison of Curves With Each Other or With Straight Lines, To Which Is Appended the Use of Geometric Proportion in the Quadrature of Infinitely Many Parabolas and Hyperbolas.*[91] Fermat's continuing study of Viète's theory of equations during this period is attested to by a 1650 treatise, *A New Use of Square Roots and Roots of Higher Order in Analysis*, which significantly simplifies Viète's technique for eliminating irrational terms from equations.[92]

During the decade 1643-1654, however, Fermat's work largely took a form appropriate to the short snatches of time left to him by the press of govern-

---

[90] E.g. Fermat to Mersenne, 26.III.1641: "Les occupations que les procès nous donnent sur la tête m'ont empêché de pouvoir lire à loisir les traités que vois m'avez fait la faveur de m'envoyer." FO.II.218.

[91] *De aequationum localium transmutatione et emendatione, ad multimodam curvilineorum inter se vel cum rectilineis comparationem, cui annectitur proportionis geometricae in quadrandis infinitis parabolis et hyperbolis usus*, FO.I.255-284; cf. below, Chap. V, §IV.

[92] *Novus secundarum et ulterioris ordinis radicum in analyticis usus*, FO.I.181-188; cf. below, Chap. IV.

mental affairs. He delved ever more deeply into the subject of number theory, in particular into methods for the solution of the equation $x^2 - py^2 = 1$ (p prime) in integers. Many of his observations on Diophantus, scratched into the margins of his copy of the *Arithmetic*, stem from this period of isolation. As parliamentary duties demanded an ever greater portion of his time and energy, his scientific correspondence waned considerably. While some ten letters of his date from 1643, the next ten years see only two letters to Carcavi, one to Gassendi, and one to Mersenne. The absence of any unidentifiable back references in correspondence after 1654 indicates that the dearth of correspondence is real and not the result of letters gone astray. When he worked at all, Fermat again worked in isolation.

D. New Correspondents, New and Old Problems (1654-1664)

Fermat resumed contacts with other mathematicians in 1654. By then, several of his old correspondents had died: Mersenne in 1648, Descartes in 1650, Etienne Pascal in 1651. Moreover, the ill feeling generated by Fermat's refusal to reveal his methods of number theory in 1643 seems to have permanently soured his relations with Frenicle and Brûlart, making a resumption of correspondence impossible. Fermat turned to new friends, of whom the first was Pascal's son, Blaise. Although the letter that opened the Pascal-Fermat correspondence is no longer extant, the young philosopher-mathematician apparently had written to Fermat, whom he knew through his father, for consultation on his studies in probability. Fermat's first letter opens *in media res* with a discussion of the division of the stakes in a dice game when it is prematurely cut off. The problem had originally been proposed to Pascal by a notorious gambler named Méré, and Pascal was pleased to learn that his thoughts on the matter were confirmed by Fermat.

The correspondence continued only a short while, ending in October 1654. Pascal had written to Fermat for a definite purpose, to wit, to check his results on probability. Fermat, however, anxious to circulate the many challenging results in number theory achieved in his period of isolation, sought to turn their correspondence to that new subject and to engage Pascal's interest by revealing some of his more striking findings. Mistaking Pascal's well-mannered reaction for enthusiasm, he wrote hopefully to his old friend Carcavi, who had again brought up the subject of publication:[93]

> I am delighted to have had opinions conforming to those of M. Pascal, for I have infinite esteem for his genius and believe him capable of succeeding in anything he might undertake. The friendship he has offered is so dear to me and so thoughtful that I believe there should be no difficulty in making some use of it in the publication of my treatises.
>
> If that did not shock you, the two of you may undertake that publica-

[93] Fermat to Carcavi, 9.VIII.1654, FO.II.299.

tion, of which I consent to your being the masters; you may clarify or supplement whatever seems too concise and relieve me of a burden that my duties prevent me from taking on. I even desire that the work appear without my name, leaving pretty much up to you the choice of any designation that will be able to mark the name of the author whom you deem your friend.

If Fermat wanted Pascal as his editor, Pascal would have none of it. By 1654, he had wearied of mathematics, which he pursued largely from habit.[94] In retrospect, it is doubtful he would have been able to fulfill the task. Pascal and Fermat were worlds apart in their basic style of mathematics. In particular, they strongly differed on the proper role of algebra; Pascal remained wedded to traditional canons of style. Number theory did not interest him in the least, as he revealed to Fermat in a curt note dated 27 October 1654:[95]

Your last letter has perfectly satisfied me. I admire your method for the division of stakes, even more so now that I understand it better. It is entirely your own and has nothing in common with mine, and it arrives easily at the same result. There! We have reestablished our understanding.

But, Monsieur, if I have competed with you in this matter, look elsewhere for someone to follow you in your numerical researches, of which you have done me the honor of sending the statements. As far as I am concerned, I confess to you that they go right past me; I am capable only of admiring them and of begging you very humbly to take the first opportunity to complete them. All our gentlemen saw them last Saturday and admired them with all their hearts. One cannot easily bear waiting for things so pretty and so desirable. Please think about them, then, and be assured that I am, etc.

Pascal's seeming interest had softened Fermat's thoughts about publication, though Fermat still refused to share the work of preparation. Pascal's rather cold letter crushed any hopes of someone else doing the job, and plans for publication again died.

But Fermat had never set much store by publication of his work. What disturbed him most about Pascal's response was his young friend's total lack of interest in number theory. As Chapter VI will show, Fermat had by this time developed a completely new approach to problems involving the properties of integers, and he was anxious to share it with other mathematicians. His notion of how to share mathematical results had not, however, changed much since the early 1640's. He still preferred to let his correspondents wrestle with the problems before revealing his solutions and methods of solution. If he did

---

[94] Emile Cailliet, *Pascal: The Emergence of Genius*, (2nd ed., N.Y., 1961), p. 106.
[95] Pascal to Fermat, 27.X.1654, FO.II.314.

not like controversy, he did enjoy intellectual combat. And so, in January 1657 he addressed "to French, English, Dutch mathematicians and those of all Europe" two difficult problems in number theory for solution.[96] Although the problems were couched in the traditional terms of determining sums of aliquot parts, Fermat was in fact looking for something entirely new from his correspondents, to wit, a proof that the solutions he offered as examples were unique under the given conditions. The so-called "first challenge" was promptly followed by a second in February; Fermat called for the complete solution of the equation $Nx^2 + 1 = y^2$ for non-square N.

The challenges evoked the response of several English mathematicians and opened a long correspondence between Fermat and Kenelm Digby, who had met Fermat personally in 1656 and who acted as spokesman and intermediary for the English. Fermat, committed to a new style of number theory, assumed in his wording of the challenges that they would be understood to call for integral solutions only. He turned back several proposed solutions because they failed to meet this criterion, thereby heating the dispute that arose. Although the English participants, John Wallis and Viscount William Brouncker, eventually arrived at their famous solution of the second challenge by means of continued fractions, they took umbrage at Fermat's none too gentle prodding on the matter and could not themselves see any value in number theory. Nor could Franz van Schooten in Leiden, who toyed with the challenge problems for a time but then abandoned them as not worth the effort. Only Fermat's earlier correspondent, Frenicle, participated enthusiastically in the affair. His continued ill feeling toward Fermat, stemming from their exchanges in the early 1640s, apparently blocked the path toward personal reconciliation, however, and Fermat and Frenicle did not resume direct correspondence at this time.

Fermat further irritated the English mathematical community by his strong criticism of Wallis' *Arithmetic of Infinites*, a copy of which Digby had given him while in Toulouse. The English may have felt that the apostle of algebraic analysis was simply being cantankerous when he objected that: "It is not that I do not approve of it. But since all of his propositions can be demonstrated in the ordinary, legitimate, and Archimedean way in many less words than his book contains, I do not know why he has preferred this style of algebraic notation to the ancient style, which is more elegant, as I hope to show him at my first leisure."[97] To back up his criticism, as well as to argue his priority in achieving most of Wallis' results, Fermat used the opportunity to polish and set down in writing the techniques of quadrature he had been developing

---

[96] "Problemata duo mathematica, tanquam indissolubilia Gallis Anglis, Hollandis, nec non caeteris Europae Mathematicis proposita a Dno. de Fermat, Regis Consiliario in Tolosano Parlamento, Castris Parisios ad Dom. Claudium Martinum Laurenderium, Doctorem Medicum, transmissa 3 nonas Januar. 1657, accepta vero 12 Kal. Febr." FO.II.332–333. For further details, see below, Chap. VI, § IV.

[97] Fermat to Digby, 15.VIII.1657, FO.II.343.

since the early 1640s. The two-thirds of his *On the Transformation and Emendation of Local Equations* that contain his new procedures for the reduction of quadrature problems to one of two basic forms was, however, set forth in the language and style of the algebraic theory of equations, a fact which must have made his criticism of Wallis seem all the more cantankerous. Anxious to gain a wide audience both for his defense of the *Arithmetic of Infinites* and as evidence that the English had carried the day in the dispute over Fermat's challenges, Wallis published the exchange of letters in the *Commercium epistolicum* of 1658.[98] Although Wallis succeeded where Descartes had failed in getting Fermat's views into print under Fermat's own name, the *Commercium* in fact contained very few letters from Fermat, and the letters it did contain revealed practically nothing of Fermat's own work either in number theory or in quadrature.

Besides his continuing studies in number theory and research into quadrature and rectification, Fermat also pursued the unfinished task of restoring the works of Pappus' "field of analysis." In 1656 he tried to repeat his success of the *Plane Loci* with his *Renewed Doctrine of Euclid's Porisms, Set Forth in the Form of an Introduction to More Recent Geometers.*[99] The recipient of this treatise was the Parisian *parlementaire* and astronomer, Ismael Boulliau, who was himself working on the same problem. Fermat's other direct correspondents during the last period of major activity were a fellow Toulousain, the Jesuit Antoine de Lalouvère, and the Jesuit Jacques de Billy.[100] Lalouvère was the only person who managed to put a major work of Fermat's into print during the author's lifetime, albeit under a pseudonym. In 1660, Lalouvère published as an appendix to his own treatise on the cycloid Fermat's *Geometrical Dissertation on the Comparison of Curved Lines with Straight Lines.*[101] In that work, Fermat showed how his methods of quadrature could be directly applied to the problem of the rectification of curves.

[98]*Commercium epistolicum de quaestionibus quibusdam mathematicis nuper habitum*, ed. John Wallis, Oxford, 1658.

[99]*Porismatum Euclideorum renovata doctrina et sub forma isagoges recentioribus geometris exhibita*, FO.I.76–84.

[100]Both men were instrumental in gaining a wider audience for Fermat's work. Antoine de Lalouvère, S.J. (?1600–Toulouse, 1664), Professor of Mathematics at the Jesuit College in Toulouse, appended Fermat's treatise on rectification (without the author's permission and under the pseudonym M.P.E.A.S.) to his own *Veterum geometria promota in septem de cycloide libris* (Toulouse, 1660), of which the first book is dedicated to Fermat. Lalouvère had already published two earlier treatises, one on quadrature (Toulouse, 1651) and one on the cycloid (Toulouse, 1658). Jacques de Billy, S.J. (Compiègne, 1602–Dijon, 1679), served as Fermat's Voltaire. Actively concerned from 1665 on with the introduction of Fermat's techniques into the secondary curriculum, while Professor of Mathematics at the Collège de Clermont (later Lycée Louis le Grand), one of France's leading secondary schools, Billy gathered excerpts from his correspondence with Fermat into a work entitled *Doctrinae analyticae inventum novum* which served as an introduction to the 1670 edition of Diophantus with Fermat's observations. Earlier, Billy had published his *Nova geometriae clavis algebra* (Paris, 1643).

[101]*De linearum curvarum cum lineis rectis comparatione dissertatio geometrica*, FO.I.211–254; see below, Chap. IV,§V.

Fermat's return to mathematical correspondence in the late 1650's concluded with a resumption of the old controversy over the law of refraction. One of Descartes' followers, Claude de Clerselier, was gathering the master's polemical scientific correspondence for publication[102] and turned to Fermat for assistance in the details of his own dispute with Descartes. Fermat replied that, while Descartes had ceded to him on the matter of the method of maxima and minima, he on the contrary remained steadfast in his objections to Descartes' law of refraction and its derivation in the *Dioptrics*. In the initial exchange with Clerselier and Rohault, Fermat largely rehearsed the arguments of twenty years earlier. But the heat of renewed battle and Cureau de la Chambre's plea for a settlement ultimately set Fermat on his last bit of concentrated scientific research. Unable to accept the idea that light could travel more quickly in a denser medium, a notion that followed directly from Descartes' theory of the nature of light, Fermat eventually settled on another principle: that a beam of light passing through one or several media takes the shortest possible path. To his great surprise, he found that the mathematical analysis of refraction on the basis of that extreme-value consideration led directly to Snell's and Descartes' sine law, a result he had always mistrusted. Fermat sent his *Analysis of Refraction* and *Synthesis of Refraction* to Cureau in 1662, and, in the covering letter, he expressed his belief that the use of his principle made his and Descartes' disagreement over the relation between the speed of light and the density of the medium mathematically irrelevant. He added, however, that Descartes' derivation still failed to agree with physical observation in that it predicted a bending away from, rather than toward, the normal in a denser medium.[103] The criticism reflects Fermat's continued lack of understanding of the principles on which Descartes' theory of light rested, and his hope that his own extreme-value principle would effect a reconcilia-

---

[102]Claude de Clerselier (1614–1686), *Lettres de M. Descartes, où sont traitées les plus belles questions de la morale, de la physique, de la médecine, et des mathématiques . . . Lettres de M. Descartes, où il répond à plusieurs difficultez qui luy ont este proposées sur la Dioptrique, la Géométrie, & sur plusieurs autres sujets* (Paris, C. Angot, 1657–1667, 3 vols.). Cf. ensuing exchange between Clerselier and Fermat in FO.II.nos. 90, 90ₐ, 93–95, 97, 99, 113–115.

[103]Fermat to de la Chambre, 1.I.1662, FO.II.457–463: ". . . j'ai trouvé que mon principe donnait justement et précisément la même proportion des réfractions que M. Descartes a établie. J'ai été si surpris d'un événement si peu attendu, que j'ai peine à revenir de mon étonnement. J'ai réitéré mes opérations algébriques diverses fois et toujours le succès a été le même, quoique ma démonstration suppose que le passage de la lumière par les corps denses soit plus malaisé que par les rares, ce que je crois très vrai et indisputable, et que néanmoins M. Descartes suppose le contraire. Que devons-nous conclure de tout ceci? Ne suffrait-il pas, Monsieur, aux amis de M. Descartes que je lui laisse la possession libre de son théorème? . . . Il resterait encore une petite difficulté que la comparaison de M. Descartes semble produire. C'est qu'il ne parait pas encore pourquoi la balle qui est poussée dans l'eau n'approche pas de la perpendiculaire, ainsi que la lumière; . . ." Fermat sent both of his treatises, *Analysis ad refractiones* and *Syntheses ad refractiones*, to de la Chambre in January and February 1662. Cf. texts in FO.I.170–179.

tion with the Cartesians proved illusory. The new principle came under attack almost immediately, but by 1662 Fermat was too old to join battle.

III. FERMAT'S STYLE OF WORK AND HIS INFLUENCE ON HIS CONTEMPORARIES

The transmission to Cureau in 1662 of the analysis and synthesis of the law of refraction marked the end of Fermat's career. In February of 1665 his death in Toulouse was announced (by Carcavi?) in the *Journal des Sçavans*: "With great sadness we have learned here of the death of M. de Fermat, Councillor of the Parlement of Toulouse. He was one of the most beautiful minds of this century, a genius so universal and of such vast extent that if all men of learning had not borne witness to his extraordinary merit, one would hardly believe all the things that must be said of him in order not to detract from his praise."[104] Because, continued the author of this *éloge*, "this journal serves principally to make known by their works those persons who have gained fame in the republic of letters, we will content ourselves here with giving a catalogue of the writings of this great man, leaving to others the task of offering him a fuller and more fitting eulogy." Of his mathematical works, only Fermat's theorems on number theory, his treatise on maxima and minima, his analytical geometry (for which, by the way, he is given priority over Descartes), his treatise *On Spherical Contacts* (an extension of Viète's *Apollonius Gallus* to spheres and planes), and the restoration of Apollonius' *Plane Loci* are mentioned. The *éloge* lays equal stress on Fermat's talents as a linguist and classicist, his conscientious service as a *parlementaire*, and his wide circle of famous correspondents.

The eulogist ended the article with the wish that Fermat's work be printed. His plea fell largely on deaf ears. At his death, Fermat's work belonged more to the history and tradition of mathematics than to its developing state. By 1665 men such as Huygens, Newton, and (somewhat later) Leibniz were in differing ways moving well beyond Fermat's accomplishments in analysis. They, their colleagues, and their followers could learn little or nothing from his analytic treatises. Only in the area of number theory could Fermat's work claim to be of value to the research mathematician, and few mathematicians were doing research in number theory in the late seventeenth century. Later, when mathematicians again focused attention on number theory, they did find Fermat's work challenging and suggestive. Hence, posterity's evaluation of Fermat came to rest more and more on his work in number theory, and his contributions to the development of mathematical analysis were gradually forgotten.

The reaction of Christiaan Huygens to Fermat illustrates the reasons for

---

[104] *Le Journal des Sçavans, De l'An M.DC.LXV. Par le Sieur de Hedouville* (Amsterdam, Pierre le Grand; 1684), 9 February 1665, pp. 79–82; passage cited is on p. 79.

Fermat's rapid fall to obscurity in the late seventeenth and early eighteenth centuries. A full generation younger, Huygens learned analysis from Fermat's treatises, both directly through copies sent to him by Mersenne and indirectly through Franz van Schooten, his teacher. He testified to this fact in a letter to Carcavi dated 1 June 1656, in which he accepted Carcavi's invitation to take up correspondence with Fermat: "Father Mersenne has honored me with his correspondence in order to stimulate me to the study of mathematics, in which he sees me naturally inclined. And he has often sent me writings by your other illustrious men, and principally by M. de Fermat, which I have begun to understand to the extent that I have profited in these sciences. Hence, from my first apprenticeship I have had marvelous esteem for this great man ...."[105]

To an extent here Huygens was being polite. For reasons to be discussed shortly, not all of Fermat's work evoked his admiration. In writing earlier to his mentor, van Schooten, Huygens remarked that he found many of the demonstrations in Fermat's restoration of the *Plane Loci* "perfunctory"; he did not feel that Fermat's effort at restoration was the equal of van Schooten's.[106] And yet, Huygens fully appreciated the importance of Fermat's method of maxima and minima and ably defended its author's priority over newer claimants such as Hudde.[107]

Huygens' correspondence with Fermat, at first through Carcavi, began in 1656 with a discussion of probability directly related to Huygens' own treatise *On the Game of Dice*.[108] He found Fermat's work pleasing. However, Fermat's further contributions to their correspondence pleased him less and less. The elderly mathematician was trying to enlist Huygens' interest in number theory, which had little relevance to the research Huygens was pursuing. When Fermat did send on some work on spirals, Huygens remarked that he was surprised that Fermat "takes pleasure in finding new curves which otherwise have no properties worthy of consideration."[109] By this time, the barrier separating the two men had become clear. Fermat delighted in pure

---

[105]Huygens to Carcavi, 1.VI.1656, FO.IV.119. Though a devoted disciple of Descartes and largely unsympathetic to Fermat (cf. above, p. 000 and n. 73), van Schooten did not hesitate to employ the latter's mathematical methods. In his commentary to Book II of Descartes' *Geometry*, published as one of the appendices to the 1659 edition of that work, he determines the normal to a conchoid "... beneficio methodi de maximis et minimis, cujus author est Vir Clarissimus P. de Fermat, in Parlamento Tolosano Consiliarius, quam Herigonus in supplemento Cursus sui Mathematici exemplis aliquot illustravit atque ibidem ad inveniendas tangentes adhibere docuit." (FO.IV.243.) It is highly likely that Huygens also learned Fermat's techniques through Hérigone's textbook.

[106]Huygens to van Schooten, 26.V.1655, FO.IV.116.

[107]Huygens to Wallis, 9.VI.1659, FO.IV.124: "Videbis item methodum illam Huddenii ad maximi vel minimi determinationem, quam tamen in solidum illi non debemus, sed primam eius inventionem Fermatio potius."

[108]Franz van Schooten, *Exercitationum mathematicarum libri quinque ... quibus accedit Christiani Hugenii Tractatus de Ratiociniis in Aleae Ludo* (Leiden, 1656–1657).

[109]Huygens to Carcavi, 26.II.1660, FO.IV.126–127.

mathematics with little thought to application. Huygens was a mathematical physicist concerned with bringing mathematical analysis to bear on problems in mechanics. Moreover, Huygens soon recognized that his friend was less and less cognizant of what was happening currently in mathematics. Fermat took to sending along papers on topics that had long since been settled. To meet Huygens' request for papers on analytical topics, Fermat was reaching back to older work rather than taking on new problems. Not only did the lack of new material dismay Huygens, but he could foresee problems in trying to publish the material. "I nevertheless fully believe," he wrote to Carcavi in 1660, "that M. de Fermat had seen no other works, since he assures us of this, but others will perhaps be more skeptical, if in publishing his work he does not cite those to whom he had shown it earlier."[110] Sadly, Huygens watched others break into print with subjects treated by Fermat, thus depriving the Toulousain of his due credit. He agreed with Fermat and Carcavi that van Schooten had been rude, if not cruel, in publishing a restoration of Apollonius' *Plane Loci* without even mentioning Fermat's paper on the same subject; van Schooten might not have used the work, but Huygens knew full well that he was aware of its existence.[111]

Throughout their correspondence, Huygens' attitude toward Fermat was one of perplexity. He knew Fermat was a great mathematician, but he could find little evidence of it in the work he was receiving. Fermat seemed out of touch, out-dated. In part, the fault was Huygens' own. As his own work diverged from that of Fermat, he began to ignore Fermat. He refused to comment on the renewed dispute concerning the law of refraction, but did indicate he thought little of Fermat's principle of the quickest path, a principle which has since become basic to the field of optics.[112] For his refusal to join the debate he drew a rebuke from Jean Chapelain:[113]

> I should tell you that M. de Fermat, Councillor of the Parlement of Toulouse and the excellent mathematician that you know, has politely complained by letter to one of his friends that he wrote you and proposed some problem for consideration, that you did not judge it worthy of your thoughts, and that he has not had any reply from you. To say something, I replied that you hoped to remain free at home and beyond all the tumult that deprives you of quiet and your books. Use this information according to your good judgment; I do not believe that you would want to neglect such an important man, who represents for us a second Viète.

[110]*Ibid.*
[111] Huygens to Carcavi, 27.III.1660, FO.IV.128–129: "Il ne se plaint pas tout à fait sans raison de M. Schooten de ce qu'il n'a pas fait mention de lui en publiant ses lieux plans. Car encore qu'il n'ait jamais vu ce que M. de Fermat en avait écrit comme il m'a assuré toujours, il l'a bien su et partant il n'aurait pas du le dissimuler."
[112]Huygens to Lodewijk Huygens, 8.III.1662, FO.IV.132–133; 19.IV.1662, FO.IV.134.
[113]J. Chapelain to Huygens, 12.VII.1664, FO.IV.135.

Huygens did not respond to this plea. Other than being polite to an old man, he felt he had nothing to gain from his correspondence with Fermat. At the news of Fermat's death, he expressed his sorrow and said he hoped for the publication of the mathematician's work, more probably for the sake of Fermat's reputation than for any value in the works themselves. His final appraisal of Fermat's later work is summed up in a 1691 letter to Leibniz apropos of Fermat's *On the Transformation and Emendation of Local Equations* as published in the 1679 *Varia opera*: "I have above set out all that I remember having seen in the posthumous works of M. Fermat, but this treatise has been published with so many mistakes and is moreover so obscure (with demonstrations suspect of error), that I have not been able to profit from it."[114] In view of what Fermat would have wanted, it is sad that Huygens' last reference to his work should focus critically on the manner in which his method of quadrature was set forth. Though proud of that method, Fermat wished by the late 1650s to be judged for other achievements. He had tried hard to engage Huygens' interest in number theory, indeed to the point of revealing to him, in the *New Account of Discoveries in the Science of Numbers* of 1659, details of methods that Frenicle and others had never been privileged to see. But it was to no avail; like most of his contemporaries, Huygens had no interest in the subtle properties of the integers.

As Huygens indicated in the passage just cited, Fermat's works themselves contained an element which deterred interest in 1665 and later. They were hard to read. Unrevised and still in the shorthand analytical style of a man who had neither time nor desire to strive for elegance, they were also couched in the language and symbolism of Viète, to whose style Fermat remained steadfastly dedicated throughout his life. The works seemed strange and unnecessarily complicated to a new generation of analysts schooled in the Cartesian model. And yet, if a well-defined French analytical school of mathematics existed in the second half of the seventeenth century and later, it was in great part due to the very works which the members of that school now found antiquated. With Descartes, Fermat had been instrumental in weaning French mathematics away from the strict Greek synthetic model.

When Fermat's papers first started flooding into Paris in 1636 and 1637, their author's use of algebraic analysis without synthetic demonstration unsettled the Parisian mathematicians. His friends decried the lack of style; his enemies refused to accept the results. Although men such as Roberval did some work in algebra, they still were committed to the Euclidean form of presentation and had not yet moved to the point of equating algebra with mathematics. They complained to Fermat, but he stood his ground. In telling Mersenne in 1636 of his research, he took account of his strange mode of presentation:[115]

---

[114]Huygens to Leibniz, 1.IX.1691, FO.IV.137.
[115]Fermat to Mersenne, 3.VI.1636, FO.II.14. My italics.

I think you will admit that these results are pretty, but I have so little opportunity to write down their demonstrations, which are among the most difficult and complex in mathematics, that I am content to have discovered the truth and to know the means of proving it whenever I shall have the leisure to do so. If I can find some occasion to go and spend some three or four months in Paris, I shall use them to set down in writing all my new thoughts in these arts, in which I will doubtlessly be much aided by your efforts.

Fermat never found the leisure time to go to Paris, nor did he ever bother to follow up his letters containing new mathematical discoveries with proofs of those discoveries. His apparent disregard for the conventions of classical mathematics disconcerted the Parisians. Mersenne's expressions of concern brought forth another revealing reply from Fermat: "You should not doubt that my demonstration offers conclusive proof, even though it seems that M. de Roberval has not found it precise. I can assure you, then, that all the propositions that I have placed in my paper are perfectly true, and even then I do not want to be believed until I have set out in writing all the demonstrations for this material."[116]

In paying lip service to the Parisians' desire for strict synthetic demonstration, Fermat was not about to change his own notions about the value of analytic mathematics. He continued to use algebra as a mode of research and exposition and spoke of his algebraic methods of finding extreme values as "a geometrical truth. . . . I maintain that my methods are as certain as the construction of the first proposition of the *Elements.*"[117] As far as the Parisians were concerned, esthetics ultimately gave way to efficiency. Fermat's analytic style might not meet their expectations of style, but his problems were baffling, his solutions enlightening and promising of still further results. His influence added heavily to the forces that moved the analytic school of mathematics from obscurity at the end of the sixteenth century to predominance at the end of the seventeenth.

In everything he did, Fermat focused on the art of analysis and on carrying out the program designed to increase its effectiveness. That is why Huygens could evince from him no real interest in the application of his techniques to physical problems. When Fermat had extended the art to the full range of inherited geometrical problems, when, that is, all that the ancients had accomplished in geometry now lay codified in the algebraic theory of equations and its precepts, Fermat turned not to mathematical physics but to number theory, which still defied such codification. Huygens could not appreciate Fermat's motivation because Huygens belonged to another, new tradition;

[116]Fermat to Mersenne, 15.VII.1636, FO.II.28.
[117]Fermat to Mersenne, II.1638, FO.II.132.

indeed, Huygens belonged to another age. Fermat for his part could not appreciate that the tradition of Viète had run its course, that in its efforts to restore the mathematics of the ancients it had in fact created something entirely new, which demanded that mathematicians stop looking toward the past and fix their eyes on an open-ended future. To have been a "second Viète" was at once Fermat's glory and his tragedy.

CHAPTER III

The Royal Road

Not to begrudge to posterity our unformed brainchildren is to some extent in the interests of the very science the products of which, at first simple and rough, grow and are are strengthened by new inventions. Indeed, it is of interest even to scholars to have a thorough view of the hidden progress of the mind and of an art in the process of advancing itself.

Pierre de Fermat

I. INTRODUCTION

In 1637 the French scientific community witnessed one of those strange coincidences once thought rare, but which the history of science has shown to be frequent: two independent minds producing simultaneously the same innovation. Early in 1637 Pierre de Fermat announced and sent on to his correspondents in Paris his *Introduction to Plane and Solid Loci* (hereafter, *Introduction*). At about the same time, Marin Mersenne received from René Descartes the galley proofs of the *Discourse on the Method* and its three *Essays,* among them the *Geometry.* [1] Both the *Introduction* and the *Geometry* set forth the same basic technique for treating geometric locus problems algebraically. Their respective authors had thought to correlate indeterminate algebraic equations in two unknowns with two-dimensional geometric curves and had invented the same framework in which to place them. Both men had invented what we today refer to as analytic geometry, the "royal road" to geometry.

In regard to priority, there is no doubt that Fermat's system was the first to come to the attention of the Parisian community; both the testimony of contemporaries and the dating of the pertinent documents attest to that fact. But when did each man first arrive at his conclusion? As this chapter will show, Fermat lacked necessary elements for his system until as late as

---

[1] *Discours de la Méthode pour bien conduire sa raison et chercher la vérité dans les sciences. Plus la Dioptrique, les Météores et la Géométrie, qui sont des essais de cette Méthode* (Leiden: J. Maire, 1637). Mersenne's relaxed attitude toward the galleys eventually led to the controversy between Fermat and Descartes over the *Dioptrics* and later the method of maxima and minima; cf. below, Chap. IV, §IV.

1635, while the *Rules for the Direction of the Mind* suggests that Descartes had already laid the foundations for the *Geometry* in the late 1620s.[2] However, the *Discourse* with its *Essays* was not published in France until early in 1638, while Descartes himself had possession of the original manuscript of the *Introduction* some months prior.[3] The fact remains that the *Geometry* was published and circulated widely, while the *Introduction* remained in manuscript until the publication of Fermat's *Varia opera* in 1679. Hence, regardless of who was the first to make the innovation, with some justice analytic geometry still bears the eponym "Cartesian."

More important and illuminating to the historian than the moot question of priority is the facet of simultaneity. What common problems and influences led both Fermat and Descartes to invent their respective systems? Although the task of the present chapter is solely to investigate in depth the background to Fermat's *Introduction*, some general observations applicable to both men may be made by way of introduction.

As Chapter II has shown, Descartes and Fermat operated against a common mathematical background, part of which they shared with the wider mathematical community as a whole, in particular the problems connected with curves. Euclid had been mastered, now it was Apollonius' turn. And to those who had read and understood the *Conics*, Pappus' *Collection* could offer the more tantalizing "special curves": the cissoid of Diocles, the conchoid of Nicomedes, the spirals of Archimedes and Menelaus.[4] The importance of such curves, especially of the conic sections, went beyond their inherent interest to the curious mathematician. Many Renaissance artists, from Alberti on, striving to reproduce accurately a three-dimensional world on a two-dimensional canvas, studied the changing images of curves under perspective collineation. Mathematicians such as Kepler came to the study of perspective through their work in optics.[5] Perspective showed how the "perfect" circle

---

[2] For details of the conceptual evolution of Descartes' algebra, see M. S. Mahoney, "The Royal Road," Chap. I, and *idem*, "Descartes: Mathematics and Physics," *Dictionary of Scientific Biography*, Vol. IV (New York, 1971), pp. 55–58.

[3] Cf. Descartes' letter to Mersenne dated 25.I.1638 in AT.I.499–504; the reference to the *Introduction* is on p. 503: "Je ne vous renvoie point encore les écrits de M. Fer[mat] *de locis planis et solidis* . . . ."

[4] Book IV of the *Collection* deals with these special curves, many of which were applied to the three famous problems of Greek geometry (to square the circle, to double the cube, and to trisect the angle). Diocles' cissoid is given by the modern equation $y^2(2a - x) = x^3$; Nicomedes' conchoid by the equation $(x - a)^2(x^2 + y^2) = b^2x^2$; Archimedes' spiral by the polar equation $\rho = a\vartheta$. Fermat took the *paradoxos grammê* of Menelaus to be a spiral of the form $\dfrac{2\pi R}{2\pi R - R\vartheta} = \left(\dfrac{R}{\rho}\right)^2$, where R is the radius of a given reference circle. For other notions of this curve, see P. VerEecke's French translation of the *Collection* (Paris, 1933), Vol. I, p. 207, n. 6. Fermat dealt with all of these curves in various of his works. His *Doctrine of Tangents* (see below, Chap. IV, §VII) derives the tangent to any point on each of the cissoid, the conchoid, the cycloid, and the quadratrix. See also his fragmentary quadrature of the cissoid in FO.I.285–290.

[5] Writers on perspective geometry, among them Kepler and Desargues, represented in the seventeenth century a distinctly synthetic geometric school of mathematics in con-

became a conic section when viewed from a different vantage point. It was to a conic section, the ellipse, that Kepler finally turned when an "oval" was needed to restore accuracy to the Copernican system. From the attempts of Galileo and others to resolve Aristotle's paradox of the wheel came the curious curve which a point fixed on a rolling wheel describes with respect to the line on which it is rolling, a curve which the French called the *roulette* and which today is known as the cycloid.[6] Similarly, while the path in space of a cannon ball falling from a tower on a rotating earth exemplified for Galileo the Platonic miracle of the circle, others (especially Fermat) with less prejudiced eye and, perhaps, greater sensitivity to the mathematical consequences of Galileo's physical principles found in that path the phenomenon of the spiral.[7]

Initially, physics and painting did not provide new curves for mathematical study in the early seventeenth century, as physics would do later. Rather, they acted to emphasize the importance of the study of curves and thereby to strengthen the already strong grip that the *Conics* of Apollonius had on the minds of mathematicians of the sixteenth and seventeenth centuries, especially after its translation into Latin by Commandino in 1566. And the *Conics*, in turn, enhanced the central importance of the treatise that placed it in the larger mathematical context of analysis, Pappus' *Mathematical Collection*. The *Conics* challenged mathematicians to understand a beautifully constructed, finished masterpiece of Greek geometric thinking. Pappus, by contrast, challenged them to be more inventive. Besides the immediate difficulties of determining a manageable geometric description of the "mechanical" or "special" curves in Book IV, there was, as outlined in Chapter II above, the more programmatic problem of restoring the works on plane, solid, and surface loci; on the porisms; and the missing last four books of the *Conics*; which

---

trast to the algebraic analysts discussed in Chapter II above. They have yet to receive adequate treatment from historians of mathematics, although their work led directly to the beginnings of projective geometry. Kepler, in particular, posited the "continuity principle" according to which the line and circle are treated as degenerate cases of conic sections, the circle as the narrowest ellipse and the line as the widest hyperbola. Like many writers on perspective, Kepler used optics as a starting point for his studies in perspective, in particular in his *Ad Vitellionem paralipomena quibus astronomiae pars optica traditur* (Frankfurt, 1604), Chap. IV of which contains the continuity principle. The family relationships of the line, circle, and conic sections were first expressed algebraically by Fermat and Descartes. (I am indebted to Prof. Albert W. Tucker of Princeton University for first calling my attention to Kepler's work on perspective.)

[6] The equation of the cycloid may be given parametrically as $x = a(\vartheta - \sin \vartheta)$, $y = a(1 - \cos \vartheta)$, where a is the radius of the reference circle. The cycloid was studied in the seventeenth century by Roberval (*De trochoide eiusque spatio*), Lalouvère (*Veterum geometriae promota in septem de cycloide libris*, Toulouse, 1660), and Blaise Pascal (*Historia trochoidis sive cycloidis*, Paris, 1658), among others. On Fermat's determination of the tangent to the cycloid, see below, Chap. IV, §VII; on his attempted quadrature, see FO.*Suppl*.87-91.

[7] See below, Chap. V, §II,A.

Pappus had cited in Book VII as being of special importance to the "field of analysis."

Here research activity in the area of curves tied in with the mathematical activity that forms the common underlying bond between Fermat and Descartes, to wit, algebraic analysis. Both men approached mathematics analytically and both used algebra as the vehicle for analysis. With specific regard to their analytic geometries, both men started with problems taken from Book VII of Pappus' *Mathematical Collection*, though significantly not with the the same problem.[8] Where the two men, and their systems, differed was in the direction toward which their efforts pointed. Descartes' goal was to illustrate by way of example a new mode of philosophizing.[9] If the *Geometry* reduces, in Book III, to a treatise on the theory of equations, it is for the same reason that the last ten *Rules for the Direction of the Mind* present the technique of setting up algebraic equations: algebra was to serve as the mathematical form of the new method of philosophy. By contrast, Fermat's goal was immediately mathematical. His *Introduction* presents in systematic detail a problem-solving technique which he fully intended to use in further mathematical research. For him the system of the *Introduction* was simply part of a single, general method, already clear in his own mind since his stay in Bordeaux, which underlay the technique of determining maxima and minima and of treating number theory analytically. By 1637, Descartes had all but finished with mathematics; Fermat was just getting started. And yet, if their paths diverged, at least they were traveling in the same country. Each man aimed at solving problems, and each man sought a systematic method for solving those problems.

Moreover, due to Fermat's and Descartes' commitment to the analytic program outlined in Chapter II, both the *Introduction* and the *Geometry* continued the effort to restore the analytic techniques of the Ancients. The two authors moved beyond their predecessors, however, in treating a new class of problems and a new class of equations. Viète, Snel, and Ghetaldi had demonstrated the efficacy of applying the algebra of equations in one unknown to the restitution of works on the section of ratios and of areas. That is, they applied algebra to determinate, metric geometric problems, the analysis of which led to determinate equations in one unknown or to determinate families of simultaneous equations. By the early 1600s algebraists were well in control of determinate algebraic systems of equations. The systematic solution of third- and fourth-degree equations had become commonplace.[10] Of growing interest to mathematicians like Viète and, later, Fermat were indeterminate equations and systems of equations in two or more unknowns, i.e., those per-

---

[8] See below, §III,B.
[9] The *Geometry* also provides the mathematical derivations of the optical "ovals" presented by Descartes in the *Dioptrics* as solutions to the general anaclastic problem.
[10] See above, Chap. I.

mitting an infinite set of solutions. The immediate source of problems involving such equations was Diophantus' *Arithmetic*, and work begun in this area by Viète was taken up by Fermat early in his career.

By the second decade of the seventeenth century, then, the raw materials lay at hand, out of which the analytic geometries of Fermat and Descartes would emerge: interest in and growing knowledge of curves, the use of algebra in the solution of determinate geometric problems, and the treatment of indeterminate equations. To Fermat and Descartes goes the credit for having seen how these three elements might be combined to form a potent, systematic heuristic device for solving both geometric and algebraic problems. The complex novelty of that achievement makes it advisable to reverse historical order by first setting forth in detail Fermat's system of analytic geometry as it is contained in the *Introduction* and then following its genesis as Fermat moved from his knowledge of algebra and his attempts to restore the *Plane Loci* of Apollonius to his new system. As the discussion below will show, the restoration of Apollonius' treatise opened up several different lines of possible development, and hence the actual endproduct must define the path of historical analysis. Once the new system was established, however, it dictated in itself extensions and applications which Fermat pursued in the years following 1636, and an examination of that corollary work will conclude the discussion.

II. FERMAT'S ANALYTIC GEOMETRY: THE INTRODUCTION TO PLANE AND SOLID LOCI

A. *The Nature of the Work: A Method and A Theorem*

Fermat's concern for general method as well as for solving particular problems caused him to set forth many of his treatises on two levels, His *Introduction to Plane and Solid Loci* is an excellent example.[11] It is titled an "introduction" as to a method, but it contains a complete and detailed proof of a well-defined theorem. Its function as an introduction is the more general of the two levels. It introduces into mathematics a brand new method which is contained in the assertion that to every indeterminate algebraic equation in two unknowns there corresponds a uniquely determined geometric curve considered as a locus. The assertion of a one-to-one correlation between geometric loci and indeterminate equations, classes of mathematical objects previously unconnected in the minds of mathematicians, provides, according to Fermat, "a general path to loci," i.e. a general technique for the resolution of locus problems. But justification of such a claim must take the form of its successful application to a more limited set of curves and equations, and in the *Introduction* Fermat seeks to prove that indeterminate equations of the

---

[11]*Ad locos planos et solidos isagoge*, FO.I.91–103. All subsequent quotations have been translated from the FO text, and the page numbers that follow them refer to it.

first and second degree in two unknowns correspond to the members of the family of curves traditionally known as the "plane and solid loci," to wit, the straight line, the circle, and the three conic sections.

Fermat begins by referring to the general problem to which the method is addressed:

> There is little doubt that the Ancients wrote many works on loci; witness Pappus who, in the beginning of the seventh book [of the *Collection*], asserts that Apollonius wrote on plane loci and Aristaeus on solid loci. But, unless we are mistaken, their investigation did not satisfy them sufficiently. This we gather from the fact that they did not express many loci in sufficient generality, as will be shown below.
>
> Therefore, we will subject this science to its own particular analysis, in order that consequently a general path to loci become clear. (p.91)

Besides the particular difficulties which his restoration of Apollonius' *Plane Loci* presented, and which will be described in some detail below, Fermat found that the theorems contained in that work were too specific. Apollonius had considered only a small segment of the conditions which determined or were satisfied by plane loci, that is, a line or a circle. Moreover, Fermat felt, the conditions themselves could often be generalized well beyond their statement by Apollonius. Fermat shared with Descartes this sense of a lack of generality in the works of the Ancients. But, while Fermat was disturbed by the restricted scope of Apollonius' theorems, Descartes was bothered by the prolixity of the theorems themselves. The Greeks' lack of a general method forced them, he complained in the *Geometry*, to compose fat volumes, ". . . where the order alone of their propositions shows us that they did not have the true method for finding all of them, but that they simply gathered together those which they had found."[12] Carrying away from their reading of the Greek geometers the same dissatisfaction, the two men came to the same conclusion: the Ancients' failure was due to insufficient analysis of locus problems. Concurring in their diagnosis, Fermat and Descartes prescribed the same remedy. In Fermat's words, the problem of loci must be subjected to "its own particular analysis."

Fermat does not explicitly discuss the general nature of this "particular" analysis. He need not. To Fermat, as to many contemporaries, "analysis" means algebraic analysis. Algebra, Viète's "art of analysis," provides a general

---

[12] *Geometry*, p. 304. Descartes' remark here contrasts markedly to his earlier belief in the "true mathematics" of the Ancients (cf. above, Chap. II, n. 14). The later attitude reveals Descartes' growing awareness that his symbolic algebra was a far more sophisticated mathematical tool than the Ancients had possessed, something that the intervening development of the methods in the *Geometry* confirmed. By the mid-1630s the analytic program was not so much restoring algebra to the analytic works of the Ancients as using the latter to help to direct the further development of the new analysts' own form of algebra.

technique for locus problems, just as for Viète and Ghetaldi it provided a general technique for the analysis and synthesis of determinate problems.[13] But the art of analysis as Viète and Ghetaldi used it does not completely suffice for the resolution of locus problems. Their section problems led, under algebraic analysis, to determinate equations in one unknown. Interpreting the results offered little difficulty; a single unknown length was sought, a single unknown length was found, given in terms of the data of the problem.

The algebraic analysis of locus problems, however, led to a new family of equations previously without geometric meaning: indeterminate equations in two unknowns. Something must be added to the art of analysis, something which makes that analysis, to use Fermat's Latin terms, *propria et peculiaris* to locus problems. To that something, Fermat turns next in the *Introduction*:[14]

"Whenever two unknown quantities are found in final equality, there results a locus [fixed] in place, and the endpoint of one of these [unknown quantities] describes a straight line or a curve."

Here, in one terse sentence, is Fermat's general assertion and his innovation. For all that it says, and Carl Boyer has termed it "one of the most significant statements in the history of mathematics,"[15] it leaves much unsaid. It embodies both a technique and an assertion. The technique is the application of algebra to locus problems, the translation of the geometric conditions into the mathematically more general language of algebra. To the immediate result of the translation one then applies what Fermat later refers to as "the precepts of the art," that is, Viète's rules for the reduction of the equation to its simplest or canonical form. The latter is the "final equality" of which Fermat speaks. Though he does not make the point explicitly, the final equality provides a criterion for classifying the problem under analysis: if it contains only one unknown, it is a determinate section problem; if it contains two unknowns, it is a locus problem. The question of what meaning to give to final equalities involving more than two unknowns is one Fermat leaves unconsidered for the moment.[16]

The assertion, then, is: a final equality with two unknown quantities corresponds to a locus. Indeed, it more than corresponds; it determines a locus. This stronger statement is what Fermat means by *fit locus loco*, a phrase which previous translators have translated "a locus results," ignoring the ablative *loco*. Fermat is laconic enough not to need further abbreviation. By the use of the ablative *loco*, he means to say that the locus which results is

[13] See above, Chap. II.
[14] FO.I.91: "Quoties in ultima aequalitate duae quantitates ignotae reperiuntur, fit locus loco et terminus alterius ex illis describit lineam rectam aut curvam."
[15] C. B. Boyer, *History of Analytic Geometry* (New York, 1956), p. 75.
[16] He takes it up later; see below, p. 123.

fixed in place by the final equality to which it corresponds. That is, the final equality contains all the information necessary to determine the locus fully. As will be clear from the exposition below of sample theorems from the *Introduction*, Fermat carefully relates the parameters of the equation to those of the curve in question; such a correlation is essential to his demonstration.

But what will he demonstrate? Certainly not the general proposition given above, for the assertion of a correspondence between loci and indeterminate equations is more a postulate than a theorem, as Fermat well realized. Such methods as the one implied by the general proposition must be judged by their efficiency rather than by proof. Fermat must test his method on a family of well known curves, and this he proposes to do:

> Whenever the local endpoint[17] of the unknown quantity describes a straight line or a circle, a plane locus results; and when it describes a parabola, hyperbola, or ellipse, a solid locus results. If [it describes] other curves, the locus is called linear. We will say nothing concerning the latter, because a knowledge of a linear locus is very easily derived, by means of reductions, from the investigation of plane and solid loci. (pp. 91–92)

The *Introduction* will deal, then, with a very familiar group of curves: the line, the circle, and the conic sections. The classification of these curves into plane and solid loci is traditional; its immediate source is Pappus. Similarly, the gross collation of all remaining curves into a single class of linear loci also stems from Pappus and classical tradition.[18] Since the *Introduction* represents a conscious transition from the old to the new, Fermat maintains these old

---

[17]The "local endpoint" (*terminus localis*))is the endpoint of the second variable, which traces out the curve determined by the equation; cf. the discussion below.

[18]Pappus discusses the classification of loci in two passages in the *Collection*. First, in Book IV, he notes: "Those problems which can be solved by means of a straight line and a circumference of a circle may properly be called *plane*; for the lines by means of which such problems are solved have their origin in a plane. Those, however, which are solved by using for their discovery one or more of the sections of the cone have been called *solid*; for their construction requires the use of surfaces of solid figures, namely those of cones. There remains a third kind of problem, that which is called *linear*; for other lines (curves) besides those mentioned are assumed for the construction, the origin of which is more complicated and less natural, as they are generated from more irregular surfaces and intricate movements." (Pappus, ed. Hultsch, iv, p. 270, lines 5–17; quoted and trans. by Heath, *History of Greek Mathematics*, II, 17). In his introductory remarks to Apollonius' *Plane Loci* in Book VII, Pappus expands on these distinctions: "Of loci in general, (1) some are *ephectic* [lit. 'fixed'], so that Apollonius says in the preface to his own *Elements* that a point is a locus of a point, a line a locus of a line, a surface of a surface, a solid of a solid; (2) some are *diexodic* [lit. 'allowing passage'], such that a line [is the locus] of a point, a surface of a line, a solid of a surface; and (3) some are *anastrophic* [lit. 'enveloping'], such that a surface [is the locus] of a point, a solid of a line. Of [loci] in the field of analysis, some are things given in position [and] are ephectic, some are called *plane* and some *solid*. Linear [loci] are diexodic [loci] of points; *surface* [loci] are either anastrophic [loci] of points or diexodic [loci] of lines.

distinctions, relying on older, intuitive geometric notions of the curves he proposes to treat. He shows how they are related to various equations. However, like Descartes and Newton, Fermat will eventually use the new system to eliminate the old distinctions as artificial and intuitive; the equations themselves sufficed for the redefinition of the curves and for their classification.

B. The Geometric Framework and the Main Theorem

The form of the main theorem of the *Introduction* depends on the procedure by which Fermat illustrates the intimate relationship between an indeterminate equation and its corresponding curve. In the work of Viète and Ghetaldi, the application of algebra to determinate problems not only provided solutions to the problems but showed also how those solutions were to be constructed. It is the latter property of algebraic analysis that Descartes takes pains to render explicit in the opening paragraphs of the *Geometry*, as Viète had done earlier in his *Canonical Recension of Geometric Constructions.*[19] For the algebraic analysis of locus problems to succeed, Fermat must provide a technique which relates the algebra to the construction of the curve and the framework within which that construction takes place. With the following statement, he fulfills that requirement and states the specific theorem that the rest of the *Introduction* is designed to demonstrate:

> The equations can easily be set up, if we arrange the two unknown quantities at a given angle—which we will usually take as a right angle—and if one endpoint of one of these [quantities] given in position is given. Provided that neither of the unknown quantities exceeds the square, the locus will be plane or solid, as will be made clear from what is said. (p. 92)

Again, much prior mathematical development lies implicit in Fermat's statement.[20] Where Descartes justifies in detail the use of line lengths as algebraic variables, that is, by showing that geometric magnitudes constitute an algebraic field, Fermat assumes it as part of the Viètan algebra he employs. He

---

Nevertheless, linear [loci] are demonstrated by surface [loci]. The loci we are treating are called plane and are in general either straight lines or circles. Solid [loci] are conic sections: parabolas, ellipses, or hyperbolas. Loci which are neither straight lines, nor circles, nor any of the said conic sections, are called linear. The loci described by Eratosthenes as 'referred to means' belong to the aforesaid according to genus, but differ from them by the peculiarity of their hypotheses. The Ancients, concentrating on the classification of the plane loci, taught them as elements [i.e. elementary material]. Ignorant of this [classification], those who followed added others, as if the [plane loci] were not [themselves] infinite in number, should someone wish to append [merely] the [loci] contained in that classification [alone]." (Pappus, ed. Hultsch, vii, p. 660.17–662.23; my trans.) It is small wonder Fermat and Descartes felt that a new classification of curves was in order!

[19] See below, p. 115.
[20] See above, Chap. II, and Mahoney, "Royal Road," Chaps. I and III.

represents the two unknowns in the equation by two varying line lengths placed at an angle to one another. Their lengths vary in accordance with the algebraic relationship of the variables which they represent. A diagram will help to clarify Fermat's technique.

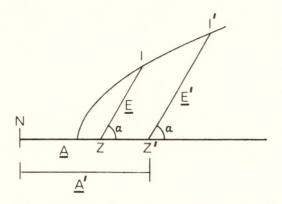

Fermat is dealing with algebraic equations of the form $f(A,E) = C$, where A and E, following the notational convention of Viète,[21] are the two unknowns and C a homogeneous constant. In the geometric picture, he takes the line NZ to represent A; NZ is "given in position," which in classical terminology means that it undergoes neither rotation nor translation, and one of its endpoints N is also "given," i.e. fixed. Hence, NZ ($= A$) changes only in length measured from N. Fermat then imagines ZI to represent E, and he joins ZI to NZ at point Z at an arbitrary, but henceforth fixed, angle α. By the construction he has set forth, ZI undergoes only parallel translation and change in length measured from Z. The "local endpoint" (*terminus localis*) is point I. It occupies varying positions in the plane as the lengths NZ and ZI change in accordance with the relationship linking A and E, i.e. $f(A,E) = f(A',E') = C$. These varying positions constitute a locus.

[21] The matter of symbolism creates some difficulties in this chapter. I resolve them by a compromise. When directly quoting Fermat, or when attempting, through paraphrase, to keep in the spirit of his style in mathematics, I use the symbolism of Viète as he did. When seeking merely to explicate the mathematics of the *Introduction*, or to discuss a logico-mathematical point, I use modern terminology and symbolism. I try when possible to avoid confusion by using distinct combinations of letters. When using Fermat's symbolism, I make, however, certain adjustments to account for the difference in language. For example, "D in A" occurring as a term in an equation makes little sense outside of the Latin context [*linea*] D in [*lineam*] A [*ducta*]; therefore, I render *in* either by the word "times" or by the symbol "·". I also use "=" rather than the word "equals," in the interest of economy and ease of reading. The expression $f(A,E)$ is a hopeless mixture of old and new, but one which I hope serves my purposes at the moment. The mathematicians of the seventeenth century lacked means for expressing a general function of one or more variables, working rather from *instar omnium* or verbal expressions. On Viète's system of algebraic notation, see above, Chap. II.

"Fermat's scheme," writes Carl Boyer, "like that of Descartes, may be characterized as an ordinate—rather than coordinate—geometry."[22] Perhaps the term "uniaxial system," corresponding to the single reference line, or axis, and in contrast to the biaxial system in common use today, would describe Fermat's construction more accurately. While both systems accomplish the same purpose and lead to the same results, they differ conceptually. In Fermat's analytic geometry, curves emerge as loci from the different positions taken by point I as the variable length ZI stands on NZ at varying distances from point N; that is, curves are generated rather than plotted. There is connected with the system an intuitive sense of motion or flow wholly in keeping with the intuition which underlies the notion of an algebraic variable. While the concept of the algebraic variable may be found sketched in the works of Viète and articulated in the opening of Descartes' *Geometry*, it receives added emphasis from the geometric correlate invented by Fermat and Descartes; from the intuitive picture of a geometric magnitude acting as an algebraic variable comes the notion of continuous variation. Calling Fermat's system uniaxial also has an historical advantage. Such a term points to the historical root of his construction, for to a large extent Fermat's choice of a framework within which to construct his loci is classical in origin, as the next section will show.

The description of Fermat's system as it has been given so far may deceive the unwary, especially when the angle between NZ and ZI is taken as right. Besides the absence of a second axis, the system differs from the modern so-called "Cartesian" coordinate system in one other important respect. Meant to illustrate algebraic relationships as they were understood at the time, the system shares the limitations of that understanding. The only "true" solutions of an algebraic equation were the positive ones; length or distance was by definition positive. Hence, Fermat's system includes only the first quadrant of the modern coordinate system. The concrete result of such a restriction is that Fermat consistently loses from one-half to three-quarters of his higher-order curves; for instance, the algebra of the system can account for only the positive half of one branch of an axial hyperbola whose center is the origin. To complete the curves, geometry must come to the aid of algebra and supply the missing branches by considerations of symmetry with respect to the axis and with respect to a line through the origin parallel to ZI.

On the basis of the system he has set up in the short introduction, Fermat devotes the body of the *Introduction* to a demonstration of the following theorem: indeterminate algebraic equations of the first and second degree in two unknowns determine curves which are members of the general family of conic sections. The examples of his demonstration will be clearer if we con-

---

[22] Boyer, p. 76.

sider first the general scheme of the proof. Fermat divides the family of second-degree equations in two unknowns into seven subfamilies, each represented by a characteristic equation. He shows for each subfamily that its members correspond to, and determine, members of one of the subfamilies—line, circle, parabola, hyperbola, ellipse—of the family of conics. The seven characteristic equations treated by Fermat are, in modern symbolism (translated from his partially verbal equations):

(I) $ax = by$ line
(II) $xy = b$ hyperbola
(III) $x^2 \pm xy = ay^2$ line(s)[23]
(IV) $x^2 = ay$ parabola
(V) $b^2 - x^2 = y^2$ circle[24]
(VI) $b^2 - x^2 = ay^2$ ellipse
(VII) $b^2 + x^2 = ay^2$ hyperbola

Proof of the correspondence takes the form of a demonstration in each case that the local endpoint of one of the variables in the equation lies on a particular curve constructed on the basis of the parameters of the equation. The seven characteristic equations treated by Fermat constitute an exhaustive division of the general equation, since, by a technique of reduction, every other second-degree equation can be reduced to one of the seven forms above and hence corresponds to the conic determined by that form.

C. *The Body of the* Introduction: *Proof by Cases of the Central Theorem*

Fermat begins with the simplest case: in his notation, the equation $D \cdot A = B \cdot E$; and in his words, the following argument:

---

[23] Fermat's solution here, restricted in each case to the positive roots, calls for a single line corresponding to the equation

$$x = y\left[\frac{\sqrt{1 + 4a} + 1}{2}\right] \quad (a > 0)$$

and ignores that corresponding to

$$x = -y\left[\frac{\sqrt{1 + 4a} + 1}{2}\right] \quad (a > 0).$$

The non-homogeneity of the equation $x^2 + xy = ay^2$ results from using the equation instead of the proportion by which Fermat actually expresses the relationship; that is, he speaks of $x^2 + xy$ and y^2 being in a given ratio, which here is denoted by a.

[24] Provided, of course, that the angle between axis and ordinate is a right angle; cf. below, n. 30. The equation of a circle in a skew system is of the form $x^2 + y^2 + axy = b^2$; note that in both the right and the skew system, the coefficients of the x^2 and y^2 terms are equal.

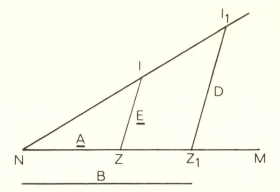

Let NZM be a straight line given in position, of which point N is given. Let NZ be equal to the unknown quantity A, and let the line ZI, erected at a given angle NZI, be equal to the other unknown quantity E. If D times A equals B times E, then point I will lie on a *straight line* given in position.

That is, as B is to D so A is to E. Therefore, the ratio of A to E is given; and the angle at Z is given. Therefore, the triangle NIZ is given in species, and angle INZ [is given]. Also, point N is given, and the straight line NZ [is given] in position. Hence, NI will be given in position. The composition is easy. (p. 92)

This argument deserves extended comment for two reasons: first, it sets the pattern for the rest of the theorems of the *Introduction*, and, second, it illustrates Fermat's habit of abbreviating his proofs almost to the point of obscurity. In the argument as Fermat sets it down, he first establishes line NZM as the axis of his geometrical framework. He then fits the equation into that framework by supposing NZ to represent the first unknown A and ZI the second unknown E. He concludes the first paragraph with a statement of the theorem he seeks to prove: if $D \cdot A = B \cdot E$, then the locus of point I is a straight line. The second paragraph of the proof is meant as an analysis of the problem, and it is a classical example of a reductive analysis using Euclid's *Data*.[25] (Fermat never cites the *Data*; he relies on the reader to recognize the material and to turn to the *Data* should he need further proof.) That is, it argues back from the given equation to a known construction: from the given equation, it follows that $A:E = B:D$. Since B and D are given constants, it follows that their ratio is given, and hence that the ratio of A to E is constant. Hence, the ratio of two sides, NZ ($= A$) and ZI ($= E$), and their included angle NZI are both given; from the *Data* it then follows that the triangle NZI is

[25]See M. S. Mahoney, "Another Look at Greek Geometrical Analysis," *Archive for History of Exact Sciences*, 5(1968), pp. 318–348; esp. p. 341.

fully determined and that side NI is given in position.[26] Here Fermat ends his analysis, leaving to the reader the task of completing it by a construction of side NI. The missing steps are easily supplied: set length $NZ_1 = B$ and length $Z_1 I_1 = D$, and then draw NI_1.

Fermat presents only the (partial) analysis of the problem, dismissing the synthesis (composition) as "easy." His use of the terminology of resolution and composition, i.e. analysis and synthesis, lends a classical tone to his proof. But, besides the fact that only the analysis is present, the argument marks a fundamental departure from classical usage. This will be clear from a consideration of what Fermat must do to prove his theorem. He must show that the given equation, when fitted into the framework of his geometric construction, uniquely determines a line. To do so, he must accomplish two tasks: first, he must provide a construction of the line on the basis of the parameters of the equation; second, he must show that point I falls on the line (NI) constructed, whatever the particular values of A and E satisfying the equation. Such a situation corresponds to an if-and-only-if proposition, which in Antiquity was almost always proved as two separate propositions, a theorem and its converse. But the classical techniques of analysis and synthesis applied primarily to ordinary if-then theorems: analysis supplied the construction or the steps of the proof by working from consequent to antecedent; synthesis comprised the rigorous logical deduction in the other direction.[27] In the above theorem, however, Fermat views resolution as providing a proof of the first part of his biconditional theorem, a proof, that is, that the equation determines a line. The synthesis, which Fermat dismisses as easy, would show that all pairs of values A, E satisfying the equation are such that when translated into the geometrical picture, the point I will lie on NI. A proof of this statement comes down to a proof that if any pair of values is to fit the requirement, then they must be so related that $A : E = B : D$; that is, they must lead to a triangle on the diagram, such as NZI, similar to the one constructed in the analysis. Why Fermat would term the proof of the converse of the theorem proved in the passage given a "composition" is readily understood. The proof he gives moves from the given ratio B/D to the construction of the triangle. The composition which he omits moves from the construction of the triangle to the originally given ratio B/D.

An important element in the proof of the first theorem is the *sotto voce* use of Euclid's *Data*. For the higher-degree equations, Fermat turns (again without explicit citation) to Apollonius' *Conics*. For example, take the proof of (IV): in Fermat's words, "if Aq equals $D \cdot E$, point I lies on a parabola." The proof begins with the basic framework involving lines NZ and ZI drawn

---

[26] Although Fermat does not state so explicitly, he includes the modern criterion of a line in his discussion by referring to the fact that angle INZ is fixed. Clearly in the case of angle NZI = 90°, INZ is the angle of which the tangent is D/B, a given ratio.

[27] Cf. Mahoney, "Another Look," pp. 321–327.

at right angles with point N fixed. Fermat draws NP parallel to ZI, and about NP as axis he describes a parabola of which latus rectum is D and the "applied sides" (here abscissas) are parallel to NZ. The assertion is that, by construction, $D \cdot \text{NP} = \text{PI}^2$, $A = \text{NZ} = \text{PI}$, and $E = \text{ZI} = \text{NP}$; hence, point I lies on the

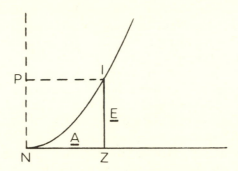

parabola. The full justification, which Fermat merely sketches out, would follow the pattern of Theorem I. The crucial step is the construction of the appropriate parabola on the basis of the parameter of the equation. Fermat merely postulates that construction. He supplies none of the steps, relying on the reader to recognize that Propositions I,52 and I,53 of Apollonius' *Conics* set forth the construction of a parabola on a given axis and with a given latus rectum. Later theorems of the *Introduction* similarly rely on the reader's knowledge of the pertinent theorems of the *Conics* (I,52–I,58), whereby he may supply the constructions (which Fermat only postulates) necessary for linking the various characteristic equations to the conic sections. Fermat also assumes that the reader will know the *Conics* well enough to realize that most of his theorems hold for any angle between the variable lengths (*Conics* I,11–14, 53, 55, 58).[28] Beyond these substantial technical points, Fermat's system owes more general aspects to the *Conics*, as will become clear below.

Following the pattern illustrated by the two above examples, Fermat links each of the seven characteristic equations given at the end of the previous section to one of the conic sections. He has a further task, however: he must show that these seven equations account for all the possible members of the family of second-degree equations. That is, he must show that the equation resulting from any combination of values for the coefficients in the general equation

$$\alpha x^2 + \beta y^2 + \gamma xy + \delta x + \epsilon y + \zeta = 0$$

may be placed under the heading of one of the characteristic equations.[29]

[28] The only exception is the circle; see below, n. 30.
[29] Any combination, that is, for which the resulting equation has real solution pairs x, y.

The "precepts of the art," i.e. Viète's algebra, supply a universally applicable technique for carrying out this distribution. Let us take, for example, the subfamily of equations corresponding to the circle. Fermat gives the characteristic equation as $Bq. - Aq. = Eq$. By an argument similar to those presented above, Fermat demonstrates that the equation determines a circle of radius B about the origin, point N. He proceeds then to account for the remaining members of the subfamily:

All equations containing Aq and Eq and A or E multiplied by given quantities may be reduced to this equation, provided that angle NZI is right[30] and, in addition, that the coefficients of Aq are equal to the coefficients of Eq.
Let $Bq - 2D \cdot A - Aq = Eq + 2R \cdot E$.
Add Rq to each [side] in order that $E + R$ replace E; it will make

$$Rq + Bq - 2D \cdot A - Aq = Eq + Rq + 2R \cdot E.$$

To these Rq and Bq is added Dq in order that $D + A$ replace this A. Let the sum of the squares Rq, Bq, and Dq equal Pq. Hence,

$$Pq - Dq - 2D \cdot A - Aq = Rq + Bq - 2D \cdot A - Aq,$$

since, by construction,

$$Pq - Dq = Rq + Bq.$$

If therefore you take A in place of this $A + D$ and E in place of $E + R$, it will make

$$Pq - Aq = Eq,$$

and the equation will be reduced to the preceding one. (p. 98)

Fermat employs the technique of reduction described here throughout the *Introduction*. It involves a change of variables with an attendant translation of the reference line;[31] the original A and E in the complex equation are re-

---

[30]The proviso that Fermat adds to his statement on the class of equations determining a circle—"provided that angle NZI is right"—receives fuller explication in Theorem VI on the ellipse. There Fermat remarks concerning equations which contain Aq and Eq on opposite sides of the equations and with opposite signs: "If the coefficients are the same and the angle is right, the locus will be a circle, as we have said. But it may be that the coefficients are the same, but the angle is not right; the locus will be an ellipse." Here the intimate relationship between the circle and the ellipse, which previous work in perspective had demonstrated geometrically, emerges in analytic form. Indeed, Fermat accomplishes as a byproduct of his efforts an analytic demonstration of the close interrelationship between the various conic sections in addition to the line and circle. The *Introduction* shows that the five curves involved are all related by the general second-degree equation. Kepler had earlier shown geometrically that the line, the circle, and the conic sections were all members of a single family of curves.

[31]That is, Fermat moves the axis and origin of his system to make them coincide with the axis and vertex of the curve. That is another reason for calling his system "uniaxial."

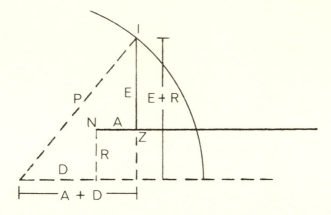

placed respectively by a suitable $A' = A + D$ and a suitable $E' = E + R$, so that the resulting equation takes the simple form of the characteristic equation for the subfamily. The new variables A' and E' are no longer measured from N, however; A' is measured from a point the length D removed along the axis from N, while E' is measured from a line parallel to the axis NZ and removed the distance R from it. Fermat's technique remains basic to analytic geometry today. Once one has shown, for example, that the simple equation $x^2 + y^2 = r^2$ corresponds to a circle about the origin, one may immediately move to the statement that the general parametric equation $(x - h)^2 + (y - k)^2 = r'^2$ also determines a circle, of which the center is now the point (h,k).

Terms such as "change of variable," "translation," and even "transformation" did not, of course, form part of Fermat's mathematical vocabulary, and, in view of the profound conceptual content they acquired during the nineteenth century, one uses them here with caution. Nonetheless, on an operational level at least, Fermat is clearly carrying out transformations of his uniaxial system. Having in each case reduced a particular example to canonical form, he supplies the construction of the curve determined by it, not only referring the curve to a new axis and (usually) a new origin, but also explicitly relating the new reference system to the old via the algebraic reduction procedures he has employed. In the example above, he accomplishes the reduction by adding a constant value to each of the unknowns; he then shows the specific parallel translation of the axis (with its fixed endpoint) determined by that additive change of variable.

Toward the end of the *Introduction* Fermat reveals his awareness of the precise geometric adjustments called for by other, more complex algebraic reduction procedures. Again, the explicit correlation between algebraic operation and geometric construction warrants the name "transformation," although the particular transformation (rotation of the axis, change of angle between axis and applicate, and multiplicative increase of variable) has not come down to the present as a standard one in analytic geometry. The equation treated represents the final form of the general second-degree equation not covered by the seven preceding theorems; Fermat calls it "the most difficult of all equations,"[32] and it contains Aq, Eq, $A \cdot E$, and data. The equation Fermat chooses as a special example of his general technique is, in modern symbols,

$$b^2 - 2x^2 = 2xy + y^2.$$

First, Fermat adjusts the equation from one containing the variables x, y to one containing $x, x + y$. In the particular example, such adjustment presents no problem:

$$b^2 - x^2 = (x + y)^2.$$

Taking, in his axial system, NZ as x and ZI as $x + y$, Fermat gets a circle of radius b about point N as a locus. But that locus is traced out by a ZI which represents $x + y$ rather than y itself; that is, the length corresponding to y is

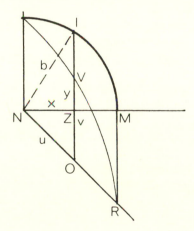

uniformly augmented by that corresponding to x. To get a length corresponding to y, Fermat must find a way to subtract the length NZ from the length ZI. He accomplishes this goal by drawing NR at an angle of 45° to NM. Then ZO drawn perpendicular to NZ also equals NZ ($= x$); and, if ZI ($= x + y$) is measured from O, then the length ZV corresponds to y. It is clear now that point V serves as the "local endpoint" of two axial systems: it is the endpoint

---

[32]FO.I.100: "Difficilima omnium aequalitatum est quando ita miscentur Aq et Eq ut nihilominus homogenea quaedam ab A in E afficiantur unà cum datis, etc.

$Bq - Aq$ bis aequetur A in E bis $+ Eq$."

The proof occupies pp. 101–102 of the text.

of the length OV (= $x + y$) measured from axis NO and inclined at an angle of 45° to that axis, and it is the endpoint of the length ZV (= y) measured from axis NZ and perpendicular to it. Hence, if one can determine the locus of point V with respect to the system NO,OV, one will also have the locus of point V with respect to the system NZ,ZV. The determination of the locus of V in the system NO,OV requires one further adjustment of the original equation. For, while OV = ZI = ($x + y$), NO is not x but $x\sqrt{2}$. Substituting then in the original equation the values $u = x\sqrt{2}$ and $v = x + y$, one gets

$$b^2 - \frac{u^2}{2} = v^2 \text{ or } 2b^2 - u^2 = 2v^2,$$

which is the equation of an ellipse. Therefore, the original equation corresponds to the same ellipse.

The solution paraphrased here displays Fermat's imaginative mathematical talents at their best. One may quickly verify that the pattern of solution underlying his treatment of the particular equation is fully general. By combining it with the reduction technique for equations containing linear terms in x and y, Fermat can determine the locus corresponding to a general equation of the form $\alpha x^2 + \beta y^2 + \gamma xy + \delta x + \epsilon y + \zeta = 0$, in which none of the coefficients is zero. The expansion of the axis and the angle between the two main axes NZ and NO will vary as a function of β and γ.

The techniques of change of variable and transformation of the geometric framework supply, then, the final element in Fermat's demonstration of the particular theorem to which he addresses the *Introduction*. By means of these techniques he can reduce any specific instance of the general second-degree algebraic equation in two unknowns to one of the characteristic equations treated in the body of the *Introduction*. He requires from his reader knowl- of the algebra of Viète, both to follow the arguments in the *Introduction* and to cope with specific equations not in the text itself. In this respect, Fermat's exposition of analytic geometry differs greatly from Descartes. Fermat treats a whole family of equations in great detail and makes the *Introduction* almost a textbook in elementary analytic geometry. But, he assumes on the part of his reader knowledge of the highly technical algebra which underlies his system. Descartes works the other way around. He gives the solution of one particular problem, Pappus' *n*-line locus problem. Many of the elementary concepts, such as the equation of the line or the circle, are left implicit in the treatment of the specific case. But, instead of setting forth an explicit reduction procedure, Descartes presents in Book III of the *Geometry* the fundamentals of the theory of equations from which the mathematically astute reader can derive techniques similar to Fermat's. The difference in treatment illustrates the different goals toward which the two men are working. Descartes considered algebra the mathematical model of a general epistemological system. The *Geometry* is presented as but an example of the method described in the *Discourse*. If the analytic geometry set forth therein is going to be used, it will not be

Descartes who uses it. The job of adapting the contents of the *Geometry* to practical use in mathematics is left to someone else. Fermat, however, is a practicing mathematician engaged in solving problems. His system is meant to be employed, and he will employ it.

The *Introduction*, Fermat promises at the outset, will point the way to a general treatment of loci. He redeems the promise at the end of the work:

> Thus we have embraced briefly and lucidly all the things the Ancients left unexplained concerning plane and solid loci, and consequently it will be clear to what locus all the cases of the last proposition of Book One of Apollonius' *Plane Loci* pertain. And in general all things which touch on this matter will be found without difficulty.
>
> But, we may be permitted to add as a crowning touch this most beautiful proposition, the simplicity of which will be immediately clear:
>
> If, given any number of lines in position, lines be drawn from one and the same point to each of them at given angles, and if [the sum of] the squares of all of the drawn lines is equal to a given area, the point will lie on a solid locus given in position. (p. 102).

The proposition is a generalized version of the three- and four-line locus problem enunciated by Pappus in his commentary on the *Conics* of Apollonius, the problem to which Descartes addresses the *Geometry* (though the manner in which Descartes generalizes the problem differs from Fermat's).[33] Fermat does not seek a general solution. Confident that it follows from what he has already presented in the *Introduction*, he provides rather an example illustrating the "path to practice." The example involves the following problem: Given two points M and N, find the locus of point I such that $IN^2 + IM^2$ is in a given ratio to the area of triangle INM. Fermat applies algebraic analysis to the

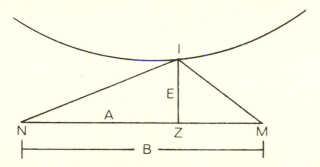

[33] Following the quotation of Pappus' long statement of the problem in his introductory remarks to the *Conics* in Book VII, Descartes sums up the problem in the *Geometry* (pp. 306–307): "La question donc qui avait été commencée à résoudre par Euclide et poursuivie par Apollonius, sans avoir été achevée par personne, était telle: Ayant trois ou quatre ou plus grand nombre de lignes droites données par position; premièrement on

statement of the proposition, letting NM = B, ZI = E, NZ = A, and angle NZI = 90°. "By the precepts of the art,"

$$2Aq + Bq - 2AB + 2Eq = DBE,$$

where $2D$ represents the fixed ratio. The nature of the locus follows on inspection: it is a circle. The construction of the locus follows from the application of the reduction technique and the comparison of the parameters with those of the characteristic equation for a circle.

With this final example, Fermat completes the *Introduction*. Having laid the groundwork relating loci and indeterminate equations, he turns to the applications of the system. Before examining those applications, however, it is necessary to go back and investigate the historical origins of Fermat's analytic geometry. Since the next section will focus on the development of specific aspects of the *Introduction* discussed in this section, it would help to review quickly the salient features so far discussed. First, Fermat employed an algebraic approach to geometric locus problems. Underlying his approach, then, is the application of algebraic techniques to geometric problems in general. Second, he recognized that the algebraic analysis of locus problems resulted in a single indeterminate equation in two unknowns, and he posited that such equations determine definite geometric curves. Third, he established a geometric framework which allowed the translation of the algebraic equations into the curves determined by them. This framework consisted of a single main axis, an origin, and a variable ordinate erected at an arbitrary angle to the axis. He related the algebraic equation to the curve by substituting the distance along the axis from the origin to the foot of the variable ordinate for one of the unknowns and the length of the variable ordinate from the axis for the other. Finally, he set forth reduction procedures embodying the techniques of change of variable and transformation of the axial system.

III. THE ORIGINS OF THE *INTRODUCTION*: APOLLONIUS' *PLANE LOCI* AND *CONICS*

A. What Needs Explanation?

Fermat concludes the *Introduction* with a hint for future historians:

If this invention had preceded the two books of the *Plane Loci* recently restored by us, rather more elegant constructions of the locus theorems

---

demande un point duquel on puisse tirer autant d'autres lignes droites, une sur chacune des données, qui fassent avec elles des angles donnés, et que le rectangle contenu en deux de celles qui seront ainsi tirées d'un même point, ait la proportion donnée avec le carré de la troisième, s'il n'y en a que trois; ou bien avec le rectangle des deux autres, s'il y en a quatre; ou bien, s'il y en a cinq, que le parallélépipède composé de trois ait la proportion donnée avec le parallélépipède composé des deux qui restent, et d'une autre ligne donnée; ou s'il y en a six, que le parallélépipède composé de trois ait la proportion donnée avec le parallélépipède des trois autres; . . . ."

therein would have resulted. And yet we do not regret even now our perhaps precocious and immature offspring. Not to begrudge to posterity our unformed brainchildren is to some extent in the interests of the very science the products of which, at first simple and rough, grow and are strengthened by new inventions. Indeed, it is of interest even to scholars to have a thorough view of the hidden progress of the mind and of an art in the process of advancing itself. (p. 103)

Fermat's words reveal the intellectual tenets of the new algebraic analytical school of mathematics, the tenets which Chapter II sought to display. He speaks of an art, mathematics, in progress, continually growing in size and power. Mathematics has for Fermat a conceptual history; it grows not only by the addition of new theorems but also by the introduction of new concepts. And the growth of mathematics reveals the growth of the mind. Mathematics serves a wider function than just the solution of mathematical problems; it exemplifies the hidden workings of the mind. The concept of mathematics as mental science received explicit treatment at the hands of Descartes, but it underlay the philosophy of most of the algebraic analysts of the day. Algebra rendered the processes of the mind visual. It thereby satisfied two desiderata: first, it enabled the mathematician to retrace his steps and present a rigorous synthesis; second, it showed the reader, conceived as a student, the path of discovery. To the algebraist, the reader who saw how a result was achieved understood it better than one who had the result forced on him in the form of a synthetic demonstration. Where the Greeks, for the sake of elegance, wiped away all traces of their path of discovery, the algebraists not only left it but made it central to their treatises. Extending this desideratum of a clearly presented path of discovery from the resolution of a mathematical problem to the progress of the art of mathematics itself, Fermat points to his restitution of the *Plane Loci* of Apollonius as part of the path leading to the innovative system of the *Introduction*. He thereby provides the historian with a part of the material needed to answer Boyer's question: "One would like to know how the transition from the analytic art of Viète to the fundamental principles of analytic geometry took place, . . ."[34]

The "fundamental principles" of Fermat's analytic geometry have been discussed in the previous section, which was concerned with them in the context of the *Introduction* itself and which viewed them more from a logico-mathematical standpoint than from an historical one. The present section examines them historically, investigating the sources of Fermat's techniques and concepts and the motives that led him to combine them into a single, original system. It traces the long and complex development that ends with the *Introduction*. Fermat has already revealed where it begins: with his restitution of the *Plane Loci* of Apollonius.

[34] Boyer, p. 74.

B. *Apollonius'* Plane Loci

Apollonius' work and Fermat's attempt to restore it have already been cited in Chapter II, which placed the restoration among Fermat's earliest mathematical efforts in Bordeaux and in the context of similar efforts carried out by a well-defined group of like-thinking mathematicians, the algebraists, indicating that it represented a link between the young Fermat and the work of François Viète. Fermat intended his restoration to be a further realization of the analytic program by adding yet another of the lost works of Greek analysis to those already restored by Viète, Snel, and Ghetaldi.[35] But he was also striking out in a new direction, one dictated by the nature of the work he sought to reconstruct. The earlier efforts of Viète and his followers had dealt with a family of closely related texts, the treatises on sections. Recognizing in Greek analysis traces of the art of algebra, they sought to bring algebra back to that analysis. Hence, they began with the algebra that they knew best, the algebra of determinate equations in one unknown. The section treatises were subject to analysis by such means because they contained problems that were obviously metric and determinate. They called for solutions in terms of a single length or area. Ghetaldi's *On Mathematical Resolution and Composition* offers examples of such problems. The only exception was the *Contacts* of Apollonius, but the theorems of that work surrendered easily to a Viète with a comprehensive grasp of Euclid's *Elements.*

Apollonius' *Plane Loci* presented, however, challenges of a new kind. Its theorems called not for point solutions, i.e. the determination of lengths and areas, but for solutions involving families of points, i.e. loci. Where the mode of application of Viète's analytic art to the restoration of the section treatises had been clear to him and to his followers, the manner in which the art might be applied to locus problems was not at first clear to Fermat. And rightly so, for at this point the analytic program began to add more to the Greek treatises than had originally been present. The relatively elementary algebra that could be applied to the section treatises because it had originally been there is another form[36] would not suffice to analyze the treatises on loci, the bases of which were at best only partially algebraic. To begin with, Fermat was forced to follow more traditional paths of geometrical analysis.

Fermat began work on the *Plane Loci* sometime during 1628. His later letters have suggested to some writers that he may have attacked the two books

---

[35]Fermat was fully conscious of his role as continuator of a tradition. His restitution of the *Plane Loci* opens with a direct reference to that tradition: "Loci plani quid sint, notum est satis superque: hac de re scripsisse libros duos Apollonium testatur Pappus, eorumque propositiones singulas initio libri septimi tradit, verbis tamen aut obscuris aut sane interpreti minus perspectis (graecum codicem videre non licuit). Hanc scientiam, totius, videtur, Geometriae pulcherrimam, ab oblivione vindicamus et Apollonium *de locis planis* disserentem Apolloniis Gallis, Batavis, et Illyricis [i.e. Viète, Snel, and Ghetaldi] audacter opponimus, certam gerentes fiduciam non alibi praeclarius quam hoc in opere, Geometriae miracula elucere." (FO.I.3.)

[36]See above, Chap. II.

of the treatise in reverse order. Writing his first letter to Mersenne in April 1636, he announced the completion of the treatise, adding that he had left an earlier, incomplete version with Prades in 1630; that early version lacked, in his words, "the prettiest and most difficult question."[37] Paul Tannery, in a footnote, identifies this proposition as Theorem I,7, but offers no evidence in support of his conjecture.[38] If Tannery's identification were correct, and if one were to assume that the last thing Fermat had been working on was the last thing he handed his depositaries, then his remark to Roberval in 1637 that "eight years ago" he gave to d'Espagnet and an unidentified person copies of his restoration of the second book would support the idea of an attack in reverse.[39] Other evidence and other assumptions, however, militate against the idea. First, as the subsequent discussion will show, "the prettiest and most difficult question" is more probably Theorem II,5 than Theorem I,7. Fermat's handling of the former theorem constitutes the most involved and sophisticated mathematical thinking in the whole treatise; the analysis is the most pronouncedly algebraic in nature, the approach most strongly hints at the ideas underlying the *Introduction*. Theorem II,5 is also the theorem proudly presented separately to Roberval in 1637 as an example of Fermat's new method; it was then "one of the prettiest propositions in geometry."[40] Second, in his initial inability to find a technique for applying the analytic art to his task, Fermat would have taken up the easiest and most directly Euclidean problems first. Propositions 2 - 4 of Book I represent just such problems; rather than providing proofs, Fermat refers in fact directly to Euclid. The early propositions of Book II, however, are comparatively difficult. Most likely, events followed this order: Fermat completed a draft of both books in 1629, but was unhappy about his treatment of Theorem II,5. By 1630, a more satisfactory handling of the theorem still eluded him. On leaving Bordeaux for Orléans, he gave an incomplete draft to Prades. He then returned to the problem of II,5 in the period from 1631 to 1636, found a way to treat it which in turn suggested a more general algebraic approach to locus problems as a whole, and put the restoration in final form sometime prior to April 1636. The *terminus ad quem* is provided by Fermat's first letter to Mersenne. Fermat's battle with and final victory of Theorem II,5 of the *Plane Loci* supplies important clues to the path of transition from the geometrical analysis of the Ancients to the algebraic geometry of Fermat. Before exploring that particular theorem, however, it will help to survey the treatise as a whole and Fermat's restitution of it.

---

[37]Fermat to Mersenne, 26.IV.1636, FO.II.5; cf. below, p. 112

[38]*Ibid*, n. 3. Tannery does offer some further evidence in FO.I.29, n. 1 (apropos of Book II of the *Plane Loci*), but it too fails to override the considerations presented above.

[39]Fermat to Roberval, 20.IV.1637, FO.II.105.

[40]Fermat to Roberval, [II.1637], FO.II.100. Fermat offers the solution of the locus problem, but not proof.

Fermat did not select the *Plane Loci* for restitution accidently. As described by Pappus, the work had several attractions. It treats the most elementary class of loci, i.e. straight lines and circles. The theorems appeared to be so close in nature to the material of the *Elements*, Pappus reports, that the Ancients taught them as an appendix to Euclid's work.[41] Moreover, the treatise receives detailed attention from Pappus. While he often gives little more than a statement of the general problem to which each treatise in the field of analysis is addressed, he sets down for the *Plane Loci* sixteen theorems representative of the classes of Locus problems treated by Apollonius. He makes clear that the theorems he lists are only representative by stating at the end of his discussion that the *Plane Loci* contained 147 theorems and 8 lemmas.[42] However, Pappus pares down the number of theorems not so much in the interests of economy as in keeping with what he takes to be the goal of analysis: the reduction of a multitude of specific theorems to a minimum of general problems and techniques. The algebraic analysts of the late sixteenth and early seventeenth centuries had adopted Pappus' goal of complete generality. Improving on Pappus and on the works in Book VII did not involve new goals as much as new techniques; algebra enabled them to analyze problems in greater generality than Pappus had attained. The *Introduction*, indeed, opens on this theme. In their investigation of loci, the Ancients, according to Fermat, failed to achieve sufficient generality in the statements of their theorems. Fermat's "particular" analysis would open a completely general approach to locus problems. The result is in fact striking: by the system of the *Introduction*, Fermat is able to condense the 147 theorems of Apollonius, already reduced to 16 by Pappus, into a single theorem which in turn also includes whatever theorems Aristaeus required for the five books of his *Solid Loci*. That single theorem is, of course, the one demonstrated by the *Introduction*.

The *Introduction* lay in the future when Fermat first turned to the *Plane Loci*, but the desire for generality that Fermat inherited from Viète and his school had already put him on a course leading to that final success. The search for generality also explains perhaps why Fermat made no attempt to go beyond the sixteen theorems cited by Pappus. Nowhere in the restored treatise is there any indication that Fermat considered it less than a complete restoration. But it was a restoration in the sense of Pappus rather than Apollonius. Fermat was no antiquarian interested in a faithful reproduction of Apollonius' original work; he was a working mathematician seeking to ferret out the analytic techniques he felt Apollonius had hidden. The *Plane Loci* was to serve as a means to an end rather than an end in itself. For the moment, Fermat would accept Pappus' reduction, but there is little doubt he meant to reduce further. He simply was not sure how to go about doing so.

The reduction of plane locus problems to their most general statement in-

---

[41] See above, n. 18.
[42] Pappus, ed. Hultsch, II, p. 670.

volved a restriction of the nature of those problems. Apollonius' *Plane Loci* contains two classes of theorems which differ considerably. The second class contains the theorems that Fermat thought of as proper locus theorems. The first class comprises the set of theorems to which Pappus was directly referring in speaking about plane loci as an appendix to the *Elements*; they are also related indirectly to the analytical treatises on sections. They form, one might say, a bridge between section problems and locus problems. Pappus, in a burst of mathematical if perhaps not linguistic genius, managed to combine them into a single proposition, Theorem I of Book I:[43]

> If two lines are drawn, either from one given point or from two, and either parallel or containing a given angle, and either having a given ratio between them or comprehending a given area, and if the endpoint of one of them lies on a plane locus given in position, then the endpoint of the other also lies on a plane locus given in position, sometimes one of the same genus, sometimes one of a different genus, and sometimes similarly situated with respect to a straight line, sometimes contrarily.

To this theorem so stated, Fermat reacts with wry understatement: "This proposition can easily be divided into eight propositions and each of the latter into multiple cases: the vagueness of the translator [Commandino] seems to be due to the lack of punctuation; indeed, even Pappus himself seems here not to have avoided vagueness due to the extreme brevity."[44] Fermat makes eight subtheorems out of the single proposition. They are the eight possible combinations of the antecedents, allowing for the case when the two lines go out from a single point in the same straight line. An example best conveys the meaning of Pappus' statement. Let A be a fixed point, and let DC be a given straight line. Consider point B to move along DC. Draw AB, which will sweep along with B changing its length as it rotates about point A. Let line AE also rotate about point A in such a way as to maintain a constant angle with AB. Further, let AE also change in length so that the ratio AE/AB is constant. The proposition asserts that point E will generate a locus given in position, i.e. a locus fully determined by the data of the problem. Here that locus is clearly a straight line inclined at the given angle to the given straight line.

But if the length AE varies in such a way that the product $AE \cdot AB$ remains constant, then the locus will be a circle passing through A. Hence, according to the particular combination of antecedents chosen, the resultant locus is sometimes the same type as the given one, sometimes a different type, and so on.[45]

---

[43]FO.I.4; Fermat takes the Latin wording of this theorem, like that of the others, from Commandino's translation of the *Mathematical Collection* (Pisa, 1588).
[44]*Ibid.*
[45]Although Newton does not cite this theorem explicitly, it clearly underlies Proposition LVII of Book I of the *Principia*.

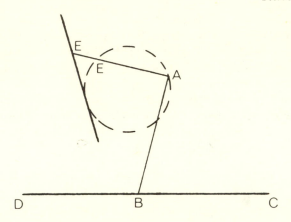

While Theorem I,1 has clear connections with the section treatises already successfully algebraicized by Viète, Ghetaldi, and Snel, Fermat apparently did not even attempt an algebraic treatment of it. In his restoration, he handles the subtheorems in strict geometric fashion. It is doubtful he could have done otherwise. For not only are the subtheorems more conveniently handled by geometric means than by algebraic, but their analytic algebraic treatment requires techniques beyond the level of the *Introduction* itself. In addition, Fermat probably took note of Pappus' remark that the Ancients considered the class of problems represented by Theorem I,1 an adjunct to Euclid's *Elements*, and hence he felt the theorem would lend little insight to the advanced analysis he was seeking to extricate from the treatise.[46] The hints which led to the system of the *Introduction* came rather from Fermat's investigation of the second class of problems included in the *Plane Loci*, those he construed as proper locus problems.

To this second class belong all the remaining theorems of the *Plane Loci*; Theorems I,2 - II,1 concern loci which are straight lines, Theorems II,2 - II,8 loci which are circles. Where the problems of the first class involved the loci of points bearing a fixed relationship to other loci (in the example above, the relation of the locus of point *E* to that of point *B*), those of the second class involve the loci of points bearing a fixed relationship to a set of given elements, themselves not loci. For example, Theorem II,1, which warrants close

46 Fermat's later work on restoring Euclid's *Porisms* suggests an alternative explanation to the one advanced above. He may have viewed the theorems of the first class of plane loci as porisms. In his *Porismatum Euclideorum renovata doctrina* . . . (1656) he remarked: "Quum locum investigamus, lineam rectam aut curvam inquirimus nobis tantisper ignotam, donec locum ipsum inveniendae lineae designaverimus; sed quum ex supposito loco dato et cognito alium locum venamus, novus iste locus porisma vocatur ab Euclide: qua ratione locos ipsos porismatum unam speciem et esse et vocari verissime Pappus subjunxit." (FO.I.82.) The example given above fits the description here of a locus on a locus.

examination, reads:[47]

> If straight lines are made to converge from given points, and if the squares of these lines differ by a given area, then the [intersection] point lies on straight lines given in position.

Fermat's solution of the theorem reveals the gradual incursion of algebraic modes of thought into his treatment of a geometrical situation:[48]

> Let two points *A* and *B* be given, and let any area less than the square [on] *AB* be given. Let *AB* be divided at *C* such that the square [on] *AC* exceeds the square [on] *CB* by the given area, and let the infinite perpendicular *CE* be erected in which any point *D* is taken. Draw *DA*, *BD*; I say that the square [on] *AD* exceeds the square [on] *BD* by the given [area].
>
> This is clear, since the square [on] AD exceeds the square [on] BD by the same [area] by which the square [on] AC exceeds the square [on] CB.
>
> If the given area is greater than the square [on] AB, point C will fall outside the line AB.

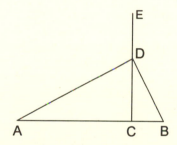

Three aspects of this solution require comment and elucidation: the general nature of the proof and the way in which it differs from standard Euclidean form; Fermat's use of the "linear algebra" of the Greek analysts; and the steps within the proof which had already been successfully algebraicized by Fermat's immediate predecessors.

Euclidean theorems are, in general, universal statements about fixed geometric relationships. The standard Euclidean proof involved a general statement followed by a particular statement, or "instantiation," referred to a specific instance of the theorem taken at random. For example, one instantiates the statement "The sum of the internal angles of a triangle is equal to

[47]FO.I.29.
[48]FO.I.29–30.

two right angles" by saying, "Let *ABC* be a triangle; the sum of its internal angles *ABC*, *BCA*, *CAB* is equal to two right angles." Only one instantiation is required, because it is sufficient to fix all relationships for the duration of the proof. Once the theorem is shown to hold for the particular instantiated figure, it is asserted to hold for all, because that figure was taken at random and without qualifications beyond those of the antecedents of the theorem. The proof follows, that is, the most common form of mathematical reasoning: what holds for *any* instance, holds for *all* instances. The Euclidean model does not, however, suffice for the demonstration of locus theorems. In Fermat's proof above, the instantiation fixes only points *A* and *B* and some area, say *M*. Construction based on these data further determines line *AB*, point *C*, and line *CE*. To complete his solution, Fermat must now show that *CE* indeed is the locus which satisfies the conditions of the theorem, and to accomplish this he must employ a second instantiative argument. He chooses any point *D* subject only to the condition that it lie on *CE*. He demonstrates that the squares of the lines joining *D* with *A* and *B*, respectively, differ by the area *M*. Now he may complete the demonstration by asserting that all points on *CE* satisfy that same condition.

Whereas centuries of practice made the single instantiation of the Euclidean proof an unconscious habit of the mathematician, the second instantiation required in a locus demonstration was too rare to be habitual. Presumably, the Greeks employed the above pattern of proof for loci, but none of their treatises on loci was extant in Fermat's time. Hence his use of a second instantiation involved a conscious adjustment of the familiar Euclidean model. And instantiation is fundamental to algebra; every algebraic equation is an instantiation. Fermat's proof contains the germ of his future system in the form of a verbal equation: ". . . the square *AD* exceeds the square *DB* by the same amount [*M*] as the square *AC* exceeds the square *CB*." By the nature of the proof itself, *AD* and *DB* are not fixed magnitudes as are *AC* and *CB*; they change in length according to the varying position of point *D* on *CE*. *AD* and *DB* are two variable lengths linked by an indeterminate equation, albeit a verbal equation. We have here, then, perhaps the first step in the "hidden progress of the mind" toward the foundation of the system of the *Introduction*.

The verbal equation itself deserves a note. It is a statement in the form of the "linear algebra" commonly used by Greek mathematicians such as Archimedes.[49] That is, Fermat refers only to a line drawing; no square areas are drawn, but rather arithmetic operations are carried out on line lengths. The shift to a symbolic algebraic statement cannot lie far beyond for Fermat, a man fully aware of the Viètan concept of an algebraic statement.

Indeed, Viète's work lies hidden in the proof. In the initial construction, Fermat posits the construction of point *C* such that $\overline{AC}^2 = \overline{CB}^2 + M$, but he

---

[49]Cf. Mahoney, "Another Look, pp. 337–340.

does not carry it out in detail. He need not. Viète and Ghetaldi have already solved myriad section problems of this type algebraically.[50] Moreover, their solutions have indicated that for any line *AB* there exist two such points, depending on whether *AC* or *CB* is taken as the larger segment. Hence, Fermat does not waste energy on a complete proof of the theorem, a proof which would take account of the plural "lies on straight *lines* given in position." Viète's algebraic techniques set both the number of those lines and their position. Further, the *diorismos* in the proof to the effect that the area *M* must be less than the square of line *AB* follows immediately from the algebraic reasoning employed by Ghetaldi in Book I of his *On Mathematical Resolution and Composition.*

Taken as a representative of the theorems of the *Plane Loci*, then, Theorem II,1 and Fermat's treatment of it go far in indicating the steps which would lead to the first important innovation of the *Introduction*, the correspondence between indeterminate algebraic equations and geometric loci. The employment of algebraic analysis itself needs no explanation; it follows from Fermat's entrenchment in the algebraic school of mathematics, and it already underlies parts of his proof. The key is the establishment of a correspondence between a new family of equations and a new class of geometric problems. The restoration of the theorems of the *Plane Loci* (with the exception of I,1) led Fermat in each case to a verbal equation such as that discussed above, an indeterminate equation linking two variable magnitudes. The successful restoration of the *Plane Loci* made clear to Fermat that the analysis (eventually algebraic) of locus problems led to such equations. The fundamental principle of the *Introduction* posits the converse and, with it, the correspondence: "Whenever two unknown quantities are found in final equality, there results a locus [fixed] in place, and the endpoint of one of these [unknown quantities] describes a straight line or a curve."

The transition from the *Plane Loci* to the *Introduction* did not, of course, follow easily the path suggested above. But the shift from geometric to algebraic reasoning did occur in the course of settling the final form of the *Plane Loci.*

C. *Theorem II,5: The Shift from Geometry to Algebra*

The demonstration of the theorem with which Fermat was still wrestling when he left Bordeaux and which he finally mastered sometime around 1635 displays the elements of the shift particularly well because Fermat apparently never "cleaned up" his presentation. In addition, the final demonstration of the theorem sheds light on the evolution of the second major innovation of

---

[50] E.g. Ghetaldi, *De resolutione et compositione mathematica* (Rome, 1630), Book I, Prob. IV (p. 22): "Datam rectam lineam in duas partes dividere, ut partium quadrata, dato quadrato differant. Oportet autem latus quadrati dati minus esse secanda."

the *Introduction*: the geometric framework or uniaxial system. The theorem is II,5:[51]

> If, from any number of given points, straight lines are drawn to a point, and if the sum of the squares of the lines is equal to a given area, the point lies on a circumference given in position.

Its form alone makes this theorem more complex and more difficult than Proposition II,1; it contains an indeterminate element lacking in the latter, the number of given points from which lines are drawn. A fully general solution of the problem must take account of this indeterminacy in the data. In addition, since more than two points are involved, Fermat must find some basic technique for linking them to one another.

Two approaches stood open to Fermat for giving a general proof of an indeterminate theorem or, more precisely, a theorem schema. The first is proof by induction. It would proceed by establishing the theorem for a minimal number of points and then by setting forth and justifying a procedure for adapting the solution to the insertion of an additional point. That is, in induction, one shows that the statement holds for n elements and then derives from the proof of the nth case a proof for the $(n + 1)$st. Although now a standard technique in number theory, proof by induction would have been, in the early seventeenth century, a geometrical approach, constructing the $(n + 1)$st case from the nth.[52] And indeed, it is the approach Fermat initially takes. In his later struggles with the theorem, however, Fermat switches to the second line of attack, proof by instantiation. That is, he provides a proof for some number n of points, but one in which the particular value of n plays no role. The proof holds as it stands for any n; hence, it holds for all n. The procedure, it would seem, is algebraic, rather than geometric, in nature. The transition, taking place in the years immediately preceding the *Introduction*, marks an important step toward the system of the latter. Hence, it is of value to follow Fermat's somewhat tortuous proof, which appears to be the piece-by-piece result of several years' work.

Fermat first attempts to lay the foundation for an inductive proof; he begins with the minimum number of points, two. Taking points A and B as given, he draws AB and bisects it at E. Taking IE as radius and E as the center, he describes circle *ION*. He shows that, for any point O on the circle, $\overline{AO}^2 + \overline{OB}^2 = 2(\overline{IE}^2 + \overline{AE}^2)$; the proof follows from a consideration of the right triangles *AZO* and *BVO*.

[51] FO.I.37.

[52] Blaise Pascal is usually credited with being the first to employ consciously the method of complete mathematical induction in his *Traité du triangle arithmétique* (cf., for example, H. G. Zeuthen, *Geschichte der Mathematik im 16. und 17. Jahrhundert*, p. 168), though W. A. Bussey ("The Origin of Mathematical Induction," *Amer. Math. Monthly*, 24(1917), 199–207) attributes the method to Maurolycus. Fermat's own method of "infinite descent" (see below, Chap. VI, §VI) is, of course, merely a reverse form of complete induction.

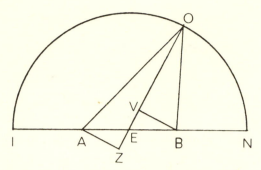

Fermat stops here, knowing that his reader will recognize an analytic proof by reduction. The step omitted calls for the determination of point I on AB such that $2(\overline{IE}^2 + \overline{AE}^2)$ equals the given area. As in Theorem II,1, Fermat again assumes that the reader will know to turn to Viète and Ghetaldi for the algebraic solution of the section problem to which he has reduced the original situation. However, in trying to add a third collinear point to the scheme, he is forced to come back to this problem himself. At first, the three-point case flows smoothly. Taking three collinear points B, D, and E, where $BD > DE$, Fermat determines a point C on the line such that $BD - DE = 3CD$. He then

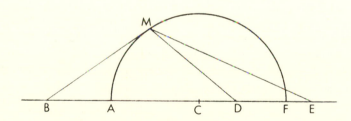

describes a circle AMF with center C and radius CA. A demonstration of the fact that, for any point M on circle AMF, $\overline{MB}^2 + \overline{MD}^2 + \overline{ME}^2 = $ const. follows with little difficulty. But at this point, the smooth flow of the proof is interrupted, for it is not clear how point A is to be determined in such a way that the constant above is equal to the given area. To carry out an inductive proof of the theorem, Fermat must establish an algorithm for determining the radius of the circle based on the radius for the case of one less point. To begin this task, Fermat returns to the two-point case. In his analysis he takes a new pair of points A and E as given and constructs the midpoint C of AE. Taking Z as the given area, he asserts that, if a circle of radius CB such that $\overline{CB}^2 = \frac{1}{2}(Z - (\overline{AC}^2 + \overline{CE}^2))$ is described about center C, then for any point Y on the circle $\overline{AY}^2 + \overline{YE}^2 = Z$. Fermat supplies neither a derivation nor a

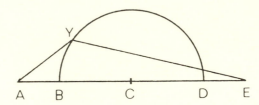

proof of the solution, saying of the latter: "We shall omit it as proved by Pappus and rather satisfactorily by others, lest we dwell further on simple matters."[53] These words seem to be an afterthought, a later addition to the above problem of determining the radius of the circle for the two-point problem, because a major break in the direction of the proof and in its style occurs right after them. One may guess that what has gone before represents the state of Theorem II,5 in 1630, when Fermat turned the text over to Prades. A determination of the radius of the circle for the three-point collinear case still escaped him.

How far had Fermat come? He could show that, for any circle, the sum of the squares of the lines joining a point on the circumference to any number of points on the diameter is a constant. Adding points and lines only increases the size of the constant. To solve even the collinear case, however, he had yet to find a rule for determining the radius and center of the circle which would make that constant equal to the given area. That rule apparently came to him later. It came to him perhaps in the following manner: the addition of another point and the corresponding line throws the circle off balance; the center and radius must be adjusted to compensate.

Whatever the general reasoning, Fermat eventually found the desired solutions and added them directly to what he already had done. He presents the solutions in the form of two lemmas "to the general method," the first of which reads:[54]

> Let there be set forth in the first, second, and third figure any number of given points A, B, C, E, and, depending on the number of points, let there be taken the *conditionary part*[55] AD of the line lengths bounded by point A and the remaining points, i.e. the fourth part in this example. Therefore, let AD be the fourth part of the line lengths AB, AC, AE; point D has a different position according as the cases vary.
>
> I say that the lengths bounded by point D and by the points on the side of A are equal to the lengths bounded by point D and the points on the side of point E.

[53] FO.I.39. The reference is to Lemma 4, or Prop. 122, of the Commandino translation.

[54] FO.I.39–40.

[55] *Pars conditionaria*: the average length. Here Fermat means the sum of the segments divided by the number of points.

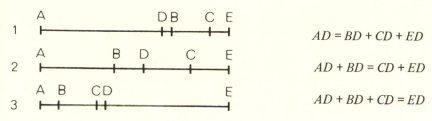

For the second figure, to take an example, the assertion is that, if AD is one-quarter of the sum of AB, AC, and AE, and D lies between B and C then $AD + BD = CD + ED$. Fermat's proof is simple and direct:

Since $4AD = AB + AC + AE$,
then $4AD - 3AD + BD = AB - AD + BD + AC - AD + AE - AD$,
or $AD + BD = CD + ED$.

The proof is also algebraic; it is an instantiative proof of a general proposition. For Fermat explicitly notes that the proof holds independently of the number of points chosen: "If five points were given, AD would be taken five times along with the four lengths bounded by the given points and point A, and hence one may proceed *ad infinitum* by the method . . . . The method is the same for however many points *ad infinitum*; it concludes the same thing however the cases may vary."[56]

All that is lacking here is the use of a parameter n to denote an indeterminate number of lengths bounded by $n + 1$ points. Fermat's demonstration scheme is algebraic reasoning in geometric garb. Hence it is proper to translate into modern symbolism:

Let $x_1, x_2, \ldots, x_n$ be given lengths.[57]

$$\text{Let } y = \frac{x_1 + x_2 + \ldots + x_n}{n} = \frac{1}{n} \sum_{1}^{n} x_i \, .$$

Lemma: If $x_j < y < x_{j+1}$, then $\sum_{1}^{j} (y - x_i) = \sum_{j+1}^{n} (x_i - y)$.

The second lemma posits the same conditions as the first, plus: Let point N be taken anywhere on the same line.

I say that [the sum of] the squares of the lengths bounded by the given points and point N exceeds [the sum of] the squares of the lengths

---

[56]FO.I.40.

[57]Stating the lemma in this fashion involves a generalization of Fermat's statement insofar as he always takes $x_1 = 0$. In addition it shifts attention away from the points which bound the segments toward the segments themselves.

bounded by the given points and point D by the square of DN taken as many times as there are given points.[58]

In the modern notation adopted above, the lemma asserts that if z is any length, then $\Sigma_n (z-x_i)^2 - \Sigma_n (y-x_i)^2 = n(z-y)^2$; the proof is obvious. Fermat, in his demonstration, again presents a proof scheme good for all values of n.

What led Fermat to these rather strange lemmas? For once, the pedagogical analyst fails his reader; Fermat gives no indication of their source in his thinking. His reasoning may be reconstructed with some certainty, however, and the reconstruction only further illustrates the extent to which algebraic habits of thought progressively creep into his treatment of geometrical problems. In addition, it illustrates the first use by Fermat of a problem-solving technique which he would employ repeatedly to great effect throughout his career and one which derives from the theory of equations.

Let us, like Fermat, take a particular number of given points as an example to facilitate our discussion. Let line AE be given containing the four points A, B, C, E, and let point O be a point for which the following equation holds:

(1) $$Z = \overline{AO}^2 + \overline{BO}^2 + \overline{CO}^2 + \overline{EO}^2,$$

where Z is a given area. From point O, draw OM perpendicular to AE. From

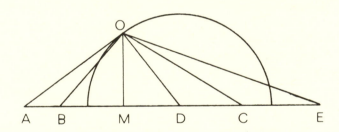

equation (1) and the Pythagorean Theorem, it follows that:

(2) $$Z = \overline{AM}^2 + \overline{BM}^2 + \overline{CM}^2 + \overline{EM}^2 + 4\overline{OM}^2.$$

From earlier portions of the Theorem II, 5, it is clear that O will lie on a circle, so we know at least the nature of the locus. Let D be the center of the circle, OD its radius. Then $\overline{OM}^2 = \overline{OD}^2 - \overline{DM}^2$. Hence,

(3) $$Z = \overline{AM}^2 + \overline{BM}^2 + \overline{CM}^2 + \overline{EM}^2 + 4\overline{OD}^2 - 4\overline{DM}^2.$$

The equation as it now stands contains two unknowns, OD and DM. It must also supply two solutions: 1. the radius OD, and 2. the length AD which determines the position of center D. The trick is to get both answers from a single equation.

The one equation indeed suffices due to the odd nature of the segment DM, which is both unknown and indeterminate. DM is unknown because of the as yet undetermined position of D. But, even with point D known, DM is an indeterminate variable quantity; equation (3) holds, whatever the position of point M, and that position varies with the changing position of point O on the locus. Hence, equation (3) must be independent of the value of DM. The next step then is to isolate terms containing DM; from the diagram:

$$AM = AD - DM, BM = BD - DM, CM = CD + DM, EM = ED + DM.$$

Hence, the substitution of these values into (3) yields:

(4) $\quad Z = \overline{AD}^2 + \overline{BD}^2 + \overline{CD}^2 + \overline{ED}^2 - 2AD \cdot DM - 2BD \cdot DM + 2CD \cdot DM$

$\qquad + 2ED \cdot DM + 4\overline{DM}^2 + 4\overline{OD}^2 - 4\overline{DM}^2.$

The middle terms of the right-hand side may be gathered together in the form $2DM(-AD - BD + CD + ED)$. The latter expression will be the only term in the equation containing a length bounded by point M. Hence, the equation will be independent of the position of M only if this term is equal to zero. But DM is clearly not always zero; hence,

(5) $\qquad\qquad\qquad AD + BD = CD + ED.$

Under what conditions will equation (5) hold? It depends on the position of point D, or rather on the length AD; for equation (5) may be rewritten from the diagram above:

(6) $\qquad\qquad AD + AD - AB = AC - AD + AE - AD$, or

$\qquad\qquad\qquad 4AD = AB + AC + AE$, or

(7) $\qquad\qquad\qquad AD = \dfrac{1}{4}(AB + AC + AE).$

That is, if point D is so chosen that the length AD is the "conditionary part" of the sum of the lengths AB, AC, AE, then equation (5) will hold. In addition, equation (4) will then read:

$$Z = \overline{AD}^2 + \overline{BD}^2 + \overline{CD}^2 + \overline{ED}^2 + 4\overline{OD}^2,$$

yielding a solution for OD in the form:

(8) $\qquad\qquad OD = \sqrt{\tfrac{1}{4}(Z - \overline{AD}^2 - \overline{BD}^2 - \overline{CD} - \overline{ED}^2)}.$

The reader will recognize first that equations (5) – (7), taken in reverse order, constitute a proof of figure 2 of Fermat's first lemma, and in addition that the same equations, combined with equations (3) and (4), constitute a proof of the second lemma.

This ingenious technique of using a single equation containing a fixed unknown and an indeterminate variable, and of eliciting two pieces of information from that equation by setting the coefficients of the indeterminate variable identically equal to zero, makes other appearances in Fermat's later attempt to restore Euclid's *Porisms* and in his *Synthesis of the Law of Refraction*.[59] Its use elsewhere, however, is not the only evidence strengthening the conjecture that it underlies the above derivation. On a more general level, it is clearly linked in origin to the sorts of research into the theory of equations—in particular, the relation of the roots of an equation to its coefficients—that Fermat had been carrying on in his investigation of maxima and minima while in Bordeaux.[60] Blocked from a geometric solution of II,5 by the peculiar nature of the problem, he may well have followed out the hint provided by II,1 to attack it algebraically. As the following chapters show, an algebraic attack meant for Fermat an attack through the theory of equations. If, in fact, this is the path Fermat pursued, it illustrates well the intrusion of algebraic thinking into his struggle with Proposition II,5 of the *Plane Loci*.

As the preceding analysis also demonstrates, the general solution of the collinear case of Proposition II,5 follows almost as a corollary to the two lemmas. Equation (7) yields the center of the desired locus, equation (8) the radius.

So far Fermat's solution of Proposition II,5 has indicated the "hidden progress" of his mind toward the algebraicization of locus problems. The remainder of the discussion will show how his treatment of the non-collinear case of the theorem lays the foundation for the geometric framework of the *Introduction*. The discussion immediately preceding has begun this task. Fermat has used one of the given points, A, not only as a reference for position but also as a reference for measurement. That is, A is serving as an "origin." The constants and parameters of the problem, including the indeterminate variable DM, are all expressed in terms of distances from A. An important extension of this procedure is the use of A and a line through it as a reference for points not on the line. A corollary to the theorem just analyzed involves precisely such an extension. It prepares the way to the non-collinear theorem by modifying further the collinear case. Until now, Fermat has been dealing strictly with the sum of the squares of the connecting lines, each taken alone. As will be clear presently, however, he must know what happens if integer coefficients are affixed to those squares. For example, given several points, say three, A, B, E, and an area Z, how does one determine the circle locus of the point M such that $2\overline{AM}^2 + \overline{BM}^2 + \overline{EM}^2 = Z$?

---

[59] In its use in the *Synthesis of the Law of Refractions*, see J. E. Hofmann, "Pierre Fermat—ein Pionier der neuen Mathematik," *Praxis der Mathematik*, VII (1965), p. 173. I should make clear that this technique is never made explicit by Fermat. It first occurred to me in trying to reconstruct the analyses of his solutions to five porisms (FO.I.78–82). My success there led me back to Pappus' discussion of the *Porisms* of Euclid and resulted eventually in the demonstration in "Another Look," pp. 346–348.

[60] See below, Chap. IV, §II.

Again, Fermat intends here a general question; he takes 2 as an *instar omnium*. Although he restricts the complication of the theorem to integer coefficients, clearly the proposition holds for any real coefficients q_i.[61] Fermat is limited, however, by the proof scheme he has instituted: "In this case, for the purposes of construction, length AD is to be taken as the fourth part of the lengths AB, AE, because in this case point A takes the place of two points. It is the same as saying: given four points A, A, B, E, to find the circle NM in which, taking any point, say M, the four squares AM, AM, BM, EM equal a given area."[62] Fermat realizes, then, the full generality of his lemmas; the choice of points is in no way restricted, even in the case of duplication. On the other hand, the use of the lemmas in the form stated by Fermat limits him to integer coefficients.[63] Algebra, as we noted earlier, begins to intrude into Fermat's geometric treatment; however the essentially geometric tenor of Fermat's restitution prevents algebra from asserting itself fully.

After this long preparation, Fermat is now in a position to tackle Theorem II,5 as stated. He begins, however, again with a particular case: given the collinear points A, B, C, E, and point Q not on AE, to find the circle MI such that if I is a point on the circle,

$$\overline{AI}^2 + \overline{BI}^2 + \overline{CI}^2 + \overline{EI}^2 + \overline{QI}^2 = Z,$$

where Z is a given area. Point Q acts, like point A in the preceding example,

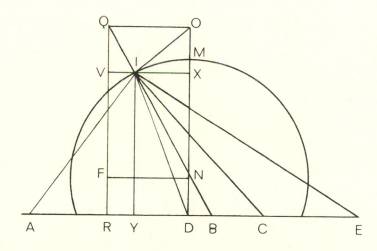

[61] Let $P_i(x_i, y_i)$ ($i = 1, 2, \ldots, n$) be n points in the plane. Suppose $P(x, y)$ to be a point on a locus determined by the condition $\Sigma q_i \overline{PP_i}^2 = M$. Then, $\Sigma q_i [(x - x_i)^2 + (y - y_i)^2] = M$, which may be expanded to the form

$$x^2 \Sigma q_i + y^2 \Sigma q_i - 2x \Sigma q_i x_i - 2y \Sigma q_i y_i + \Sigma q_i (x_i^2 + y_i^2) = M,$$

the equation of a circle, the center and radius of which are directly computable.
[62] FO.I.44.
[63] The reason is simple: the number of points can only be an integer.

as a weighting factor, tending to pull the center of the circle toward it. Fermat breaks up this pull into orthogonal components. Using AE as a reference line, he constructs QR perpendicular to AE and sets

$$AD = \frac{AR + AB + AC + AE}{5};$$

this step accounts for the horizontal effect of point Q on the center of the circle. If DO is erected perpendicular to AE at D, the center will lie on DO. Just as the horizontal component AR of Q's position with respect to A has its effect on the position of point D, so too does its vertical component $QR = DO$. But point O is only one point; the four points A, B, C, E also have their effect. With reference to the center-line DO, the latter work as though located at point D, the foot of AE now considered perpendicular to DO.[64] Hence, the center N of the circle satisfying the locus results from the equation $DN = \frac{1}{5} DO$. The importance of the preceding corollary becomes clear. The equation determining the radius also reflects the added weight of the four points on AE. If NM is that radius, then the following relationship holds:

(9) $\quad Z = \overline{AD}^2 + \overline{RD}^2 + \overline{BD}^2 + \overline{CD}^2 + \overline{ED}^2 + 4\overline{DN}^2 + \overline{NO}^2 + 5\overline{NM}^2.$

The proof follows from use of the Pythagorean Theorem and Lemma 2. Fermat explains the general proof scheme in a subsequent paragraph, illustrating how each specific case may be handled *mutatis mutandis*.

Note in particular here the manner in which A becomes a reference point, not only for those points given on the line AE but also for those assumed not to be on it. The position of each point not on the reference line is treated in terms of its horizontal distance from A and its vertical distance from the reference line. In addition, the position of each point with respect to the center and its distance from the center is expressed in terms of horizontal and vertical components. For example, point B is referred to center N via the components BD, DN. The coefficient 4 of $\overline{DN}^2$ in equation (3) results from the fact that all four collinear points A, B, C, E have equal vertical components DN.

Fermat next puts the final touches on his proof scheme: "But since many

[64] This portion of the proof, beginning with the construction of the perpendicular QR, is most interesting because it appears to be entirely unmotivated by what has gone before. One will look in vain for missing mathematical steps in the proof. Rather, the motivation may have come from the work in geostatics that Fermat was carrying on at this time, i.e. the early 1630s. Assume for the moment that line AE represents a balance arm, points Q, A, B, C, E, equal weights. If AE is in equilibrium, then from the statics of Archimedes it follows that weight Q acts along the perpendicular to the balance arm; that is, Q has the same effect that it would have if placed at the foot R of the perpendicular to the balance arm AE. Then D will be the fulcrum, or balance point. Now turn the diagram sideways and view DO as a balance arm in equilibrium. Weight Q acts as if placed at the foot O of the perpendicular to DO, and points A, B, C, E, since they all lie perpendicular to DO, act as if their combined weight were located at D. Hence, the fulcrum N lies $\frac{1}{5}$ of the distance DO from D $\left(4 \cdot \frac{1}{5} = 1 \cdot \frac{4}{5} \right)$.

cases will emerge from the diverse positions of the assumed line containing two or more points, while the remaining points take different positions on either side of some assigned line, [and] although each case has its own particular shortcuts, it helps to have a model for general proofs and constructions."[65]

Take any number of points A, B, C, D, E, F, either in the same line or in different ones. "Take any line SR in the same plane, such that all the given points are on one side of the line SR." Erect perpendiculars from the given points to SR: AG, BH, CI, DK, EM, FN. Then set GL equal to the condi-

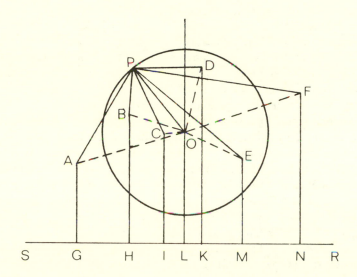

tionary part of (GH, GI, GK, GM, GN). Erect LO perpendicular to SR and equal to the conditionary part of (AG, BH, CI, DK, EM, FN). The circle satisfying the locus will have its center at O, and its radius OP will be determined by the equation

$$Z = \overline{AO}^2 + \overline{BO}^2 + \overline{CO}^2 + \overline{DO}^2 + \overline{EO}^2 + \overline{FO}^2 + n\overline{OP}^2,$$

where Z is the given area and n the number of points, here six.

While an equation involving two unknown variables does not appear in Fermat's final solution, as one does in Proposition II,1, the variables do appear in the course of the proof. In the penultimate figure above, they are the orthogonal lengths XI, YI. The purpose of Fermat's derivation is to convert the equations containing these variables and determining the center and the radius to ones containing only the fixed terms of the problem.

From the consideration of Fermat's restitution of Apollonius' *Plane Loci*, it

[65]FO.I.47.

should be clear that the treatise provided both ample motivation and rich inspiration for Fermat's development of an analytic geometry. As exemplified by Propositions II,1 and II,5, the *Plane Loci* suggested a correspondence between indeterminate equations and geometric loci. The statement of problems and the patterns of proof, especially the need to express theorem and proof schemes, invited the introduction of algebraic techniques and modes of expression. The necessity of determining the relative position of points in terms of distance led to the use of horizontal and vertical (orthogonal) components with respect to a fixed point and a fixed line. In short, Fermat's work on the *Plane Loci* brought together the ingredients for analytic geometry, ingredients which, to a large extent, prescribed their own recipe. One particular theorem inspired the cook, as he himself said in a 1636 letter to Roberval: "I have omitted the principal application of my method, which is for finding plane and solid loci. It has served me in particular for finding that plane locus that I earlier found so difficult: If from any number of given points straight lines are drawn to a point, and the sum of the squares of the lines is equal to a given area, the point lies on a circumference given in position."[66]

The theorem should look familiar: it is Proposition II,5. As we noted above, Fermat was probably referring to this theorem, rather than I,7, when he said to Mersenne earlier in 1636: ". . . I have completely restored Apollonius' treatise *On Plane Loci*. Six years ago I gave to M. Prades, whom you perhaps know, the only copy I had of it, written in my own hand. It is true that the prettiest and most difficult problem, which I had not yet solved, was missing. Now the treatise is complete in every point, and I can assure you that in all of geometry there is nothing comparable to these propositions."[67]

Fermat had, then, to break away from traditional geometric techniques in order to overcome the difficulties which the proof of Proposition II,5 presented. The theorem, in turn, suggested the path he might follow. Yet, the final form of the proof had to be couched in terms of the classical treatise he intended to restore. Hence, the demonstration omits any hint of a new method, except insofar as it excites the admiration and curiosity of the reader as to how Fermat arrived at so intricate a proof.

In addition to the outward form, one major element of the system of the *Introduction* still distinguishes it from its adumbration in Proposition II,5 of the *Plane Loci*. Although the latter employs a fixed point and fixed line, to which other points and lines are referred, it does so in an *ad hoc* fashion quite different in intent from the systematic use of the uniaxial framework in the *Introduction*. Moreover, the reference system of II,5 would, on the face of it, seem to point in the direction of a coordinate system similar to our present one, but conceptually incompatible with the uniaxial system. Hence, the *Plane Loci* may well satisfy the quest for a motivating factor, but one must

[66]Fermat to Roberval, 22.IX.1636, FO.II.74; Fermat quotes the theorem in Latin from the *Plane Loci*.
[67]Fermat to Mersenne, 26.IV.1636, FO.II.5.

look elsewhere for the source of Fermat's uniaxial system. Apollonius supplies this missing link in another work, the *Conics*.

D. Apollonius' Conics and Viète's Algebra: A System and Its Equations

The *Conics* figures prominently in the corpus of Greek analytical works. Indeed, Apollonius himself stressed in his introductions to the various books of his treatise the manner in which his theorems might be employed in the analysis of problems, especially in the determination of *diorismoi*.[68] By discussing the *Conics* in the context of the *topos analyomenos* (*Collection*, Book VII), Pappus emphasized the work's analytic nature. Hence, in addition to the information it supplied about the properties of the conic sections, the *Conics* served mathematicians such as Fermat as a prime example of Greek geometrical analysis.

Pappus' discussion of the *Conics* brings out one aspect of the work whose importance was not lost on Fermat. He credits Apollonius with having introduced the standard names of the conic sections: ellipse, hyperbola, parabola. In addition, Pappus indicates that the source of these names was the analytic field of the application of areas. In seeking the most general characteristic relationships obtaining for each section, that is, its *symptômata*, Apollonius found that their verbal statement could be construed as one of the three basic forms of the problem of applying an area to a given line length.[69] Take as an example his derivation of the *symptôma* of the parabola. Proposition I,11 of the *Conics* runs in paraphrase: Let *ABG* be an axial triangle of the circular cone of which the vertex is *A* and the base the circle about the diameter *BG*. If a plane perpendicular to the plane of $\triangle ABG$ and parallel to one of its sides, say *AG*, cuts the cone, the resulting curve in the surface of the cone is, by definition, a parabola. Let the intersection of the cutting plane with the circular base of the cone be the line *DE*, and that of the cutting plane with the axial triangle be the line *ZH*. Although *DE* is by construction perpendicular to *BG*, it is not necessarily perpendicular to *ZH*, since the cone may be oblique, when only one of the infinite axial triangles would be perpendicular to the base. Apollonius has already shown that no restrictions in this regard are necessary, since *ZH* bisects the intersection line of the cutting plane with any plane parallel to the base. To derive the fundamental property of the parabola, Apollonius now imagines any plane parallel to the base to cut the cone; let *MN* be the diameter of the resulting circular section. In the plane of this latter section, let *KL* be drawn parallel to *DE*, intersecting *ZH* at *L*. From the properties of a circle, it is clear that

$$\overline{KL}^2 = ML \cdot LN.$$

[68]Cf. Mahoney, "Another Look," pp. 327–329, 336–337.
[69]The substance of the following discussion is an abbreviation of the brilliant reconstruction of E. J. Dijksterhuis in his *Archimedes I* (Groningen, 1938).

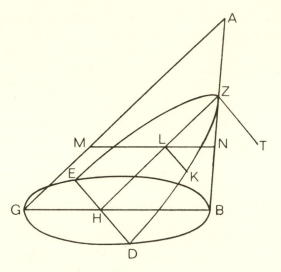

By similar triangles,

$$\overline{KL}^2 = ZL \cdot \frac{\overline{BG}^2 \cdot ZA}{AB \cdot AG} \; .$$

Whereas *KL* and *ZL* are variable lengths dependent on the placement of the cutting plane *MN* within a given cone, the second factor in the product on the right is fixed with respect to the cone and the placement of the cutting plane *ZH*. Hence, it may be represented by a fixed length *ZT*. The *symptôma* of a parabola, then, is

$$\overline{KL}^2 = ZL \cdot ZT \; .$$

A verbal expression in the Greek form would be: to apply an area ($\overline{KL}^2$) to a line (*ZT*). *ZL* is the unknown "side" of the rectangle whose other side is *ZT* and whose area is equal to *KL*^{2}. From the Greek word *parabolē* for such exact application comes the name which Apollonius now applies to the curve characterized by the *symptôma*. The constant side *ZT* Apollonius calls the parameter of the ordinates, and he erects it perpendicular to *ZH* at *Z* for the sake of geometric exposition. Similar derivations for the other two conic sections lead to the *symptômata*

(H) $$\overline{KL}^2 = ZL \cdot ZT + \frac{ZT}{ZZ'} \cdot \overline{ZL}^2 \quad \text{and}$$

(E) $$\overline{KL}^2 = ZL \cdot ZT - \frac{ZT}{ZZ'} \cdot \overline{ZL}^2 \; ,$$

where again *ZT* and *ZZ*, (the transverse axis) are fixed by the choice of the cutting plane *ZH*. The ancient doctrine of the application of areas supplies the problems related to these equations. Equation (H) corresponds to Euclid VI,29: To a given straight line to apply a parallelogram equal to a given rectilineal figure and exceeding by a parallelogrammic figure similar to a given one. Here, ZT is the given straight line, and $\overline{KL}^2$ may be transformed into the given rectilinear figure by making of it a rectangle the ratio of whose sides is ZT/ZZ'. From the Greek *hyperbolē*, meaning "excess" comes the name for the curve characterized by *symptôma* (H). Similarly, *symptôma* (E) corresponds to Euclid VI,28, which is the same as VI,29 with the word "exceeding" replaced by "deficient." From *elleipsis*, "deficiency," comes the name of the curve.

Apollonius' major contribution to the study of conic sections was, in Pappus' view, his establishment of the correlation between the curves and the analytic field of the application of areas. Practitioners of the analytic program of the seventeenth century discerned algebra in that field. In his *Canonical Recension of Geometric Constructions*, Viète demonstrated how certain related geometric constructions corresponded directly to second-degree determinate algebraic equations.[70] While his original intent had been to use those geometric constructions as constructive solutions of the corresponding equations, he also succeeded thereby in translating geometric algebra into symbolic algebra. Fermat now followed his lead. Apollonius provided the *symptômata* of the conic sections and linked them, as Pappus emphasized, to the application of areas. Fermat now took the next step and translated those *symptômata* into symbolic algebra. The geometric *symptômata* became algebraic equations, not Viète's determinate equations but Fermat's indeterminate equations in two unknowns. In addition, Apollonius' technique for deriving the *symptômata* became the basis for the uniaxial system of the *Introduction*. The Greek geometer expressed those *symptômata* in terms of: 1) a variable line segment ZL measured along a fixed line ZH from a fixed point Z; 2) a variable segment KL erected at an arbitrary angle to ZL; and 3) certain fixed segments.

Fermat's efforts to prove Proposition II,5 of the *Plane Loci* would seem to have suggested to him that an algebraic approach to loci required a system for referring points and lines to one another in terms of distance. He found that system in the *Conics*. What Fermat brought, or rather (in keeping with the nature of the analytic program) brought back, to Apollonius' scheme was algebra. Specifically, he recognized that ZL and KL were unknown variables linked by an indeterminate equation. Indeed, he algebraicized Apollonius'

---

[70]*Opera*, ed. van Schooten, pp. 229–239. Viète's goal in this work is to relate the operations of algebra, i.e. addition, subtraction, etc., to those of geometry, much in the way in which Descartes does later. Where Descartes, however, establishes a way to maintain multiplication as a closed operation, Viète retains the classical notion of a product of two lines forming an area. In his work, Viète employs various corollary constructions based on the golden section to arrive at constructive solutions of the equations $x^2 + bx = d^2$, $x^2 - bx = d^2$, and $bx - x^2 = d^2$; the first and last would correspond in order to equations (H) and (E) above within the field of the application of areas.

theorems, turning the *symptômata* into equations. His task was made easier by the fact that he adopted Apollonius' basic geometric framework as his own.

E. The Question of Timing: 1635

If the elements of Fermat's thought that led to the analytic geometry of the *Introduction* now seem reasonably clear, the question of precisely when he took the step from the *Plane Loci* to the *Introduction* remains open. The concluding paragraph of the *Introduction* suggests strongly that the step coincided roughly with the completion of the *Plane Loci*, and Fermat's early letters to Mersenne in turn indicate that the latter reached completion sometime late in 1635 or early in 1636. But both sources are vague in their phrasing and do little more than hint. Fortunately, there is one more piece of evidence which, when linked to the above, allows a rather firm dating of the transition.

In 1635 Fermat dispatched to Beaugrand solutions to the three- and four-line locus problems.[71] Of these, the *Demonstration of the Three-Line Locus* is still extant.[72] In it, Fermat shows that, if *AM*, *MB*, *BA* are three straight lines given in position, and if from point *O* lines *OE*, *OI*, and *OD* are drawn

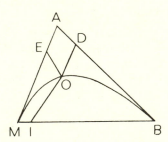

to intersect *AM*, *MB*, *BA* respectively at given angles, and if the ratio $(OE \cdot OD)/OI^2$ is constant, then point *O* lies on a conic section. Only the analysis of the solution is given, in keeping with the style of the *Introduction*. The manner of proof is essentially that of the theorems of the *Introduction*, with the exception that Fermat makes no use of Viète's algebra. Again, however, the problem had classical origins and hence, like the *Plane Loci*, called for a proof in the classical geometric style. Indeed, the source of the problem is Pappus' discussion of Apollonius' *Conics*, and its demonstration requires command of that text.

The *Demonstration*, then, represents firm evidence that in 1635 Fermat had

---

[71] Cf. above, Chap. II.
[72] *Loci ad tres lineas demonstratio*; text in FO.I.87–89.

extended his study of loci beyond the straight line and circle of the *Plane Loci* to encompass the conic sections. More important, he was treating conic sections *as loci*. It is probably at this point in time that Apollonius' *Conics* made its crucial contribution to the genesis of the *Introduction*. Once it did so, the *Introduction* was complete; Fermat had his "own particular analysis" of loci. The time was 1635, just before he was introduced to the mathematical community in Paris.

IV. EXTENSIONS OF THE SYSTEM OF THE *INTRODUCTION*: *THE INTRODUCTION TO SURFACE LOCI*

The two previous sections have established the nature of Fermat's system of analytic geometry and its genesis in Fermat's thought from his first struggles with the *Plane Loci* in 1628 to his final insight gained from the *Conics* in 1635. By 1636 the system was firmly settled in Fermat's mind, and he turned to other matters. The system of the *Introduction* became part of his method, a useful item in his own analytical toolbox. Hence, one finds traces of its influence throughout Fermat's later research. Its direct influence manifests itself in two areas in particular. First, in the years immediately following the circulation of the *Introduction*, Fermat tried to extend the system to three dimensions. In 1643, he sent to Carcavi his *Introduction to Surface Loci*,[73] in which he attempted, with little success, to describe surfaces of revolution and translation in the same analytic manner as he had the conic sections in the *Introduction*. In another direction, he employed analytic geometry in the graphical solution of determinate equations of the third and fourth degree, which was the subject of an appendix to the *Introduction* written shortly after its appearance. This work led, in turn, to an algebraic treatise, the *Tripartite Dissertation*, in which Fermat sought to emend, on the basis of the curves appropriate to their graphic solution, Descartes' classification of algebraic equations. As work in this area progressed, it came more and more to impinge on Fermat's research in the areas of the theory of equations and the determination of maxima and minima. In particular, he became less interested in the geometric picture of curves and more interested in the equations which determined them. The equations became for Fermat complete

---

[73]*Isagoge ad locos ad superficiem*; text in FO.I.111–117. The work is dedicated to Carcavi and dated Toulouse, 6 January 1643: Fermat announced its dispatch in a letter to Mersenne dated 13.I.1643: "J'envoyai par le dernier courrier mon *Isagoge ad locos ad superficiem* à M. de Carcavi, de laquelle il ne manquera pas de vous faire part. Vous y trouverez des propositions aussi belles que la Géométrie en puisse produire, et, bien que mon discours soit concis, il m'a semblé que je n'en devais pas dire davantage, s'agissant d'une méthode générale de laquelle les exemples et l'usage peuvent être infinis. Je m'imagine même que cette matière n'a pas été exactement connue des anciens; car, qu'y a-t-il dit dans tous leurs Livres en fait de lieux qui vaille la proposition suivante, par exemple? [Fermat here gives the theorem discussed below.] Et toutefois la construction en est aisée par ma méthode, et non seulement cette proposition a été par moi découverte, mais la voie générale pour en trouver infinies." (FO.II.245–246.)

"pictures" in themselves of the curves, as increased facility in the techniques of an algebraically based "differential calculus" permitted Fermat to extract from the equation alone important information about the nature and behavior of the curve, as, for example, its tangent at any given point, the area of any segment, its points of inflection, etc.

Fermat first announced his research into a three-dimensional analogue of the *Introduction* in a 1638 letter to Mersenne.[74] In that letter, he noted a corollary to Proposition II,5 of the *Plane Loci* which, in his words, ". . . allonge infiniment l'étrivière au lieu plan." Given the conditions of Proposition II,5, that is, a variable point and any number of given points in a plane linked in such a way that the sum of the squares of the lines joining the variable point to each of the given points is a constant, then not only is the locus of the variable point a circle in the plane, but it is also a sphere of which that circle is a great circle. Fermat offered no proof, but indicated that he had been working on material of a similar nature for some time: "Finally, I have found many marvelous things on the subject of surface loci, but I cannot tell you about them all at once."[75]

Again, then, Fermat had turned to Pappus' introduction to Book VII of the *Collection* for yet another work to restore. Now he sought to rescue Euclid's *Loci ad superficiem*, or *Surface Loci*, from oblivion. But here he faced a particularly difficult task, in that Pappus gives very little information about Euclid's text. In fact, in his introduction to his Book VII, he offers only the title of the work and, later on in the book, two lemmas, both of which concern conic sections.[76] Fermat had first to decide what surface loci were.

The ancient authors could offer Fermat only confusion. Proclus spoke of "loci constructed on lines" and "loci constructed on surfaces," and seemed to imply that lines and surfaces respectively themselves constituted the loci. But Pappus, in Book IV, spoke of surface loci as if they were space-curves lying in the surfaces of solids, such as the cylindrical or hemispherical helix. Health conjectures that, "However this may be, Euclid's *Surface-Loci* probably included such loci as were cones, cylinders and spheres."[77] Certainly, this is the interpretation which Fermat placed on the term *locus ad superficiem*.

Fermat worked on the problem from 1638 until January 1643, when he dispatched the finished copy to Carcavi. In the interim, he offered Mersenne

---

[74] Fermat to Mersenne, 22.X.1638, FO.II.173–174.

[75] *Ibid*.

[76] Pappus, ed. Hultsch, Vol. II, pp. 1004–1014, for the lemmas. The second lemma is Pappus' statement of the focus-directrix property of the conic sections: The locus of a point the distance of which from a given point is in a given ratio to its distance from a fixed straight line is a conic section, which is an ellipse, a parabola, or a hyperbola according as the given ratio is less than, equal to, or greater than unity. (Heath, *History of Greek mathematics*, I, p. 440).

[77] Heath, *History of Greek Mathematics*, I, pp. 439–440, whence comes the above discussion on the conflicting reports about surface loci.

examples of his work in his letters; these examples all dealt with the sphere. As the title indicates, the final version made no pretense of being a faithful reconstruction of Euclid's text. Freed from the hindrance of any prescribed theorems or contents, Fermat could give full vent to the spirit in which the ancient analytic treatises were being restored. They were to serve the present rather than to bring back the past. Hence, the *Introduction to Surface Loci* presents from the outset a general technique for the treatment of a certain class of surfaces. And that technique is couched in modern terms, as Fermat's short introduction illustrates:[78]

> The demonstration of surface loci to be presented complements the *Intro-duction to Plane and Solid Loci*. The Ancients only hinted at this demon-stration, but neither taught it by general precepts nor even at least outlined it by some noble example, unless they placed it perhaps in those long buried monuments of geometry in which so many great findings of the Ancients lie with the roaches and worms or have altogether disappeared.
>
> Nevertheless, a most brief dissertation will make clear that a general method for this material will not be lacking. If we should have the leisure, we will some day make known to many the single inventions which up to now we have treated summarily in geometry.
>
> Therefore, nothing prevents our seeking in plane, spherical, or conical, cylindrical, conoidal or spheroidal surfaces those same *symptômata* which we looked for and proved in curves, if we prefix [to our investigation] lemmas constitutive of the single loci of this type.

Fermat will make no attempt, then, to reconstruct Euclid's text or meth-ods. His approach will be algebraic from the outset. He will seek to express certain surfaces in the same manner as he has plane curves, by algebraic equa-tions expressing their *symptômata*. Even with his successful treatment of plane curves as experience to build on, Fermat faces difficulties. He has no model from which to operate, no hint of how algebra might be applied.

To begin with, Fermat's uniaxial system itself constitutes a conceptual hindrance to its extension to three dimensions. While one moving ordinate may rest comfortably in the plane while sliding along the axis, a second ordinate jutting out into space and moving simultaneously in two directions would be in a precarious situation. The mind, indeed, boggles at such a bal-ancing act. Far more comfortable to the imagination is the framework of of three fixed, mutually orthogonal, axes emanating from a fixed origin. But the latter concept follows from the notion of a biaxial, or coordinate system, which, as has been noted earlier, is not part of the analytic geometries of either Fermat or Descartes. Hence, in his *Surface Loci*, Fermat has no three-dimensional axial system. Rather, he has a multitude of plane systems, each of which contains a plane section of the solid under consideration.

[78] FO.I.111.

Fermat's approach is at once strange and inherently extremely difficult to execute successfully. It is an excellent example of the extent to which analytic techniques and concepts in the early seventeenth century had not yet caught up with geometrical intuition. Fermat begins the *Introduction to Surface Loci* with a group of six lemmas "constitutive" of the surfaces he takes to make up the class of surface loci. The six lemmas may be schematized in a single statement as follows:[79]

If a surface is cut by any number of planes *in infinitum* and the common section(s) of all the infinite number of cutting planes and the said surface is (are):

1) a straight line
2) a circle
3) sometimes a circle, sometimes an ellipse, but nothing else
4) sometimes a circle, sometimes an ellipse, sometimes a parabola or hyperbola, but nothing else
5) sometimes straight lines, sometimes circles, sometimes ellipses, sometimes parabolas or hyperbolas, but nothing else
6) sometimes straight lines, sometimes circles, sometimes ellipses, but nothing else,

the surface first posited will be:

1) a plane
2) a sphere
3) a spheroid
4) a parabolic or hyperbolic conoid (paraboloid or hyperboloid)
5) a cone
6) a cylinder.

To these six commonly known surfaces, Fermat adds a seventh by generalizing the notion of a cylinder to include solids generated by the rectilinear translation through space of a parabola or a hyperbola.[80]

Before trying to understand what Fermat has in mind with such an involved preliminary lemma, which, incidentally, is derived from the definitions and lemmas of Apollonius' *Conics* and Archimedes' *Conoids and Spheroids*, it would be good to examine the first proposition of Fermat's treatise. It reads: "If from any number of given points in whatever planes straight lines are

[79]FO.I.112.

[80]FO.I.112–113: "Quia tamen saepissime occurrunt loci in quibus sectiones sunt lineae rectae, parabolae aut hyperbolae et nihil praeterea (quod ipsa statim quaestionis analysis indicabit), conveniens est et necessaria omnino huic disputationi *nova cylindrorum constitutio, in quibus bases inter se parallelae sint parabolae aut hyperbolae, et latera, bases hujusmodi connectentia, sint lineae rectae, inter se parallelae,* ut accidit in cylindris communibus. Ita enim fiet, ut nulla omnino cylindrorum hujusmodi per planum sectio det circulos aut ellipses, eruntque aut scaleni aut recti ad imitationem communium, prout analysis topica propositae quaestionis exposcet."

drawn to one point such that the sum of the squares of all the lines is equal to a given area, the point to which they are drawn will lie on a spherical surface given in position."[81]

Fermat begins by assuming some plane given in position and asks what the locus of points satisfying the conditions of the theorem will be on that plane. The answer results easily on the basis of the *Plane Loci*. For let points A, C, E be the given points of the problem lying arbitrarily in space. From each of these points drop the perpendiculars AB, CD, EF, respectively to the plane. Let I be a point on the plane so linked to A, C, E, that

$$\overline{AI}^2 + \overline{CI}^2 + \overline{EI}^2 = \text{the given area.}$$

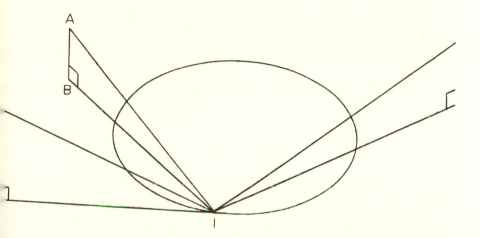

Since A, C, E are given points, and the plane containing I is assumed given, the lengths AB, CD, EF, and their squares, are also given. Hence, by application of the Pythagorean Theorem,

$$\overline{BI}^2 + \overline{DI}^2 + \overline{FI}^2 = \text{some given area,}[82]$$

and, by *Plane Loci* II,5, I lies on a circle. The same result, except for the change in the constant area, will obtain however the plane assumed given is chosen, provided of course that the plane contains at least one point I satisfying the conditions of the theorem. Hence, by lemma 2, the surface locus of points satisfying the conditions of the theorem is a sphere.

The remaining theorems of the *Introduction to Surface Loci* all follow the same pattern of demonstration, though they involve some of the other surfaces mentioned in the lemmas. Fermat's treatment clearly relies heavily on

[81] FO.I.113.
[82] The right-hand side of the equation is, more precisely: the given area $- \overline{AB}^2 - \overline{CD}^2 - \overline{EF}^2$.

the *Introduction to Plane and Solid Loci*. Imagining the surface locus cut by some plane, and dropping perpendiculars from the given points in space to the plane, he is able to reduce the three-dimensional problem to a two-dimensional one, that is, to a locus problem involving co-planar points. By this technique, any second-degree problem determining a surface may be treated as a series of second-degree plane curves which Fermat's *Introduction* has rendered easily manageable. Fermat's technique of reduction explains, then, the presence of the intuitively geometric lemmas with which his *Surface Loci* opens. The lemmas are needed to move from the determination of the possible plane sections of the surface to the surface itself.

The use of the lemmas, however, detracts from what would seem to be Fermat's goal of an algebraic geometry of three dimensions. The reduction to a series of plane problems and the construction of the surface from them enables Fermat only to determine the nature of the surface. That is, he can establish correspondence but not determination. The theorem above clarifies this point. It states that the sphere satisfying the locus problem is given in position. But Fermat has shown in his demonstration only that the surface is a sphere. Nowhere does one find an algorithmic determination of the center of the sphere or its radius in terms of the given points. Fermat could accomplish this determination, but not algorithmically; Euclid or Theodosius,[83] not Viète, would supply the solid geometry by which Fermat could determine the center and radius of the sphere on the basis of its circles. Nor, given the level of his mathematical development, is Fermat on the right track toward the fully algebraic treatment of surfaces. In modern terms, his method calls for the determination of a surface from its traces on arbitrary planes intersecting it. The problem challenges the modern mathematician with all the refinements of analytic geometry at his command.

Be that as it may, Fermat's three-dimensional system lacks even the foundation of a solid analytic geometry. For, in the *Introduction to Surface Loci*, one looks in vain for an equation which determines, or even corresponds to, the surface in question; nowhere is there an equation of the form $x^2 + y^2 + z^2 = r^2$ for the sphere, for instance. The only equations are those determining plane curves. Fermat means literally what he says in the opening paragraphs of the work; he will seek the *same* characteristic properties (*symptômata*) for surfaces as for plane curves.

Fermat's attempts to extend the system of the *Introduction* remained incomplete and open-ended throughout his career. He never returned to the

[83] Theodosius' *Three Books of Sphaerics* was relatively well known in the sixteenth and seventeenth centuries. The first Latin translation appeared in Vienna in 1529. In 1558 a Latin version by Maurolycus (Messina) and the *editio princeps* of the Greek (Paris) both appeared. Conrad Dasypodius (Rauchfuss) published a Greek-Latin edition at Strassburg in 1572. The modern edition is by Ernst Nizze, *Theodosii Tripolitae Sphaericorum libri tres* (Berlin, 1852).

question of surface loci directly. After 1643, he became increasingly involved in the theory of equations, quadrature, and number theory. Perhaps the increased insight which time and work in algebra brought to him led him to reconsider the matter of the algebra of surfaces. For, in 1650, in the context of an algebraic treatise on the elimination of variables, he stated the foundation of what would be solid analytic geometry:[84]

> In order that you may understand more clearly, there are some problems which involve only one unknown and which can be called *determinate*, in order to differentiate between them and locus problems [*problemata localia*]. There are others which have two unknown quantities and which cannot be reduced to a single one; these are locus problems.
>
> In the former problems we seek only one point; in the latter, a line. But, if the proposed problem admits three unknown quantities, a problem of this type investigates not a point alone, nor a line alone, but a whole surface peculiar to the question; then emerge surface loci, and so on. Just as in the first case the data suffice for determining the question, so in the second case one datum is absent for determining the question, and in the third only two data can complete the determination.

As this statement indicates, by 1656 the relation between algebra and geometry had come full circle. A latent algebra had dictated much of the form and techniques of the Greek analytic treatises. As later mathematicians tried to cope with these treatises in terms of pure geometry, they became hopelessly entangled. With the return to algebra, led by the practitioners of the analytic program, many of the problems straightened themselves out. So it was for Fermat. Tied to geometry in 1643 (for another example, see the section to follow), he broke loose from the bonds in the context of a purely algebraic treatise.

In one more sense, too, the relation between geometry and algebra had come full circle by 1650. In translating into geometrical terms the algebraic techniques of the Babylonians, the Greeks had established a correspondence between geometric dimensionality and the degree of an algebraic expression. The conceptual revolution in algebra, begun by Viète and completed by Descartes, freed algebra of this hindering correspondence and made algebraic equations unidimensional. Fermat restored a correspondence between dimension and algebraic equations; he characterized it, however, not by the degree of the equation but by the number of unknowns in the equation. Analytic geometry and the history of *n*-dimensional manifolds since the last century attest the fruitfulness of Fermat's adjustment of the link between algebra and geometry.

[84]*Novus secundarum et ulterioris ordinis radicum in analyticis usus,* FO.I.186–187. For further discussion of this text, see below, Chap. IV, §VII.

V. USES OF THE SYSTEM OF THE *INTRODUCTION:* GRAPHIC SOLUTION AND CLASSIFICATION OF EQUATIONS

The very title of Fermat's *Introduction* points to its main purpose and to the desires that motivated it: to establish a technique for bringing symbolic algebra to bear on the complex of classical geometrical problems involving loci. The concluding paragraph of the treatise further emphasizes its original motivation. But like Descartes, Fermat quickly recognized that his new technique had implications extending well beyond the problems it was specifically designed to handle. The same system of algebraic geometry that used algebra to solve locus problems made possible the inverse application of loci to the solution of algebraic problems. The same system that employed the geometric properties of the conic sections (including the circle and straight line) as a unified family of curves to establish the essential unity of the family of quadratic equations could in turn use the algebraic properties of determinate and indeterminate equations to extend and subdivide the ancient classification of curves. Even if one leaves aside for the moment the new techniques by which Fermat analyzed the equation of a curve for such intrinsic properties as its tangent or its quadrature (see Chapters IV and V below), his system of algebraic geometry, though designed as a tool, suggested as a mathematical system in itself new problems and new possibilities that transcended its roots in Pappus and Apollonius. Here again one encounters the fascinating theme of Fermat's mathematical development: techniques originally stimulated by and addressed to problems within an ancient tradition quickly became the foundations of a new and entirely different tradition. Often, the transition was so subtle that it escaped Fermat's notice. The foremost proponent of Viète's analytic program at times fell victim to the erroneous belief on which it was founded. The assumption that symbolic algebra represented a means toward restoring the analysis of the Ancients often masked crucial points of incompatibility between the algebraic and geometric approaches to mathematics and emphasized the one to the detriment of the other.

A. *The Graphical Solution of Equations: Geometry vs. Algebra*

Soon after the *Introduction* began to circulate in Paris in 1637, it was joined by an *Appendix to the Introduction to Loci, Containing the Solution of Solid Problems by Means of Loci.*[85] Hence, the idea of applying the system of the *Introduction* to the graphical solution of equations occurred to Fermat very shortly after the completion of the system itself, if indeed any

---

[85]*Appendix ad isagogen topicam continens solutionem problematum solidorum per locos*, FO.I.103–110. All quotations from Fermat in Section A are translated from that text and will not be further identified. On the joint circulation of the *Introduction* with the *Appendix*, cf. Fermat to Mersenne, II.1638, FO.II.134: "Je serai bien aise de savoir le jugement de M[rs] de Roberval et de Pascal sur mon *Isagoge topique* et sur l'*Appendix*, s'ils ont vu l'un et l'autre."

time at all separated the main work from its appendix.[86] As the *Appendix* itself shows, the stimulus to make such an extension already lay in the classical Greek sources that provided Fermat with his problems and in Viète's algebra from which he derived his techniques of solution. In the *Appendix*, however, their combination represented to some extent a union of incompatibles.

Fermat's Greek sources offered many examples of the solution of determinate problems by means of curves. Archytas of Tarentum reportedly employed three intersecting surfaces of revolution to double the cube (i.e. to construct two mean proportionals), and Eutocius reports the early use of the intersection of an hyperbola and a parabola to solve the same problem as posed in Archimedes' *Sphere and Cylinder* II,4. In the introductory remarks to Chapter VI of his *On the Emendation of Equations*, Viète reemphasized the thrust of these examples by proposing that for equations having irrational (*asymmetrae*) roots, "it is for the geometer rather than the arithmetician to set them out (*exhibere*) accurately.[87] Once Fermat had established the intimate relationship between curves and indeterminate equations, it required little to translate Viète's proposal for the geometric solution of algebraic equations into definite procedures. To do so was the purpose of the *Appendix*:

> Now that the method for finding loci is clear, it remains to inquire in what way the solution of solid [i.e. cubic] problems can be most elegantly derived from what has been said above. To do this, we must restrict that freedom of the unknown quantities to overstep their limits; for in locus problems there are infinitely many points by which the question proposed is satisfied.

Though, as shown above, Fermat's clear insight into the relationship between dimensionality and the number of unknowns—or, at least, an explicit statement of that insight—still lay in the future, nonetheless the link between a geometric locus and the infinitely many solutions of an indeterminate algebraic equation served as the very foundation for the method of the *Introduction*. Moreover, Fermat knew from his study of Viète that the reduction of an indeterminate equation in two unknowns to a determinate equation in one required a supplementary condition in the form of a second indeterminate equation.[88] The translation of this fact into the system of the *Introduction*

---

[86] In Descartes' case, of course, there is no question of separation; graphical soi.tior forms an integral part of the *Geometry*.

[87] *Opera*, ed. Schooten, pp. 140–141: "nam etsi radices sint asymmetrae, exhibebuntur ea methodo veris proximae, accurate autem exhibere, est Geometrae potius quam Arithmetici."

[88] Again, an explicit statement of this point is first found in that same later text, *Novus secundarum et ulterioris ordinis radicum in analyticis usus* (1650; cf. above, n. 84). But there it serves as assumed foundation for a method of elimination of radicals and is treated as something well known.

reiterated in different terms what Fermat's classical sources had already told him: "Therefore the question is most easily made determinate by means of two local equations, since indeed the two loci given in position intersect one another, and the intersection point [being] given in position forces the question from the infinite into prescribed limits." In sum, the reasoning that had gone into the *Introduction* made it clear that the correspondence between indeterminate equations and their loci carried over to systems of determinate equations and the intersection of curves.

The general problem to which Fermat addressed the *Appendix*, however, began with a single determinate equation, and it remained for him to develop a technique for incorporating it into an indeterminate system. Again, he could look to Viète for guidance. The use of a second, auxiliary unknown is an outstanding characteristic of Viète's approach to the theory of equations. The three methods to which Fermat most often referred and from which he derived his most fruitful ideas—*syncrisis, anastrophe*, and *climactic paraplerosis*[89]—all employ such an unknown. Moreover, in each case, the introduction of the second unknown achieves a reduction in the degree of the equation under study. Faced with the basic problem of using the quadratic curves of the *Introduction* to solve cubic and quartic equations,[90] Fermat took, then, the step one might expect. In the first problem of the *Appendix*, to solve the equation $x^3 + bx^2 = bc$, he set each side of the equation equal to bxy, "in order that by division of this solid, on the one hand by x and on the other by b, the matter is reduced to [quadratic] loci."

Introduction of the auxiliary variable y transforms, then, the determinate equation into a system of indeterminate equations of reduced degree:

$$x^2 + bx = by$$

$$c = xy.$$

From the *Introduction* it is clear that the first equation determines a parabola and the second an hyperbola. Their point of intersection—for Fermat, they have only one, for the reasons set forth above (pp. 82-83)—determines a fixed value for y common to them both, and through that a fixed value for x that is a solution to the original equation.

Fermat here again trusted to the previous training of the reader to provide the reasoning just set forth. In the *Appendix*, he merely determined that the intersection of the two curves will fix the endpoint of the y-ordinate in position and then repeated the formula that threatened to drive his corre-

---

[89]For the technical details of *syncrisis* and *anastrophe*, see below, Chap. IV, §II,A; for *climactic paraplerosis*, see below, §V,B.

[90]Note the continued influence here of the ancient classification of curves and problems. "Solid problems" are problems reducible to cubic equations, where the degree of the equation still carries dimensional connotations. Quartic equations are not, properly speaking, solid problems, but are treated as such because the method of graphical solution is the same for both degrees. In a sense, then, Fermat is already thinking along the lines of a new classification at the same time that he preserves the old.

spondents to distraction: *et est facilis ab analysi ad synthesin regressus.* Showing that the value of x corresponding to the common y-ordinate was in fact a solution of the original equation would constitute part of that synthesis, and Fermat expected his reader to know where to look for the proof, to wit, to Viète's theory of equations.

Rather than provide that *regressus ad synthesin* in any detail, Fermat was content to generalize the procedure of his one example for the graphic solution of cubic equations: "The method for all cubic equations is not dissimilar. For having placed on one side all solids containing A [x] and on the other the given solid (with or without solids containing A or A^2), one can put together an equality similar to the one above." Expecting a certain amount of problem-solving ingenuity from his readers, Fermat deliberately left the general procedure somewhat vague with regard to how one is to divide up the terms of the original equation. Clearly, however, the procedure always works if one splits the general equation $x^3 + ax^2 + bx = c$ into the two equations

$$x^3 + ax^2 + bx = pxy \text{ or } x^2 + ax + b = py$$

$$c = pxy,$$

where p is chosen to simplify as much as possible the resulting equations.

Fermat then went on to show in the *Appendix* that the same basic procedure also works for quartic equations, moving as was his habit from a particular example to a general procedure. Given the equation $x^4 + bx + cx^2 = d$, one separates the quartic term from the others in the form $x^4 = d - cx^2 - bx$ and equates each side to the common value c^2y^2. The resulting system of equations reduces the original problem to the intersection of the parabola $x^2 = cy$ and the circle $(d - bx)/c^2 - x^2 = y^2$. The general procedure follows immediately from this example via Chapter I of Viète's *On Emendation*, by which the cubic term of a general quartic $x^4 + bx^3 + cx^2 + dx + e = 0$ can be eliminated and the equation reduced to the form treated in the example.

The motivation that gives the *Appendix* its shape and thrust emerges clearly from the concrete example that Fermat pursues next. Harking back to Eutocius' commentary on Proposition II,4 of Archimedes' *Sphere and Cylinder*, Fermat proposes to find two mean proportionals between two given magnitudes by means of a graphical solution. Taking b and c ꞏꞏꞏꞏ x as the greater mean, and z as the smaller, he transforms the original statement of the problem, $b:x = x:z = z:d$, into the equation $x^3 = b^2d$. The transformation of this single equation into the system of equations $x^3 = bxy$ (or $x^2 = by$) and $b^2d = bxy$ (or $bd = xy$) provides a solution for x via the intersection of a given parabola and a given hyperbola.[91] Raising the cubic equation to a quartic via

[91] Note here how the goal of presenting a systematic method leads Fermat to move all the way to a single cubic equation and then back to a system of quadratic equations, rather than generating that system directly from the original continued proportion. His choice of two separate designations for the second proportional, $C(z)$ and $E(y)$, manifestly serves this goal.

127

multiplication by x, i.e. to $x^4 = b^2 dx$, makes possible its solution by intersecting parabolas; i.e. $x^4 = b^2 y^2$ (or $x^2 = by$) and $b^2 dx = b^2 y^2$ (or $dx = y^2$). Just as Descartes, then, brought his *Geometry* to a close with the problem of finding mean proportionals, so too Fermat in the *Appendix* directed the power of his algebraic geometry to that same problem.

But if Fermat was addressing a classical complex of problems in the *Appendix*, he was also taking aim at a more recent one. He was interested less in the specific solutions to the problems than in the "elegance" with which he was able to derive them, an "elegance" that Viète's solution techniques lacked: "Forget, therefore, Viète's *climactic parapleroses* by which he reduces quartic equations to quadratics by means of cubic equations of a squared root. For, as has been shown, quartic and cubic equations can henceforth be solved with the same elegance, ease, and brevity; nor, do I think, can they be solved any more elegantly." To Viète's *climactic paraplerosis*, which reduced an equation like $x^4 + bx = c$ to the form $x^2 + xy = (b/2y) - (y^2/2)$, where y is a root of the equation $(y^2)^3 + 4c(y^2) = b$, Fermat would oppose the technique of graphical solution presented in the *Appendix*. It was a geometer's choice, as Fermat's next point clearly shows: "In order that the elegance of this method be clear, here is the construction of all cubic and quartic problems by means of parabola and circle."

Again Fermat generalizes on a specific example: Let $x^4 - cx = d$, or $x^4 = cx + d$. Add to each side of the equation the terms lacking to make the left-hand side a perfect square of the form $(x^2 - b^2)^2$, i.e. let

$$x^4 - 2b^2 x^2 + b^4 = cx + d + b^4 - 2b^2 x^2,$$

for some b. Letting $2b^2 = n^2$, set, on the one hand,

$$x^2 - b^2 = ny$$

and, on the other hand,

$$b^4 + d + cx - n^2 x^2 = n^2 y^2.$$

The first equation yields a parabola and the second a circle. The extension of the procedure followed in the example to any quartic equation rests only on the mathematician's ability, employing the methods of Viète's theory of equations, to rid that equation of its cubic term. Since any cubic equation can be raised to a quartic, cubic equations are subsumed under the same general procedure.[92]

What has been gained by this new approach to solving equations? Fermat answers the question in his concluding words: the general procedure shows quite clearly the futility of trying to solve such problems as finding two mean proportionals or trisecting the angle by plane means (i.e. circle and straight-edge). Despite the specific examples solved in the *Appendix*, Fermat thus

[92] One must also take care in the second indeterminate equation that the x^2-term and the y^2-term have contrary signs. Fermat gives an example to show how this condition can always be fulfilled.

ultimately comes down on a metamathematical point: cubic and quartic problems necessarily involve the use of solid curves (i.e. at least one conic section) in their solution. Viète's theory of equations, together with its translation through the *Introduction* into geometrical terms, leaves no doubt in the matter.

The new approach has its price, however. Throughout the *Appendix* Fermat loses many more roots of equations than he finds. Indeed, he never finds more than one. He does not, because he neither expects nor demands more than one. That is what makes the metamathematical tone of his final point so strange in relation to the rest of the treatise. The multiplicity of roots for higher degree equations represents an indispensable foundation for the Viètan theory of equations on which the methods of the *Appendix* are based, and the irreducibility of cubic and quartic equations to quadratic means of solution rests ultimately on that theory of equations. But Fermat ignores such multiplicity through most of the *Appendix* because he is operating in the framework of Pappus' classical geometrical analyst. He seeks answers, not metamathematical theorems, and for him only the real, positive, geometrically constructible root counts. The methods of the *Appendix* determine that root; Fermat does not require that they do more.

Just as the "elegance" of Fermat's graphical method of solution reveals the geometrical bias that motivated it, so too the final *caveat* speaks to geometers. And yet the *Appendix* ends on an algebraic note. In showing the uniform solvability of cubic and quartic equations by means of circle and parabola—and not, therefore, by purely plane means—Fermat broached a larger subject to which his reading of Descartes' *Geometry* would force him to return, to wit, the classification of algebraic problems by means of the curves required to solve them, and vice versa.

B. *The Tripartite Dissertation*

The lack of order and uniformity in the *Appendix* makes its title particularly appropriate. It represents an afterthought to the *Introduction*, a catch-all in which to record several immediate, but non-essential corollaries to the new system of algebraic geometry. Fermat surely realized that he was merely scratching the surface of the graphic solution of equations and of the classificatory possibilities suggested by it. In 1636 and 1637, however, he had many different things of a fundamental nature to communicate to his new friends in Paris; it was not the time to fill in details, nor, given the novelty of his results, was there outside stimulation to do so. The years immediately following produced that stimulation. Receipt of a copy of Descartes' *Geometry* in December 1637 and the subsequent controversy with Descartes over the method of maxima and minima[93] introduced to Fermat a mathematical peer

---

[93]For the details of this controversy, which decisively shaped Fermat's mathematical career following 1638, see below, Chap. IV, §IV.

and rival of whom he had previously known nothing. The ambiguous outcome of the controversy made him painfully aware of the fact that Roberval and his circle did not constitute the only group of mathematicians in Paris that could and did judge his efforts. Fermat not only faced Descartes, he also faced the "Cartesians," for whom the *Geometry* robbed Fermat's results of their novelty. By the early 1640s, the fundamental similarity of Fermat's and Descartes' various methods had shifted the emphasis from foundations to details, and the shift was reinforced by the Cartesians' stress on systematic generality over flashes of insight. To fight the Cartesians on their own ground, Fermat transformed the *Appendix* into a full-scale investigation of the graphic solution of equations, the *Tripartite Dissertation on the Solution of Geometric Problems by the Most Simple Curves Properly Suited to Each Class of Problems.*[94]

In the *Tripartite Dissertation*, Descartes and his *Geometry* play the dual role of hero-villain. Nothing emerges so clearly from the treatise as Fermat's profound respect and admiration for his rival,[95] yet it is couched in the form of an attack on the classification of curves in Book II of the *Geometry* and on the treatment of mean proportionals at the end of Book III, matters in which Descartes seems to have taken some pride. Fermat had to attack Descartes as Galileo had to attack Aristotle in the *Dialogue Concerning the Two Great World Systems*. In each case, it is the followers, not the leader, who are the villains, and in each case their villany arises out of blind devotion. The errors of the leader open the only avenue of approach, as Fermat indicates at the very beginning of the *Tripartite Dissertation:*

---

[94]*De solutione problematum geometricorum per curvas simplicissimas et unicuique problematum generi proprie convenientes dissertatio tripartita*; the text in FO.I.118–131 is taken directly from the *Varia opera*, pp. 110–115. It is difficult to set a precise date of composition for the treatise. It does not itself bear a date or contain any readily datable material, nor does Fermat discuss it in any of his correspondence. Since it directly discusses Descartes' *Geometry*, it postdates December 1637, when Fermat first received a copy of that work. The language of the treatise would seem to indicate that Descartes was still alive when it was written; for example, Fermat says *Cartesium . . . hominem esse* rather than *hominem fuisse*, and, with few exceptions, all other verbs referring to Descartes are in the present tense. Hence, 1650 would be a reasonable *terminus ad quem*. As will become clear from comparisons made in the discussion to follow, however, both the mathematical content of the treatise and the specific audience to which it was addressed would argue for a more narrow time span between the years 1641 and 1643. But nothing in that discussion depends on this more precise dating. The earlier date would make the *Animadversio in Geometriam Cartesii* of 1657 (see below, p. 137) a condensation of an already completed work rather than an outline for a work to be written, as Hofmann would have it ("Neues über Fermats zahlentheoretische Herausforderung von 1657," *Abh.d.Pr.Akad.d. Wiss., Math.-naturw.Kl.*, Jhrg.1943, Nr.9, Berlin, 1944, p. 39).

[95]Cf. Fermat's concluding words to Part II: "Veritatem enim tantum inquirimus et, si in scriptis tanti viri alicubi delitescat, eam libenti statim animo et amplectemur et agnoscemus. Tanta me sane, ut verbis alienis utar, hujus portentosissimi ingenii incessit admiratio, ut pluris faciam Cartesium errantem quam multos κατορθοῦντας.

Although it might seem paradoxical for someone to say that in matters mathematical Descartes is only a man, to see that it is true the subtler Cartesians should consider whether Descartes' classification of curves into certain classes or degrees does not contain an error and whether those curves should not be classified more probably and more easily according to the true laws of geometric analysis. We feel it will be evident to everyone or (if that is too general) at least to mathematicians and analysts that we do not pursue this matter to the detriment of such a famous man; for the truth is of concern to Descartes and to all Cartesians, and there is particular merit in pursuing truth vehemently even though it may at times contradict our wishes.

Introduced at the very outset, this theme runs as a thread through the rest of the treatise.

In the context of the ancient tradition to which both men (but Fermat in particular) were responding, Descartes' error was no light matter. The proper classification of geometrical problems enabled the mathematician to avoid a serious offense against traditional mathematical esthetics, to wit, the use of a more complex solution than necessary. Implicit in Fermat's claims for the "elegance" of his method in the *Appendix*, the canon of simplicity becomes explicit in the *Tripartite Dissertation* in response to what Fermat perceives as its violation in the *Geometry*. To expose that violation in the realm of curves defined by indeterminate equations, Fermat starts from what he takes to be common ground in the realm of determinate equations. He and Descartes agree on the classification of the latter: determinate equations of degree $2n$ and $2n - 1$ form an irreducible common class.

From the justification of this basic classification emerges Fermat's most subtle tribute and debt to Descartes, for Fermat appeals to the theory of equations, to the "precepts of the analytic art," for which Viète had represented Fermat's source and authority. Now, for the first time, Descartes joins Viète in that role; throughout the *Tripartite Dissertation* the two men exercise their authority in tandem, and from Fermat's standpoint it is Descartes who is thereby honored. Though not apparent at the outset, Descartes has earned the honor. As Fermat progressively refines the methods he first discussed in the *Appendix*, the extensions and changes in style increasingly reveal the direct influence of Descartes' advanced and more lucid treatment of the theory of equations in Book III of the *Geometry*. It is, therefore, ironic that Fermat's faith in the classification of determinate equations should turn out to be misplaced; on strictly algebraic grounds the classification has no justification.

Fermat, however, thought the justification lay implicit in the techniques of his two authorities:[96]

---

[96] Cf. *Géométrie* (1637), pp. 383–387.

The reason why two adjacent powers, although they are of different degree, constitute nonetheless only one genus of problems is because quadratic equations can be reduced to simple or linear [*laterales*] equations by a simple method known to both ancients and moderns, and hence can be resolved without difficulty by circle and straightedge. Moreover, with the aid of a new method, set forth by Viète and Descartes, quartic [*quadrato-quadraticae*] equations, or equations of the fourth degree, can be reduced to equations of the third degree, or cubic equations. It is to this task that Viète addressed that subtle *climactic paraplerosis* peculiar to him; it may be found in Chapter VI of his book *On Emendation of Equations*. Descartes uses a not dissimilar technique to the same end, though he uses different words to describe it.

Viète's method of *climactic paraplerosis* formed the basis of Fermat's graphic solutions in the *Appendix*. In the transfer to the *Tripartite Dissertation*, however, it undergoes a subtle and interesting shift in emphasis. To see that shift, and to understand the extension Fermat wants to make of the method itself, it will help to set down here one detailed example of Viète's procedure. In the words of Problem I of Chapter VI of *On Emendation*, Viète seeks "to reduce a quartic equation containing a linear term to a quadratic equation by means of a cubic equation of a squared root." The example he gives is $x^4 + bx = c$.[97] Transposing the linear term to the other side of the equation and adding the terms necessary to make the left-hand side a perfect square, Viète transforms the original equation into $(x^2 + \frac{1}{2}y^2)^2 = c - bx + x^2y^2 + \frac{1}{4}y^4$, where y is an auxiliary unknown. To complete the reduction to a quadratic equation, Viète must now find a value for y that makes the expression on the right-hand side a perfect square. To do so, he borrows a leaf from Diophantus' *Arithmetica*;[98] the trick lies in eliminating all terms containing x, and the rule for squaring a polynomial provides the means. Consider the square of the expression $(b/2y) - xy$, i.e. $(b^2/4y^2) + x^2y^2 - bx$. Setting that square equal to the above expression yields a cubic equation in y^2; that is,

$$\frac{b^2}{4y^2} = c + \frac{1}{4}y^4, \text{ or } b^2 = 4c(y^2) + (y^2)^3.$$

As a beneficiary of the ingenuity of Tartaglia and Cardano, Viète is in a position in fact to solve that cubic equation, which is of the form $u^3 + au = b$. By substituting the solution, say $y = m$, into the expressions $(b/2y) - xy$ and $x^2 + \frac{1}{2}y^2$, he has then reduced his original quartic equation to a quadratic, to wit:

[97]*Opera*, ed. Schooten, p. 141. Because Chap. II has already discussed the issue of Viète's symbolism and his concept of homogeneity in some detail, and because these issues are irrelevant to the mathematical point in question, I shall make free use of the more familiar modern notation.

[98]I.e. the method of "single" and "double" equations; for greater detail, see below, Chap. VI, §III, B.

$$x^2 - \frac{1}{2}m^2 = \frac{b}{2m} - mx.$$

The specifically problem-oriented thrust of the *Appendix* focused Fermat's attention there on the resulting quadratic equation. He wished to contrast the ease with which his graphic solution reduced a quartic equation to the intersection of two quadratic curves with the complexity of Viète's reduction via a cubic equation. In the *Tripartite Dissertation*, however, it is precisely the algebraic necessity of solving a cubic equation that Fermat wants to point out. Hence, he here speaks of the reduction of a quartic equation to a cubic equation, rather than to a quadratic. Despite the seeming shift from geometry to algebra signified by this change of emphasis, however, the geometrical bias of the *Appendix* maintains its effect on Fermat's thinking and works to vitiate the conclusions he goes on to draw from Viète's procedure.

Mindful of the concluding theorem of the *Appendix*, which insisted on the geometrical irreducibility of cubic problems, Fermat infers a corresponding algebraic irreducibility of a cubic equation. It is, after all, the reducibility of a quartic equation to a cubic and the irreducibility of the latter that forms the basis for uniting the two in a single genus distinct from the genus of quadratic and linear equations, i.e. that forms the basis of Fermat's perceived common ground with Descartes. Yet the desire to assert the correspondence of irreducibility, conditioned perhaps by the technique of graphic solution toward which Fermat is heading, blocks out a fact that Fermat surely knows from his study of Viète:[99] Tartaglia's classic solution of the cubic equation is based precisely on a reduction of the cubic equation to a quadratic! Hence, there is no algebraic basis for separating Fermat's first two *genera* of equations, but rather only the geometric impossibility of determining cube roots by use of circle and straightedge.

In large part, Viète's casuistic presentation of the theory of equations reinforced Fermat's geometrical bias, if indeed it was not a source of that bias. By separating *climactic paraplerosis* (reduction from quartic to cubic) in Chapter VI of *On Emendation* from *duplicate hypostasis* (reduction from cubic to quadratic) in Chapter VII, Viète established an apparent distinction which his general description of the former method could only work to stress: "It has already been shown above that quadratic, quadratoquadratic, cubocubic equations, and further equations rising by alternate degrees, can be reduced by climactic paraplerosis if not by anastrophe."[100] The classification of equations into pairs of even and odd degree, then, already lay implicit in Viète's work. Fermat had given it geometric expression in the *Appendix*, and a remark in the *Geometry* moved him to make it explicit in the *Tripartite Dissertation:*[101]

[99]Cf. Viète, *De emendatione*, Chap. VII, Prob. I.

[100]*Opera*, ed. Schooten, p. 141.

[101]*Géométrie* (1637), p. 323: "Au reste ie mets les lignes courbes qui font monter cete equation iusques au quarré de quarré, au mesme genre que celles qui ne la font

In a similar manner (though with somewhat more difficulty) the Viètan or Cartesian analyst will reduce a cubocubic equation to a quadratocubic, i.e. an equation of the sixth degree to one of the fifth degree. And, because in the aforesaid cases, in which there is only one unknown quantity, equations of even degree can be reduced to the next lower odd degree, Descartes has confidently asserted on p. 323 of the French version of his *Geometry* that exactly the same thing holds true of equations in which there are two unknown quantities.

Descartes' sin, according to Fermat, lies in that transfer of the method of reduction from determinate to indeterminate equations. It will not work, he goes on to argue; indeed, analysts will find that a similar reduction of indeterminate equations is impossible. How, for example, would one go about reducing the quartic $x^4 = a^3 y$ to a cubic; how does one apply *climactic paraplerosis* here? One cannot, and hence one cannot classify curves on the same basis as determinate equations.

What Fermat never questions in the *Tripartite Dissertation* is the algebraic validity of the classification of determinate equations. Yet, on what basis, other than the mere implied assurance of Viète and Descartes, can Fermat justify his claim of the existence of a general method for reducing an equation of degree $2n$ to one of degree $2n - 1$? Despite Viète's claim for his method, he never applies it to anything other than quartic equations; Descartes nowhere actually reduces a sextic equation to a quintic. Nor does Fermat set forth the details of that extended reduction. He does not for two reasons (other than the fact that no such general reduction procedure exists). First of all, his confidence in Viète, combined with his own experience in the technique of setting polynomial expressions equal to a square,[102] provides an intuitive basis for believing that just as a quartic may be reduced to a quadratic by means of a cubic, so too a sextic may be reduced to a cubic by means of a quintic, an octic to a quartic by means of a septic, and so on. It is for him less a question of general proof than of the experience and skill of the algebraist in selecting the proper coefficients and terms in the expression to be squared; Part II of the *Tripartite Dissertation* attests to his own considerable skill in that regard. But second, and more important, Fermat's commitment to the $(2n, 2n - 1)$ classification does not really derive from the reduction procedure, but rather from the technique of graphic solution that forms

---

monter que iusques au cube. & celles dont l'equation monte au quarré de cube, au mesme genre que celles dont elle ne monte qu'au sursolide. & ainsi des autres. Dont la raison est, qu'il y a reigle generale pour reduire au cube toutes les difficultés qui vont au quarré de quarré, & au sursolide toutes celles qui vont au quarré de cube, de façon qu'on ne les doit point estimer plus composées."

102 See below, Chap. VI, §III, B. The fact that Fermat was engaged in such research during the early 1640s supports the argument for placing the composition of the *Tripartite Dissertation* in that same period.

the core of his treatise. What really links equations of degree $2n$ and $2n - 1$ in Fermat's mind is not the reducibility of the former to the latter, but their common solution by means of curves of degree n at most. Despite all the references to algebraic techniques of reduction, geometric construction forms the backbone of the *Tripartite Dissertation* as it does that of the *Appendix*. In both works, algebra serves geometry to the detriment of the former. They are not the only examples in which Fermat's geometric point of view blocks algebraic or analytic insight.[103]

The essentially geometric framework of Fermat's treatise begins to emerge clearly as he continues in his attack on Descartes. For just as classification itself is addressed to simplicity of solution, so the most telling sign of faulty classification is that it leads to unnecessarily complex solutions. Hence, robbing Descartes' classification of curves of its algebraic foundation will not suffice to reveal his transgression against canons of elegance. To do that, Fermat introduces the main theorem of the *Tripartite Dissertation*: "Let there be proposed a problem in which the unknown quantity ascends to the third or fourth power. We will resolve it by conic sections, which are of the second degree. But if the equation ascends to the fifth or sixth power, then we set forth the solution by curves of the third degree. If the equation ascends to the seventh or eighth power, we will set forth the solution by curves of the fourth degree, and so on *ad infinitum* by a uniform method."

Although Descartes also requires only cubic curves to solve fifth- or sixth-degree equations, Fermat then points out, he believes it necessary to resort to fifth- or sixth-degree curves in order to solve seventh- or eighth-degree equations. "Certainly," he continues, "he offends pure geometry who takes for the solution of any problem curves that are too complex and of higher degree, ignoring more simple and appropriate curves. For both Pappus[104] and more recent writers have often maintained that it is no light transgression (*peccatum*) in geometry when a problem is solved by improper means (*ex improprio genere*). In order that this not happen, one must correct Descartes and restore various problems to their proper and natural places (*sedes*)." Before carrying out that correction and restoration, however, Fermat puts the cap on his criticism of Descartes' classification. Not only does it lack algebraic validity, it also contradicts Descartes' own geometrical construction meant to generate curves of increasing complexity.

Descartes introduces that construction on p. 319 of the *Geometry* as a "way of separating all curved lines into certain classes." Taking the perpendicular lines *GA* and *AK* as fixed, he imagines the indefinite straightedge *GL* to rotate about the fixed point *G*, while *L* always remains in the line *AK*. He then links the rotating straightedge to a generating curve *CNK* defined with re-

---

[103]See below, Chap. IV, p. 204.
[104]*Mathematical Collection*, Book IV, Prop. 59; ed. Hultsch, p. 270, l. 27ff. (Identification by Tannery.)

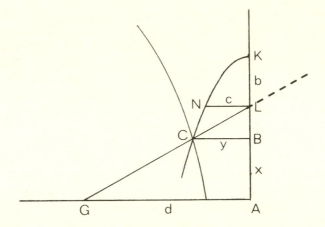

spect to AK as axis and point K as origin by holding the segment KL constant and allowing it to slide along AK. The intersection point C of the generating curve and the straightedge traces a new curve defined with respect to axis AK and origin A. In symbols, if $GA = d$, $AB = x$, $CB = y$, $KL = b$, $NL = c$, and $BK = u$, then a generating curve $y = f(u)$ will define the generated curve $y = f(b + xy/(d - y))$.

Descartes asserts on p. 322 of the *Geometry* that the generated curve will belong to the next higher genus of curves than the one to which the generating curve belongs. By his classification of curves given on p. 323, the classification to which Fermat so strongly objects, his assertion would imply that a curve of the first or second degree would generate one of the third or fourth degree, and a curve of the third or fourth degree one of the fifth or sixth. Fermat's counterexample is as simple as it is telling: let $y^3 = u$, i.e. let the generating curve be a simple "cubic parabola." Then the curve generated will be given by the equation $y^3 = b + xy/(d - y)$, which is a quartic equation. But a fourth-degree curve belongs to same class as a third-degree curve, not to the next higher class.

The single counterexample suffices to destroy the generality of Descartes' assertion; a more detailed algebraic demonstration of its invalidity would be superfluous. Moreover it would detract from the methodological point Fermat is pursuing. For the same construction that Descartes uses to classify curves can also be interpreted as a technique of graphic solution, and as such is guilty of the transgression cited above. That is, even if one grants for the moment the validity of the relation between generating and generated curves, Descartes' construction translates into the graphic solution of a problem of degree $2n$ or $2n - 1$ by means of the intersection of a straight line and a curve of degree $2n - 2$ or $2n - 3$. Fermat's counterexample emphasizes the transgression by showing that, as a technique of graphic solution, Descartes' construction would employ a cubic curve to solve a fourth-degree problem!

By now, however, Fermat has gone beyond beating a dead horse. He is flailing at an imaginary target. Whatever the classificatory deficiencies of the above construction, Descartes never intended it to serve as his method of graphic solution. Book III of the *Geometry* gives ample evidence that he was perfectly well aware of the main proposition Fermat has just introduced, i.e. that determinate equations of degree $2n$ and $2n - 1$ can be solved by curves of degree n at most; the point emerges clearly and obviously from his theory of equations. It is unclear whether Fermat failed to see this or saw it and chose to ignore it; a later condensed version of the *Tripartite Dissertation* suggests the former alternative.[105] One way or the other, Fermat clearly felt that the first part of his treatise had to push Descartes' error to the limit in order that his own method of graphic solution, with its implications for the classification of curves, receive an attentive hearing from the Cartesians, "to whom it seems a remedy to ignore any errors, especially those of Descartes, to the prejudice of the truth."

The second part of the *Tripartite Dissertation*, then, leaves Descartes' *Geometry* to present the details of Fermat's own method. He proposes to show in detail how any equation of degree $2n$ or $2n - 1$ can be solved graphically by means of curves of degree n at most. In keeping with his usual style of presentation, and that of his main sources, Viète and Descartes, Fermat describes his general procedure in words and then elucidates the description on the basis of concrete examples. If done with care, a translation of the general procedure into symbols will not do violence to Fermat's verbal description at the same time that it reveals more clearly to the modern reader what the procedure entailed.

The procedure begins with an equation of the canonical form

$$(1) \qquad x^{2n} + a_1 x^{2n-1} + \cdots + a_{2n-2} x^2 = r.$$

Though Fermat relates explicitly how to raise an equation of degree $2n - 1$ to one of degree $2n$ (by simple multiplication through by x), he refers the reader to Viète and Descartes for the details of carrying out the more important step of removing the term containing x in order to reduce any given equation to the initial canonical form.[106] Borrowing, then, a leaf from Viète and Diophantus, he introduces an expression containing an auxiliary unknown together with powers of x. This expression, when squared and set equal to the left-hand side of equation (1), is designed to produce a resulting equation in two unknowns of degree n. That is, he introduces an expression of the form

$$(2) \qquad x^n + b_1 x^{n-1} + \cdots + b_{n-3} x^3 + cxy,$$

---

[105] The *Animadversio in Geometriam Cartesii* (1657) repeats the charge *verbatim*.

[106] Both Viète (*On Emendation*, Chap. I) and Descartes (*Geometry*, Book III) show how to remove from an n-th degree equation the term containing the $(n - 1)$th power of the unknown by altering the roots to make its coefficient zero. Given an equation $x^m + ax^{m-1} + \ldots + px + q = 0$, one can substitute $1/u$ for x to make p the coefficient of u^{m-1} and then eliminate it by the procedures of Viète and Descartes.

where the particular value of c is arbitrary and the values of the various b_i are chosen in such a way that, when the square of expression (2) is substituted for r in equation (1), the terms containing the $n - 2$ highest powers of x cancel out.[107] For example, $a_1 = 2b_1$, $a_2 = b_1^2 + 2b_2$, $a_3 = 2b_1b_2 + 2b_3$, and so on; since this criterion for selecting the b_i involves $n - 2$ equations in $n - 2$ unknowns, it poses no theoretical problems. Since x appears in every term of expression (2), x^2 will appear in every term of its square, as it does in every term of the polynomial in equation (1). Hence, after cancellation of the $n - 2$ highest terms leaves an equation of degree $n + 2$, removal by division of the x^2 common to every term will yield finally an equation of the form

(3) $\quad p_1 x^n + p_2 x^{n-1} + \cdots + p_0 = q_1 x^{n-1} y + q_2 x^{n-2} y + \cdots + q_{n-2} y^2$

where, *inter alia*, $p_0 = a_{2n-2}$ and $q_{n-2} = c^2$.

Hence, the substitution of the square of expression (2) for the constant r in equation (1) has yielded the indeterminate equation of a curve of degree n. A second such curve follows immediately by setting expression (2) equal to $\sqrt{r}$; i.e.

$$x^n + b_1 x^{n-1} + \cdots + b_{n-3} x^3 + cxy = \sqrt{r}.$$

The point of intersection of the two curves will determine a value of x which is a root of the original equation. Fermat's own first example will help to elucidate this abstract procedure in concrete terms. Let $x^6 + bx^5 + cx^4 + dx^3 + mx^2 = n$; "all problems that ascend to the fifth or sixth power can be reduced to this form, for it involves nothing other than raising a fifth power to a sixth or to free the latter from the last term containing x or the linear unknown (*latere*), all of which Viète and Descartes have amply treated." Let $x^3 + bxy$

---

[107]Fermat's verbal description of the general procedure reads: "Modus autem operandi talis est. Data quaevis aequatio in qua unica tantùm reperitur ignota quantitas reducatur 1° ad gradum elatiorem sive parem: deinde ab adfectione sub latere omnino liberetur, quo peracto remanebit aequatio inter quantitatem cognitam vel homogeneum datum ex una parte, & aliquod homogeneum incognitum cujus singula membra à quadrato lateris incogniti adficientur ex una parte, ex altera homogeneum istud incognitum aequetur quadrato cujus latus effingendum eo artificio ut in aequatione ipsius quadrati cum homogeneo incognito elatiores quantum fieri poterit lateris ignoti gradus evanescant. Cavendum etiam ut singula lateris quadratici sic effingendi homogenea à radice vel latere ignoto adficiantur, & ultimum tandem ex illis à secundâ etiam radice incognita adficiatur. Orientur tandem beneficio divisionis simplicis ex una parte, & extractionis lateris quadrati ex altera, duae aequationes linearum curvarum problemati dato convenientium constitutivae, & earum intersectio solutionem problematis exhibebit ea qua dudum usi sumus in solutione problematum per locos methodo. . . . Notandum porro in problematibus quae ad nonam aut decimam potestatem ascendunt, ita effingendum latus quadrati ut in eo sint quatuor ad minus homogenea quorum beneficio evanescant tres elatiores lateris ignoti gradus. In problematibus autem quae ad undecimam aut duodecimam potestatem ascendunt latus effingendi quadrati constare debere quinque ad minus homogeneis, ita formandis ut eorum beneficio quatuor elatiores lateris ignoti gradus evanescant. Perpetuâ autem & facillimâ methodo, haec lateris quadrati effingendi forma per solam & simplicem divisionem vel applicationem ut verbis geometricis & in re purè geometricâ utamur expediri Analystae experiendo deprehendent, & characterum + & – variatio nullum methodo praejudicium est allatura."

serve as the indeterminate expression to be squared and substituted for n. Then,

$$x^6 + bx^5 + cx^4 + dx^3 + mx^2 = x^6 + 2bx^4y + b^2x^2y^2, \quad \text{or}$$
$$bx^3 + cx^2 + dx + m = 2bx^2y + b^2y^2,$$

which defines a third-degree curve. Similarly $x^3 + bxy = \sqrt{n}$ defines a third-degree curve, and the intersection of the two yields a root for the original sixth-degree equation.

A comparison of the above technique of graphic solution with Fermat's demonstration in the *Appendix* that the parabola and circle suffice to solve graphically all equations of the fourth degree reveals clearly the straightforward way in which the main content of the *Tripartite Dissertation* grows out of the earlier memoir. For the two procedures are distinguished only by the increased complexity of the technique employed in the later one to "complete the square" of the higher degree polynomial. Both procedures derive from Chapter VI of Viète's *On Emendation*, the earlier one as a direct geometric translation of *climactic paraplerosis*, the later one as an extension of Viète's method via Diophantus' technique of "single and double equations" (see Chapter VI, §III, B). In overall plan of attack, little had changed in Fermat's treatment of graphic solution in the interval separating the *Appendix* from the *Tripartite Dissertation*. If, however, Fermat could now extend the applicability of his approach, it was due to increased sensitivity to the possibilities of algebraic manipulation raised by his reading of Descartes' *Geometry*. In a sense, the *Tripartite Dissertation* reveals that debt more by what it omits than by what it includes. Underlying the verbal description and examples in Part II is Fermat's assumption that his readers share both the sensitivity and the technical competence to carry out the steps he outlines; he assumes, that is, that his readers have also read and understood Book III of the *Geometry*.

Nonetheless, whatever his debt to Book III, Fermat devotes Part III of the *Tripartite Dissertation* to an attack on Descartes' treatment of mean proportionals, which concludes that book and which constitutes the material which Descartes so graciously left to posterity as crumbs from a feast. The peculiarly simple nature of the equations that express the problem of finding a given number of mean proportionals between two given magnitudes—the largest of n mean proportionals between a and b is given by the root of $x^{n+1} = a^n b$— makes possible more than a mere halving of the degree of the problem to be solved. To take Fermat's first example, the determination of six mean proportionals between a and b, i.e. the solution of the equation $x^7 = a^6 b$, can always be solved by means of two fourth-degree equations, using the technique of Part II. But in this special case, a further reduction is possible. Just set x^7 and $a^6 b$ each equal to $x^4 y^2 b$; then the solution emerges from the intersection of the two cubic curves $x^3 = by^2$ and $a^3 = x^2 y$. Similarly, the largest of twelve mean proportionals is determined by the intersection of either of two sets of lower-degree curves: (1) $x^5 = by^4$ and $a^3 = x^2 y$, or (2) $x^4 = by^3$ and $a^4 = x^3 y$.

It would be a mistake, however, to see the thrust of Part III as aimed at graphic solution of equations or the theory of equations. For the central problem of the reduction of the equation $x^{n+1} = a^n b$ to curves of the lowest possible degree lies outside the realm of algebra. It is essentially a problem in number theory, a subject in which Fermat was becoming increasingly engaged at the time he wrote the *Tripartite Dissertation*.[108] The uniform general technique of the reductions presented in Part III involves splitting the original equation into two equations of the forms $x^{n+1} = x^p y^q b$ and $a^n b = x^p y^q b$. The possibility of extensive reduction of degree, especially for large values of n, lies in the second equation. While the degree of the first equation decreases arithmetically and is bounded by the value of q, that of the second may, by proper choice of p and q, decrease geometrically. That proper choice emerges from the partition of n into two numbers p and q such that either q represents the greatest common divisor (*GCD*) of n and p or the *GCD* of n, p, q is maximized. As such, the choice is still indeterminate; e.g. $72 = 63 + 9 = 64 + 8 = 36 + 36$. But the reduction of the first equation introduces a determinative element, for to achieve maximum arithmetic reduction p should be as large as possible under the conditions of partition. Hence, the correct partition of 72 is 64,8, whence the determination of 72 mean proportionals reduces to the intersection of two ninth-degree curves via the equations $x^{73} = x^{64} y^8 b$, or $x^9 = y^8 b$, and $a^{72} b = x^{64} y^8 b$, or $a^9 = x^8 y$.

Except for a series of specific examples, Fermat does not go into very much detail concerning these special cases of reducibility, nor does he pick up and deal with the problem of partition, although it is clearly on his mind: "One should note that the form of the equation must often be changed in order that its homogeneous terms be subject to an easy division by aliquot parts; this need be said only once." Rather, he has shifted to the realm of number theory in order to press home once more Descartes' transgression against the canons of mathematical simplicity. For, he notes, "it is clear from the aforesaid that one can assign between the degree of a problem and the degree of the curves solving it a ratio greater than any given ratio. When the Cartesians see this, I don't doubt they will subscribe to the necessity of our admonition and our emendation."

Fermat's proudest achievement in number theory provides the vehicle for this final assault on Descartes: "Moreover, since I have found that numbers of the form $2^{2^n} + 1$ are always prime numbers and have long since signified to analysts the truth of this theorem,[109] I will derive from it without any diffi-

---

[108] See below, Chap. VI, §III, B.

[109] Fermat first announced his conjecture in a letter to Mersenne on Christmas Day 1640 and repeated it in his correspondence with Blaise Pascal in 1654. It is hard to sense just how much time is signified by the Latin phrase *jam dudum*, and it is interesting, perhaps even significant, that the reduction of equations by means of the theorem does not appear in the *Observation on Descartes' Geometry*. For further details concerning the conjecture, see below, Chap. VI.

culty a method with the aid of which we construct a problem of which the degree bears to the degree of the curves serving for its solution a ratio greater than any given ratio."

Ironically, Fermat's misplaced faith in the primality of the numbers $F_n = 2^{2^n} + 1$ has basically nothing to do with the validity of the point he wishes to make here. That point is merely that the determination of $F_n - 1$ mean proportionals can be solved by means of curves of degree F_{n-1} and that, by choice of a suitably large value of n, the ratio of $F_n - 1$ to F_{n-1} can be made to exceed any preassigned value. Its validity rests on the peculiar property of the numbers F_n that $F_n - 1 = (F_{n-1} - 1)^2$. Hence, given the equation for $F_n - 1$ mean proportionals,

$$x^{F_n} = a^{F_n - 1} b,$$

one can partition $F_n - 1$ into $(F_{n-1} - 1)(F_{n-1} - 2)$ and $F_{n-1} - 1$. Since $(F_n - 1) - (F_{n-1} - 1)(F_{n-1} - 2) = (F_{n-1} - 1)^2 - (F_{n-1} - 1)(F_{n-1} - 2) = F_{n-1} - 1$, the first solution curve is given by the equation

$$x^{F_{n-1}} = y^{F_{n-1} - 1} b.$$

The second indeterminate equation, after extraction of the $(F_{n-1} - 1)$-*th* root, reduces to

$$a^{F_{n-1} - 1} = x^{F_{n-1} - 2} y.$$

Nonetheless, if the primality of F_n does not enter into the computational foundations of Fermat's procedure, it does play a role in the larger classificatory context of that procedure. For, in Fermat's scheme of things, "in equations in which only one unknown quantity is found on the one side, the exponent of that pure power must be a prime number in order to designate the degree of the problem. For, if that exponent is a composite number, the problem immediately reduces to the degrees of the numbers that measure it [i.e. its prime divisors]."

The *Tripartite Dissertation* ends, then, on the classificatory note with which it started. And yet, it leaves open the very question it poses: if Descartes' classification of curves is incorrect, what is the correct one? Only in a later condensed version of the treatise, a letter to Kenelm Digby published as an appendix to Frenicle's solution of the challenge problems of 1657 (see Chapter VI, §IV) and entitled *Observation on Descartes' Geometry*,[110] does Fermat provide the answer:

The simple equation, i.e. one in which there is only one unknown, determines separate species of problems. Equations of the first and second

---

[110]This short memoir, for the most part made up of extracts from the *Tripartite Dissertation*, represents one of the two major papers of Fermat uncovered since DeWaard's publication of the *Supplement* to FO in 1922. Both papers were edited and published by J. E. Hofmann in his "Neues über Fermats zahlentheoretische Herausforderungen von 1657," *Abh.d.Preuss.Akad.d.Wiss.,Math.-naturw.Kl.*, Jrg. 1943, Nr.9, Berlin, 1944. They had originally appeared as appendices to Frenicle's *Solutio duorum problematum* (1657); for further details, see below, Chap. VI, §III, D.

degree constitute the first species of problems, equations of the third and
fourth degree the second, equations of the fifth and sixth degree the third,
and so on in that order *ad infinitum.*

The twofold equation, i.e. one in which there are two unknown quantities, serves for the production and constitution of curves by means of infinitely many points. Lines of the first degree, i.e. straight lines, arrogate to
themselves the first species of lines; lines of the second degree, i.e. the
circle, parabola, hyperbola, and ellipse, the second; cubic curves or curves
representing the third power the third; quadratoquadratic curves the fourth,
and so on. That is, we construct each problem by curves of the same species.

And so the graphic method of solution first adumbrated in the *Appendix* now
reaches fruition. It becomes the foundation for the classification of both determinate and indeterminate equations. In doing so, it reveals on a deeper,
attitudinal level the continuing influence on Fermat of the classical sources
that originally stimulated the *Introduction.* For the thrust of Pappus' analysis
had been to solve problems, and Viete canonized that thrust in the motto that
concludes his own *Introduction* and serves as the title of Chapter II above. So
too, the fully articulated classification of the *Tripartite Dissertation* honors
this tradition; it is addressed to solving problems.

CHAPTER IV

Fashioning One's Own Luck

Hos cupiam similes tentando excu-
dere sortes
Fermat

I. INTRODUCTION

As one product of Fermat's first spurt of creative mathematics, the algebraic geometry of the *Introduction* required some six or seven years to reach fruition. Once Fermat had all the pieces, however, he welded them together into a finsihed treatise that contained a clear exposition and justification of his new technique. If the *Introduction* failed to attract the attention it deserved from the mathematical community at large, it was not due to any lack of clarity, but rather to the almost simultaneous appearance of Descartes' *Geometry*, which combined basically the same technique with a brilliant reformulation of the theory of equations. As one of the *Essays* accompanying the *Discourse on Method*, the *Geometry* was a *tour de force* calculated to put any similar work in the shadows, and, in the case of Fermat's *Introduction*, it succeeded.

But Fermat's creative period in Bordeaux had other offspring, in particular his methods for determining the extreme values of algebraic polynomials and for drawing the tangent to any point of an algebraic curve. A memoir describing these methods began circulating in Paris in 1636. Entitled *Method for Determining Maxima and Minima and Tangents to Curved Lines* (hereafter, *Method*),[1] it lacked the clarity

---

[1] Critical edition in FO.I.133–136. The holograph, now lost, circulated widely among contemporary mathematicians, many of whom made their own copies of it. Van Schooten's copy is found in MS Groningen 110, ff.6v–7r. Mersenne took a copy to Italy with him in 1644, which accounts for the copies in MS Florence, ff.83r–84, and in MS Ricci, ff.37–39. Fermat's holograph probably served as the basis for the published version in the 1679 *Opera varia*. It is difficult to date the treatise precisely. Descartes received the original from Mersenne early in January 1638, which means that it must have been sent from Paris sometime in December 1637. Tannery assumed that Mersenne got it from Carcavi (cf. AT.I.482-3; FO.II.126, n.2), who had brought it to Paris when he

of exposition of the *Introduction* and perhaps for that reason alone did not succumb to the *Geometry* but became instead the focus of a bitter dispute that eventually involved most of the mathematicians in Paris. Brought thus to the attention of a wide audience, it also became for both contemporaries the posterity Fermat's central work on the method of maxima and minima.

The *Method* was ill suited to that role. Its lack of clarity was in part deliberate. Fermat had originally written it sometime around 1629, purposely restricting his presentation to the statement of an algorithm, a mechanical procedure bereft of any justification or theoretical foundation. Indeed, Descartes opened the dispute with an attack on its adequacy as an algorithm, particularly with respect to the determination of tangents. Only in the course of the dispute did Fermat find himself forced to provide the proof of validity he had omitted from the *Method*. Only some ten years after first setting down the bare algorithm did he record the reasoning that lay behind its invention. Hence, to understand the development, both historical and conceptual, of the method of maxima and minima and the method of tangents derived from it, one must look beyond the *Method* to Fermat's other, later accounts of that method. But doing so requires that one separate the chronological order in which Fermat recorded his thoughts from the order in which they occurred to him. This requirement takes on added importance from the fact that, at about the same time that Fermat was providing the original justification for the algorithm of the *Method*, he was also revising the foundations of his method in the light of new insights gained from his reading of Descartes' *Geometry*.

Fermat's extant works contain four major papers devoted to the method of maxima and minima and several others dealing with the method of tangents.[2] With respect to the dates of composition of the first four, the *Method* is the earliest. A second, untitled paper, beginning "Volo meâ methodo . . . ," was probably written in the midst of the dispute, sometime around April 1638.

---

moved from Toulouse in 1636. But, in a letter to Roberval dated 22.IX.1636, Fermat wrote: "Sur le sujet de la méthode *de maximis et minimis*, vous savez que, puisque vous avez vu celle que M. Despagnet vous a donné, vous avez vu la mienne que je lui baillai, il y a environ sept ans, étant à Bordeaux." At no time during the period 1637–1638 is any text other than the *Method* mentioned as a source of Fermat's method. Hence, by Fermat's own testimony, it seems likely that the *Method* was composed while he was still in Bordeaux; was deposited with Despagnet, who then, at the request of Fermat in 1636, turned it over to Roberval, or at least made it available to him; and then, toward December 1637, was turned over to Mersenne for transmission to Descartes. The only hint provided by the actors themselves is Descartes' reference to "un Conseiller de ses amis qui vous l'a donné pour me l'envoyer." Despagnet was a *conseiller* of the Parlement of Bordeaux; neither Roberval nor Carcavi held that office, though Carcavi had held it previously.

[2] Cf. FO.I.136–169, which includes eleven memoirs other than the *Method;* Fermat's correspondence for the years 1637–1638 in FO.II; and FO.*Suppl.* 72–86, 120–125. Two of these memoirs occupy a central place in the present chapter. Material from the others will be found in appropriate footnotes.

The third paper, *Analytic Investigation* [*of the Method of Maxima and Minima*] (hereafter, *Analytic Investigation*),[3] has been dated 1643 or 1644; for reasons to be given below, however, 1639 or 1640 seems a more likely date. The fourth, a letter to Pierre Brûlart de Saint-Martin, bears the date 31 March 1643. Yet, with respect to the ideas contained in these papers, the *Analytic Investigation* undoubtedly records the earliest form of the method of maxima and minima. Indeed, we will argue that that is precisely the purpose for which Fermat wrote it, to wit, to present the original theoretical foundations of the algorithm of the *Method*. The unfortunate aftermath of the dispute with Descartes explains why he should have done so at such a late date. The memoir "Volo meâ methodo . . ." retains its role as an explication of the *Method*, though it must now be viewed against the background of the *Analytic Investigation*. The letter to Brûlart, however, as the last account of the method, also represents an entirely new approach to it, one which illustrates the growth of Fermat's ideas after his exposure to a wider mathematical world.

By contrast, the method of tangents derived from the method of maxima and minima presents no such difficulties of chronological order. Rather, it offers the difficulty of looking back over the watershed constituted by the *Introduction* to see how Fermat first came upon the algorithmic presentation in the *Method* at a time when he could not yet express curves in terms of equations. If the *Analytic Investigation* provides the original foundations of

---

[3] Ed. FO.I.147–153. There the memoir bears the title "Methodus de maxima et minima," provided by the editors. Of the memoir Tannery says, "Cet important morceau a été conservé par une copie de Mersenne, aujourd'hui perdue elle-même, mais dont il subsiste deux transcriptions de la main d'Arbogast: l'une au net (Manuscrit du prince Boncompagni), l'autre en brouillon (Bibl. Nat., *Fonds français*, 3280, nouv. acq.), qui a servi à M. Ch. Henry pour le texte qu'il a donne: *Recherches sur les manuscrits de Pierre de Fermat* (Rome, 1880), pages 180–183," (FO.I.147, n. 2). Since Tannery wrote that in 1891, two other copies have come to light, one in MS Florence, ff.93^v-96^r, and another in MS Ricci, ff.47^v-51^v. Both of these copies bear the title given in the text above, and hence one may assume that this was the original title. The presence of the treatise in two Italian MSS. is explained by Mersenne's journey to Italy in 1644-1645. Showing his usual lack of common sense, Mersenne carried the unique original with him; writing to Torricelli from Rome on 10 January 1645 (FO.IV.85), he warned, "Quaero te vero ne perdatur illa charta syncriseos et anastrophes, ne, si pereat exemplar primum, illo semper tractatu careamus." The tract apparently got lost in the mails for a time, because Torricelli replied he had never seen the copy, but then in February acknowledged receipt of it. Torricelli turned it over to Ricci for copying. Cavalieri also saw the memoir at this time and may have been responsible for the Florence copy.

Mersenne's journey sets the only reliable *terminus ad quem* for the treatise. The conclusion of the treatise sets a *terminus a quo*; indirect reference to the dispute between Fermat and Descartes places its composition after the summer of 1638. Fermat wrote two letters to Mersenne in 1639 on the subject of syncrisis (see below) which forms the basis of the *Analytic Investigation*. Moreover, he had, by 1643, an entirely new version of his method (see below). These two pieces of evidence, combined with documentation of pressure being placed on Fermat during 1639 by Descartes and Mersenne to demonstrate the method, suggest that the *Analytic Investigation* was probably written and sent to Paris sometime around 1639 or 1640.

the method of maxima and minima, it says nothing of the method of tangents. It does not because Fermat felt he had already clarified the relations between the two methods in the course of the dispute with Descartes. Several papers written during the period 1637–1638 contain that clarification, though all of them treat curves in the form of equations. Hence, though the historical problems posed by the derivative method are no less difficult, they are at least relatively straightforward.

The confusion of chronological and conceptual order in Fermat's papers on the method of maxima and minima, however, makes the narrative argument of this chapter a complex one. The sources must serve a dual purpose: first, to give an historical account of the origins and development of the method, together with its corollary, the method of tangents; and, second, to justify the manner in which the sources themselves are here used to establish that account. The *Analytic Investigation* forms the crux of the argument and makes it complex. For the argument rests in part on the theory that Fermat wrote that paper to record old ideas and that, although it was written sometime around 1640, it in fact represents the foundations of the method as conceived in the late 1620s. The technical and personal issues raised by the dispute with Descartes provide strong support for that interpretation, but those issues in turn require a clear understanding of both the *Method* itself and its theoretical underpinnings. Hence, the argument turns back on itself. To break the vicious circle, the discussion to follow will begin with an exposition of the *Analytic Investigation* and the *Method*. Having established the procedures and concepts of Fermat's original method, it will then examine the Fermat-Descartes controversy in some detail, giving equal attention to the technical issues and to the personal feelings involved in that affair. The controversy will then provide justification for placing the *Analytic Investigation* at the very beginning of Fermat's thoughts on the subject of maxima and minima, at the same time that it sets up the context for his subsequent alteration of the theoretical foundation of the method.

Broader questions than mere chronology of ideas are involved here. One unhappy historical result of the elevation of the *Method* to the rank of prime expression of Fermat's method of maxima and minima has been a systematic misinterpretation of its theoretical foundations. Taken alone, its vague and laconic description comfortably conforms to an anachronistic reading in terms of concepts developed long after Fermat's work. Almost every secondary account of it sees in it some form of infinitesimal calculus or the method of limits. By contrast, the *Analytic Investigation* shows clearly that it contains no such thing, that it rests on purely finite algebraic concepts derived from the theory of equations. Clarity on this point then places the two later versions of the method of maxima and minima in a similarly different light, for they too are devoid of infinitesimal or limit considerations. Moreover, clear insight into the conceptual foundations of the *Analytic Investigation* and its derivative *Method*, combined with a more accurate chronology of those foundations, unravels a further source of confusion prevalent in the secondary literature about the method of maxima and minima. Historians have tended to view Fermat's method as being of a piece and hence have tried to reconcile the various versions of the method contained in his works. They have had difficulty doing so, and for good reason. As will become clear below, the versions are not all reconcilable; they represent different views that Fermat held at different times. Those few historians, moreover, who have recognized differences in Fermat's accounts have been misled by an incorrect dating of the content of the *Analytic Investigation*. All have failed to discern the real and essential novelty of Fermat's last version of the method, that contained in his letter to Brûlart of 1643.[4]

To return to the beginning of the story, however, we must go back to Fermat's earliest research in Bordeaux at the end of the 1620s.[5] He himself takes us back there in his *Analytic Investigation*, written, as we shall presently see, to blunt Descartes' accusation that the methods of the *Method* were the lucky result of trial and error.

II. THE ROOTS OF AN EQUATION AND THE ROOTS OF A METHOD

A. *Viète's syncrisis*

The sting of Descartes' charges that Fermat operated *à tâtons* and *par hasard* in obtaining mathematical results put the Toulousain in an uncommonly open, discursive mood when he sat down to write the *Analytic Investigation*. For one of the very few times in his career, he looked to his memory for the substance of what he had to say. Thinking about his inven-

---

[4] In a very real sense, secondary accounts of Fermat's method of maxima and minima and his method of tangents go all the way back to the Fermat–Descartes controversy. Every modern history of mathematics that treats either the seventeenth century or the invention of the calculus, or both, has discussed Fermat's methods. Such accounts tend to be largely expository and more or less misleading, depending on the extent to which the author reads infinitesimals or limits into Fermat's procedures. Two more detailed investigations attempt to probe the conceptual foundations of the method, as well as the problem of the proper chronology of concepts: Heinrich Wieleitner, "Bemerkungen zu Fermats Methode der Aufsuchung von Extremwerten und der Bestimmung von Kurventangenten" (with a mathematical note by J. E. Hofmann), *Jahresbericht der deutschen Mathematiker-Vereinigung*, 38 (1929), pp. 24–35; and Per Strømholm, "Fermat's Methods of Maxima and Minima and of Tangents. A Reconstruction," *Archive for History of Exact Sciences*, 5 (1968), pp. 47–69. Neither Wieleitner nor Strømholm succeeds, in my opinion, in his attempt, and much of what follows below is greatly at variance with each and both of those articles. Rather than carry on a debate with these authors, I shall leave it to the reader to compare the various interpretations.

[5] Cf. above, n. 1. Also "Méthode expliquée": "Je désire seulement qu'il [Descartes] sache que nos questions *de Maximis et Minimis* et *de Tangentibus linearum curvarum* sont parfaites depuis huit ou dix ans et que plusieurs personnes qui les ont vues depuis cinq ou six ans le peuvent témoigner." (VI.1638, FO.II.162).

tion of a method of determining extreme values, he recalled: "While I was pondering Viète's method of syncrisis and anastrophe and was exploring more accurately its use in discovering the structure of correlate equations, there came to mind a new method to be derived from it for finding maxima and minima, by means of which some doubts pertaining to *diorismos*, which have caused trouble to ancient and modern geometry, are most easily dispatched."[6]

By Fermat's own testimony, then, the method of maxima and minima derived from two sources, Viète's theory of equations and the ancient problem of *diorismos*.[7] Pappus' *Mathematical Collection* provided the main source for the latter. That Viète's *Analytic Art* should provide the key to the *Collection* was an article of mathematical faith for Fermat and other followers of Viète's analytic program.[8] Following Fermat's hints, we must look to Viète's method of syncrisis and then to the nature of Fermat's encounter with *diorismos* in Book VII of the *Collection*, if we seek to discover the origins of the method of maxima and minima.

Chapter XVI of Viète's *On Recognition of Equations*[9] presents a method for combining two similar equations in order to obtain expressions for the constants of the equations in terms of their roots. The method and the text to which it belongs form part of Viète's brilliantly original, but (as befits a professional lawyer) frustratingly casuistic theory of equations.[10] Fond of coining Latin neologisms from Greek, Viète called the method *syncrisis*,

---

[6] *Anal.Inv.* (FO.I.147).

[7] A *diorismos* is, in classical Greek geometry, a subsidiary limiting condition that must be added to the statement of a problem in order to guarantee its solvability in general terms. The simplest example in Euclid's *Elements* (I,22) requires that, for any three lines set for construction into a triangle, the sum of any two must be greater than the third. For further details, see Mahoney, "Another Look at Greek Geometrical Analysis," *Archive for History of Exact Sciences*, 5(1968), pp. 327-329.

[8] See above, Chap. II.

[9] *Francisci Vietae Opera Mathematica in unum volumen congesta ac recognita Operâ atque studio Francisci à Schooten Leydensis Matheseos professoris* (Leiden, Elzevier, 1646), pp. 104ff. *De recognitione aequationum* was the first of two treatises forming the work *De Aequationum recognitione et emendatione. Tractatus duo quibus nihil in hoc genere simile aut secundum huic aevo hactenus visum* first printed in Paris in 1615 under the editorship of Viète's Scots student Alexander Anderson. The work was probably written sometime during the early 1590s. For a brief précis of the treatise, see Frédéric Ritter, "Francois Viète (1540-1603), inventeur de l'algèbre moderne. Essai sur sa vie et son oeuvre," *La revue occidental* (1905), pp. 389-401.

Anastrophe is the subject of Chap. III of *De emendatione aequationum* (*Opera*, 1646, pp. 134-138). It employs a means very similar to syncrisis to move from the solution of an equation like $bx - x^3 = c$ to that of $x^3 - bx = c$. Despite Fermat's reference to *anastrophe*, the technique of *syncrisis* formed the basis of his method of maxima and minima and hence it alone is discussed in the present chapter.

[10] A detailed history of the theory of equations during the sixteenth and seventeenth centuries ranks high on the list of desiderata in the history of mathematics. As I have suggested in Chap. II above, the development of that theory may well have been the main thrust of mathematical research at the time and formed the contemporary link between the analytical treatises of the Greeks and the symbolic algebra of the moderns.

meaning "combination," "composition," or "comparison," and defined it as "the collation with one another of two correlate equations in order to discover their structure."[11] "Two equations are understood to be correlate," the text continues, "when they are both similar and furthermore consist of the same given magnitudes, both in terms of smaller degree and in the wholly constant terms. Nevertheless, the roots are different, because either the formulas of the equations by their nature may be satisfied by two or more roots or the quality, or sign, of the terms differs."[12] Viète made further distinctions concerning such pairs of equations, but they are superfluous to the present discussion.

Like most innovators, Viète failed to take full advantage of his own brainchild. Words confuse and obscure, where use of symbolic notation provides clarity. Descartes saw this clearly in his *Geometry*. He offered contemporaries relief from the verbosity of the *Analytic Art*; his notation can do us the same favor without disguising the substance of Viète's method. Viète uses the following example to explicate his method of syncrisis:

B *parabola* in A *gradum*−A *potestate* aequatur Z *homogeneae*,

that is,

$$b^{n-m}x^m - x^n = c^n,$$

where the constants b and c carry superscripts only to denote their dimensionality in keeping with the law of homogeneity.[13]

This equation may serve as the first of a pair of correlate equations if we take the x to denote one of the roots of the equation. If we then take $y < x$ as a second root, the second correlate equation is $b^{n-m}y^m - y^n = c^n$. To carry out syncrisis we equate the left-hand sides of the two equations:[14]

$$bx^m - x^n = by^m - y^n,$$

or

$$b(x^m - y^m) = x^n - y^n.$$

We may now solve directly for b:

$$b = \frac{x^n - y^n}{x^m - y^m}.$$

Substitution of this value for b into one of the original equations then yields

[11] *Opera* (1646), p. 104: "Syncrisis est duarum aequationum correlatarum mutua inter se ad deprehendendum(!) earum consitiutionem collatio."

[12] *Ibid.*, p. 105: "Due autem equationes correlatȩ intelliguntur, quum ambe similes sunt, & praeterea iisdem datis magnitudinibus constant, sive adfectionum parabolis, sive adfectionum homogeneis. Radices tamen ideo diversae sunt, quoniam vel ipsae aequationum formulae de duabus pluribusve radicibus ex sui constitutione sunt explicabiles, vel in iss diversa est adfectionum qualitas, seu nota."

[13] See above, Chap. II.

[14] Now that the point regarding homogeneity has been made, we can adopt the more convenient style of notation.

an expression for c:

$$c = \frac{x^n y^m - x^m y^n}{x^m - y^m} .$$

The method of syncrisis shows, then, how the constants of the one original equation are related to two of its roots.

The far-reaching implications of the above procedure emerge clearly from the example *in specie* that Viète provides. In the case where $n = 2$ and $m = 1$, syncrisis yields:

$$b = x + y$$

$$c = xy,$$

that is, in this particular quadratic equation, the coefficient of the x term is equal to the sum of the roots of the equation and the constant term to their product. It is a simple and direct matter to extend the method for equations of higher order—namely by taking as many correlate equations as there are roots—and thereby to generate the elementary symmetric functions of those equations.

B. Pappus and the Uniqueness of Extreme Values

The keystone of Viète's method of syncrisis was the assumption, at the time rapidly becoming fundamental to algebra, that every equation of degree greater than 1 has more than one solution and that, indeed, an nth degree equation may have as many as n roots.[15] Syncrisis revealed how those several roots were related to the constants of the equation. As Fermat "pondered" Viète's method, this basic multiplicity of roots may have triggered a reaction to another text he appears to have been studying carefully. For hand in hand with Viète's art of analysis went the ancient treatise to which it was mainly addressed, Book VII of Pappus' *Mathematical Collection*.[16] In the *Analytic Investigation* Fermat refers to Proposition 61 of

[15] Any proof of the Fundamental Theorem of Algebra, of which Gauss's was the first, assumes the existence of the field of complex numbers as well as the field of real numbers. Such fields themselves are commonly defined in terms of families of solutions to equations. Historically it appears that such fields were constructed precisely so that an nth-degree equation could have n roots, and such construction represents a conscious inventive effort by mathematicians rather than the "discovery" of a property inherent in equations. In Viète's time and for some time thereafter, mathematicians contented themselves with the possibility of as many as n roots but did not require that every equation have exactly n roots. Descartes was among the first to postulate n roots and to fill out the absence of real roots with those he called *imaginaires*. Cf. *Géométrie* (Lieden, 1637), p. 380. For other contemporary statements of the theorem, see Hofmann, *Gesch. d. Math.*, I, p. 166.

[16] Critical Greek-Latin edition by Friedrich Hultsch, *Pappi Alexandrini Collectionis quae supersunt* (Berlin, 1877); French trans. Paul VerEeck (Paris, 1933). Fermat primarily used the Latin translation by Commandino (Pisa, 1588), though he had access to a Greek text, perhaps the one Mersenne cites as available in Paris (cf. *Universae*

that Book and to Commandino's commentary on it. From the importance Fermat attaches to his analytic solution of the problem posed by the proposition—the solution is supposed to crown the memoir—it seems more than likely that he in fact devised the method of maxima and minima in seeking it.

Proposition 61 of Book VII of the *Collection* reads:[17]

Given the three lines AB, BC, CD, if one sets rectangle ABD to rectangle ACD as the square on BE to the square on EC, the ratio is unique and the least [ratio] of rectangle AED to rectangle BEC. And I say it is the same

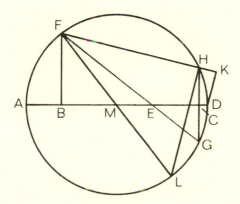

[ratio] as that of the square on AD to the square on the excess by which the straight line that can contain the rectangle AC·BD exceeds that which can contain the rectangle AB·CD.

Pappus' theorem, couched in the terminology of Greek geometry, is clarified by Fermat's statement of the problem to which it offers a solution:[18]

Let BDEF be a straight line, in which points B,D,E,F are given. The point N

---

geometriae mixtaeque mathematicae synopsis, Paris, 1644; Preface). To keep our discussion in historical context, we will cite and translate from the Commandino edition.

Book VII of the *Collection* dealt exclusively with geometrical analysis and was the focus of seventeenth-century interest in Pappus' work. See above, Chap. II.

[17]Pappus, *Collection*, trans. Commandino (Bologna, 1660), p. 295: "Tribus datis rectis lineis AB, BC, CD si fiat ut rectangulum ABD ad rectangulum ACD, ita quadratum ex BE ad quadratum ex EC, singularis proportio, & minima est rectanguli AED ad rectangulum BEC. Itaque dico eandem esse, quae est quadrati ex AD ad quadratum excessus, quo recta linea, quae potest rectangulum contentum AC, BD excedit eam, quae potest contentum AB, CD." Commandino here follows Pappus in the common Greek notation *rectangulum* ABD, meaning the rectangle with adjacent sides AB and BD.

[18]*Anal. Inv.*, FO.I.151. Format follows here the classical notation for geometrical products: "rectangle DNE" is the product of the line segments DN and NE.

is to be taken between points D and E such that the rectangle BNF has the minimum ratio to the rectangle DNE.

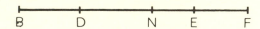

What caught Fermat's eye was the phrase "unique and the least." Commandino had already singled out the phrase in his commentary to the proposition. He cited the Greek, ὅ μοναχὸς λόγος καὶ ἐλαχιστός ἐστιν ὁ ὑπὸ αεδ πρὸς τὸ ὑπὸ βεγ, and then added, "One cannot satisfactorily perceive what is signified by these words, nor what one should understand by *monachos* and *epitagma* in these lemmas, since we lack the books of Apollonius in which they are written.[19]

Fermat understood. *Epitagma* he took to mean "condition" in the same sense as *diorismos*. *Monachos* means "unique," "singular"; *elachistos* means "least."[20] Linguistically there was no problem. But then the mathematical sense of the phrase dawned on him, perhaps after he had already attacked the problem algebraically. The translation of the problem into algebraic symbols leads to a quadratic equation. And a quadratic equation normally has two roots, corresponding geometrically to two section points which satisfy the demands of the problem. Fermat took the leap called for by Pappus. There is only one minimum ratio. Hence, there is only one section point at which the line can be divided to yield that ratio. Hence, the quadratic equation that expresses the condition of the minimum can have only one root.

Now the questions that were to lead to his method of maxima and minima lay clear in his mind: how can one manipulate an equation so as to yield a

[19]Pappus, trans. Commandino, p. 296: "Singularis proportio, & minima est rectanguli AED ad rectangulum BEC. Graecus codex ὅ μοναχὸς λόγος καὶ ἐλαχιστός ἐστιν ὁ τοῦ ὑπὸ αεδ πρὸς τὸ ὑπὸ βεγ, quibus verbis quid significetur, quidque per monachos, & epitagma in his lemmatibus intelligat, satis percipi non potest, cum Apollonii libris careamus, in quibus ea conscripta sunt."

[20]For the meanings of these terms in Greek mathematics, see entries under each in Charles Mugler, *Dictionnaire historique de la terminologie géométrique des Grecs* (Paris, 1958–59). In a memoir written sometime in the late 1630s (De Waard says April 1638; FO.*Suppl.*xvi), beginning "Volo meâ methodo . . ." (FO.I.140-147, under the title *Ad eamdem methodum*), Fermat noted (p. 142): "Ut pateat hujus methodi certitudo, desumam exemplum e libro Apollonii *De determinata sectione*, qui, ut refert Pappus initio septimi libri, difficiles determinationes habebat; et eam quae sequitur difficilimam esse existimo, quam ut inventam supponit Pappus septimo libro, nec enim illam veram esse demonstrat, sed, ut veram supponens, alias inde consequentias deducit. Hoc loco Pappus vocat minimam proportionem μοναχὸν καὶ ἐλάχιστον, *minimam et singularem*, ideo scilicet quia, si proponatur quaestio circa magnitudines datas, duobus semper locis satisfit quaestioni, sed, in minimo aut maximo termino, unicus est qui satisfaciat locus: idcirco Pappus vocat *minimam et singularem*, id est unicam, proportionem omnium quae proponi possunt minimam. Commandinus hoc loco dubitat quid per intelligat Pappus, et veritatem quam modo explicui ignoravit."

unique root, and, since a quadratic equation has two roots, what happens to the second root in the case of a maximum or minimum? The second question engaged his attention first. To ease his investigation, Fermat turned to the simplest and best known example of an extreme value, Euclid VI,27, the *diorismos* that limits the Greek geometric solution of quadratic equations. Conveniently, it too involved a quadratic equation.

Fermat recorded his investigation in the *Analytic Investigation*. The problem of dividing a line AB (= b) into two segments such that their product is a given quantity translates algebraically into the quadratic equation

$$bx - x^2 = c.$$

Geometrically, the section problem requires a *diorismos*, since the value of the product cannot be arbitrarily large. That *diorismos* involves a maximum, and Fermat already knew what it was; by Euclid VI, 27 the product of the segments cannot be greater than $\frac{1}{4} AB^2$.[21] The corresponding value for x in the algebraic formulation is $b/2$. "The point satisfying the proposition cuts the given straight line in half," he says, "and the maximum rectangle is equal to the fourth part of the square of B; from no other section of this straight line will there result a rectangle equal to the fourth part of the square of B."[22] But the problem is still a quadratic one, and there should be another solution. Where is it? Fermat looked at the general problem again.[23]

But, if it is proposed to cut the same straight line B under the condition that the rectangle contained by its segments is equal to a plane Z (which is to be supposed less than the fourth part of the square on B), then two points satisfy the proposition, and the point of the maximum rectangle lies between them.

What about these two points, or two roots? Applying Viète's method of syncrisis, Fermat determined that their sum is always equal to B, which of course is obvious from the geometrical picture. But now the sum of the roots interested Fermat less than their difference. What happened to the difference between the roots as Z approached the maximum value? Maybe then Fermat could discover what happened to the second root at the maximum value.[24]

[21]Euclid, trans. Heath (*The Thirteen Books of Euclid's Elements*, Repr., N.Y., 1956), VI, 27: Of all the parallelograms applied to the same straight line and deficient by parallelogrammic figures similar and similarly situated to that described on the half of the straight line, that parallelogram is greatest which is applied to the half of the straight line and is similar to the defect.
[22]*Anal. Inv.*, FO.I.148. In quoting directly from Fermat, we preserve his Viètan notation, translated into lower case in the preceding discussion.
[23]*Ibid.* The last part of the statement is not proved; Fermat takes the fact to be intuitively obvious.
[24]*Ibid.*, pp. 148–149. In regard to the discussion on pp. 156–157 below, it is important to remember that *evanescere* held no special meaning for Fermat. It was the common

If, in place of plane Z, another plane is taken which is greater than plane Z but less than one fourth of the square of B, then the straight lines A and E [i.e. x and y $(=b - x)$)] will differ from each other by less than the ones above, since the points of division will approach more closely to the point constitutive of the maximum rectangle. And, as the rectangles of the divisions increase, the difference of these A and E will decrease, until with the last division of the maximum rectangle it disappears, in which case a μοναχός or unique solution will hold, since the two quantities will be equal, i.e. A will equal E.

The geometrical situation made clear what Fermat could not discern from the equation itself. Even for the maximum value of c (i.e. Z) the equation $bx - x^2 = c$ had two solutions, but the two solutions appeared to be one because they were equal. Geometrically, the two segments determined by those solutions were of equal length, and hence their endpoints coincided at the midpoint of the line. Once Fermat had this insight, the result of syncrisis also yielded the extreme value almost by inspection. Since the sum of the segments, or roots, is always equal to b, in the case of the maximum each root was equal to $b/2$ and their product to $b^2/4$. In Fermat's own words,[25]

Since, therefore, in the above two correlate equations,[26] by the method of Viète, B will equal $A + E$, if E is equal to this A (which appears to hold always at the point constitutive of a maximum or minimum) then, in the proposed case, B will equal $2A$; that is, if the straight line B is cut in half, the rectangle contained by its segments will be a maximum.

Fermat now had a method for determining maxima and minima. For he now saw how to manipulate an equation to obtain a repeated root. To use another example from the *Analytic Investigation*, suppose one must determine $\max(bx^2 - x^3)$.[27] One begins by generalizing the problem to that of

---

Latin word for "to disappear." Only when used by Newton, did the word take on special mathematical significance, i.e. something akin to "approach to a limit."

At this point, the problem of notation again obtrudes on our narrative. Throughout his memoirs on maxima and minima and on the rule of tangents, Fermat remained loyal to the notation of Viète. But that notation hides more from the modern reader than it reveals. As always, a compromise is necessary. When quoting directly from Fermat, I shall keep his Viètan notation; while discussing the mathematics of his method, however, I shall employ the modern Cartesian x, y notation. The reader will have no difficulties with changing notation if he bears in mind that Fermat's A is *always* rendered by x, his E by y, throughout this chapter. Otherwise, with few exceptions, which will be noted, the only other change in notation is the shift from upper-case to lower-case consonants to denote constants.

[25]*Anal. Inv.*, FO.I.149.

[26]I.e., $bx - x^2 = c$ and $by - y^2 = c$, where c (i.e. Z) now is taken to be the maximum rectangle.

[27]The particular example may not have been chosen at random. It belongs to the famous *diorismos* attached to Archimedes' solution of the problem in *Sphere and*

solving the equation $bx^2 - x^3 = c$ for *any* c less than that maximum. The equation will have at least two roots, say x and y. By syncrisis,

$$bx^2 - x^3 = by^2 - y^3$$
$$b(x^2 - y^2) = x^3 - y^3$$
$$bx + by = x^2 + xy + y^2.$$

This relationship holds for any two roots of the original equation. Moreover it holds independently of the particular value of c. But, at the maximum value of c, the roots will be equal to one another; hence, for that maximum value of c, $2bx = 3x^2$, or $2b = 3x$.[28]

Cylinder, II, 4; that solution and its *diorismos* received extended attention from Eutocius in his commentary to the proposition. Cf. Mahoney, "Another Look," pp. 339–340.

[28] Two interesting mathematical sidelights arise here which, though they would interrupt the narrative above, should not be ignored. First, as a minor point, one should note that, although Fermat in the particular problems he treats has from syncrisis a direct expression for the maximum itself (e.g. $c = xy$, where $x = y = b/2$), he never explicitly computes that maximum. That is, he determines what value of x will maximize $bx - x^3$, but nowhere in his memoir, or in any other memoir, does the reader learn what the maximum value itself is. The reason for this should be clear. Fermat is dealing with geometrical construction problems, where what is required is the section point or line segment that yields the maximum or minimum, not the extreme value itself. The beauty and simplicity of his method lie in the fact that it yields the desired section point or line length directly.

Second, there is about the method an unfortunate ambiguity that Fermat nowhere resolves. Take his solution of max $(bx^2 - x^3)$. Note that he moves directly from the equation $2bx = 3x^2$ to the final result $2b = 3x$ by discarding the solution $x = 0$. To him, of course, that last solution seemed obviously extraneous, since it led to no value at all for the solid to be constructed. What Fermat failed to see is that $x = 0$ yields the *minimum* value for the expression (albeit a local minimum). Similarly, the final equation for the maximum of $bx - x^3$ is $b = 3x^2$, to which Fermat simply says, "which equation will give the desired maximum solid." It will if one takes the positive root of that equation; the negative root, however, will yield a minimum. Again, the original geometrical purpose of the derivation dictates the choice of root. A negative value for x makes no geometrical sense. The final equation for the minimum proportion mentioned above does force Fermat to consider the question of two solutions. Of the equation $DZB + A^2D - A^2Z + A^2B = 2DZA$ he says (FO.I.152): "Nec morabitur Analystam ultimae istius aequalitatis ambiguitas: nec prodet quippe se, vel invito, latus utile. Imo et in aequationibus ambiguis quae plura duobus habent latera, non deerit solitum ab utraque hac nostra methodo, sagaci tantisper Analystae, praesidium."

Fermat, then, dismisses the problem posed by two or more solutions to the final equation obtained by his method. In the cases he treats, the geometry of the situation comes to his aid. But the geometry blinded him apparently to a serious ambiguity in his conception of the method. He never makes clear whether he takes "uniqueness" to mean that *all* the roots of the equation stating the maximum condition are equal (which, for equations of the third and higher degrees is not true) or merely two of the roots are equal (which is correct). His method works, because whichever of the two bases he had in mind, he in fact equated only two of the roots. That is, in generalized form, if M is a maximum of minimum value of $P(x)$, then the equation $P(x) - M = 0$ is of the form $(x - a)^2 R(x) = 0$, where a is the value of x yielding the extreme value and $R(a) \neq 0$. Fermat's method itself obtains the value of a. Depending, however, on how many local extreme values $P(x)$ has, there will be several a's. Fermat never seems to have discerned this fact clearly. Indeed, even in the most sophisticated version of his method, Fermat never saw the distinction between absolute and local extreme values, with the result that

We must pause here for a moment to emphasize an important aspect of Fermat's line of thought. Secondary accounts of Fermat's method generally interpret the passage cited on page 154 as a consideration of a limit. In an obvious sense, it is just that. As the value of Z approaches the maximum, the two roots A and E approach equality, or their difference approaches 0. But the careful reader should blind his modern mathematical eye. Whatever the similarities of this passage with modern discussions of approach to a limit, the differences have greater weight. In the first place, Fermat does not speak of the difference between the roots approaching 0; at the maximum, the difference *is* 0. Secondly, and most importantly, one must be clear about what Fermat is trying to discover in the passage. *His* question is: what happens to the other root when, for the maximum value of Z, the equation appears to have only one root? By looking to the geometrical picture in this passage and following the two roots as Z approaches its maximum value, he finds the answer: the other root is still there, but it is equal to the main root. It may be a bad pun, but the roots of Fermat's method of maxima and minima lie in the domain of the finite theory of equations and not in any consideration or introduction of infinitesimals or limits.

The genesis of Fermat's method explains an obvious contradiction inherent in it. After dividing by $A - E$, Fermat sets $A = E$. He has divided by 0! To save the algorithm that Fermat derived from his original method and set down in the *Method*, modern mathematics must rely on the sophisticated theory of limits of Cauchy. Fermat's understanding of the situation was more naïve. The method relied on a false assumption[29] and on the full generality of Viète's theory of equations. In order to apply the method of syncrisis in the case of a maximum or minimum, he falsely (though intentionally) assumed that the equation had at least two *different* roots. He held that the false assumption would be vindicated by the fact that the relationships discovered by syncrisis were fully general and held for all particular values of the parameters of the equation. That is, in the equation $bx - x^2 = c$, the relationships $b = x + y$ and $c = xy$ held, *whatever the particular values of b and c*. And, in the event that c is the maximum value of the expression $bx - x^2$, Fermat knew that the relationship $b = x + y$ still held true, despite the fact

---

the proof of the method in the Letter to Brûlart (see below) is vitiated by that very distinction. His failure to see the distinction, and hence to react to it explicitly, makes it impossible to resolve for ourselves the ambiguity surrounding his conception of the method.

[29]The false assumption, or *regula positionis falsae*, represented a standard problem-solving technique from the time of the Babylonians on up to cossist algebra. In its simplest form, it takes a linear equation $L(x) = Q$ and posits an arbitrary solution a. If $L(a) \neq Q$, then the correct solution results from the proportion $x : a = Q : L(a)$. There was also a technique of "double false position" for more difficult problems. Fermat, of course, uses the technique in an entirely different sense here; indeed, he uses it meta-mathematically. His false assumption is a counterfactual assumption about the the nature of the roots of an equation.

that then $x = y$, or $x - y = 0$. In his own mind, Fermat saw no division by 0. The division by $x - y$ belonged to the *general* examination of the equation; setting x equal to y belonged to the consideration of a *particular* case. Seeing his way clearly past this hurdle, Fermat easily took the next step in the development of his method, a step that only emphasized the apparent paralogism.

C. Improving Syncrisis and the Method

"Exploring more accurately" Viète's method of syncrisis within the theory of equations per se, Fermat discovered a minor inconvenience which that theory imposed on the analyst when applying it to the solution of equations. For, besides exposing the relationships inherent between the roots of an equation and its constants, syncrisis could perform a corollary service. In a note written in 1639, but again clearly relating mathematical research carried out years earlier, Fermat described that corollary service and a minor emendation he had made to it, which in turn caused an emendation in his original method of maxima and minima.[30] To use his example, suppose that[31]

$$abx - bx^2 - x^3 = c$$

and that one root of this equation is $x = r$. Then r satisfies the equation, that is,

$$abr - br^2 - r^3 = c.$$

By the method of syncrisis, one may set

$$abx - bx^2 - x^3 = abr - br^2 - r^3$$

or

$$ab(x - r) = b(x^2 - r^2) + (x^3 - r^3).$$

Following division by $x - r$, the last equation reduces to

[30]Contained in letter Fermat to Mersenne, 1.IV.40, FO.II.XXVIII[bis], pp. 187–188. The body of the letter contains number-theoretical material and, when printed in the *Varia opera* of 1679 (pp. 173–176), it did not contain this note on syncrisis. The editors of FO clearly indicate that the note derives from a different source from that of the body of the letter, to wit, from the manuscript of Vicq d'Azyr (later Boncompagni), but they follow Vicq in placing it in the letter. DeWaard, however, (CM.IX.243), questions whether the note does not properly belong in the letter written in February 1639: "Fermat avait promis cette méthode à la fin de sa lettre du 20 février 1639. Le fragment actuel aurait-il été écrit dans l'original sur un feuillet detaché, que Vicq d'Azyr aurait mis ici à tort, et qui devrait être placé peu après cette promesse?" In all likelihood, DeWaard is correct. The note does not fit the rest of the 1640 letter; indeed, it interrupts the flow of that letter. FO records only one letter from Fermat for the whole of 1639, which suggests that much may be missing from this year, including the letter in which this note had its original place.

[31]The equation as given in FO.II.187 reads: $bda - ba^2 - a^3 = z$. The editors note that as such it probably does not stem from Fermat, who would have written $BDA - BAq - Ac$ aeq. Z. In shifting to Cartesian notation, I have also altered Z to c and D to a, in the first case to prevent confusion of a constant with an unknown and in the second to avoid the distraction of the combination dx.

$$ab = bx + br + x^2 + xr + r^2,$$

and, since r is given, one now has a quadratic equation which can be solved to yield the two remaining roots.

In theory this technique is easily applied, but Fermat saw a practical difficulty. Take the equation

$$x^3 - 9x^2 + 13x = \sqrt{288} - 15,$$

for which one of the roots, $3 - \sqrt{2}$, is given. The direct application of syncrisis involves dividing the equation

$$x^3 - 9x^2 + 13x = (3 - \sqrt{2})^3 - 9(3 - \sqrt{2})^2 + 13(3 - \sqrt{2})$$

by $x - 3 + \sqrt{2}$, which Fermat viewed as unnecessarily difficult. As he noted in the letter to Mersenne from which this example is taken, "the means that Viète used to solve similar problems, which he calls *syncrisis* in his treatise *On Recognition of Equations*, is defective and does not say everything."[32] He added: "I will give a general method for all similar solutions, which succeeds with no trouble and which does not have the defects of Viète's method, which is quite cumbersome due to the divisions [it entails], particularly for somewhat difficult examples like the one in question which common analysts cannot solve by syncrisis."

In the letter from which the first example above is taken,[33] Fermat redeemed his promise of an emendation. If r is one solution of the equation, he argued, then,

there are two lines equal to $A\,[x]$ and unequal to $N\,[r]$. Posit that one of these two lines is $N + E\,[r + y]$ and now form the equation as if $N + E$ were A; we will have

$$abr + aby - by^2 - y^3 - br^2 - 2bry - 3ry^2 - r^3 - 3r^2y = c.$$

Since $x = r$ is one solution,

---

[32] Fermat to Mersenne, 20.II.39, FO.II.179–181. This letter was written in response to one from Mersenne, now lost, in which Mersenne transmitted to Fermat (as he had also done to Descartes) three algebraic problems posed to him by a certain Dounot and his friends. No trace remains of the original proposal or of Mersenne's letter. In his letter Fermat repeats them in the following form: "Pour première question, proposé: $1C - 6N$ égal à 40 et la valeur d'$1N$, 4, et encore $1C + 4N$ égal à 80, où N est encore 4, ils demandent la méthode pour trouver la racine en pareilles questions sans aller à tâtons. . . . Ils proposent ensuite $1C - 8Q + 19N$ égal à 14, et après avoir déterminé que le problème est ambigu et donné trois valeurs de la racine, savoir 2, $3 - \sqrt{2}$, $3 + \sqrt{2}$, ils ajoutent: *Qui dederit quartam solutionem, portento erit simile*. . . . Voici la dernière question: $1C - 9Q + 13N$ aeq. $\sqrt{288} - 15$. *Quaeritur $1N$. Hoc problema recipit tres solutiones quarum exhibimus primam, scilicet* $3 - \sqrt{2}$, *quae satisfacit exacte. Si reliquas duas dederim, ero illis magnus Apollo.*" With a patience all his own, Fermat separated the wheat from the chaff in these proposed problems. To the first, he reminded Dounot and company that methods for solving cubic equations already lay at hand and warned them that failure to use them only led to a waste of valuable time and effort.

[33] Cf. n.30.

$$abr - br^2 - r^3 = c;$$

therefore,

$$aby - by^2 - y^3 - 2bry - 3ry^2 - 3r^2y = 0.$$

Dividing now by y, "which is a simple division and not a composite one like that of Viète and the others," one obtains

$$ab - by - y^2 - 2br - 3ry - 3r^2 = 0.$$

This quadratic equation then yields a value for y which, added to r, gives the desired second root. A repeated application, Fermat noted, would yield the remaining outstanding root.[34]

"I don't want to make too much of this, nor have I said everything in giving you this one example; it is merely to facilitate the operation [of the method]."[35] At the time, i.e. 1639, Fermat had indeed not said everything about his emendation of Viète's method of syncrisis; he had not yet committed the *Analytic Investigation* to paper, or at least had not yet dispatched it to Paris. But the emendation itself dated back to 1629[36] and the same adjustment that simplified application of syncrisis had also simplified the method of maxima and minima based on it. Fermat's improvement on Viète's method gave the final and characteristic form to Fermat's own new method. The *Analytic Investigation* describes the effect.

After deriving $\max(bx^2 - x^3)$ by his original procedure, Fermat noted; "Nevertheless, because this practice of division by binomials is rather laborious and often intricate, it has seemed convenient to compare the roots of the correlate equations to each other by their difference so that, in this way, the whole job is accomplished by a single application to (i.e. division by) this difference."[37]

Suppose now one must find $\max(bx - x^3)$. According to the method as outlined so far, one would take the correlate expression as $by - y^3$. But, Fermat says, "since E (just like A) is an uncertain quantity, nothing prevents its [i.e. the second root's] being called $A + E$." That is, the syncrisis of the two correlate equations assumes a slightly different form:

$$bx + by - x^3 - 3x^2y - 3xy^2 - y^3 = bx - x^3.$$

[34] Again (cf. n.28), Fermat seems unwilling to consider both solutions of a quadratic equation as equally valid. The quadratic equation above will yield two solutions for y; each one added to r will yield the two outstanding roots. A repeated application of the method is superfluous. Fermat's call for such a second application is a good indication that he had his method and its foundation on uniqueness in mind when writing this letter to Mersenne. The method had put him in the habit of taking only single solutions to quadratic equations.

[35] FO.II.188.

[36] It must have, since the emendation became the basis for the form of the *Method*, which Fermat himself dates from that time. Also, cf. Fermat to Mersenne, 26.IV.36, FO.II.5: "J'ai trouve aussi beaucoup de sortes d'analyses pour divers problèmes tant numériques que géométriques, à la solution desquels l'analyse de Viète n'eût su suffire."

[37] *Anal. Inv.*, FO.I.149–150.

Fermat's emendation now requires an additional step, to wit, cancellation of terms common to both sides of the equation. The result is

$$by - 3x^2y - 3xy^2 - y^3 = 0,$$

or

$$by = 3x^2y + 3xy^2 + y^3.$$

Instead, now, of dividing through by $x - y$, the earlier expression of the difference of the roots x and y, Fermat divides by the explicit difference y of the roots $x + y$ and x; that is,

$$b = 3x^2 + 3xy + y^2,$$

"which is the structure of two correlate equations of this sort." Fermat goes on:[38]

> To find the maximum, the roots of the two equations should be equated to one another, in order to satisfy the precepts of the aforesaid method, from which this latter [method] has taken both its mode of operation and its very reasoning.[39]
> Therefore A and $A + E$ are to be equated to one another; hence E will give nothing. Since, therefore, B, by the already determined structure of the correlate equations, is equal to $E^2 + 3A^2 + 3AE$, then all homogeneous quantities containing E should be eliminated as representing nothing. There will remain $B = 3A^2$, which equation will give the desired maximum solid.

Thus, the apparent division by a quantity equal to 0, somewhat masked in Fermat's original conception of the method, obtrudes directly on the sensitivity of the reader. One takes the expression to be maximized, sets it equal to another expression in which $x + y$ has been substituted for x in the original, cancels common terms, divides by y, and sets $y = 0$. Though more obvious now, the difficulty had the same resolution in Fermat's own mind. In the application of syncrisis to the original equation expressing the general problem to be maximized, y represented a real difference between the roots. It only became 0 when one took the particular instance of a maximum (or minimum). Making it 0 did no violence to what had previously been obtained, since syncrisis yielded fully general relationships irrespective of any particular values of the constants in the equation.

In possession of his new method, Fermat could now attack the *diorismos* of Pappus VII, 61. The actual solution follows exactly the procedures of the

---

[38]*Ibid.*

[39]This statement alone should suffice to refute Wieleitner's claim that the *Analytic Investigation* represented the last version of the method of maxima and minima. The verb *desumpsit* leaves no ambiguity that the use of A and E as different roots preceded the use of A and $A + E$ as the roots, and the latter usage is that of the *Method*. Similarly, Strømholm's attempt to place the *Analytic Investigation* and the *Method* in entirely different traditions within Fermat's work also breaks apart on this one sentence.

original method and is distinguished only by the relatively large number of terms involved in the calculations. It serves to climax the *Analytic Investigation*. Through the solution of that problem, Fermat hoped to show the overall effectiveness of his technique. In one important way, it does differ from the other examples he offered in displaying his method. The problem it solves requires that one determine a minimum proportion. The details of the derivation offer no further insight into Fermat's method of maxima and minima, but the applicability of the method to proportions may have opened the way to another innovation, Fermat's method of drawing tangents to curves. Before turning to that second offspring of the method of syncrisis, however, let us follow the first one step further in its development.

D. *Diophantus' Adequality, or How to Hide a Method*

So simple (once he had it) yet so powerful, Fermat's method gave him a distinct advantage over his colleagues in the solution of problems involving extreme-value *diorismoi*. And besides learning mathematics per se in Bordeaux, Fermat also had learned that mathematics was a competitive business. His behavior at several points in his career conflicted with the Baconian ideal of scientific community to which he was so fond of alluding. Indeed, that ideal never did completely subdue his competitive streak, as the debate with English mathematicians in the mid-1650s shows.[40] If it had been up to Fermat in the early years of his career, he might never have set his method of maxima and minima to paper. But he was under pressure to do so, certainly from the mathematicians in Paris in 1636 and probably from those around him in Bordeaux in the late 1620s.[41] Only his desire to reveal as little as possible of the reasoning behind the method explains the form in which he set it down in the first part of the *Method*.

The *Method* is brief and vague. It poses more questions than it answers. All the refined reasoning presented above was reduced to the following:[42]

[40] See below, Chap. VI, §IV.

[41] Fermat announced his methods to Roberval in a letter dated 23.VIII.36 (FO.II.56): "J'ai trouvé beaucoup d'autres propositions géométriques, comme la restitution de toutes les propositions *de locis planis* et autres; mais ce que j'estime plus que tout le reste est une méthode pour déterminer toutes sortes de problèmes plans ou solides, par le moyen de laquelle je trouve l'invention *maximae et minimae in omnibus omnino problematibus*, et ce, par une équation aussi simple et aisée que celles de l'Analyse ordinaire." Roberval must have requested to see these results, because Fermat later (22.IX.36; cf. n. 1) referred him to Despagnet for copies. Writing in April 1637, Roberval thanked Fermat for the restored theorem (II,5) from Apollonius' *Plane Loci* and suggested that, if Fermat would forward some of his other papers, the "Little Academy" that had been formed around Roberval would be willing to publish them (FO.II.103). In the meantime, he and Fermat had been corresponding about problems of center of gravity, to which Fermat had been applying his methods, and Roberval repeatedly admitted his inability to solve the problems and his desire to see Fermat's derivations.

[42] *Methodus*, FO.I.133–134. According to the editors of FO (cf. FO.II.133, n. 1), the words in brackets in the FO Latin text are taken from the copy of the *Method* included in the *Analytic Investigation* (FO.I.153, where, however, only the presence of the frag-

The whole doctrine of finding maxima and minima is based on two expressions in symbols and this single rule: Set some term of the question to be A (either a plane, or a solid or length, so as to be of the same dimension as the proposition to be satisfied) and, having found the maximum or minimum in terms involving a degree (or degrees, as the case may be) of A, posit further that the same term as before is $A + E$ and find again the maximum or minimum in terms involving A and coefficient degrees, as the case may be, of E. *Adequate*, as Diophantus says, the two homogeneous expressions equal to the maximum or minimum and, having removed common terms (which done, all homogeneous quantities on either side will contain E or degrees of it), apply all terms to E or to a higher degree of E, until one of the homogeneous quantities, on one side or the other, is completely freed from affection by E. Remove then on both sides the homogeneous quantities in any way involving E or degrees of E, and set the remaining homogeneous quantities equal, or, if nothing remains on one side, equate rather the negative terms to the positive, which comes down to the same thing. The solution of this last equality will give the value of A, which known, the maximum or minimum will become known by repeating the traces of the foregoing resolution.

Far more than the complex terminology of Viète's algebra makes this verbal description confusing to the historian. It even perplexed Fermat's contemporaries, who understood that terminology. All of the lucid and ingenious reasoning that had led Fermat to the method was excluded; only the dangerous paralogism mentioned above remained, now emphasized to an even greater degree by the absence of the counteracting reasoning. Fermat offered an algorithm, nothing more.

The algorithm was meant for analysts, and it began in true analytical fashion. Set up an expression in terms of the unknown x and the data of the condition to be maximized or minimized. From that expression derive a second by substituting $x + y$ for x. Set the two expressions *adequal*. What is *adequal*? It was Fermat's path over a serious hurdle posed by his desire to conceal the foundation of his method. What was the hurdle?

Stepping back for a moment, one will recognize the two expressions that Fermat has just set up as the left-hand sides of the two correlate equations derived from the equation expressing the general, non-extreme construction

---

ment is noted and the reader is referred back to the present text) and need not have appeared in the original. They give no reason for this conjecture, however. The final sentence may make no sense to the modern reader, but as Chap. II has shown, the notion that one could obtain a rigorous synthetic demonstration of an analytic derivation figured strongly in seventeenth-century justifications of the analytic approach to mathematics, for example in the treatise *De resolutione et compositione mathematica* of Marino Ghetaldi (Rome, 1620); cf. further, Mahoney, "Royal Road," Chap. III, pp. 206–213.

problem. To use the first example from the *Analytic Investigation*, which also appears first in the *Method*,

$$bx - x^2 = c_1$$

$$b(x + y) - (x + y)^2 = c_1,$$

where $c_1 < \max(bx - x^2)$. In the original conception of the method, however, x, y, and c_1 were at first *variable* quantities, dependent on one another through the relation $c_1 = x(x + y)$. Only after syncrisis, when Fermat took account of the particular nature of the roots in the case of an extreme value of c_1, did x become the *unknown* root corresponding to $\max(bx - x^2)$. In the algorithmic form of the method, however, x represented that unknown from the start, which meant that y had to be zero, if indeed there were only one such unknown. To avoid going into a detailed justification of setting the left-hand sides of the above two equations equal to one another, which would have been the normal next step in syncrisis but would also have thrown attention on the dubious nature of y, Fermat required a gimmick, preferably one in some way related to the true bases of the method. He found it in Diophantus' *Arithmetic*.[43]

In Proposition 9 of Book V of that work, Diophantus wants to partition 1 into two fractions such that when a given number is added to each of them the results are squares. If the given number is 6, the problem reduces to that of partitioning 13 into two squares differing by less than 1.[44] Diophantus begins by taking one half of 13, i.e. $6\frac{1}{2}$, and finding the smallest fraction which, added to $6\frac{1}{2}$, makes a square. Adding $\frac{1}{400}$, he obtains $(\frac{51}{20})^2$, which multiplied by 2 yields $13\frac{2}{400}$. Hence, he must add a little to $(\frac{51}{20})^2$ and subtract a little from it in such a way that the resulting numbers both remain squares and add up to 13. How he does this need not concern us here. In Proposition 11, where he employs the same technique, he calls it παριότης, which Xylander rendered into Latin as *adaequalitas* and Heath has translated "near-as-possible equality."[45]

Like Fermat's method of maxima and minima, Diophantus' technique was based on a false assumption.[46] That is, Diophantus assumed for the moment

[43]Or, more precisely, in the edition and translation of the *Arithmetica* published by Xylander (Wilhelm Holtzmann) in Basel in 1575 and republished with a commentary by Gaspar Bachet de Meziriac in Paris in 1621. Fermat's personal copy of the Bachet edition held in its margins most of his results in number theory (see below, Chap. VI), and the reedition of the work by Samuel de Fermat in 1670 included his father's marginalia. Strangely, those marginalia for the pertinent propositions in Diophantus make no mention of *adaequalitas* or the use Fermat made of it in the method of maxima and minima.

[44]I.e. let x and $1 - x$ be the numbers. Then $x + 6 = m^2$ and $1 - x + 6 = n^2$, or $13 = m^2 + n^2$, where $m^2 - n^2 = 2x - 1 < 1$.

[45]Cf. T. L. Heath, *Diophantus of Alexandria: A Study in the History of Greek Algebra* (2nd ed., Cambridge, 1910; repr. N. Y., 1964), pp. 95-98, 207ff.

[46]Fermat explained his concept of adequality somewhat more clearly in a memoir (FO.I.140-147) which opens, "Volo mea methodo . . ." (or, in a French version that also circulated, "Je veux par ma méthode . . .") and which De Waard (FO.*Suppl*.xvi) dates April 1638. In determining the value of $\max(bx^2 - x^3)$, he set up the two correlate ex-

that the desired fractions were equal, in order then to determine their actual difference. Also like Fermat's method, the technique was purely finitistic, involving neither infinitesimals nor approaches to a limit. The technique was, therefore, close enough to what Fermat had in mind. Using it, he could hide the foundation of his method behind a classical authority. He merely turned Diophantus' reasoning around. Fermat's method assumed for the moment that the roots were unequal, in order then to determine the conditions under which they were equal. In both cases, the men meant "temporary equality." And so, in the *Method* Fermat *adequated* the left-hand sides of the two correlate equations. "For the time being," the roots were considered unequal, thus making the next steps, Fermat's emended form of syncrisis, possible.

Whatever confusion the introduction of this Diophantine term wrought among Fermat's contemporaries—and its original meaning and intent clearly escaped Mersenne if no one else[47]—it has certainly led historians of mathematics astray. For into it they have read the pseudo-equality of the differential calculus; for them it becomes one more peg on which to hang a quasi-modern interpretation of Fermat's method. It cannot, however, provide that service. Fermat's method was finitistic, and so too was his use of the term *adequality*. Indeed, only the finitistic origins of his method explain the one remaining divergence of the algorithm from the original method.

In the last step of the part of the algorithm corresponding to syncrisis, Fermat wrote, "apply all terms to E or to a higher power of E until one of the homogeneous quantities, on one side or the other, is completely freed from affection by E." Simply put, divide all the terms of the adequality by y, and keep dividing by y until at least one term no longer contains it. This passage has embarrassed the modernizers of Fermat, since it does not fit the modern

---

pressions and then added: "Id comparo primo solido $Aq.$ in B - $Ac.$, *tanquam essent aequalia, licet revera aequalia non sint*, et hujusmodi comparationem vocavi adaequalitatem, ut loquitur Diophantus (sic enim interpretari possum graecam vocem παρισόης qua ille utitur)." Following division by E, he again returned to the theme of the counterfactual nature of adequality: "Deinde utrimque deleo homogenea quae afficiuntur ab E: superest ex una parte B in A bis, et ex alia $Aq.$ ter, inter quae non amplius facere oportet, ut antea *comparationes fictas* et adaequalitates, sed veram aequationem." This memoir contains two of the extreme-value problems treated in the later *Analytic Investigation*, plus the determination of the tangent to an ellipse. Unlike the latter treatise, however, the memoir continues to employ adequality instead of revealing the actual foundations of the method.

As indicated in the last section of this chapter, the term "adequality" did eventually take on a new meaning. Though originally coined in the sense of "feigned equation," the term appears in Fermat's treatise on quadrature (see below, Chap. V, §IV) in the sense of "approximate equality" or "equality in the limiting case"; there Fermat says (FO.I.257), "adaequetur, ut loquitur Diophantus, aut *fere aequetur* (emphasis added)." Unlike the method of maxima and minima, the method of quadrature did not rest on equation-theoretic foundations. The inequality of a curve and the polygon inscribed in it was a real inequality.

[47] See the version of the method of maxima and minima published in Pierre Hérigone's *Cursus mathematicus* (*Supplément*, Paris, 1644), pp. 59–69 (discussed below, p. 199) for which Mersenne is usually given credit.

algorithm for the derivative of an expression. In fact, in the problems Fermat worked out, the proviso of repeated division by y was unnecessary. But, thinking in terms of the theory of equations, Fermat could imagine, even if he had not experienced, cases in which the adequated expressions contained nothing less than higher powers of y. Syncrisis, a finite algebraic method, would allow repeated division by y in order to free some term of it.[48]

Thus, in the *Method*, the carefully thought out method described in the *Analytic Investigation* became a bare algorithm. Stripped naked of its original foundation, it did not, however, go out into the world unclothed. It donned, rather, the nondescript cloak of adequality and acquired thereby an appearance of poverty of mathematical reasoning that deceived contemporaries and posterity alike. For adequality only threw attention on the difficulty that Fermat's emendation of syncrisis had already exacerbated, the apparent division by zero.

Two other aspects of the *Method* added to the controversy that attended its circulation among members of the mathematical community. First, the example that Fermat provided was the relatively simple, already well-known maximum of Euclid VI, 27. No outstanding problem was solved, no breakthrough made that could attest to the power of the method. Second, Fermat attached to the *Method* a description of his technique for determining tangents to curves, with the claim that that technique derived from the method. Here, he even omitted the algorithm, and the example he provided solved an already well-known problem: to draw a tangent to any point on a parabola. But the method of tangents involved other difficulties besides the absence of some brilliant solution to an outstanding problem. Its presentation left vague the connection between it and the method of maxima and minima. Since the method of tangents formed the initial core of the controversy between Fermat and Descartes, we must look back at the origins of that method before turning to the controversy itself and its effects on Fermat's thoughts about extreme values.

III. OF DUBIOUS PARENTAGE: THE METHOD OF TANGENTS

There is for the method of tangents no text like the *Analytic Investigation*, which dispels the shadows surrounding its genesis and original foundations. For reasons that will become clear below,[49] Fermat's subsequent, more detailed accounts of the method represent advances in his own thinking made later on the basis of clearer insight into the relation of his analytic geometry

[48] The proviso of repeated division by the difference of the roots appears also in the memoir cited in n. 46: "Hac divisione peracta, si omnia homogenea dividi possunt per E, iteranda erit divisio per E, donec reperiatur aliquod ex homogeneis quod hujusmodi divisionem non admittat, id est, ut Vietaeis verbis utar, quod non afficiatur ab E." The *Analytic Investigation*, perhaps because it follows the particular problems treated so closely, includes no such proviso.

[49] See below, pp. 167–168 and 190–191.

to the method. One finds the earliest account of the method—earliest both in terms of written composition and in terms of content—in the second part of the *Method*. With the help of the understanding of the origins of the method of maxima and minima gained above, however, that memoir proves to be revealing enough. Quoted in full, Fermat's method of tangents in the earliest version reads:[50]

> To the above method [of maxima and minima] we have reduced the determination of tangents to given points in any curve.
>
> For example, given parabola *BDN*, its vertex *D*, axis *DC*, and point *B* on it, to draw straight line *BE* tangent to the parabola and meeting the axis in point *E*.

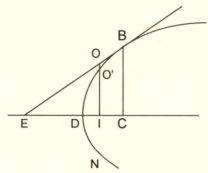

> Taking any point in straight line *BE* and from it drawing ordinate *OI*, and from point *B* ordinate *BC*, the ratio of *CD* to *DI* will be greater than the ratio of the square on *BC* to the square on *OI*, because *O* is outside of the parabola. But, due to similar triangles, as *BC* squared is to *OI* squared so *CE* squared is to *IE* squared. Therefore, the ratio of *CD* to *DI* will be greater than that of the square on *CE* to the square on *IE*.
>
> But, since point *B* is given, the applied line *BC* is given; hence point *C* is given, as well as *CD*. Therefore, let *CD* be equal to a given *D*. Let *CE* be *A*; let *CI* be *E*.
>
> Then, *D* has a greater ratio to $D - E$ than has A^2 to $A^2 + E^2 - 2AE$. And, multiplying means and extremes by each other, $DA^2 + DE^2 - 2DAE$ will be greater than $DA^2 - A^2E$. Adequate, therefore, according to the above method; with common terms removed, $DE^2 - 2DAE$ will be adequal to $-A^2E$, or, what is the same thing, $DE^2 + A^2E$ will be adequal to $2DAE$. Let

[50]*Methodus*, FO.I.134–136. Fermat could be very careless about confusing notation, and this problem is an excellent example. The letters *D* and *E* denote points in the diagram as well as unknown line lengths. The editors of FO distinguished between the two uses typographically by italicizing the letters in their algebraic role, but Fermat certainly did nothing of the sort. Since part of the narrative of the present chapter focuses on the confusion wrought by Fermat's presentation in the *Method*, it has seemed best to leave the problem of notation alone here. The careful reader should have no real difficulty separating the two uses of the letters.

everything be divided by E; hence, $DE + A^2$ will be adequal to $2DA$. Remove DE; hence, A^2 will be equal to $2DA$, and thus, A will be equal to $2D$. Hence we have proved that it is the double of CD, which in fact is the case.

As in the case of the method of maxima and minima, it is a simple matter to verify that the procedure given by Fermat works. Indeed, as in the case of extreme values, the method of tangents all too easily translates into the modern idiom of the differential calculus. Far more interesting historically is the question of why Fermat thought it worked and how he came upon it.

The successful resolution of Pappus VII, 61 may have provided Fermat with the crucial insight. His method applied to proportions as well as quantities, and curves were defined by proportions.[51] Could he find an extreme value involved in drawing a tangent? Consider his account of the method of tangents. For any point O, other than B, on the tangent, the ratio OI^2/DI is greater than the ratio BC^2/CD, since, for O' on the parabola, $BC^2/CD = O'I^2/DI$ and $OI > O'I$. Hence, the ratio BC^2/CD is, in a sense, a minimum value for ratio OI^2/DI. Treat it as such by the method of maxima and minima. Solution of the problem requires the determination of the length CE; let it first be denoted by x. Since, for the purpose of analysis, the problem is assumed solved, the triangle BCE is fixed, and hence so too is the ratio $BC/x = m$. Therefore, from the triangle $BC = mx$, and from the parabola $BC^2/CD = m^2x^2/d$. Now set $CE = x - y$ (or $x + y$; the result will be the same), whence $DI = d - y$ and $BC^2/CD = m^2(x - y)^2/(d - y)$. Following the precepts of the method of maxima and minima, adequate the two expressions for BC^2/CD:

$$\frac{m^2x^2}{d} \approx \frac{m^2(x - y)^2}{d - y}$$
$$x^2(d - y) \approx d(x - y)^2,$$

which is precisely the equation arrived at in the *Method*. The remainder of the derivation follows the method of maxima and minima step for step. One has, then, here a plausible reconstruction of the original path along which the method of tangents was derived from the method of maxima and minima, to wit, by treating the ratio BC^2/CD as a quasi-minimum. The method of tangents can be generalized beyond its application to the parabola, but an express *caveat* is necessary first.

Present secondary accounts of the method of tangents tend to overlook a small, but crucial detail which is of utmost importance in understanding the future development of that method. Note that, in the passage quoted above from the *Method* as well as in the reconstruction that follows it, Fermat (and we) employ the classical Greek expression of the parabola. A parabola is a curve such that, for any two points B and O' on it, $BC^2/CD = O'I^2/DI$. This is Apollonius' *symptôma* with the mean terms alternated. It was Fermat's only

---

[51] See below.

means for expressing the properties of curves before his invention of an analytic geometry in the *Introduction*. And that invention, as was shown in Chapter II, came about sometime around 1635, several years after the date Fermat assigned to his invention of the method of tangents. When the need arose during the dispute with Descartes to explain the method of tangents in greater detail, Fermat did not hesitate to adjust the method to take account of the new possibility of expressing curves in terms of equations. In particular, as noted in Chapter III, the invention of analytic geometry radically altered mathematicians' concept of curves and vastly increased the number of curves suitable for mathematical investigation. Only when one was certain that every curve, and any curve, had some algebraic equation uniquely corresponding to it and implicitly containing all its properties, could one generalize, in the full sense of the word, any algebraic method of tangents. Hence, Fermat's discussions of the method of tangents after 1635 or 1636 can offer the historian little help in working out the original method. For Fermat naturally and tacitly made the shift from proportional to equational expressions of curves and may not himself have realized how profound a conceptual shift accompanied the latter. In attempting to reconstruct a generalization of the original method, a generalization that might explain Fermat's confidence in that method, we must bear in mind the very different understanding of curves attendant upon it.

Most of the curves for which Fermat had the defining properties in 1629 were, like the parabola, concave with respect to their diameters. Hence, for those curves, the various ordinates $O'I$ were less than the segments OI drawn to the tangent, and one could treat the problem in terms of determining a minimum. For convex curves, the $O'I$ would be greater than the corresponding OI, in which case the main defining ratio would be treated as a maximum. Now,

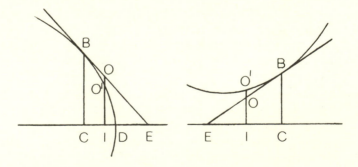

letting the ratio $f(BC)/g(CD) = k$, where k is some constant of proportionality, serve as the defining ratio of the curve (in the case of the ellipse, for example,

$BC^2/(CD \cdot CD') = k$),[52] Fermat needed merely to take $BC = mx$ from the triangle and work from the adequality

$$\frac{f(mx)}{g(CD)} \approx \frac{f(m(x - y))}{g(CD - y)}$$

to determine the tangent to any curve for which he knew the defining proportion. Again, in the case of the ellipse, the initial adequality would read

$$\frac{m^2 x^2}{d(r - d)} \approx \frac{m^2 (x - y)^2}{(d - y)(r - d + y)}.$$

Fermat's method of tangents, then, was indeed fully general and did derive from his method of maxima and minima. Unfortunately, he failed in the *Method* to make either point clear. Contenting himself with the single example of the tangent to a parabola, he quickly drew the *Method* to a close:[53]

> The method never fails. Indeed it can be extended to many quite pretty questions, for with its help we have found the centers of gravity of figures bounded by curved and straight lines and of solids, and many other things about which we will perhaps speak elsewhere, if leisure permits.
>
> Concerning the quadrature of areas contained by curved and straight lines, as well as the ratio of the solids resulting from them[54] to cones of the same base and height, we have already corresponded at length with Lord de Roberval.

Fermat had succumbed to the pressure of colleagues to put his methods of maxima and minima and of tangents down on paper, to give them some hint of the techniques he was employing to provide such brilliant solutions to the problems posed to him. But he succumbed in less than 600 words, and there he in all probability meant to let the matter rest. When, a short year later, Descartes let loose a storm of criticism of Fermat's methods, the cherished illusion that the *Method* would suffice as an account of these methods was the first thing to crumble.

---

[52] Although it is not generally the case that plane curves can be expressed in such a closed form involving separated variables, all the curves with which Fermat was familiar in 1629 (most notably the conic sections), and most of those he dealt with later, can be so expressed. Exceptions like the cycloid were transcendental curves that required adjustment of the method for even more fundamental reasons.

[53] *Methodus*, FO.I.136. These last two paragraphs of the *Method* discuss, of course, work that followed the invention of the methods discussed in it. Indeed, the last paragraph refers directly to correspondence that took place during the fall of 1636 and the spring of 1637. Whether one should take this to mean that the memoir was first composed at that late date is questionable. The arguments made in n. 1 above still obtain. It is entirely possible that Fermat had someone, perhaps Mersenne, add these last two paragraphs to his original copy so as to bring Descartes, to whom it was to be sent, up to date on the progress of the method.

[54] I.e. the corresponding solids of rotation.

IV LOOKING UNDER THE BED: DESCARTES VS. FERMAT, 1637–1638

The sketchiness of Fermat's *Method for Determining Maxima and Minima and On Tangents to Curved Lines* demanded of its readers an uncommon amount of good will. It required at the very least that they begin its study under the assumption that Fermat was correct and that the method indeed worked. In the hands of someone not well disposed to it or its author, however, the memoir was easy prey for criticism and misunderstanding. And, at the end of 1637, no one could have been less well disposed than René Descartes.

With the publication earlier that year in Leiden of his *Discourse on Method* and its three *Essays*, Descartes had set out on his crusade to reform philosophy and had begun to assume a belligerent stance toward anything or anyone that contradicted his belief that he alone had the key to knowledge, that in the realm of mathematics his *Geometry* contained everything to be said on its subject. Pleased that someone might learn from his work,[55] he rejected any notion that such a "student" might in fact uncover something entirely new, something not derivable from that work. The *Geometry* enjoyed his most jealous protection, for the new method of mathematics had been the source of the larger philosophical method of the *Discourse*. To attack it, to correct it, or to find something not already in it, was to impugn Descartes' whole program. To claim to have achieved similar results independently of, or earlier than, Descartes was to question the uniqueness of the mission communicated to him in the close winter quarters at Ulm in 1619.

By the end of December 1637, Fermat had done just that. He had questioned the validity of Descartes' derivation of the laws of reflection and refraction in the *Dioptrics*, he had expressed doubt that mathematics alone could lead to physical truths, and he had claimed to have invented some years earlier mathematical methods that either duplicated those of the *Geometry*, or were more effective, or indeed had no counterpart in Descartes' work. His claims were made in ignorance of the effect they would have on Descartes, of whom he had known nothing until the spring of 1637, when he received a copy of the *Dioptrics* from Beaugrand.

Fermat was also ignorant of the context into which Descartes would set those claims. He knew nothing of the overt and anonymous attacks that had already begun on Descartes and his work, attacks coming mainly from the

---

[55]Examples of this attitude are legion in Descartes' correspondence. To Alphonse Pollot, for example, he wrote: "Je tiens à grand honneur que vous veuillez prendre la peine d'examiner ma *Géométrie*, et je vous garde l'un des six exemplaires qui sont destinés pour les six premiers qui me feront paraître qu'ils l'entendent." (12.II.1638, *Alquié*.II.20.) Perhaps the best example, however, is provided by the closing words of the *Geometry* itself: "Et j'espère que nos neveux me sauront gré, non seulement des choses que j'ai ici expliquées; mais aussi de celles que j'ai omises volontairement, afin de leur laisser le plaisir de les inventer."

very group of men in Paris with which Fermat had so recently come into correspondence. He did not know that his mentor, Beaugrand, had accused Descartes of plagiarizing Viète, nor that Beaugrand was anxious to repay Descartes in kind for his withering criticism of Beaugrand's *Geostatics*. Fermat was unaware of the personal feud between Roberval and Descartes. Hence, he wrote his critique of the *Dioptrics* and asked that Descartes be shown his mathematical papers with no sense that Descartes would read them as stemming from a protégé of his most hated enemies. Unwittingly, Fermat had become a member of what Descartes viewed as a conspiracy to destroy the *Discourse* and its *Essays*.

Only against the background of this perceived conspiracy can one understand the vitriolic fury of Descartes' response to Fermat and the zeal with which he strove subsequently to destroy Fermat's reputation. Only by knowing Fermat's own ignorance of the circumstances can one understand his perplexity over that response. Few scientific debates in history reveal so much of the personalities of the participants, or the extent to which personal factors can influence rational discourse. Most interesting is the fundamental contrast in personalities that emerges from this debate: Descartes learned nothing from it, even though he was in error; Fermat, even though he was correct, gained new mathematical insights that led him to revise his methods and sharpen his tools. These new insights make the debate an integral part of the history of Fermat's methods of maxima and minima and of tangents. Hence the debate must intervene in our narrative as it intervened in the development of Fermat's thought.

The storm broke in January 1638, but it had built up over the previous six months. Its formation explains its fury. For Fermat, the incident began innocently. In April or May of 1637, Beaugrand, now *secrétaire du Roi*, showed him a copy of a work recently submitted for licensing, the *Dioptrics* by René Descartes. Fermat did not know that Beaugrand had obtained the copy by highly questionable means[56] and that he had the right neither to show the

---

[56] Descartes had made an enemy of Beaugrand through his brutal criticism of the latter's *Geostatics*. Seeking revenge, Beaugrand used his position as *secrétaire du roi*, one of the duties of which was to oversee the issuance of licenses to print and distribute books in France, to gain hold of the printer's copies of the *Dioptrics*. Whether he stole them himself or had them stolen is not clear; one way or the other he circulated them among his friends, presumably hoping thereby to embarrass Descartes by exposing the work to criticism before it appeared publicly. Mersenne did his best to smother the results of this indiscretion, but he failed. Fearful of Descartes' wrath—Descartes had entrusted the task of procuring a license to him—Mersenne revealed the details a small bit at a time. When Descartes finally knew the full story, he resolved never again to have anything to do with Beaugrand. Cf. Descartes to Mersenne, 22.VI.1637, AT.I.390–391: "Pour l'auteur de la *Géostatique*, il n'a pas fait, ce semble, un trait d'honnête homme, d'avoir retenu la *Dioptrique* en la façon que vous me mandez. Et je m'étonne, puisqu'il en fait si peu d'état, de ce qu'il a pris tant de peine pour la voir avant les autres, et qu'il a même en quelque façon négligé son honneur pour cet effet." Also, Descartes to Mersenne, 1.III.1638, AT.II.25: "... je m'étonne de ce que vous daignez encore parler à [Beaugrand], aprés le trait qu'il vous a joué. Je serais bien-aise d'en apprendre encore une fois

work to Fermat nor even to possess it himself. Engaged at the time in correspondence with Roberval and the elder Pascal over matters of quadrature and determination of centers of gravity,[57] Fermat probably paid little attention to Descartes' treatise, until he received from Mersenne a letter (now lost) requesting that he forward any comments on it directly to Mersenne. Mersenne, for his part, had written to prevent any pre-publication discussion of the treatise, lest Descartes discover just how careless the Minim friar had been with the copies entrusted to him.[58]

Eager to please Mersenne, Fermat felt called upon to say something. Though he had some criticism to make about Descarte's efforts in optics, he had no intention of provoking anyone. He began his reply to Mersenne in what he thought was an innocuous way:[59]

You have asked for my judgment of M. Descartes' treatise on *Dioptrics*. On the one hand, surely the little time M. Beaugrand gave me to peruse it would seem to free me from the obligation to satisfy you exactly and point by point. Besides, since the material itself is very subtle and very thorny, I do not dare to hope that unformed and not yet well digested thoughts could give you any great satisfaction. But, on the other hand, when I consider that the search for truth is always praiseworthy and that we often find what we are seeking by groping about in the shadows, I felt you would not think ill of me if on this subject I tried to outline my ideas for you, which, since they are obscure and halting, I will perhaps clarify some other time if the bases of my thought meet approval or if I do not change my own mind.

Descartes was clearly a stranger to Fermat, for no one who knew Descartes or his work could have so begun a letter that he was likely to see.[60] Only from

---

l'histoire au vrai, car vous me l'avez mandée à diverses reprises, et diversement, en sorte que je ne sais ce que j'en pourrais dire ou écrire assurément, en cas qu'il se présentât occasion de l'en remercier selon son mérite."

[57]See below, Chap. V, §II.

[58]Descartes already had a fairly good idea. Writing on 27 April 1637, he warned (AT.I.364): "Au reste, je remarque par vos lettres que vous avez fait voir ce livre à plusieurs sans besoin, et au contraire que vous ne l'avez point encore fait voir à Monsieur le Chancelier, pour lequel seul néanmoins je l'avais envoyé, et je désirais qu'il lui fût présenté tout entière." M. le Chancelier was Pierre Séguier, who was absent from Paris at the time (cf. AT.I.365). His absence may explain in part how Beaugrand managed to gain hold of copies of the work.

[59]Fermat to Mersenne, IV or V.1637, FO.II.106–107. The FO version bears the conjectured date ⟨Septembre, 1637⟩, but, according to De Waard (CM.VI.247), that is the month in which Mersenne forwarded the critique to Descartes.

[60]It is not entirely clear that Fermat in fact meant the letter to be forwarded to Descartes, for the latter noted in his letter of 18.I.1638 (*Lettres de Descartes*, ed. Clerselier, Vol. III, #55; see below, p. 176): "Je vous renvoie l'original de sa démonstration prétendue contra ma *Dioptrique*, parce que vous me mandiez que c'était sans le sçeu de l'auteur que vous me l'aviez envoyé." Fermat could not, however, have been so naïve as to believe that Mersenne would not forward the substance of his criticism, if not the letter itself. Though Mersenne's initial request to Fermat is lost, it is more than likely

the depths of his innocence could Fermat have criticized one of the *Essays* on the grounds that ". . . we often find what we are seeking by groping about in the shadows (*à tâtons et parmi les ténèbres*)." In fact, the phrase *à tâtons* would come back to haunt him.[61] The picture of the author of the *Dioptrics* groping about in the dark for the truth of his subject would alone have sufficed to ruffle Descartes' mane. But, having slipped the lock on the lion's den, Fermat blithely walked in.[62]

He referred to the efforts of Alhazen, Vitelo, and Maurolycus to establish a mathematical law of refraction and noted that now M. Descartes was trying his hand at it. He outlined Descartes' theory of light as an inclination to motion and located the crucial assumption that one may investigate inclination to move via the mathematical examination of actual motion. Fermat could not assent to this assumption because of the great difference between potency and act. Moreover, he observed, the instantaneous transmission of light precluded analysis of it by analogy to motion in time. But, of course, mathematics could not increase one's understanding of physical matters anyway.

These general reactions out of the way, Fermat turned to Descartes' proof of the law of reflection and tried to show that it really offered no proof, since by the author's choice of components of motion the result was built into the assumptions. Hence, "of all the infinite ways of dividing the determination to motion, the author has taken only that one which serves him for his conclusion; he has thereby accommodated his means to his end, and we know as little about the subject as we did before." If Descartes had not yet found a convincing method for handling this difficult material, nonetheless Fermat saw promise. "We must seek the truth in common," he concluded, and he would be willing to lend his hand. Hoping that Mersenne would find his remarks sufficient, Fermat closed his letter with some references to other works he had been sent, including Galileo's *Two New Sciences*, promising to react to them at some later date.

Fermat might not have known Descartes, but Mersenne did. And Mersenne feared the reaction such a letter might call forth from the temperamental philosopher. He hesitated for several months in forwarding the letter, but finally did so in September, after having received a cordial letter from Des-

---

that he there indicated his intention to transmit all comments about the *Dioptrics* to its author.

[61] Cf. Descartes to Hardy, VI.1638, FO.IV.48–51; in reference to the method of tangents, Descartes remarks: ". . . il est fort vraisemblable que M. de Fermat . . . ne l'a trouvé qu'à tâtons. . . ." See also below, p. 180.

[62] In the interests of maintaining the mathematical focus of the present chapter, most of the technical details of the debate over the *Dioptrics* are here omitted. For Fermat's efforts in this area, see below, Appendix I. The intricacies of Fermat's and Descartes' debate over the law of refraction, as well as the origins and development of the latter's theory of light, are brilliantly laid out in A. I. Sabra's *Theories of Light from Descartes to Newton* (London, 1967).

cartes asking that all criticism of the *Dioptrics* be sent for reply.[63] At the time, September 1637, Mersenne had in fact little to fear in the case of Fermat's critique. For Fermat was not unknown to Descartes, who had already seen something of his work.

At the same time that Fermat first saw the *Dioptrics*, Descartes had received from Mersenne Fermat's restitution of Proposition II,5 of Apollonius' *Plane Loci*, which Fermat had included in one of his letters to Roberval.[64] As Chapter III has shown, that result was, by 1636 at the latest, already obsolescent in view of the development of the *Introduction*. But Descartes did not know that, and Fermat's habit of sending old work to correspondents created in this instance a first impression on Descartes that shaded all of his future estimations of Fermat's abilities. Descartes' observation on the proposition, sent to Mersenne in May 1637 reeked of condescension:[65]

You also send me a proposition of a mathematician, a Councillor of Toulouse, which is very pretty and which has very much pleased me. For insofar as it is quite easily solved by what I have written in my *Geometry*, where I give the general method for finding not only all plane loci but also all solid loci, I expect that this Councillor, if he is an open and honest man, will be one of those to make the most of the work and that he will be one of those most capable of understanding it. For I will tell you honestly that I feel there will be very few people who will understand it.

At their first indirect meeting, therefore, Descartes had seen in Fermat not a rival, but a prospective convert to his methods. He reacted to Fermat's first letter on the *Dioptrics* in that light.

Fermat's critique did not shock Descartes. Clearly, Fermat had failed to grasp the novelty of Descartes' new method and to seize the insights afforded by it. For example, in paraphrasing the opening arguments of the work, Fermat had translated the phrase *bien aisé à croire* into *probablement*. "When I say that something is easy to believe," Descartes lectured, "I do not mean to say that it is only probable, but that it is so clear and so evident that there is no need for me to stop and demonstrate it."[66] In Descartes' view, Fermat's

---

[63] Descartes to Mersenne, 25.V.37, AT.I.378-379: "Enfin, parce que mon explication de la réfraction, ou de la nature des couleurs, ne satisfait pas à tout le monde, je ne m'en étonne aucunement; car il n'y a personne qui ait eu encore assez de loisir pour les bien examiner. Mais lorsqu'ils l'auront eu, ceux qui voudront prendre la peine de m'avertir des défauts qu'ils auront remarqués, m'obligèrent extrèmement, principalement s'il leur plait de permettre que ma réponse puisse etre imprimée avec leur écrit, afin que ce que j'aurai une fois répondu à quelqu'un serve pour tous."

[64] See above, n.41.

[65] Descartes to Mersenne, 25.V.37, AT.I.#76.

[66] Descartes to Mersenne for Fermat, 5 or 12.X.37, FO.II.#23 (cf.CM.VI.#629). De Waard set the proper date for the letter through that of the covering letter addressed to Mersenne, which says of Fermat's criticism: "J'ai été bien-aise de voir la lettre de Monsieur de Fermat, et je vous en remercie; mais le défaut qu'il trouve en ma démonstra-

other objections likewise pointed not to faults in the *Dioptrics*, but to Fermat's failure to understand the argument. Fermat had not heard the articles of faith, so Descartes shouted them again and dispatched his reply early in October.

The misunderstanding present at the beginning of the controversy presaged its vehemence and recriminations. Fermat had known nothing of Descartes and in his ignorance had criticized the *Dioptrics* in a manner totally unsympathetic to Descartes' basic goals. Descartes had known of Fermat only through the one proposition he had seen, a proposition he took to be entirely representative of the man's talents and accomplishments. It had shown Descartes only that Fermat had the brains to benefit from reading the *Geometry*. When Descartes received Fermat's critique of the *Dioptrics*, he had no inkling of Fermat's other mathematical papers, far more representative of his mathematical abilities, then lying in Paris.

Descartes discovered the true stature of his opponent toward the end of December 1637. Earlier that month, before replying to Descartes' letter, Fermat had copies of the *Method* and the *Introduction* turned over to Mersenne for transmission to Descartes.[67] The extant sources do not reveal why he did so, but the fact that, at about the same time, he had received a copy of the *Geometry* raises the suspicion that he sent his two papers to document his independence, if not indeed his priority, in obtaining results very similar to Descartes'.[68] In transmitting the works, Mersenne gave Descartes to understand that Fermat had been surprised at the absence of any treatment of extreme values in the *Geometry* and wanted to show Descartes how to deal with such problems. Throughout the affair, however, Mersenne showed a rare talent for making a bad situation worse, and whatever he may have said to Descartes can offer little indication of Fermat's intent.[69] Whatever the par-

---

tion n'est qu'imaginaire, et montre assez qu'il n'a regardé mon traité que de travers. Je répons à son objection dans un papier séparé."

[67] Descartes acknowledged receipt of the *Method* on 18.I.1638 and of the *Introduction* on 25.I.1638; of the latter he said to Mersenne: "Je ne vous renvoie point encore les écrits de Monsieur Fermat *De locis planis et solidis*, car je ne les ai point encore lus; et pour vous en parler franchement, je ne suis pas résolu de les regarder, que je n'aie vu premièrement ce qu'il aura répondu aux deux lettres que je vous ai envoyées pour lui faire voir." (*Alquié*.II.16.)

[68] Descartes certainly took this to be Fermat's intention, and it may have been so phrased by Mersenne (his letter to Descartes on this occasion is no longer extant). See the preceding note and the passage on p. 178. Cf. also Descartes to Mydorge, 1.III.1638 (AT.II.16): "La troisième est un écrit latin de M. de Fermat *De maximis et minimis*, qu'il m'a fait envoyer, pour montrer que j'avais oublié cette matière en ma *Géométrie*, et aussi qu'il avait une façon pour trouver les tangentes de lignes courbes, meilleure que celle que j'ai donné."

[69] For one thing, Fermat emphasized his refusal to have his letters published from the very beginning of the controversy, an attitude that ill fits a desire to compete with Descartes or assert priority over him. In light of Descartes' demand that all criticism of his work be published together with his rebuttals (see n.63), this steadfast refusal must have upset both Mersenne and Descartes. In December 1637, for example, Fermat wrote upon

ticular reason for Fermat's action, the result seems clear: in Descartes' eyes Fermat lost his amateur status. What Descartes had previously discerned as a modicum of scientific talent was beginning to take shape in his mind as part of a sinister conspiracy to destroy his work, as he put it, "to smother his brainchild at its very birth."[70]

What else could Descartes have thought? On the heels of a critique of the *Dioptrics* that questioned the very foundations of his new method of natural philosophy came now Fermat's claim to have invented himself major portions of the *Geometry*, indeed to have obtained results reaching beyond the scope of that treatise. Surely Fermat's latest communication represented an enormous threat to Descartes' program. Indeed, a more sinister thought may also have entered Descartes' mind. How could someone who, as late as the previous spring, had produced only such a meager display of talent as the geometrical restitution of *Plane Loci* II,5 have achieved what had just arrived in the mail? Mersenne had been careless with the printer's copies of the *Dioptrics*. Had he let the *Geometry* wander about also? If Descartes never dared openly to suggest plagiarism, one must nevertheless note that in the ensuing controversy he always tried to read the *Method* in terms of the *Geometry*, as if the latter had somehow been the source of the former.

Charges of plagiarism would not be necessary in this case, for Descartes felt—as Talleyrand would put it some time later—that a blunder was worse than a crime. To plagiarize correctly something one understands is a crime; to plagiarize incorrectly what one does not understand is stupidity. Descartes apparently decided to base his case on Fermat's mathematical stupidity. On 18 January 1638 he took pen in hand to demolish the upstart in Toulouse. The critique of the *Method* had to go to Fermat via Mersenne, and the covering letter to Mersenne that accompanied it reveals something of Descartes' frame of mind.[71] He wrote that he was returning the original copy of Fermat's first letter on the *Dioptrics* but not that of the *Method*, because: "... I thought that I should retain the original and content myself with sending you a copy, principally because it contains faults which are so apparent that he might perhaps accuse me of having imagined them if I do not retain his own handwriting to defend myself."

---

receiving Descartes' first reply, that he would continue the debate, "... [ne] point par envie ni par émulation ..., mais seulement pour découvrir la vérité ..., je ne désire pas que mon écrit soit exposé à un plus grand jour que celui qui peut souffrir un entretien familier, de quoi je me confie à vous (FO.II.116)." Again, in February 1638, he wrote (FO.II.133): "Quoiqu'il en soit, je ne me pique pas d'être cru que par ceux qui le voudront, et vous proteste que j'aimerais mieux prononcer: *Jamjam efficaci do manus scientiae*, que de souffrire que rien de ce que je vous ai envoyé soit imprimé sous mon nom, ce que je vous prie d'empêcher par le pouvoir que vous avez sur tous ces messieurs qui se mêlent de cette étude." The Latin phrase is taken from Horace, *Epodes*, XVII, 1 (identification by eds. of FO).

[70] Descartes to Mersenne, 29.VI.38, AT.II.#126. For the pertinent passage, see below, p. 192.

[71] Descartes to Mersenne, 18.I.1638, *Clers.*, #55.

Descartes also retained, at least for the moment, the remnants of charity and good will. He was still willing to believe that Fermat had some talent in mathematics—albeit less than Fermat might think—and that he had let his reputation for knowing some algebra get the best of his good sense. That came, naturally, from being influenced by people who really weren't equipped to judge such matters. Descartes named no names in his letter, but he clearly had Roberval and Beaugrand in mind. Mersenne's crowning act of tactlessness (of which more later) would confirm Descartes' suspicion that the Parisian mathematicians were behind Fermat's "attack" on the *Discourse* and its *Essays*.

Well and good, Descartes would make one more effort to bring Fermat back to his senses, but only one more:

> I would be happy to know what he will say, both about the letter attached to this one, where I respond to his paper on maxima and minima, and about the one preceding, where I replied to his demonstration against my *Dioptrics*. For I have written the one and the other for him to see, if you please; I did not even want to name him, so that he will feel less shame at the errors that I have found there and because my intention is not to insult anyone but merely to defend myself. And, because I feel that he will not have failed to vaunt himself to my prejudice in many of his writings, I think it is appropriate that many people also see my defense. That is why I ask you not to send them to him without retaining copies of them. And if, even after this, he speaks of wanting to send you still more papers, I beg of you to ask him to think them out more carefully than those preceding; otherwise I ask you not to accept the commission of forwarding them to me. For, between you and me, if when he wants to do me the honor of proposing objections, he does not want to take more trouble than he did the first time, I should be ashamed if it were necessary for me to take the trouble to reply to such a small thing, though I could not honestly avoid it if he knew that you had sent them to me.

Fermat may not have been beneath contempt for Descartes, but he was certainly on the borderline. Beyond what the letter to Mersenne says explicitly, one must note the paranoia implicit in Descartes' words. Only with regard to the *Dioptrics* did Fermat ever attack Descartes or require him to defend himself. Nowhere at all did Fermat denigrate Descartes. All the correspondence that passed between the two is accounted for. What grounds could Descartes possibly have had for suspecting that Fermat had done so "in many of his writings"? The answer lies, most probably, in the veiled references to a conspiracy in Paris. Fermat was doomed to share by association the guilt Descartes threw on the Parisian establishment.

In criticizing the *Method* in the other letter he wrote on 18 January,[72]

[72] Descartes to Mersenne for Fermat, 18.I.1638, FO.II.#25.

Descartes dosed his vehemence with high-handed sarcasm. "I would prefer," he began,

> to say nothing about the paper you have sent me, because I could not say anything that would be to the advantage of the person who wrote it. But, because I recognize that it is same person who earlier tried to refute my *Dioptrics* and because you inform me that he sent the paper after having read my *Geometry* and having been surprised that I had not found the same thing—that is, as I interpret things, with the intention of entering into competition and of showing that he knows more about the subject than I—and then also because I learn by your letters that he has the reputation of being very learned in mathematics, I feel obligated to reply to him.

Descartes would later insist that a method of maxima and minima did in fact lay implicit in the *Geometry*; it could be derived from the method of tangents therein. For him, however, the main problem had been that of drawing a tangent to a curve. In this initial attack on the *Method*, however, Descartes left aside the method of maxima and minima and concentrated on Fermat's technique for determining tangents. He focused on what he took to be the treatise's (and the method's) two main weaknesses: first, he argued that the method of tangents did not operate in accordance with the method of maxima and minima from which it supposedly derived, and secondly he insisted that the method of tangents as it stood could not serve as a method, since its algorithmic application to other curves led to the same result as for the parabola.

Let us, he began, employ Fermat's method of maxima and minima step-for-step in drawing the tangent to a parabola. What is the result? We need a maximum, and that maximum is surely the tangent *BE* to the parabola *BDN*. For,

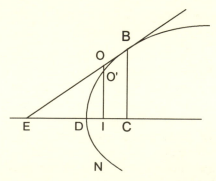

if *BE* is tangent to the parabola, then it is the greatest line that can be drawn from point *E* to the parabola. Following the algorithm of the *Methodus*, Descartes set $EC = x, BC = b$; whence, $BE^2 = x^2 + b^2$. He then took $x + y$ in place of x (or $x - y$; the result would be the same, he argued) and calculated the corresponding value for *BE*. If, when *BC* = *b*, *CD* = *d*, then the latus rectum of the parabola is b^2/d. And since $EC = x + y$ and the corresponding new value for *CD* is $d + y$, then $BC^2 = (b^2 d + b^2 y)/d$. Therefore,

$$\frac{b^2 d + b^2 y}{d} + x^2 + 2xy + y^2 = BE^2 = x^2 + b^2.$$

From this equation, the steps of the *Method* lead directly to the final result $b^2/d + 2x = 0$, "which does not give the value of the line *x*, as the author assures us; consequently, his rule is wrong."

Descartes made no effort to interpret the answer he had derived from his application of the method of maxima and minima to the length *BE*. Had he done so, he would perhaps have recognized that the point he had determined on the axis did in fact lead to a construction of the tangent. For, if *EC* = $-b^2/2d$, then *E* is the intersection point of the *normal* to the parabola at point *B* with the axis.[73] He had, as Fermat would later show clearly, determined not a maximum but a minimum, and that minimum made possible the construction of the tangent via the normal. Indeed, with less than his usual candor, Fermat would later argue that the determination of this minimum was precisely the direct link between the method of maxima and minima and that of tangents. But of this, more in its place.[74] For the moment, it suffices to note that Descartes believed he had refuted by *reductio ad absurdum* any connection between the two methods. He turned then to the method of tangents itself as it was presented in the *Method*.

Even if, in the case of the parabola, the method seemed to lead to the right result, it was a result that everyone already knew. It was also a result that did not really follow from the argument. For, Descartes maintained, one could insert in place of the word "parabola" at each occurrence the word "hyperbola," or "ellipse," or "circle," and the argument, including the inequalities, would maintain its consistency while yielding exactly the same answer, *EC* = 2*d*, a false answer for any curve other than the parabola. Therefore, Descartes would alter Fermat's claim slightly: *semper fallit methodus.*[75]

Descartes had a keen sense of irony. One by one, he took Fermat's criticisms of the *Dioptrics* and turned them against the *Method*. In this case, Fermat's supposed derivation of the tangent to the parabola proved nothing, for Fermat had chosen the one curve for which the argument held. He had, that is, adapted his *medium* to his conclusion. Surely the desire to render such poetic justice led Descartes to the misunderstanding underlying his

---

[73] Given Descartes' own approach to tangents via normals, his failure to recognize the mathematical sense of this solution seems even more surprising.

[74] See below, pp. 189–190.

[75] Descartes suggests changing the passage in the *Method* that reads: "ergo CE probavimus duplam ipsius CD, quod quidem ita se habet. Nec unquam fallit methodus." He would have it read: "non ideo sequitur CE duplam esse ipsius CD, nec unquam ita se habet alibi quam in parabole, ubi casu et non ex vi praemissarum verum concluditur: semperque fallit methodus."

criticism, or perhaps led him to twist deliberately Fermat's intent. For he was only half-right, and one suspects he knew it. The inequality $BC^2/OI^2 < CD/DI$ does obtain as a general property of other conic sections besides the parabola, e.g. for the ellipse. One could, therefore, follow the letter of Fermat's algorithm by inserting the word "ellipse" for "parabola" at each occurrence of the latter without destroying the consistency of the initial steps of the derivation. But one could use this fact only as a criticism of the completeness of the algorithm as an algorithm. If "method" meant a mechanical procedure that only an idiot could fail to follow, then Fermat's description of his procedure fell short of a method. But only slightly, for one would have to be little more than an idiot not to discern in the above inequality the *symptôma*, i.e. the characteristic defining property, of the parabola. Only someone determined to fault the *Method* would refuse to grant that the substitution of "ellipse" for "parabola" required also the replacement of $BC^2/OI^2 < CD/DI$ by $BC^2/OI^2 < (CD \cdot CD')/(DI \cdot ID')$, the corresponding inequality based on the *symptôma* of the ellipse. But Descartes had no intention of granting Fermat any favors. He would play the idiot; he would interpret the procedures of the *Method* to the letter and thereby hopefully destroy it.

If then Descartes' first objection to the *Method* had some validity, the second sprang mostly from malice. As Section III above makes clear, the original connection between the method of maxima and minima and the rule of tangents lacked mathematical rigor. Fermat had tied the two together more from intuition than from clear mathematical insight. Moreover, the *Method* did little more than assert a direct connection. One could, even if one were friendly to the treatise, make the connection as Descartes did, though a semblance of good will would probably lead one to discern the validity, albeit indirect, of the result or, if one saw no sense in the result, to the conclusion that one had applied the method incorrectly. One could, as we did in Section III, reestablish the original connection, though to take the trouble would mean to believe in Fermat's results. Descartes had every right to insist on clarification on the first issue he raised. But the second objection was transparently malevolent and tried to make of the *Method* something it was never meant to be, to wit, a completely stated algorithm.

In the concluding portions of his critique, Descartes shifted from sarcasm to condescension. The lawyer from Toulouse would have done better to have read the *Geometry* more closely before grabbing his pen. For, if Fermat was genuinely interested, Descartes would show him how to patch up his treatise and thereby also demonstrate that the *Geometry* in fact contained all this material, else how could Descartes know how to correct Fermat? With the carrot, Descartes applied the stick and made a charge that would stick to Fermat even after the debate had run its course. It was the charge that later provoked the *Analytic Investigation*: "For, first of all, his rule (i.e. the one he takes pride in having found) is such that with no work and by accident one can easily fall into the path that one must take to find it."

"With no work and by accident"—that was Descartes' reply to Fermat's suggestion that "we often find what we are seeking by groping about in the shadows." Coming from the author of the *Discourse*, no other characterization of a man's mathematics could have been more devastating.

What was the path into which Fermat had fallen? It was "nothing other than a false assumption, based on the manner of proof by *reductio ad absurdum*." Descartes held such a line of reasoning to be the "least valued" and "least ingenious" of all ways of mathematical thought. By contrast, his own was drawn "from knowledge of the nature of equations," a knowledge present *only* in the *Geometry*. Knowledge of the best sort, it was *a priori*.

In part, Descartes' mathematical acumen had asserted itself through the hypercritical haze. Fermat's method did derive from a false assumption, to wit, the assumption that, at the maximum or minimum, the equation had two different solutions. But *reductio ad absurdum* did not enter the picture; Fermat's method rested no less than Descartes' on a profound knowledge of the "nature of equations." Descartes had no monopoly on the theory of equations.[76]

Blinded, however, by just such a sense of monopoly, Descartes challenged Fermat to try his hand at a really difficult problem: find the tangent to the curve $x^3 + y^3 = pxy$.[77] He foolishly predicted that Fermat would not be able to do so and suggested that Fermat study carefully the method of the *Geometry*, where, he remarked offhandedly, his opponent would also find the proper method for determining maxima and minima. Descartes himself had not dealt with this topic explicitly simply because he could not take the time to solve every particular problem that lay before him. He had estab-

[76] As his detractors were fond of pointing out. In the years following publication of the *Geometry*, Descartes had to defend himself against charges of having plagiarized Viète or Harriot or both. Even if such charges were unfounded, and even if Descartes' version of the, theory of equations constituted a marked improvement over that of his predecessors, a bit more homework on his part prior to publication might have toned down his claims to unprecedented novelty and originality.

[77] Descartes to Mersenne, 18.I.1638, FO.II.129–130: "Puis, outre cela, sa règle prétendue n'est pas universelle comme il lui semble, et elle ne se peut étendre à aucune des questions qui sont un peu difficiles, mais seulement aux plus aisées, ainsi qu'il pourra éprouver si, après l'avoir mieux digérée, il tâche de s'en servir pour trouver les contingentes, par exemple, de la ligne courbe BDN, que je suppose etre telle qu'en quelque lieu de sa circonférence qu'on prenne le point B, ayant tiré la perpendiculaire BC, les deux cubes des deux lignes BC et CD soient ensemble ègaux au parallélépipède des deux mêmes lignes BC, CD et de la ligne donnée P."
It is interesting to note as a sidelight that neither Descartes in posing the problem nor Fermat in solving it set down anything approximating an accurate drawing of it. So far had both men gone in the direction of complete algebraic analysis of curves that they were both able to dispense with accurate drawings. The equation told them everything they wished to know. Roberval was responsible for the first accurate representation of the curve, which he called the *noeud du ruban* or *galand*. Others better disposed toward Descartes called the curve the *folium Cartesii*, and this name has become common. Roberval's diagram enabled Fermat in a later treatment of the problem (Fermat to Mersenne, 22.X.1638, FO.II.#35) to recognize the multivalued nature of the function and to distinguish between tangents to opposite branches.

lished the Method; he should have to do no more. Besides, finding tangents posed no great difficulties. Instead of vainly trying to show that he knew more than Descartes, Fermat would do better to tackle the problems cited in the *Geometry* as still outstanding. Even if he should succeed in doing so, he would only accomplish this by use of Descartes' methods and thus have no claim to having excelled him. In short, Descartes spoke as one who had run down mathematics, conquered it, and eaten the lion's share; the rest he would leave to Fermat and the other hyenas.

Descartes dispatched his letter to Paris and confidently awaited Fermat's humble surrender. He would have to wait for some time, for Mersenne's lack of tact and sensitivity knew no limits. Instead of properly forwarding the letter to Fermat, Mersenne gave it to Roberval and Pascal. Perhaps he only meant to spare Fermat's feelings, perhaps he wanted to discover whether he had valued Fermat too highly, perhaps he was simply lacking in good sense. Whatever his reason, he badly misjudged the situation. Fermat needed no help in defending his methods, and he certainly did not need the help of Roberval and Pascal. Descartes' veiled references to these two members of the Parisian scientific establishment in the covering letter of 18 January should have sufficed to deter the Minim friar from taking such a step. Receiving a reply to his letter from these two only confirmed what Descartes had mistakenly assumed, perhaps from the outset: that they were operating the strings and that Fermat was merely a puppet. He never accepted their rationale for intervening, to wit, that they wished to mediate in a dispute between two men they respected and whom they knew about equally well. Nor, finally, would Descartes accept their mediation.[78]

The intervention of Roberval and Pascal delayed by six months Fermat's direct response to Descartes' critique of the *Method*. Moreover, by trying to see more in Descartes' second objection than the obvious misunderstanding or misinterpretation that prompted it (though they did clearly locate that misunderstanding) and by their own inability to establish the proper connection between the method of maxima and minima and the rule of tangents, Roberval and Pascal forced Descartes to harden his unreasonable stand. Where a relatively unknown Fermat might have succeeded in soothing the ruffled feelings of Descartes and in clarifying the *Method*, Roberval and

[78] Descartes [to Mydorge], "Réponse à un écrit des amis de M. de Fermat," 1.III.1638, AT.II.13: "Quant à ceux qui ont écrit le papier auquel j'ai répondu en celui-ci, vu qu'ils ont voulu être les avocats de ma partie, en une cause la moins soutenable de son côté qu'on puisse imaginer, j'espère qu'ils ne voudront pas être mes juges, ni ne trouveront mauvais que je les récuse, aussi bien que quelqu'un de ses amis. Car enfin je ne connais à Paris que deux personnes au jugement desquels je me puisse rapporter en cette matière, à savoir M. Mydorge et M. Hardy. Ce n'est pas qu'il n'y ait sans doute plusieurs autres qui sont très capables, mais ils me sont inconnus; et pour ceux qui se mêlent de médire de ma Géométrie sans l'entendre, je les méprise." Roberval, for his part, assured Fermat in a letter dated 1.VI.1638 that he had entered the debate together with Pascal merely out of concern for the truth. That letter, especially its opening paragraph, is filled with Baroque

Pascal, who enjoyed Descartes' fervent enmity, only succeeded in widening the gulf between the two greatest mathematicians of the age and ultimately in hurting Fermat.

The four letters that passed between Descartes and the Parisians offer insight more into the highly refined vituperation of French baroque gentility than into the issue at stake.[79] Unable themselves to clarify the connection between Fermat's two methods and to discern the preeminent correctness of the result Descartes had derived in his first objection, Roberval and Pascal fell into an acrid debate with Descartes over whether the line *BE* could in any way be interpreted as a maximum. *In fine*, they said no, and he said yes. In regard to Descartes' second main point, Roberval and Pascal demanded a little bit of common sense on Descartes' part, enough, that is, to see that the inequality in the derivation stemmed from the *symptôma* of the parabola and had to be changed when the curve under investigation was altered.[80] Descartes for his part continued to insist on Fermat's exact wording and on the lack of method in the *Method*.[81] But even the above *précis* lends more clarity to the participants' positions than they in fact possessed, for they were constantly clad in personal insults and accusations that kept alive an argument that had no reason to exist in the first place. Indeed, from the outset, neither side would address the other directly. Roberval and Pascal wrote through Mersenne, and Descartes, who feared that any letter sent to the Parisians directly would be falsified, addressed his remarks to Claude Mydorge, with whom he pleaded not to let the originals out of sight, but to deliver only copies to Roberval and Pascal.[82]

---

hyperbole, which should not perhaps be taken to indicate as close a relationship between Fermat and Roberval as it might otherwise suggest (letter in FO.II.#29).

[79]An example: (Descartes) "J'admire que l'écrit . . . ait trouvé des défenseurs . . ." (Roberval and Pascal) "Quand M. Descartes aura bien entendu la méthode . . . alors il cessera d'admirer que cette méthode ait trouvé des défenseurs et admirera la méthode même . . ."

[80]See Roberval's extended discussion of this point in his letter to Descartes of April 1637, AT.II.107–113. Having made his mathematical point, Roberval could not resist adding some invective: "Si quelqu'un voulait dire qu'au moins la Méthode serait défectueuse, en ce que l'Auteur n'avertit point qu'il faut raisonner par des propriétés spécifiques, nous lui répondons que ceux qui mêlent de raisonner, ne doivent point ignorer cette condition, qui est de pure Logique, laquelle il suppose être connue par ceux qui liront son Traité, autrement il les renvoie aux écoles, pour y apprendre à raisonner, et les avertit qu'ils ne se mêlent point de reprendre ses Écrits, qu'ils n'entendent bien la Logique et le sujet dont il traite."

[81]Cf., for example, Descartes to Mersenne, 1.III.1638, AT.II.#112.

[82]Descartes to Mersenne, 1.III.1638, AT.II.26: "Je viens à la seconde [lettre], où vous me mandez avoir differé d'envoyer ma réponse *De maximis et minimis* à M. de Fermat, sur ce que deux de ses amis vous ont dit que je m'étais mépris. En quoi j'admire votre bonté, et pardonnez-moi si j'ajoute votre crédulité, de vous être si facilement laissé persuader contre moi par les amis de ma partie, lesquels ne vous ont dit cela que pour gagner temps, et vous empêcher de la laisser voir à d'autres, donnant cependent tout loisir à leur ami pour penser à me répondre. Car ne doutez point qu'ils ne lui en aient mandé le contenu; et si vous l'avez laissée entre leurs mains, je vous prie de voir s'ils n'en auraient

The intervention of Roberval and Pascal reached a dead end in April. With his response to their first defense of Fermat, Descartes had asked Mydorge in a covering letter to consider the case himself and to ask also the opinion of Claude Hardy. On the same day, 1 March 1638, Descartes wrote in addition to Mersenne, asking that the Minim solicit the opinion of Girard Desargues.[83] He kept up the debate with Roberval and Pascal for one more letter, but, feeling that his two opponents were only stalling for time for Fermat to collect his wits, Descartes informed his other correspondents in more detail about what he took to be the failures of Fermat's methods. He "corrected" these failures and, in doing so, showed just how well in fact he did understand the method of maxima and minima.

A quick look at Descartes' own methods shows just how closely the two men stood in their use of the theory of equations and how, therefore, Descartes could have gone so far in interpreting Fermat's method in terms of the *Geometry*. Book II of the *Geometry* contains the following procedure for

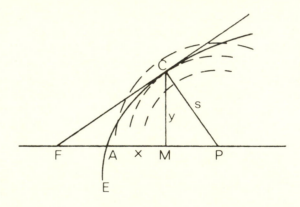

---

point effacé ces mots: *E jusques a*, et mis en leur place: *B pris en*. Cars ils me citent ainsi en leur écrit, pour corrompre le sens de ce que j'ai dit, et trouver là-dessus quelque chose à dire; mais s'ils avaient changé quelque chose dans le mien (de quoi je ne veux pas les accuser), ils seraient faussaires, et dignes d'infamie et de risée. J'envoie ma réponse à M. Mydorge, et je l'ai enfermeé avec la lettre que je lui écris, afin que, si vous craignez qu'ils trouvassent mauvais que vous lui eussiez faites voir plûtôt qu'à eux, vous puissiez par ce moyen vous en excuser." To Mydorge, Descartes wrote on the same day (AT.II.15): ". . . je vous prie d'en retenir une copie avant que l'original leur soit mis entre les mains par le Révérend Père Mersenne." Descartes' almost paranoid distrust of his adversaries was fully unwarranted. Roberval and Pascal had not changed his text any more than Fermat had changed the text of the *Dioptrics* (cf. above, p. 174); they saw no difference between drawing a line from *E* to *B* and drawing a line from *B* to *E*. Moreover, they did not send Fermat the contents of Descartes' critique. Fermat remained in the dark for another two months.

[83] Descartes to Mydorge, 1.III.1638, AT.II.22: "Je vous prie que Monsieur Hardy ait aussi la communication des pièces de mon procès." Descartes to Mersenne, 1.III.38, Alquié.II.37: ". . . je serais bien aise que M. Desargues les vît, s'il lui plaît d'en prendre la peine; mais il ne faut point faire voir un papier sans l'autre, et pour cela je voudrais qu'ils fussent tous écrits de suite en un même cahier."

determining the normal to a curve at any point.[84] Let CE be a curve related to the axis $AM(=x)$ by some (algebraic) relation $F(x,y) = 0$. Let $P_{(v,0)}$ be the intersection point of the axis with the normal to the curve at C, and let $PC = s$. Then,

$$s^2 = y^2 + (v - x)^2,$$

which yields either

$$y = \sqrt{s^2 - x^2 + 2xv - v^2} \text{ or } x = v - \sqrt{s^2 - y^2}.$$

Choosing the more convenient of these two equations, one can then eliminate from $F(x,y) = 0$ one of the unknowns, thereby obtaining a new equation of the form $G(x,s,v) = 0$ or $H(y,s,v) = 0$. In either case, the situation mathematically is the following. For any fixed value of v, the different values of s produce a family of circles about P. If CP is the desired normal, then the corresponding values of v and s will define a circle that is tangent to the curve at point C. For all other values of v and s, the circle will either cut the curve in two points or not cut it at all. Hence, in the case of the one correct circle, the two intersection points have coalesced to one. In the equation $G(x,s,v)$, as in the equation $H(y,s,v) = 0$, the correct values of s and v would so render the equation that two of its roots would be equal. That is, if $G(x,s,v)$ is of degree n, then in the case of the circle tangent to the curve at $C_{(x_1,y_1)}$ it must be of the form

$$(x - x_1)^2 (x^{n-2} + g_1 x^{n-3} + g_2 x^{n-4} + \cdots + g_{n-2}).$$

By comparing corresponding coefficients, one has n simultaneous equations to determine the n unknowns s, v, $g_1, g_2, \cdots g_{n-2}$, and therewith the solution to the original problem.

The futility of any debate between Fermat and Descartes becomes tragically clear when one recalls from Section I above that Fermat's method of maxima and minima rested on the principle that, if M is an extreme value of the expression $P(x)$, then the equation $P(x) = M$ is of the form $(x - x_1)^2 R(x) = 0$, where x_1 is the value of x for which $P(x)$ is maximized (minimized). Both men had come to the same conclusion about the form of an equation with repeated roots. They were, in fact, merely raising different structures on the same foundation.

If Descartes discerned the similarity, he did not reveal the fact. Instead, he offered his "corrections." His understanding of Fermat's method is presented in greatest detail in his letter to Mersenne dated 3 May 1638.[85] After making an inessential emendation to the beginning of Fermat's rule,[86] Descartes

---

[84] *Géométrie* (Leiden, 1637), pp. 341–351.
[85] Descartes to Mersenne, 3.V.1638, FO.II.#27.
[86] "Premièrement donc à ces mots: et inventâ maximâ, il est bon d'ajouter: vel aliâ quâlibet cujus ope possit postea maxima inveniri. Car souvent, en cherchent ainsi la plus grande, on s'engage en beaucoup de calculs superflus." Descartes apparently had in mind here his own approach to the tangent through the normal.

zeroed in on the new term *adequate*. Where Fermat had merely said, "Adequate . . . the two homogeneous expressions equal to the maximum or minimum . . . ," Descartes insisted on a more exact description of what was happening. "One must say, 'Adequate in such a way that the quantity to be found by this equation is one when it is referred to the maximum or minimum, but one resulting from the two that could be found by the same equation and that would be unequal if they were referred to something less than the maximum or greater than the minimum.'"[87] Fermat himself could hardly have wished for a better, more succinct definition of what he had meant by "adequation." Descartes then went on to show how Fermat should have determined the tangent to the parabola.

Let *BDN* be the given parabola, of which *DC* is the diameter; it is required to draw from the given point *B* the straight line *BE*, which meets *DC* at point *E* and is the greatest line that can be drawn from the same point *E* to the parabola (that is, outside of the parabola, as those who are not voluntarily deaf will well understand, whence I call it the greatest).

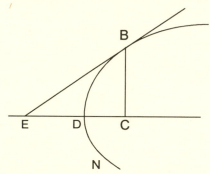

I take *b* for *BC* and *d* for *DC*, whence it follows that the latus rectum is b^2/d, and, without stopping to seek the maximum, I look only for the square of *BC* in terms other than those that are known, taking *A* for the line *CE* and thereafter taking *A* + *E* for the same:

By similar triangles (and in *x*, *y* notation), $\dfrac{x}{b} = \dfrac{x+y}{BC}$; hence $BC = \dfrac{bx+by}{x}$ or $BC^2 = \dfrac{b^2x^2 + 2b^2xy + b^2y^2}{x^2}$. But BC^2 must also be equal to $\dfrac{b^2(d+y)}{d}$, since

B lies on the parabola. Equating these two expressions leads directly to the correct result. Descartes then landed what he thought would be the deathblow to his opponent: "Now it should be noted that this condition which

[87] "Adaequentur tali modo ut quantitas per istam aequationem invenienda sit quidem una cum ad maximam aut minimam refertur, sed una emergens ex duabus quae per eandem aequationem possent inveniri essentque inaequales, si ad minorem maximâ vel ad majorem minimâ referrentur."

was omitted is the same one that I explained on p. 346 as foundation for the method that I used to find tangents, and that it is all the whole foundation on which M. Fermat's rule *should be* founded. So that, having omitted it, he makes it apparent that he found his rule only by *trial and error* (*à tâtons*), or at least that he did not clearly perceive its principles."

The two italicized phrases deserve close attention. For Descartes apparently wanted desperately to attach the second phrase to Fermat's mathematics,[88] and so he ventured to suggest what should be the foundation of Fermat's method. The irony could not have been lost on Mersenne. He already knew that what Descartes insisted should be the foundation of Fermat's method in fact was the foundation. Fermat had already told him; on 20 April 1638 Fermat had written:[89]

> Beyond the paper sent to Roberval and Pascal, and to supplement what is too concise there, M. Descartes should know that, after having drawn the parallel that intersects the tangent and the axis or diameter of the curved lines, I first give it the name it should have by virtue of having one of its points on the tangent; this is accomplished by the rule of proportions derived from the two similar triangles. After having given a name, both to our parallel and to all the other terms of the problem, in the same way as in the parabola, I again consider this parallel *as if* the point that it has on the tangent were in fact on the curved line, and, according to the specific property of the curved line, I compare this parallel by *adequality* to the other parallel drawn from the given point to the axis or diameter of the curved line.
>
> This comparison by *adequality* produces two unequal terms which in the end produce (according to my method) the equality that gives us the solution of the problem.

Written independently of one another, the letters of Descartes and Fermat contain only distinctions without differences. Wherein, then, did the two men disagree?

Their disagreement ultimately rested on the point that Descartes had made at the beginning of his letter of 18 January. Fermat was content, as Descartes had correctly discerned, to employ a method based on a clear and evident "false position." He was willing to use the phrase "as if," to move from a counterfactual assumption to a factual result. Descartes tried to avoid such procedures entirely. Both methods derived from the same source in the theory of equations, to wit, that an equation with a repeated root x_1 must be of the form $(x - x_1)^2 R(x) = 0$ but they used that source in entirely different ways, as a close comparison of the two methods of tangents will show.

[88]So much so that he repeated the phrase verbatim in his letter to Hardy written a month later (FO.IV.48–51).

[89]Fermat to Mersenne, 20.IV.1638, FO.II.#26. Regarding the inclusion of the note as an appendix to this letter, see FO.II.137, n. 1.

The lack of substance in their controversy became clear to both men practically from the minute they entered into direct correspondence. Fermat clarified all points in the two papers he sent to Mersenne toward the end of June or the beginning of July.[90] He had composed the first, a memoir entitled "The Method of Maxima and Minima Explained and Sent to M. Descartes by M. Fermat," before he saw Descartes's letter of 3 May. In it he attempted to codify his method of tangents in a clearer way. Whatever the curve, he noted, one begins by drawing the same standard diagram, and one denotes the various unknowns in the following standard way: $BA = b, BC = d,$

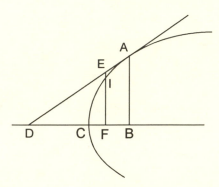

$BD = x.$ "We take an *arbitrary* (*à discrétion*) point, such as E, on the tangent, from which we draw EF parallel to AB, and we let line BF be y." Then, $CF = d - y$, and $FE = (bx - by)/x$. "Even though line FE is unequal to the ordinate drawn from point F to the curve, I nevertheless consider it *as if* it were in fact equal to the ordinate and consequently compare it by *adequality* to the line FI according to the specific property of the curve." That is, in modern terms, if $y = f(x)$ is the curve, then the tangent is determined by employing the method of maxima and minima to the equation

$$\frac{bu - bv}{u} = f(x_1)$$

where u is the subtangent, v the decrement, and x_1 the abscissa of the point to which the tangent is to be drawn.

Up to this point, Fermat had done nothing more to his original method than to refine the algorithm. Adequality had begun to take on a life of its own.[91] Originally designed to mask the reasoning presented in Section I, the Diophantine term permitted the extension Fermat was now making, while maintaining a close correspondence between the steps of the algorithm and the more rigorous reasoning on which the method was based.

[90]FO.II.#30,31. Fermat may have already sent the memoir "Volo mea methodo . . ." to Roberval in April (see above, n.46). If so, no one mentioned it in the correspondence that followed.
[91]See below, p. 213.

After justifying the method's utility by solving the problem Descartes had set for him on 18 January—to find the tangent to $x^3 + y^3 = pxy$—Fermat came to the question of the exact relationship between the method of maxima and minima and the rule of tangents. And here his mind took a sudden turn. Instead of offering the original argument, something akin to the reconstruction offered in Section III above, Fermat offered the following:

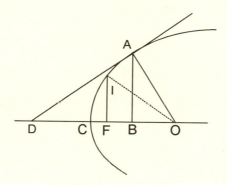

Given the curve AC as in the diagram and point A on it, one seeks not a maximum but a minimum, to wit: "We seek, then, the point O on the diameter such that the line OA is the shortest that can be drawn from point O to the curve. Having found point O by the method, join the two points O and A by the line OA and draw the line AD perpendicular to OA." After showing how one proves that AD so drawn is indeed the tangent, Fermat moved to the more important question of how his method of maxima and minima yields point O. He again took the parabola as an example: given parabola CIA and point A on it, find point O such that OA is the shortest of all lines joining O to the parabola. Establishing his notation, Fermat set $BC = d$, $BA = b$, the latus rectum $= c$, and $OB = x$; whence $OA^2 = x^2 + b^2$. If now $OF = x + y$, then $FI^2 = cd - cy$, and $OI^2 = OF^2 + FI^2$, or

$$OI^2 = x^2 + y^2 + 2xy + cd - cy.$$

By adequation, $OA^2 \approx OI^2$, or

$$x^2 + b^2 \approx x^2 + y^2 + 2xy + cd - cy.$$

The term x^2 drops out on both sides, as do b^2 on the left and cd on the right—since for the parabola $b^2 = cd$—and there remains

$$y^2 + 2xy \approx cy,$$

or, by application of the method, $2x = c$. Consequently, x will be equal to one-half the latus rectum, and the tangent follows immediately.

The shift from the original method of tangents to the above derivation is as surprising as it is sudden. For it is clearly *ex post facto*. It could not have been the original method for the simple reason that its steps do not follow the algorithm of the *Method* or that which opens Fermat's paper. To offer the determination of the normal by the method of maxima and minima completely severs the algorithm from the rigorous demonstration of the method. What led Fermat to take such a radically new approach? One can imagine that Descartes himself showed the way. In the first place, Fermat has in effect only repeated here what Descartes did in the letter of 18 January to disprove the *Method* by *reductio ad absurdum*. For compare the results. Descartes applied the method of maxima and minima to the line *BE*, taking the tangent as a maximum. When he reached the result $(b^2/d) + 2x = 0$, he could find no sense in it. When Fermat applied the same method to the line *OA*, he found that the distance from the foot of the ordinate to the point of intersection of the normal with the axis was equal to one-half of the latus rectum. He was looking for his result, of course, and Descartes was not. Why was he looking for it? Probably because since first inventing his own method of tangents he had read the *Geometry* and found that Descartes approached the problem of tangents through the normal and did so by determining the normal as a minimum. What better way could Fermat find to show the simplicity and efficacy of his method of maxima and minima than to apply it to the very technique Descartes had employed?

Hence, despite the new tack taken in the "Méthode expliquée," it seems unlikely that Fermat had in fact changed his mind. The method of tangents did rest on the determination of a minimum, but it was the quasi-minimum discussed in Section III and not the normal. The attempt to ground the method of tangents in the determination of the normal clearly stemmed from Fermat's desire to show Descartes how much simpler his method of maxima and minima made Descartes' own method of determining tangents.[92]

Nonetheless, the controversy with Descartes had significantly altered Fermat's view of his own method of tangents. For, as noted above, in the course of time the term "adequality" had begun to become more than a mask for the real reasoning behind the method. The tortuous argument presented in Section III had become simpler by virtue of taking *OI* as counterfactually equal instead of as representing a maximum or minimum. But Fermat's rule of tangents had become simpler for a more important reason. The original method, born of Fermat's research during the late 1620s, had been tortuous because it still operated from the ancient proportional definition of a curve. Such a definition had made it necessary to speak in terms of the inequality $BC^2/OI^2 < CD/DI$. Since 1635 or 1636, however, Fermat had a new manner of expressing a curve, and his "Méthode expliquée" and the letter to Mersenne of 20 April show the fruits of that new means of expression. The *Introduc-*

---

[92] He pointed out in the "Méthode expliquée" that if Descartes' method of normals were applied to a curve defined, say, by the equation $by^9 + by^7 + by^5 + by^3 + by = x^{10} - x^9d - x^7d + x^5d - x^3d - xd$, simply removing the radicals would make the final expression all but intractable.

tion had made it possible to describe curves in terms of *equations;* equations had become the *propriétés spécifiques* of curves.[93] And so the original rule of tangents, still preserved in the *Method*, found a new and simpler form of expression in turn. In the equation $F(x, y) = 0$ defining the curve, substitute $(bu - bv)/u$ for y and x_1 for x and proceed by the method of maxima and minima. What was lost through this simplification was the original close connection between the method of maxima and minima and the rule of tangents. Both became, in a sense, applications of the technique of adequality, which thereby began the transition from codeword to mathematical concept.[94]

The method of maxima and minima itself underwent a fundamental revision in the course of the controversy, but before investigating this revision we must follow out the last stages of the debate. Fermat concluded his "Méthode expliquée" by citing several of the problems in the *Geometry* that could be treated using his methods. But, he said, "that merits a separate discourse, and, if he agrees, we will confer on the matter when he wishes. I want him to know only that our propositions concerning maxima and minima and tangents to curved lines were completed eight or ten years ago and that several persons who saw them five or six years ago can testify to that fact."

Descartes, for his part, seems to have grown gradually aware of his basic mistake. Desargues, whom he had asked to referee the debate, had written through Mersenne in April[95] to tell Descartes that, as far as he could determine, most of the dispute rested on a misunderstanding. He agreed that Fermat's wording failed to make clear that the inequality had to be changed for each curve and that therefore the method did seem less than general. But, by the same token, Fermat clearly meant the change to be made. "Hence," he summed up, "M. des Cartes is right, and M. de Fermat is not wrong." He also agreed that Fermat had unambiguously asserted that the rule of tangents derived from the method of maxima and minima. But Desargues also warned Descartes that he (Desargues) had not yet had a chance to examine the entire

---

[93] The term *proprietas specifica* does not appear in Fermat's work until his letter to Mersenne of 20.IV.1638 (FO.II.#26). It is used repeatedly by Roberval and Pascal in their exchange of letters with Descartes, though there it refers to the *symptômata* of Apollonius and serves as the Latin translation of that Greek word. Fermat's use of it in the April letter and thereafter, however, refers to the algebraic equations defining the curves in accordance with the method of the *Introduction*, and hence is slightly different from that of Roberval and Pascal. Given the absence of communication during this period between Paris and Toulouse, it is hard to imagine that Fermat got the term from his Parisian colleagues. Yet it is also difficult to discern the source of the term in his own work, unless he simply was matching his translation of the *symptômata* into algebraic equations with a translation of the technical term from Greek into Latin. Of fundamental importance to the nascent concept of a function, the term simply appears suddenly in Fermat's writings and begins to have its effect, as here in the reformulation of the original method of tangents.

[94] See below, pp. 210–213.

[95] Desargues to Mersenne, 4.IV.1638, FO.II.39–47. Ironically, this letter of intervention into someone else's argument is the sole extant autograph letter of Desargues. It is preserved in the Bibliothèque de Lyon.

Method and so could make no judgment on how clear that connection might be.

It also seems to have become clear to Descartes that the absence of any reply from Fermat to his letter of 18 January was in fact due to Fermat's ignorance of its content. When Fermat did see that letter, he wrote to complain of its rude tone.[96] Even before he had seen the "Méthode expliquée," Descartes answered that complaint in mild words, and his explanation showed some uncertainty about the conspiracy of the existence of which he had been so sure in January:[97]

> I beg him most humbly to excuse me and to consider that I did not know him. Rather, his *De maximis* came to me in the form of a cartel on the part of him who had already tried to refute my *Dioptrics* even before it was published, as if to smother it before its birth, having had a copy of it that had not been sent to France for that purpose. Hence it seems to me that I could not have replied to him in words any softer than I used without evidencing some sort of laxity or some sort of weakness.

Growing awareness of Fermat's true relationship to Roberval and Pascal, the always calm tone of Fermat's own letters, and the hints contained in those letters that Fermat and he did not in fact disagree on the essentials all tended to cool Descartes' temper and induce in him a bit more humility toward Fermat's achievement.

Receipt of the "Méthode expliquée" and the covering letter which responded to additional points raised in Descartes' letter of 3 May ended the dispute for Descartes. On 27 July he wrote to Fermat and could only admit that, "seeing the last method that you use for finding tangents to curved lines, I can reply to it in no other way than to say that it is very good and that, if you had explained it in this manner at the outset, I would not have contradicted at all."[98]

Descartes did admit his concern over finding tangents by a minimum rather

---

[96] Fermat first wrote in February, after receiving word from Mersenne that Descartes was unhappy both about Fermat's critique of the *Dioptrics* and about Fermat's own *Method*. About the first point he was not surprised: ". . . parce que les choses de physique peuvent toujours nous fournir de doutes et entretenir les disputes." But he was indeed surprised that Descartes should find anything with which to disagree in the *Method* (Mersenne had mentioned only the existence of the letter of 18.I.1638, nothing of its content): ". . . puisque c'est une vérité géométrique, et que je soutiens que mes méthodes sont aussi certaines que la construction de la première proposition des *Eléments*. Peut-être que les ayant proposées nuement et sans démonstration, elles n'ont pas été entendues ou qu'elles ont paru trop aisées à M. Descartes, qui a fait tant de chemin et a pris une voie si pénible pour ces tangentes dans sa *Géométrie*." Nonetheless, Fermat was willing to stop the affair right then if Descartes was so sensitive to criticism. Again, in his letter to Mersenne of 20 April, he repeated his offer to remain silent if Descartes was getting too upset.

[97] Descartes to Mersenne, 29.VI.1638, AT.II.#26.

[98] Descartes to Fermat, 27.VII.38, FO.II.#32.

than a maximum—he still could not see that Fermat's method was indifferent to the nature of the extreme value—but he had to admit that Fermat was right, whatever the particular details.

V. THE AFTERMATH: PROCEEDING BY TOUCH

Unfortunately, with Descartes as an opponent, it did not suffice to be right. Indeed, it hurt more than it helped. Although Descartes addressed Fermat in the letter of 27 July in the most complimentary terms, as he did also in another letter written toward the end of the year,[99] Descartes had too much at stake to retire to his study and lick his wounds. He had argued from the position that Fermat's method did not work because Fermat did not have the insights of the *Geometry*. But he now saw clearly that Fermat indeed had those insights and that he had gained them independently of the *Geometry*. Moreover, Fermat's method not only worked, but it was clearly simpler to apply than Descartes' own. An honest man would have to admit defeat to his opponent.

Descartes' stubborn pride, however, prevented him from making his admission public. Two letters to Constantine Huygens, one written during the middle of the controversy and one sometime after it, show no trace of its outcome.[100] The second letter merely claims the total victory that the first promised. As late as 1640, Descartes still wrote to Mersenne to speak of Fermat's basic inadequacy as a mathematician and thinker.[101] The charges made

[99] Descartes to Fermat, 11.X.1638, FO.II.#34. The letter, the opening sentence of which is quoted above, Chap. II, p. 58, contains the most effusive praise, including approval of Fermat's determination of the tangent to a cycloid, "une preuve très assurée."

[100] Descartes to Constantine Huygens, 9.III.1638, Alquié.II.42-45: "Il y a de plus un conseiller de Toulouse qui a un peu disputé contre ma *Dioptrique* et ma *Géométrie*; puis quelques amis qu'il a à Paris lui ont voulu servir de seconds, mais je me trompe fort, si lui ou eux se peuvent dégager de ce combat sans confesser que tout ce qu'ils ont dit contre moi est paralogisme." Of course, Fermat disengaged himself from combat by forcing Descartes to admit his own error. Yet, in another letter to Huygens, written on 19.VIII (hence, following the letter of 27.VII to Fermat), Descartes did nothing but repeat his charges of a conspiracy made in his letter of 29.VI to Mersenne; he gave no indication whatsoever of the outcome of the dispute, preferring apparently to let Huygens believe that the prediction of 9.III had proved accurate.

[101] Descartes to Mersenne, 23.VIII.1638, FO.IV.60-64: ". . . la démonstration prétendue de la roulette envoyée par M. Fermat . . . [est] . . . le galimatias le plus ridicule (! cf. above, n. 99)." The letter also contains Descartes' own resumé of the whole dispute. Fermat's derivation of the tangent to the cycloid, as Descartes well knew, was correct. Cf. also Descartes to Mersenne, 28.X.1640, Alquié.II.264-272: "M. de Zuylichem [Constantine Huygens] m'a envoyé quatre traités, que vous lui avez fait copier. . . . Le 3 est de M. Fermat pour les tangentes, où le premier point n'a rien de nouveau, et le suivant qu'il dit que j'ai jugé difficile [*folium Cartesii*], n'est aucunement résolu. Et bien qu'en l'exemple qu'il donne de la roulette [cycloid], le *facit* vienne bien, ce n'est pas toutefois par la force de sa règle; mais plutôt il parait qu'il a accommodé sa règle à cet exemple." Fermat's solution to both the problems mentioned were correct and direct results of his method (see below, pp. 212–213). Descartes again wrote to Mersenne on 4.III.1641, asking that no copy of the *Meditations* be sent to Fermat: ". . . je crois bien qu'il sait des mathématiques, mais en philosophie j'ai toujours remarqué qu'il

by Descartes lingered over Fermat's reputation like smoke over a battlefield; they masked the victor.

Even though Descartes admitted that Fermat's rule worked, he continued to demand a demonstration, and from his letters one can gather that Mersenne and others took up the cry.[102] It was against this background that the *Analytic Investigation* took shape. By 1639, Fermat had found a new version of his method of maxima and minima, but it would not do for his immediate needs.[103] He saw clearly that he would have to reveal the process by which he had come upon his original method. The letter from Mersenne cited on p. 158, containing the algebraic problems posed by Dounot, presented the right opportunity. Their solution involved the application of Viète's method of syncrisis. The problem containing surds gave Fermat an opportunity to show Mersenne and others his ingenious emendation of Viète's method. And, since Fermat was on the subject of syncrisis, he might as well put together the demonstration of maxima and minima that he had derived from his investigation. "While I was pondering . . . ," he began. We have seen the main content of that demonstration.

But we have not seen the conclusion of the *Analytic Investigation*, and that conclusion indicates the purpose for which it was written. After deriving the solution to the problem of Pappus VII,61, Fermat added: "Confidently, then, as earlier, so we now assert for always that the legitimate, but not fortuitous (as it has seemed to some) determination of maxima and minima is contained in this unique and general decree: . . ."[104]

That "unique and general" decree was a verbatim copy of the body of the *Method*. To his reiteration of his 1636 paper, Fermat added a challenge: "If there are still those who claim our method to be due to luck, *hos cupiam similes tentando excudere sortes* (I would like them to fashion similar luck *à tâtons*)."[105] *A tâtons*, for that is what Fermat meant here by the word

---

raisonnait mal." The most famous of all slights, of course, was the remark made to van Schooten; see above, Chap. II.

[102]Descartes to Mersenne, 20.II.1639, *Alquié*.II.126-128: "M. de Beaune me mande qu'il désire voir ces petites observations sur le livre de Galilée que je vous ai envoyées; et puisque vous lui avez fait voir notre dispute de M. Fermat et de moi, touchant sa règle pour les tangentes, je serais bien aise qu'il vît aussi ce que j'en ai une fois écrit à M. Hardy, où j'ai mis la démonstration de cette règle, laquelle M. Fermat n'a jamais donnée, quoiqu'il l'eût promise, et que nous l'en ayons assez pressé, vous et moi." Fermat had indeed promised the demonstration in a letter to Mersenne on 15.VI.1638 (CM.VII. 285-286). Assuring Mersenne and Descartes that the method was correct and simpler than Descartes', he promised: "Il ne me reste qu'à faire voir sa démonstration à quoi je donnerai mon premier loisir."

[103]Namely, the demonstration sent to Brûlart; see below, § VI.

[104]*Anal. Inv.*, FO.I.153.

[105]*Ibid.* To this Fermat added a challenge problem to close the memoir: "Qui hanc methodum non probaverit, ei proponatur: Datis tribus punctis, quartum reperire, a quo si ducantur tres rectae ad data puncta, summa trium harum rectarum sit minima quantitas." Fermat himself never recorded his solution to the problem, if indeed he had one. J. E. Hofmann has traced the attempts at solution made by Italian mathematicians who read the memoir when it accompanied Mersenne to Italy (see above, n. 4): "Uber die

tentando. He had not merely been lucky, he did not operate *à tâtons*, his method was not fortuitous. He had worked it out systematically on the basis of Viète's theory of equations. The *Analytic Investigation* showed how he had done it, back in the late 1620s.

VI. LEARNING NEW TRICKS: THE LETTER TO BRÛLART

The simplicity of Fermat's methods struck no sympathetic chord in Descartes. Engrossed in defending the *Discourse* and then in preparing the manuscript of the *Meditations*, the prelude to his *magnum opus*, the *Principles of Philosophy*, that old dog had neither time nor patience for new mathematical tricks. Besides, the discovery of worthy competition and the obvious loss of monopoly over algebraic analysis had probably taken some of the pleasure out of mathematics for Descartes. But Fermat had no wider intellectual horizons than mathematics, and his controversy with Descartes brought him new insights which he hastened to employ.

In the letter to Mersenne on 3 May, Descartes had observed in connection with his "correction" of the procedure for determining the tangent to a parabola that: "Beyond that, one should form two equations and show that one finds the same thing in supposing EI to be $A + E$ as when one supposes it to be $A - E$. For without that the reasoning behind this operation is incomplete and proves nothing."[106] Descartes had meant, of course, that Fermat had failed to show that it made no difference on which side of BC one took OI. Fermat, however, had read the objection in terms of his method of maxima and minima as a whole, and he had replied in his letter of the end of June that: ". . . it is not necessary to say that one must carry out two operations, one for $A + E$ and the other for $A - E$, because one alone suffices for the construction, even though the demonstration [of the method] that I have not yet given derives its principal foundation from the fact that $A + E$ yields the same thing as $A - E$."[107]

One cannot be sure on the basis of this one hint whether Fermat had already found his new version of the method, and that therefore the "demonstration that I have not yet given" differed from that provided by the *Analytic Investigation*. For, in one sense, the demonstration in the *Analytic Investigation* did rest on the foundation he cited above, though one would find it hard to call it the "principal foundation." Fermat's emendation of Viète's syncrisis consisted in replacing the roots x and y with the roots x and $x + y$, where y then became the difference between those roots. It clearly made no difference whatever to the result of the method if one took instead x and $x - y$. Had Descartes seen the *Analytic Investigation* instead of the

geometrische Behandlung einer Fermatschen Extremwert-Aufgabe durch Italiener des 17. Jahrhunderts," *Sudhoffs Archiv*, 53(1969), pp. 86–99.
[106]FO.II.145.
[107]Fermat to Mersenne, end of June or beginning of July 1638, FO.II.152.

Method, he would hardly have made his remark. Of course, he probably would have said nothing at all.

Fermat may not, however, have been referring to the *Analytic Investigation* when he wrote the above. He may have been referring to a demonstration which he finally set down on paper and sent to Pierre Brûlart de Saint-Martin on 7 April 1643.[108] He sent it through Mersenne, and in the covering letter he returned for the first time since June 1638 to Descartes' remark that $A + E$ and $A - E$ "yield the same thing."[109] One must, he insisted, be clear about what the phrase itself means:

There could still be some equivocation concerning what I have said [in the letter to Brûlart], to wit, not only that $A - E$ should· yield the same equation as $A + E$, but also that, if $A + E$ yields less than A, $A - E$ should also yield less than A. For it appears at first that, if $A - E$ yields the same equation as $A + E$, it is infallibly the case that, the one yielding less than A, the other will similarly yield less than A. Such, however, is not the case, as I think I have explained by the example that I have appended.

But, to remove all equivocation, when I say that $A - E$ should yield the same equation as $A + E$, I mean that, in following my method by the positing of $A - E$, one should find A equal to the same quantity as if we employ $A + E$ by the same method. But, when I add as a second condition that if $A + E$ yields less than A, $A - E$ should also yield less than A, I mean that if, by the positing of $A + E$, the homogeneous quantities that represent the maximum are less than the homogeneous quantities that represent the maximum when A alone is posited, then in the same way, by the positing of $A - E$ the homogeneous quantities that represent the maximum should be less than the homogeneous quantities that represent the maximum when A alone is posited.

One may immediately understand, then, by "yield the same thing" that the same final answer results from the method independently of whether one takes an increment or a decrement. But Fermat pushed the idea somewhat further. The answer should be the result of the same equation. That is, his method found the extreme value of some polynomial $P(x)$ by deriving from

[108] Fermat to Brûlart de Saint-Martin, 31.III(?).1643, FO.*Suppl.* 120-125. The letter, preserved only in the Florence MS. (fols. 113v-115r), was first discovered by Giovanozzi and published by him in *Archivio di Storia della Scienza*, 1(1919), pp. 137-140. It represents perhaps the most important discovery of new material after the publication of FO, though Tannery and Henry did note (FO.II.253, n. 2) that such a letter had been written. Fermat had known Brûlart since 1640, when the two men began exchanging results in number theory (see below, Chap.VI, §III), and had promised him a proof of the method of maxima and minima sometime early in 1643. He finally sent it, along with a covering letter, to Mersenne on 7.IV.1643. In the MS. itself, the letter bears the date 31.V, but the MS. is only a copy of the original letter, and De Waard gives a compelling argument (FO.Suppl.120-121) for viewing *maii* as a misreading of *martii*.

[109] Fermat to Mersenne, 7.IV.1643, FO.II.253-254.

it, through the use of another unknown y, some polynomial $P_1(x)$ which, set equal to 0, yields the value of x for which $P(x)$ attains an extreme value. For the method to give "the same equation" meant, then, that, whether one took an increment $x + y$ or a decrement $x - y$, the resulting $P_1(x) = 0$ had to be the same. Moreover, the increment and the decrement had to "give the same thing" in another sense: if $P(x + y) \gtreqless P(x)$, then so too $P(x - y) \gtreqless P(x)$. The importance of the second condition became eminently clear in the accompanying letter to Brûlart, to which we will turn shortly.

Fermat never wasted words. Feeling that he owed Mersenne these final thoughts on a matter raised by Descartes, he closed the matter as far as Mersenne was concerned: "There you have what I thought I should say to you on this subject. For to make the matter entirely clear and perfectly demonstrated would require a complete treatise, which I will not refuse to write when I can find sufficient leisure to do so."

Regrettably, Fermat never found that leisure to write his treatise. Had he done so, one might then have been able rightly to speak of him as the founder of the differential calculus. For what he did record of his thought in this regard, as found in his letter to Brûlart, shows that his mind was wandering toward the questions and methods that lay at the heart of the differential calculus and that he was beginning perhaps to realize the difficulties inherent in such a new approach to mathematical problem-solving. The letter also shows how carefully Fermat did read the *Geometry* when he finally did obtain a copy, for Descartes' more sophisticated theory of equations makes its presence clear.

Fermat opened his letter to Brûlart with an explicit statement of the foundations that underlay his method of maxima and minima. This time he would hide nothing. The first principle he had already discussed in the *Analytic Investigation*: an extreme value is unique. In general, the conditions to be maximized or minimized, when expressed in a general equation, allow several solutions, but for the maximum or minimum value the equation has a repeated solution. The line of argument and the particular example given, the division of a line into two segments forming a maximum solid ($\max(bx^2 - x^3)$), follow closely the exposition of this principle in the *Analytic Investigation*. In the letter to Brûlart, however, the argument takes on a slightly different twist by the particular implication of this first principle that Fermat chose to emphasize. He summed up the first principle of his method by noting: "Hence one must seek a unique point, on either side of which all the terms of the problem are either always greater or always less than that which is produced by the point sought." That is, to use modern terms, if M is the maximum value of $P(x)$, then there is a unique x_0 for which $P(x_0) = M$. Any other point x_1 yields a value for $P(x) < M$.[110] Moreover, the equation

---

[110]The argument only holds, of course, within a limited neighborhood of x_1, since several local maxima or minima may be involved. Fermat, however, knew nothing of the

$P(x) = M_1 < M$ has more than one root x_1. Both the *Analytic Investigation* and the *Method* (especially via adequality) had laid stress on the equal values of $P(x)$ yielded by the two roots bracketing the desired unique x_0. But such an approach proves tortuous even for the most sympathetic reader and leaves uncomfortably vague the relationship between these roots and the single desired root itself. The new approach of the letter to Brûlart, which concentrates on the roots rather than the particular values of $P(x)$ and uses both an increment and a decrement in the roots, now opened a straighter path to the method by placing emphasis on the inequalities in $P(x)$ determined by the neighboring points.

The second foundation, too, had in a slightly different form underlain the earlier version of the method. The method had originally derived from an examination of the relationship between the roots and the coefficients of the equation that expresses the conditions of the maximum or minimum, and that relationship could be determined only under the assumption that the equation had at least two distinct roots. The point determinative of the extreme value could only be established via points lying on either side of it. In the letter to Brûlart, Fermat modified his statement of the second foundation so as to stress the manner in which both an increment and a decrement were involved in establishing the required relationship:

> Hence, it is necessary to compare the unique point with those that can be imagined on either side [of it]. That cannot easily be done by positing one [equation] alone, because if, for example, we call the line that gives us the unique point A, we must add to it or subtract from it another quantity in order to seek the ratio between the unique point and those on either side of it. Hence, in order to make the comparison with another point taken arbitrarily on one side of the unique point, we can call the line that yields it $A + E$; and, similarly, to make the comparison with another point taken on the other side of the unique point, we can call the line that yields it $A - E$, forming the first by addition and the second by subtraction. Then we must find a method by means of which $A + E$ and $A - E$ yield the same term representing A, in order that the said A represent the middle point. Everything on these two sides exceeds or falls short depending on whether we seek the maximum or the minimum.

Fermat had clearly changed his mind about his method in an essential way. He had always believed notation to be an arbitrary matter. He continued in this belief. But, as the new version of the method indicates, he began to realize the importance in exposition of a choice of notation that made clear what was going on. In the earlier versions, x and $x + y$ had been used to de-

---

distinction between local and global extreme values, taking all to be global. See above, n. 28, and below, p. 204.

note the points bracketing the point determinative of the maximum, and then x had been used again to denote the maximum point. Such denotation unfortunately suggested, as the version of the method printed in Hérigone's *Course in Mathematics* amply demonstrates,[111] that y was equal to zero all along. By indicating through explicit notation that both an increment and a decrement were involved, Fermat had already taken a large step toward clarifying the foundations of his method. The conditions that the method must meet in order to conform to Descartes' objection, or rather to Fermat's clarification of Descartes' objection, could be stated more precisely. If x is the value determinative of the extreme value, then the method must yield that x regardless of whether one uses $x + y$ or $x - y$ to find it. Also the value of the expression giving the conditions to be maximized or minimized, when $x + y$ and then $x - y$ are substituted for the unknown, must in each case be less than the value for x as argument in the case of a maximum, or greater than that value for a minimum.

As Fermat had noted in his letter to Mersenne, however, the method must yield not only the same final x but also the same equation of which that x is the solution. This extended condition requiring the same equation became, in the next two paragraphs of the letter to Brûlart, the new keystone of the method itself:

> Now it appears that my method yields the same equation for $A + E$ as for $A - E$, which experience and reason will show you right away. For $A - E$ always yields the same terms as $A + E$ and differs only in that the odd powers bear contrary signs, which does not change the equation.
>
> Hence, it is clear that $A + E$ yields the same equation as $A - E$ by my method. But this does not entirely suffice, for, if it were only a question of finding the same equation using $A + E$ as using $A - E$, we could equally well take the two terms that contain, say, $Eq.$ or $Ec.$, etc., as those which only contain E, and equate them to one another. But this will not work. Hence, besides the preceding condition, which requires that $A + E$ and $A - E$ yield the same equation, we must add another condition which requires that, if $A + E$ yields less than A, $A - E$ also yields less than A, and similarly, if $A + E$ yields more than A, $A - E$ also yields more than A.

Only the example given in the text makes clear what Fermat had in mind in these two paragraphs. He again took a problem that he had already treated in the *Analytic Investigation*: to divide a line such that the solid formed by one of the segments multiplied by the square of the other is a maximum. If x is one of the segments and b the given line, then the expression for the solid is $bx^2 - x^3$. If x_0 is the value of x for which the expression attains a maximum,

---

[111]In the edition of the *Method* that Mersenne adapted for the 1644 *Supplément* to Hérigone's *Cursus mathematicus* (p. 59ff.), he simply set E (i.e. y) equal to zero at the outset and then went on to operate with it as if it denoted a real quantity. The result is a method containing a patent division by 0.

then

$$bx_0^2 - x_0^3 = M,$$

where M is the maximum solid constructible under the given conditions. From the principle of uniqueness, it follows that any other values of x, say $x_0 \pm y$, will give

$$b(x_0 \pm y)^2 - (x_0 \pm y)^3 < M.$$

Fermat then moved to the implications of this last statement.

First, in what way do $x_0 + y$ and $x_0 - y$ "give the same equation"? Here Fermat borrowed a technique from Book III of the *Geometry*. The two arguments may be substituted for x in the original expression, the binomials expanded, and the results ordered according to powers of y. The result will be two polynomials in x_0 and y which differ only in the signs attached to the odd powers of y, the coefficients of the powers of y being the same in each case:

$$(bx_0^2 - x_0^3) + y(2bx_0 - 3x_0^2) + y^2(b - 3x_0) + y^3$$
$$(bx_0^2 - x_0^3) - y(2bx_0 - 3x_0^2) + y^2(b - 3x_0) - y^3.$$

The task is to use these expressions in conjunction with the basic inequality that serves a principle for the method to arrive at an equation which yields the value of x_0. The value of the original expression $bx^2 - x^3$ will be less than the maximum M for all arguments other than x_0. Hence, the values for $x = x_0 + y$ and $x = x_0 - y$ will both be less than M. From the above expressions it is clear that such can be the case only if the coefficient of the y term is identically equal to 0. For, if it is greater than 0, the value for $x_0 + y$ will be greater than M; if it is less than 0, the value for $x_0 - y$ will be greater than M. Therefore,

$$2bx_0 - 3x_0^2 = 0,$$

or, discounting the extraneous root $x_0 = 0$,

$$x_0 = \frac{2}{3}b.$$

The equation that yields x_0, then, arises from setting the coefficient of the y term equal to 0. There is only one such equation, since the expansions of the original expression for the arguments $x_0 \pm y$ differ only in the signs attached to the odd powers of y, not in the coefficient expressions themselves. And that is what Fermat meant when he asserted that $x + y$ and $x - y$ "yield the same equation."

But that is only one sense of the assertion that $x + y$ and $x - y$ yield the same thing. For, if one were concerned only to get an equation, one could choose the coefficient expression of any of the powers of y. The other sense of the assertion establishes the justification for choosing the coefficient expression of y. For to say that both arguments yield the same thing must also be taken to mean that if the value of the original expression for $x + y$ is less than the value for x (i.e. the maximum value) then so too is the value for $x - y$, a condition that can only obtain if the coefficient of the y term is identically equal to 0. Used in this sense, the concept of "giving the same thing" enabled Fermat to supplant nicely the notion of adequality with a technique that shifted attention to the inequalities on which the method was now based.

Although the other coefficient expressions do not contribute to the determination of the value of x_0, they are not useless to the analyst, as Fermat went on to note. In order to conform to the basic inequality, it is also the case that the coefficient of the positive y^2 term must be less than 0 for a maximum, or greater than 0 for a minimum.[112] The modern reader will recognize immediately that that coefficient is the second derivative of the original expression (divided by 2) and that Fermat therefore clearly set forth in the letter to Brûlart of 1643 what is now called the second-derivative criterion for extreme values.[113]

---

[112] FO.*Suppl*.125: "La raison principale de ceci est que les deux termes marqués par y^2, étant en plus grande raison que ceux qui sont measurés par les plus hautes puissances au dessus de y^2, ils serviront de clef pour déterminer la plus grande ou la plus petite. Car si le terme marqué + est moindre que le terme marqué −, en ce cas la proposition aboutira à la recherche de la plus grande; que si le terme marqué + est plus grande que le terme marqué −, en ce cas la question sera de la plus petite. Que si nous employons x − y, les deux termes measurés par y^2 auront chacun le même signe." Regarding Fermat's opening remark here, see below, p. 203.

[113] Apparently, it never occurred to Fermat to ask what would result from setting the coefficient of the y^2 terms equal to 0, nor did he recognize that the solutions of that equation would yield the curve's points of inflection. His one discussion of points of inflection, contained in his *Doctrine of Tangents* (see below, § VII), points to the conceptual source of the omission. There (FO.I.166ff.) he briefly outlines the problem "of investigating by the art the points of inflection at which the curvature changes from convex to concave, and conversely." He asserts without proof that the tangent drawn to a point of inflection forms with the axis an angle less than that formed by the tangent to any other point on the curve. Hence, if in the diagram point H marks the point of inflection of the curve *AHFG*, "we seek first by the above method [of tangents] the *property of the tangent* to any point on the curve. Having found this, we seek by the doctrine of maxima and minima the point H for which, when perpendicular *HC* and tangent *HB* have been drawn, the ratio of *HC* to *CB* is minimized; for, that being the case, the angle at B

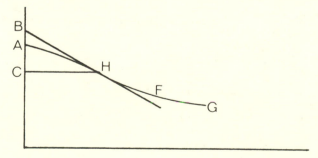

But what of the final y^3 terms in each expression? Fermat deals with them in concluding remarks that, on the one hand, show how his new version of the method made the difficulties inherent in his new approach to mathematics more evident than had previous versions while, on the other hand, presentation of the method via a relatively simple example acted to mask those difficulties. Fermat's account in the letter invites generalization beyond expressions of the third degree. Take an algebraic polynomial $P(x)$ of degree n. Substitution of the arguments $x + y$ and $x - y$ for x would yield:

$$P(x + y) = P(x) + \sum_{i=1}^{n} y^i P_i(x)$$

$$P(x - y) = P(x) + \sum_{i=1}^{n} (-y)^i P_i(x).^{114}$$

Clearly, for, say, a maximum the dual inequality $P(x \pm y) < P(x)$ will be maintained for all values of y only if $P_1(x) = 0$ and $P_2(x) < 0$. And the roots of $P_1(x) = 0$ will be the values for x that maximize $P(x)$. But, if contrariety of sign is what leads Fermat to set the polynomial coefficient of y equal to zero, what of the other $P_i(x)$ for odd i in the above expressions? They will bear contrary signs, but one cannot set them equal to zero. For the zeros of, say, $P_3(x) = 0$ will not coincide with those of $P_1(x) = 0$, and the latter have already been fixed. Hence, whatever one does about $P_1(x)$ and $P_2(x)$, the remaining terms $(\pm y)^i P_i(x)$ threaten to upset the simultaneous inequalities. How does one meet this problem? How did Fermat reason it away?

Perhaps one should ask first whether Fermat even saw the problem. For it is a problem of proof, not a problem of practice. And proofs came hard to

---

will be minimized." At the same time that one recognizes the validity of the derivation, one will see the fundamental lack of connection between it and the coefficient expression of the y^2 terms in the letter to Brûlart. The *property of the tangent* found by Fermat's method of tangents is an equation of which the roots represent the length of the subtangent measured along the axis from the foot of the ordinate to the point of tangency. In the above diagram, the property of the tangent yields length CB; since point C is given, CB in turn determines point B. Only after point B has been thus determined does Fermat then form an expression for the tangent of angle B and minimize it by the method of maxima and minima; and he does so only in this one specific instance. By contrast, the polynomial coefficient of the y^2 terms in the letter to Brûlart directly expresses the derivative of the tangent of angle B. Accustomed to thinking of the property of the tangent in terms of the subtangent, Fermat would understandably have had difficulty in recognizing such a different form of that property in the polynomial coefficient, and there is no indication he ever did recognize it. His failure to do so is another clue to the subtle yet fundamental differences between Fermat's method of tangents and modern differential calculus.

[114]The technical validity of Fermat's method lies, of course, in the fact that $P_i(x) = \dfrac{d^i P(x)}{dx^i} \cdot \dfrac{1}{i!}$. Also, for algebraic polynomials, $P_n(x)$ is always a constant.

Fermat. His talent lay in the direction of innovative solutions to particular mathematical problems. His intuitive genius often succeeded, as in the case of the method of maxima and minima, in abstracting the core of a particular solution technique and in generalizing it to make it applicable to a whole class of problems. But fully general, rigorous proofs were not his forte. Throughout the letter to Brûlart, one reads Fermat's basic message: it has to be this way, because the method works in practice. In problem after problem, he had applied the method and obtained the correct result. And the reasoning he had applied to obtain an answer had its firm foothold in the concept of the uniqueness of extreme values. On that concept he had sought, in both the *Analytic Investigation* and the letter to Brûlart, to found a fully general method and a convincing proof of its validity. His own conviction of the method's general applicability, indeed of its validity, required no proof; it resulted from extrapolation from the many problems he had solved. Did Fermat see the difficulty raised above? Yes and no. The answer depends on how one poses the difficulty.

Fermat did consider the question of the contribution of the terms containing higher powers of y. One of the major reasons for choosing the coefficient expression of y to set equal to 0 lay in the fact that that expression outweighed the others in its effect on the value of the whole:

> The reason for this is because the terms measured by the lowest power of E always have a greater ratio between them than those that are measured by $Eq.$ or by $Ec.$, etc., and those that are measured by $Eq.$ have a greater ratio between them than those that are measured by $Ec.$, $Eqq.$, etc. As in this example, taking $A + E$ and equating the two terms measured by E alone, we have, on the one side, $2BAE$ and on the other, $3Aq.E$. Now, $2BAE$ has a greater ratio to $3Aq.E$ than (taking the two terms measured by $Eq.$) $BEq.$ has to $3AEq.$. The reason for this is because analytic [i.e. algebraic] multiplication doubles B in the preceding equation, which is simple [i.e. linear]. If, then, we equate $2BAE$ to $3Aq.E$, then $BEq.$ will be less than $3AEq.$

> We can prove from this that all the terms bearing the sign + will be less than those that bear the sign –. And the last power of E, which always stands alone and which here is $Ec.$, will not change the order of the equation, no matter what sign it bears, something which we will see clearly from inspection alone.

Clearly, then, Fermat here addressed the question about the possible effect of higher-order terms on the inequalities serving as foundation for the method. But he did so in a certain way. He focused on the coefficient expressions and argued that the relationships between x and the data determined by setting the coefficient expression of y equal to 0 ensured further that, in the case of a maximum, any positive contributions to the value of the original expression would be offset by the negative ones, thus preserving the inequalities. Fermat saw the answer (and hence the question itself) in the signs of the coefficient expressions. His argument was vague, to say the least; it asserted more than it proved. And, when he came to the highest power of y, he gave up trying to prove. It would have no effect on the inequality, whatever its sign, "as will appear clearly to us from inspection alone."

In retrospect, one wants to say that Fermat stopped at the very point at which he should have started. For it is the size of this final y term that controls everything. One must place limits on its size. But, from the invention of the method of maxima and minima to the letter to Brûlart, Fermat never did so. He always left the value of the increment or decrement arbitrary: "Hence, in order to make the comparison with another point taken *arbitrarily* (*à discrétion*) on one side of the unique point . . .". Why is Fermat's argument so unconvincing? Because it is not valid. The value of y cannot be arbitrary; one need only choose $|y| > b$ to destroy the inequalities. Indeed, the modern mathematician demands more. One must keep y within an arbitrarily *small* range.

The modern mathematician makes this demand, however, for a reason entirely foreign of Fermat's way of thinking about maxima and minima. The value of y must remain within a δ-neighborhood of x because extreme values may be local rather than global. That is, an expression may have more than one extreme value, or it may have more than one value of x for which it attains a given extreme value. In short, extreme values are not unique! Yet the uniqueness of extreme values was an *idée fixe* for Fermat; it guided his thought on maxima and minima completely, and he could not free himself of it. Why? The answer lies in his style of mathematics. Fermat was a problem-solver.

All the problems Fermat had solved with his method served to reinforce the conviction he had originally drawn from Pappus. All had unique maxima or minima, either because they were quadratic problems to start with (and hence indeed had a single extreme value) or because they were cubic problems for which one of the extreme values was 0 and hence irrelevant to the *geometrical* situation of the problem. In none of his writings, however, did Fermat ever attack a quartic problem, a problem which might have confronted him directly with two relative maxima or minima. He never considered a problem which could have shown him that his use of an arbitrarily large increment and decrement raised the possibility that the y interval might reach all the way into the other extreme value and hence destroy his basic inequalities. And because, as Descartes correctly observed, Fermat worked from the particular to the general, he therefore never really saw the question raised above.

Fermat's method of maxima and minima remained finitistic from beginning to end. He never placed limits on the size of the interval y, and he never employed infinitesimals or limit procedures. He did not do so, because he never encountered the need to do so. His notation alone shows that the increment never entered the domain of the infinitesimal. If some historians have read an infinitesimal into Fermat's method of maxima and minima, they have done so as a result of mixed translation of Fermat's notation into modern notation. For his unknowns, Fermat always followed the notation of Viète. The first was always A, the second E, etc. Direct translation into Cartesian notation requires that A be replaced by x, E by y, and so on, as has been done throughout this chapter. But most accounts of Fermat's method do only half the job. A becomes x, but E remains e (or worse, is translated into h, or even Δx).[115] To the modern reader, however, x usually denotes a finite variable in the domain of real numbers, while e (or h or Δx) denotes an infinitesimal quantity ranging over the immediate neighborhood of 0. To the modern mind they are variables of two quite different sorts. A and E, however, meant to Fermat variables of the same sort, as do x and y to the modern reader. For that reason x and y have been used throughout this chapter to translate A and E. The earliest forms of a true infinitesimal calculus, those of Newton and Leibniz, both indicated the essential difference between the finite variable and the infinitesimal increment through notation. Newton first used x and o, and then x and $o\dot{x}$; Leibniz employed x and d and then x and dx. But, throughout his career, Fermat used A and E in his method of maxima and minima and his rule of tangents. To him, E was the same sort of quantity as A. Both were finite algebraic unknowns. As far as these methods were concerned, Fermat's mind was locked in the theory of equations and hence closed to any use of infinitesimal quantities.

Unlike the new foundation Fermat had offered for the method of tangents in the "Méthode expliquée," the new demonstration of the method of maxima and minima did not sever the connection between the algorithm of the *Method* and its justification. The coefficient of the y term, $P_1(x)$, is the same whether one takes an increment or a decrement. The most direct path to that coefficient is the one outlined in the *Method*. Given $P(x)$, expand $P(x + y)$. From $P(x + y)$ subtract $P(x)$. Divide the result by y, then drop all terms still containing y. The algorithm remained the same; only the foundations had changed.

VII.   FINE TUNING: THE PATH TOWARD QUADRATURE AND RECTIFICATION

Though the basic algorithm remained the same, Fermat was too acute a problem-solver not to vary its application when the conditions of a problem warranted. When applied to certain special curves (see below), the method of tangents raised the problem of determining the extreme values of expressions

---

[115]Even though Boyer in his *History of the Calculus* (N. Y., 1959) retains Fermat's original E in describing the method, he goes on to note (p. 156): "The procedure which Fermat here employed is almost precisely that now given in the differential calculus, except that the symbol Δx (or occasionally h) is substituted for E." But certainly the issue is far more than merely typographical!

containing surds. In an *Appendix to the Method of Maxima and Minima*, written in April 1644, Fermat pointed the way past this problem by a refinement of the method. Again, the new technique derived from his earlier study of Viète's theory of equations, and again Fermat revealed its algebraic foundations only after having set forth the technique itself. In this case, Carcavi learned of the foundations in 1650 through receipt of Fermat's *New Use of Roots of the Second and Higher Orders in Analysis*.[116] There Fermat dealt with the problem of reducing higher-order systems of simultaneous equations to determinate equations in one unknown. The example he gives is the system

$$x^3 + y^3 = c^3$$
$$ax + y^2 + by = f^2,$$

from which seeks to eliminate the unknown y. Clearly, the direct solution of either equation for y will involve surd expressions, something Fermat would like to avoid.

To do so, he first separates the variables in each equation to get

$$c^3 - x^3 = y^3$$
$$f^2 - ax = y^2 + by,$$

and then combines the two equations in a proportion

$$\frac{c^3 - x^3}{y^3} = \frac{f^2 - ax}{y^2 + by}.$$

Reduction of the proportion in turn into an equation yields

$$c^3 y^2 - x^3 y^2 + c^3 by - bx^3 y = f^2 y^3 - axy^3,$$

in which each term contains y and which therefore may be divided through by y. The result is a mixed equation in x and y involving powers of y no

[116]*Novus secondarum et ulterioris ordinis radicum in analyticis usus.* Text in FO.I. 181-188. Fermat's letter of 20.VIII.50 to Carcavi makes clear reference to the treatise (ma méthode générale pour le débrouillement des *asymmetries*), announcing that it had been dispatched by the previous messenger. The letter goes on to speak of the applications of the method to the determination of tangents to curves of which the equations contain algebraic surds. The points raised there merely repeat what Fermat had already communicated to Mersenne and others some six years earlier; they are discussed below, p. 208ff. Fermat discusses them again in the letter to Carcavi in order to emphasize once more the superiority of his method of tangents over that of Descartes, whose departure for Sweden in 1649 had still not cooled the passions fired by the dispute of 1638.

The treatise itself is a mixed bag. In addition to the method of elimination and its corollary, both discussed above, it also treats in general terms what Fermat calls *deficient* and *abundant* equations. The former are indeterminate equations in two or more unknowns, and their solutions take the form of loci of increasing dimension; cf. above, Chap. III, p. 123. The latter are systems of equations in which the equations outnumber the unknowns. Fermat notes they occur most frequently in the mathematical analysis of physical problems, and he uses the method of elimination as a technique for depressing the degree of the equations to arrive at a simple and direct solution without surds. The treatise ends with a discussion, but not the solution, of the problem: Given an ellipse and a point off its plane, to cut the conical surface, of which the vertex is the given point and the base the given ellipse, by a plane in such a way that the section is a circle.

greater than 2. Repetition of the procedure, combining now the quadratic equation of the original pair with the reduced equation, yields a fourth equation which, following division through by y, contains only first powers of y. One may now solve a linear equation directly to obtain an expression for y in terms of x, and that expression in turn, when substituted for y in one of the original equations, will yield the determinate equation in x corresponding to the system and involving no algebraic surds.

Such a reduction procedure would seem at first glance to offer no advantages over the more direct solution of one of the original pair of equations, since in depressing the power of y it raises that of x. In Fermat's example, the resultant determinate equation contains the lead term $a^2 x^8$. The appendix to Fermat's treatise, however, which directly follows the exposition of the above example, reveals the procedure's special virtues. Consider, for example, the equation

$$\sqrt[3]{ax^2 - x^3} + \sqrt{x^2 + cx} + \sqrt[4]{b^3 x - x^4} + \sqrt{ex - x^2} = k.$$

Here, Fermat notes, none of Viète's procedures addressed to the elimination of algebraic surds will work.[117] His new method of elimination just presented, however, will turn the trick. To show how, he operates on the simpler equation

$$\sqrt[3]{cx^2 - x^3} + \sqrt[3]{x^3 + a^2 x} = c.$$

He first rearranges the equation in the form

$$c - \sqrt[3]{x^3 + a^2 x} = \sqrt[3]{cx^2 - x^3}$$

and then sets the left-hand side equal to the expression $c - y$. The original determinate equation is thus transformed into a system of equations in two unknowns,

$$(c - y)^3 = c^3 - 3c^2 y + 3cy^2 - y^3 = cx^2 - x^3$$

and
$$y^3 = x^3 + a^2 x.$$

Treating this system according to the reduction procedure introduced in the body of the treatise, Fermat is now able to derive a single determinate equation in x that contains no algebraic surds.

It is against the background of the above algebraic technique for the elimination of surds from equations that one must read Fermat's *Appendix to the*

---

[117]*Ibid.*, p. 184: "Superiori methodo debetur perfecta et absoluta asymmetriarum in Algebraicis expurgatio; neque enim symmetria climactismus Vietaea [*On Emendation of Equations*, Chap. V], quae unicum hactenus ad asymmetrias fuit remedium, efficax satis et sufficiens inventa est." It is striking testimony to the continued influence of Viète on Fermat that, in the wealth of Descartes' theory of equations in the *Geometry*, Fermat could find nothing more advanced than what Viète had devised to handle the problem of algebraic surds. Also striking is the continued pattern of development even in the refinements of Fermat's method of maxima and minima; Viète's theory of equations remains the starting point.

Method of Maxima and Minima,[118] for surely the former is the source of the latter. The opening words of the *Appendix* point directly to the method of elimination: "Because surds often occur in the course of problems, the analyst will not hesitate to make use of cubic expressions or, if need be, of ones of higher order. Indeed, with their help, multiple and intricate ascent [in degree] will usually be avoided." The first example he gives shows clearly the application of the new algebraic technique to the method of maxima and

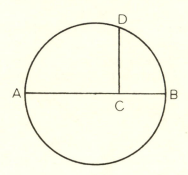

minima. Suppose, in circle *ABD*, one wishes to determine max $(AC + CD)$. Setting $AB = b$ and $AC = x$, one has for *CD* the expression $\sqrt{bx - x^2}$. Hence, algebraically, the problem is to find max $x + \sqrt{bx - x^2}$. Immediate and direct application of the algorithm of the *Method* would lead to the adequality

$$x + \sqrt{bx - x^2} \approx x + y + \sqrt{b(x + y) - (x + y)^2}.$$

Removal of the radicals involved by the usual means would raise the equation to a quartic. Instead, Fermat suggested the following dodge. Proceed analytically and assume the problem to be solved; that is, let z be the desired maximum and consider the equation

$$z = x + \sqrt{bx - x^2}.$$

Then,

$$z - x = \sqrt{bx - x^2} ,$$

or

$$z^2 - 2zx + x^2 = bx - x^2 .$$

Having removed the irrationality, "one should carry out a transposition such that the maximum degree of *O* (here z) occupies alone one side of the equation, so that by this means one can determine the maximum, to which the technique leads."[119] Rearranging, then, one has

---

[118]Text in FO.I.153–158. Previously unedited and extant only in a copy in the hand of Arbogast, it has been tentatively dated by De Waard (FO.*Suppl*.xvi) 21.IV.1644.
  [119]*Ibid*., p. 154.

$$z^2 = bx - 2x^2 + 2zx \,.$$

Now maximize the expression on the right-hand side according to the standard method:

$$bx - 2x^2 + 2zx \approx bx + by - 2x^2 - 2y^2 - 4xy + 2zx + 2zy$$

$$2y^2 + 4xy \approx by + 2zy$$

$$2y + 4x \approx b + 2z$$

$$4x = b + 2z \,.$$

Fermat then takes the value of z resulting from this last equation and substitutes it into the original equation in x and z:

$$x + \sqrt{bx - x^2} = 2x - \frac{b}{2}.$$

He then solves for x. To show just how effective the adjustment to his method is, Fermat follows up this example with one taken from Archimedes' *Sphere and Cylinder*. The problem underlying Proposition 15 of Book I of that work is to find the surface of the greatest cone inscribed in a given sphere. The algebraic analysis of the problem leads to the expression

$$\sqrt{bx^2 - bx^3} + bx - x^2 \,.$$

To maximize the expression Fermat follows the procedure just outlined above.

With the presentation of his solution to this last problem, Fermat had come full circle with his method of maxima and minima. In the memoir of April 1644, his last memoir on the subject, Fermat offered the solution to the very first problem he had posed to Mersenne in his first letter to the Minim in 1636.[120] Despite the furor of the dispute with Descartes, despite even the attempt at a new demonstration of the method in the letter to Brûlart, Fermat's actual solution techniques had not changed since his invention of the method in the late 1620s; even the problems remained the same. Indeed, in this regard, it is interesting to note that, for all its ingenuity, the above refinement of the method represents a step backward from the methods of the letter to Brûlart. For the proof technique of that letter pointed toward the use of series expansions of functions, an approach which makes possible a far simpler solution of the problem of quadratic surds. Application of the generalized binomial theorem to the expressions under the radicals removes entirely the need to raise the degree of the expressions or to introduce an auxiliary unknown. The fact that the generalized binomial theorem still lay

---

[120]Cf. FO.II.6. Fermat had earlier sent Mersenne the solution of the second problem posed there: To find the greatest cylinder inscribable in a given sphere, cf. *Problema missum ad Reverendum Patrem Mersennum 10^a die November 1642*, FO.I.167-169, which is solved by the same technique as that of the *Appendix* and hence places the invention of the technique back at least two years.

years in the future does not alone explain why Fermat did not follow such a path. More important to understanding Fermat and his conception of his method is the point that the use of the series expansion was a *proof* technique; the method of proof employed in the letter to Brûlart never affected the practice of the method. That practice had been established in Bordeaux in 1629. It never changed.

If the practice of the method of maxima and minima remained static, that of the rule of tangents did not. Its application required refinement in keeping with the growing number of sophisticated curves Fermat's own analytic geometry had brought into existence. Although he gave no examples in the *Appendix* itself, Fermat did emphasize there the importance of the refinement of the method to the problem of determining tangents.[121] An earlier (ca. 1640) memoir indicates the sort of curves that Fermat was analyzing and in which he was meeting the difficulties the technique of the *Appendix* was intended to resolve. That untitled memoir, beginning "Doctrinam tangentium antecedit . . ." (hereafter, *Doctrine of Tangents*),[122] contains the most powerful, sophisticated version of the method of tangents Fermat could devise. Meant to show that the procedures outlined in the *Method* and further clarified and systematized in the "Méthode expliquée" applied not only to algebraic curves but also to those we now call "transcendental," the *Doctrine of Tangents* also lent to adequality a new meaning that would bear fruit later in Fermat's work on quadrature and rectification. In the memoir, Fermat determines the tangents to any point on each of the four special curves: the cissoid, the conchoid, the cycloid, and the quadratrix.

"The curved lines," he began, "to which we seek the tangents, attain their specific properties either via straight lines only or via curves involving various sorts of straight and curved lines."[123] Here, for the first time, Fermat's reading of the *Geometry* affected his understanding of the nature of curves and, in turn, his sense of the general applicability of the original method of tangents. For Descartes' distinction between "mathematical" and "mechanical" curves is here mirrored precisely in a distinction between curves that can be defined by an indeterminate algebraic equation in two unknowns corresponding to variable rectilinear segments and curves that require for their definition

---

[121]*Appendix*, p. 157: "Elegantius tamen et fortasse magis γεωμετρικῶs quaestiones de maxima et minima speciales tangentium beneficio resolvuntur, licet et ipsae tangentes ab universali methodo deriventur. . . . Et generalis ad inventionem maximae et minimae geometrica est quaestionum ad tangentes abductio; nec ideo minoris facienda universalis methodus, quum ejus ope et maxima et minima et ipsae tangentes indigeant." The covering letter to Carcavi placed even greater emphasis on the application to tangents; cf. above, n. 116.

[122]Text in FO.I.158–167. One of the very few texts of which an autograph copy exists (Bib. Nat., *Fonds fr.* 3280, 112–117), the *Doctrine of Tangents* was first published in the *Varia*, pp. 69–73. De Waard dates the text 1640 (FO.*Suppl.*xvi); Mersenne was circulating it by late October 1640 (cf. FO.II.218, n. 2).

[123]*Ibid.*, p. 159.

arc lengths on other curves.[124] For the former curves, the rule of tangents first presented in the *Method* sufficed. In the *Doctrine of Tangents* Fermat added nothing to that rule except to provide finally the general algorithm that neither the *Method* nor the "Méthode expliquée" had attempted to phrase in words:[125]

> We consider in the plane of any curve two straight lines given in position, of which one is called, if you will, the diameter, the other the applicate. Then supposing the tangent to a given point in the curve to have already been found, we consider the specific property of the curve, no longer [however] in the curve [itself], but in the tangent to be found, and, having removed the homogeneous quantities (which the doctrine of maxima and minima calls for), we finally get an equation which determines the intersection point of the tangent with the diameter, and therefore the tangent itself.

In this short paragraph, Fermat encapsulizes years of mathematical research and achievement. The whole new system of the *Introduction* lies in the first sentence. The second sentence puts into words the procedures so painstakingly set forth in the "Méthode expliquée," as a comparison with the discussion above will quickly show. They apply to any curve of the class Descartes called "mathematical" and we call "algebraic," and Fermat's determination of the tangents to the cissoid, the conchoid, and the *folium* of Descartes,[126] the first three curves treated in the *Doctrine of Tangents*, differs from his earlier applications of those procedures only in the algebraic complexity of the "specific properties" of the curves. In the case of the cissoid, an algebraic surd is involved.

But "Roberval's curve," as Fermat called the cycloid, differed in kind from the cissoid and conchoid. Its definition involved an arc length. No matter, said Fermat, it too falls before the subtlety of the method of tangents. Fall it does, but not without its own subtle effect on that very method. Something new has happened to the concept of adequality in the following passage:[127]

---

[124]Cf. Descartes, *Géométrie* (Leiden, 1637), p. 315ff: "... prenant comme on fait pour géométrique ce qui est précis & exact, & pour méchanique ce qui ne l'est pas; & considérant la géométrie comme une science, qui enseigne généralement à connaître les mesures de tous les corps, on n'en doit pas plutôt exclure les lignes les plus composées que les plus simples, pourvû qu'on les puisse imaginer être décrites par un mouvement continu, ou par plusieurs qui s'entresuivent & dont les derniers soient entièrement reglés par ceux qui les précèdent (p. 316)." In his subsequent discussion, Descartes insists that the relation of the motions is exact only if it can be expressed algebraically.

[125]*Doctrine of Tangents*, p. 159.

[126]The *folium* of Descartes is, of course, the curve $x^3 + y^3 = pxy$, to which Descartes proposed that Fermat should find the tangent during the course of their dispute. The cissoid of Diocles and the conchoid of Nicomedes are among the so-called "special curves" discussed by Pappus in Book IV of the *Mathematical Collection*. Both are algebraic curves, the equation of the first being $y^2 = \dfrac{(a-x)^2}{a+x}$, that of the second, $\left(x - \dfrac{ax}{a+y}\right)^2 + y^2 = k$.

[127]*Doctrine of Tangents*, p. 162.

As long as the homogeneous quantities involve only straight lines, they are sought and designated according to the preceding formula. Indeed, for the purpose of avoiding irrationalities, sometimes, if you will, the applicates to the tangents found by the above method should be taken in place of the applicates to the curves themselves. Finally, (and this is worth the trouble) portions of the tangents already found should be taken in place of the portions of the curve subtending them, and the adequality can proceed as we have set forth above: it will satisfy the proposed problem with no difficulty.

The shift in concept hinted at in the "Méthode expliquée" was continuing its course. Adequality, originally borrowed from Diophantus to condense (and hide) the counterfactual reasoning of the method of maxima and minima, became in the "Méthode expliquée" the counterfactual assumption of equality between the ordinate to the curve and the distance along the same line from the axis to the tangent. Now, however, it made a giant step toward the notion of approximate equality.

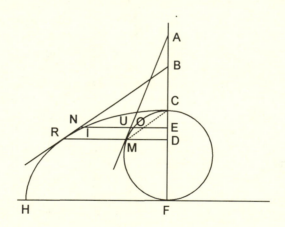

Look at Fermat's analysis of the tangent to the cycloid. Let circle *COMF* be the referent circle of cycloid *HRIC*. Given point *R* on the cycloid, draw tangent *RB*, where *B* is the intersection point of the tangent with the diameter of the curve. The specific property of the cycloid is that ordinate *RD* is the sum of ordinate *DM* of the circle and arc *CM* of the same circle. Draw the tangent to point *M* of the circle by the standard method.[128] Now, taking $DB = x$, $DA = b$, $MA = d$, $MD = r$, $RD = c$, arc $CM = n$, and *DE* (*pris à discretion*) $= y$, determine the length of *EN* by the standard use of similar triangles. $EN = \dfrac{cx - cy}{x}$. Treat *EN* as if it were equal to *EI*, the ordinate to the cycloid drawn

[128] That is, simply draw the perpendicular to the radius drawn to M.

212

from point E. That is,

$$EN = \frac{cx - cy}{x} \approx OE + \overarc{CO} = OE + \overarc{CM} - \overarc{MO}.$$

In the face of the arc lengths CM and MO, the method of tangents as presented so far would seem to be helpless, but here Fermat took the next bold step in the development of his method: "But, in order to reduce these three terms to analytic terms, [and] to avoid irrationalities according to the above warning, let us take in place of straight line OE the straight line EU applied to the tangent, and in place of curve MO let us take the portion MU of the tangent, which subtends this MO."[129]

In algebraic terms, $MU = yd/b$, $EU = (rb - ry)/b$, and CM is given (since point R is given) as n; hence, the final adequality reads:

$$\frac{cx - cy}{x} \approx \frac{rb - ry}{b} + n - \frac{yd}{b}.$$

From this point, the method of tangents follows its usual course, and leads to the final result

$$\frac{r + d}{b} = \frac{c}{x}.$$

Fermat interprets this result geometrically and shows that it is equivalent to drawing RB parallel to chord CM in the referent circle.

Not only, then, does adequality serve to posit a counterfactual equality between the ordinate to the curve and the ordinate to the tangent, but, in this final version of the method of tangents, Fermat extended its use to cover the counterfactual equality of an arc length along a curve to the segment of the tangent subtending it. And, where the difference between counterfactual equality and approximate equality (or equality in the limiting case) was clear in the *Analytic Investigation* and the *Method*, that clarity disappears in the *Doctrine of Tangents*. Early in the 1640s, the notion of adequality, originally born of an attempt to conceal the purely finite algebraic foundations of the method of tangents, began to suggest applications to a new class of problems, applications that teetered on the thin line separating the realm of the finite from that of the infinitesimal. Fermat apparently never crossed that line in his method of maxima and minima and of tangents, perhaps because he never returned to the subject after 1644. But, during the same period of the early 1640s, another class of problems increasingly began to demand his attention. Treatment of the problem of the quadrature of curves by classical and semi-classical means had by then reached a dead end. To secure a breakthrough, Fermat turned to his notion of adequality and, in doing so, finally crossed the line into the realm of the infinitesimal. He never looked back, however, to bring the method of maxima and mimima with him.

[129] *Doctrine of Tangents*, p. 163.

CHAPTER V

Archimedes and the Theory of Equations

It is not that I do not approve of it; but, since all of his propositions can be demon-strated in the ordinary, legitimate, *and* Archimedean manner *in many fewer words than his book contains, I do not know why he has preferred this style of algebraic notation to the ancient style, which is more convincing and more elegant, as I hope to show him at my first leisure.*
Fermat to Digby[1]

I. INTRODUCTION

Fermat's research into the quadrature and rectification of curves followed the same basic pattern of development as his analytic geometry and his method of maxima and minima. Fermat began with old problems inherited from the Greek mathematicians, to which he applied the new algebraic techniques inherited from Viète. But that algebra did more than supply a vehicle for new solutions. It affected in subtle and profound ways the very manner in which Fermat read the ancient authors. For, by its nature and conception, Viète's *Analytic Art* provided or suggested solution techniques that possessed greater generality than the particular problems to which they were applied. As a result, Fermat was able to generalize the problems themselves and to establish close ties among problems that the ancients had viewed as unrelated.

For example, Fermat's effort to reconstruct the lost text of Apollonius' *Plane Loci,* a work limited to circles and straight lines as loci, led to the analytic methods of the *Introduction.* Yet the applicability of those methods extended far beyond the problems that had inspired them. They encompassed all algebraic curves in the plane. Through the *Introduction* the ancient tripartite distinction between plane, solid, and linear loci gave way to the unifying concept of curves defined by indeterminate equations in two unknowns. With this tripartite distinction disappeared also the three-dimensional

[1] 15.VIII.1657, FO.II.343.

bounds of Greek geometric intuition. The world was soon filled with an infinity of curves unimagined by the ancients.

The analytic geometry of the *Introduction* did more, however, than merely increase the number of mathematical curves in the world. It altered the very meaning of the word "curve" by fundamentally revising the mathematician's intuition of a curve. No longer was a curve "given" or "known" in terms of the intersection of a plane with some familiar solid such as a cone, or by some easily visualized composition of moving lines. The *Introduction* showed that a curve was "given" by the indeterminate equation to which it uniquely corresponded. This basic conceptual shift belies the superficially geometric style of much of Fermat's work, especially in the realm of quadrature and rectification. For it makes little essential difference whether one denotes variable line lengths by single literal symbols or by the traditional combination of letters denoting the endpoints. When Fermat defines a curve such that the cube of the ordinate bears a fixed ratio to the square of the abscissa, he is treating curves algebraically.

To Fermat, algebra meant more than the use of literal symbols for the unknowns and the employment of certain solution techniques for equations. Algebra meant the "analytic art," and the "analytic art" meant the study of equations themselves. Through the influence of Viète, concrete solutions yielded in importance to the determination of relations among problems and hence among their solutions. The focus had shifted from the solution of a particular problem to the question of how many other problems were also solved in the process. Equations contained information about more things than their mere solution. Suitably compared and deciphered, they told of the most intimate of mathematical relationships.

In the *Introduction*, for instance, Fermat showed that the equation of a curve contained all the information required for its construction. The method of tangents, however, showed that the equation could be made to yield more about the curve. For, by a particular application of the method of maxima and minima to the equation of the curve, one could construct the tangent to any given point. In this way too, one might add, the method of maxima and minima, originally inspired by a single problem in Apollonius' *Determinate Section*, proved more general than its inspiration. It not only unified the greater part of the ancient *diorismoi* under one systematic approach; it also established a hitherto unsuspected relationship between extreme values and tangents to curves. Hence, in its application to the equations of curves it became one of the keys for unlocking the secrets contained in these equations.

Following the *Introduction*, then, the study of equations and the organization of mathematical problems according to their equations increasingly occupied Fermat's attention. Viète's own work had set the style. His theory of equations, applied to determinate equations and hence to geometric point constructions, had unified and systematized the informal "reduction analysis" of the Greek geometers in this one area.[2]

In particular, it had made possible the systematic study of solvability and hence constructibility. Fermat's refinement and extension of Viète's methods, in the *Introduction* and the method of tangents, now brought locus constructions—indeed, curves in general—within the aegis of the analytic art. Nowhere does this trend emerge so clearly, both as achieved result and as guide to research, as in Fermat's mature work on the quadrature and rectification of curves. For there he sought and was able to show that the equation of a curve sufficed to determine whether a segment of the curve was quadrable—that is, whether that segment was equal to a rectilinear area expressible in closed algebraic form—and, if so, to determine its area. In the realm of the rectification of curvilinear segments, he established a uniform reduction analysis which algebraically transformed the problem of rectification to that of quadrature.

The very algebra that made all this possible, the "analytic art" of Viète, paradoxically prevented Fermat from discerning the profound conceptual changes his own thinking underwent in the course of his work. Just as he wondered in 1637 how Descartes could fail to be convinced by the method of maxima and minima, a method "as evident as the first proposition of Euclid's *Elements*,"[3] so too in 1657 he could write to Digby and imply that his methods of quadrature, in contrast to those of Wallis, remained true to the "ordinary, legitimate, Archimedean way" of mathematics. For Fermat shared Viète's belief that algebra was basically a symbolic language for expressing and analyzing the content of Greek geometry and arithmetic. He was apparently unaware that the use of algebra might lead, as in fact it did, to the introduction of new concepts alien to the ancient geometry to which it was applied. Yet, "adequality" represents just such a new concept, and adequality is the concept that separates the method of maxima and minima and the methods of quadrature from Euclid and Archimedes. It is not, however, the same adequality in each case. In the method of maxima and minima, adequality means temporary or counterfactual equality, conceived in a purely finitistic way.[4] But, as the last chapter suggested in its conclusion, during the years between 1637 and 1644, adequality shifted in concept more and more toward an approximate equality or equality in the limiting case. In its algebraic, operational use, it remained the same, but in the manner in which it was applied to the analysis of mathematical situations and in the sorts of reasoning it permitted or suggested, it underwent a fundamental change. And yet the

---

[2] On Viète's work, see above, Chap. II; on "reduction analysis" in classical Greek geometry, see Mahoney, "Another Look at Greek Geometrical Analysis," *Archive for History of Exact Sciences* 5(1968), 331–337.

[3] See above, Chap. IV, p. 192.

[4] See above, Chap. IV, §II.

change was so subtle that Fermat could still believe that he had basically altered nothing in the old Archimedean approach to problems of quadrature.

In Fermat's work on quadrature and rectification, then, the historian may follow the partly unconscious introduction of the concepts basic to the new infinitesimal calculus that lay on the horizon. Besides revealing another aspect of Fermat's development as a mathematician, the present chapter may serve as a case in point for the manner in which fundamentally new ideas sneak into a science.

As in the case of the analytic geometry and the method of maxima and minima, Fermat's mature work on quadrature and rectification had its crude beginnings. It began with problems and techniques inherited from the ancients, in this case from Archimedes. Even though the first tentative application of algebra to this inherited material brought forth striking new results and achieved a generality unknown to the ancients, Fermat's work prior to 1637 still addressed only a limited class of problems. Only during the 1640s and 1650s did he achieve the general methods he was seeking.

How he did so may have to remain a mystery. In his early letters to Mersenne and Roberval, Fermat left more than enough clues to reconstruct the course of his research on quadrature and the determination of centers of gravity. After 1638, however, he ignored the subject in his correspondence, except to claim priority in the achievement of specific results. Through Mersenne the mathematical world learned of some of Fermat's later accomplishments, though nothing of his methods.[5] Only after his friend Lalouvère published Fermat's treatise on rectification in 1660 did Fermat indicate in his letters that he had been doing additional research on the subject.[6] Fermat's major treatise on quadrature, written sometime around 1657 or 1658, remained among his private papers until its posthumous publication in the *Varia opera* of 1679. Not having circulated, it could raise neither protest nor defense, as it most certainly would have done. Hence, it never became a subject of Fermat's correspondence and thereby of further clues regarding its conception and development. Controversy supplied the crucial evidence and insights of Chapter

---

[5] The *Seconde Partie* (1637) of Mersenne's *Harmonie universelle* ("Nouvelles observations physiques et mathématiques"), p. 2, discusses Galileo's problem of the path of a cannonball falling from a tower *(Two World Systems*, Second Day), referring anonymously to Fermat's assertion of the spiral nature of that path and to his work on the quadrature of spirals. In 1644 Mersenne's *Cogitata physico-mathematica* again took up the Galilean problem, but now named Fermat as the discoverer of the spiral path. Through both the *Cogitata* and Mersenne's visit to Italy that same year (1644), Torricelli learned of Fermat's quadrature of the "higher parabolas" (see below, 244ff.) and in 1646 he sent his own work on the quadrature of hyperbolas to Fermat for comment. The ensuing correspondence, apparently no more than a single exchange of letters, is no longer extant. Fermat, however, referred to it in a letter of 20.IV.1657 to Digby as evidence of his priority over Wallis in the quadrature of the curves in question; cf. FO.II.337–8: "bien que la quadrature tant des paraboles que des hyperboles infinies ait été faite par moi depuis fort longues années et que j'en aie autrefois entretenu l'illustre Torricelli, je ne laisse pas d'estimer l'invention de Wallisius, qui sans doute n'a pas su que j'eusse préoccupé son travail."

[6] See below, §V.

IV. Chapter V can look to no controversy and is poorer for the fact. A remark of Huygens' in a letter to Leibniz in 1691 makes the situation more regrettable. To Huygens, at least, the treatise on quadrature seemed to make no sense.[7] Yet, on careful examination, it makes brilliant sense; Leibniz had followed the same pattern of thought in his own research on quadrature.[8] The reason for Huygens' quandary—he was, after all, one of Fermat's most sympathetic correspondents—lay in a combination of obsolescent or vague terminology and confusing printing errors that attended the published form of Fermat's treatise. One can only wish that Fermat had still been alive to enlighten his distinguished successor.

But Fermat was not alive. So the historian must take the finished treatises on quadrature and rectification, explicate their contents, and try to fit them into the patterns of thought and research suggested by Fermat's other work. To provide a foundation, he must go back again to Fermat's activity in Bordeaux and Toulouse prior to his introduction to the Parisian mathematicians and discover the nature of the work on quadrature and centers of gravity on which Fermat later built so impressive a structure.

II. FROM SPIRALS TO CONOIDS

In the course of correspondence with Mersenne and Roberval during 1636, Fermat supplied a complete inventory of the results he had thus far achieved in the realm of quadrature. At the start he posed them as problems, asking that Roberval and others try their hand before Fermat sent on the solutions. Toward the end of the period, however, Fermat began outlining his solution techniques in response to those of Roberval. One may mark the end of this first period of research and publication by the treatise *On the Determination of Centers of Gravity by the Method of Maxima and Minima*, sent to Roberval via Mersenne in February 1638, just before the controversy with Descartes eclipsed all other matters.

The first results Fermat announced concerned spirals. Working from the model of Archimedes' *On Spirals*, which investigates the properties of the curve $\rho = \alpha\vartheta$, Fermat defined a new spiral,

$$\frac{2\pi}{\vartheta} = \left(\frac{R}{\rho} \right)^2 ,$$

---

[7] Huygens to Leibniz, 1.IX.1691, FO.IV.137: "J'ai recherché là dessus ce que je me resouvenais d'avoir vu dans les oeuvres posthumes de M. Fermat, mais ce traité est imprimé avec tant des fautes, et de plus si obscur, et avec démonstrations suspectes d'erreur, que je n'en ai pas su profiter."

[8] Compare, for example, the transformation techniques of Leibniz' *Analysis tetragonistica ex centrobarycis* (25.X.1675, trans. Child, *Early Mathematical Manuscripts of Leibniz*, 65–90) with those of Fermat's *Treatise on Quadrature* presented below, 254ff.

and gave the area under that spiral for any number of turns.[9] In a memoir he sent to Paris at this time, Fermat also presented the quadrature of the "Galilean spiral,"

$$\frac{R}{R - \rho} = \left(\frac{\alpha}{\vartheta} \right)^2,$$

where α is a given central angle. Other papers, no longer extant, but reported by Mersenne, apparently extended the last result to cover all spirals of the form

$$\frac{R}{R - \rho} = \left(\frac{\alpha}{\vartheta} \right)^n.^{[10]}$$

At about the same time that the memoir on the Galilean spiral, to which we shall presently turn, arrived in Paris, Roberval and Etienne Pascal also received the first indication that Fermat's results in quadrature extended beyond spirals. In a letter of 23 August 1636 Fermat included the following problem.[11]

Let *AB* be a parabola with vertex *A*, and let the figure *DAB* be rotated about the fixed line *DA*. It will describe the Archimedean parabolic conoid, the

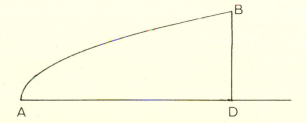

ratio of which to the cone of the same base and vertex is 3:2. But, if the figure *DAB* is rotated about the fixed line *DB*, a new conoid results; one seeks its ratio to the cone of the same base and vertex. We have found the ratio to be 8:5; the matter did not lack difficulty. Indeed, we have also found the center of gravity of the same conoid.

---

[9] The equations in polar coordinates are, needless to say, anachronistic. Both Archimedes and Fermat gave verbal definitions of their curves in terms of generation by a combination of rotational and rectilinear motions. Here and in what follows, however, use of the equations accurately describes the content of their achievements while saving words, and the dangers of conceptual anachronism are slight.

[10] Cf. Fermat to Mersenne, 3.VI.1636, FO.II.12–14, as well as the reports by Mersenne cited in n. 5 above.

[11] FO.II.55. On Fermat's derivation of the center of gravity of this figure, see below, 240ff.

A month later, apparently in response to the enthusiasm expressed over the above result, Fermat indicated the true extent of his research. In speaking of his "method" Fermat noted first that it served to find "propositions similar to those on the conoid which I sent in my last letter." For example, he went on:[12]

> I have squared infinitely many figures composed of curved lines; as, for example, if you would imagine a figure like the parabola but such that the cubes of the applicates [ordinates] are proportional to the lines they cut off from the diameter [i.e. the abscissas]. This figure approaches the parabola and differs only in that, whereas in the parabola one takes the ratios of the squares, I take in this figure that of the cubes; it is for that reason that M. de Beaugrand, to whom I showed the proposition, calls it a "solid parabola." Now, I have demonstrated that this figure is in sesquialterate [3:2] proportion to the triangle having the same base and altitude. You will find in examining it that I have had to follow a path other than that of Archimedes in the quadrature of the parabola and that I would never have solved it by the latter means.

Almost in passing, then, Fermat defined a new class of curves unknown to the ancients. For the solid parabola $y^3 = kx$ he referred to was only one representative of the class of "higher parabolas" $y^q = kx^p$, about which he would have much more to say in the future. Mentioning here only the one example, Fermat then went on to generalize his cubature of the "new conoid," giving the volume for any segment cut off from it by a plane drawn parallel to the base. "In the demonstration," he added, "besides the help I have derived from my method, I have used inscribed and circumscribed cylinders."[13]

Fermat's results impressed Roberval and his circle, though Roberval quickly discerned through his own efforts at solution that the variety of Fermat's accomplishments was more apparent than real. Even if Roberval could not furnish the fully general results and methods demanded by Fermat, he did discover the essence of Fermat's methods of quadrature and cubature and recognized that it relied on a powerful command and application of theorems about the sums of powers of the natural numbers.

Fermat himself had, in fact, provided the crucial hint in a letter to Mersenne apparently dealing with unrelated matters. Sometime during September or October 1636, Fermat responded to Mersenne's request for a solution to the problem of finding the sum of the cubes of a series of numbers in arithmetic progression.[14] Fermat not only solved the problem but added further that he could give the sum for any power of the numbers in any arithmetic progression. As an example he offered a formula for the sum of the fourth powers of

---

[12] Fermat to Roberval, 22.IX.1636, FO.II.73.
[13] *Ibid.*, p. 74.
[14] Letter in FO.II.63–70; for a discussion of its contents, see below, 230ff.

the integers. Whether Roberval needed or took this hint, he nevertheless wrote back to Fermat on 11 October 1636 to say that he had corroborated Fermat's results, adding:[15]

> Unless I am quite mistaken, I have found the same means as you, using lines parallel to the axis and segments of these lines taken between the parabolas and the line tangent to the same parabolas at the vertex; those segments follow the natural sequence of the square numbers or the cube numbers, etc. Now, the sum of the square numbers is always greater than the third part of the cube which has for its root the root of the greatest square, and the same sum of the squares with the greatest square removed is less than the third part of the same cube; the sum of the cubes is greater than the fourth part of the quadratoquadrate and, with the greatest cube removed, less than the fourth part, etc.

Expressed in symbols, Roberval's method, which he strongly suspected to be Fermat's also, rested on the theorem:

$$\sum_{i=1}^{N} i^m > \frac{N^{m+1}}{m+1} > \sum_{i=1}^{N} i^m \quad \text{(for all } m, N\text{)}.$$

Somewhat disappointed, it would seem, at the speed with which his Parisian correspondent had uncovered his method, Fermat acknowledged Roberval's success, while he questioned the full generality of Roberval's demonstration. He replied on 4 November: "You have also used the same means as I in the quadrature of solid parabolas, quartic parabolas, etc. *ad infinitum*. But you take something to be true of which you could not possibly have the precise demonstration, to wit, that the sum of the squares is greater than the third part of the cube . . . ."[16] From there Fermat went on to offer hints toward obtaining that "precise demonstration," that is, toward obtaining the formulas for the sums of the powers of the integers. Even without the proof, however, the basic theorem put Roberval in a position to emulate any and all of Fermat's results. Together, the two men explored the limits of their common method. To their disappointment, they quickly reached that limit. By the time Fermat wrote to Roberval on 16 December 1636, he knew how far the method would go and that it would not suffice to solve all the quadrature problems he faced. In the years following he undertook to find a fully general method of quadrature.

Despite its limits, Fermat's early method of quadrature represented an advance over the method of the ancients. Moreover, it provided the conceptual framework for Fermat's later methods; indeed, the very limits of the early

---

[15] Roberval to Fermat, 11.X.1636, FO.II.81–82.
[16] Fermat to Roberval, 4.XI.1636, FO.II.83–84.

method dictated the nature of those later methods. Hence, it will help to have a more detailed understanding of the former method. Fermat's work on quadrature up to 1636 had two distinct aspects: first, his success in obtaining the general formulas for the sums of powers of the integers; second, the application of these sums to the geometrical problems of quadrature and cubature. The two aspects had, in a sense, always been inseparable. One finds them united in Fermat's ancient model, Archimedes' *On Spirals*. In another sense, however, they are distinctly separable in Fermat's work prior to their renewed unification in the work before us. Whether historically the second aspect precipitated the first by posing a need, or the first made the second possible by posing a solution, is an issue that must, given the lack of evidence, remain moot. For present purposes, it may help to let the second motivate the first.

A. The Archimedean Model

You will also see my spirals, of which the demonstration will be immediately known to you (for it is similar to that of the new figures that I have squared or to which I have found equal cones), and you will grant me that these propositions shed not a little light on geometry.[17]

As this remark in the December letter to Roberval suggests, all of Fermat's early efforts at quadrature were shaped by the particular model from which he chose to work. For, unlike the majority of his predecessors and contemporaries, Fermat did not follow the methods of Archimedes' *Quadrature of the Parabola*.[18] Instead, he focused his attention on the less frequently studied treatise *On Spirals*.[19] There, in the course of twenty-eight propositions concerning various properties of the spiral $\rho = a\vartheta$, Archimedes had shown that the area of the first turn of the spiral (i.e. from $\vartheta = 0$ to $\vartheta = 2\pi$) is equal to one-third of the area of the circle of radius $2\pi a$, adding that the same ratio holds for any sector of the first turn of the spiral with respect to the corresponding sector of the referent circle. As the "moderns" were wont to com-

---

[17] Fermat to Roberval, 16.XII.1636, FO.II.93–94.

[18] In the *Quadrature of the Parabola*, Prop. 20ff., Archimedes begins with the inscribed triangle having the same base and vertex as the parabolic segment and constructs a series of inscribed polygons by repeating the procedure for each remainder segment. What is summed, then, is a series of triangles, not rectangles.

[19] The treatise was available to Fermat in various editions: the *editio princeps* of Archimedes' works (Basel, 1544), the Latin translation of Commandino (Venice, 1558), or the new Greek edition of Rivault (Paris, 1621). The latter seems the most likely source of Fermat's knowledge of the text. Fermat quoted from the introduction to *On Spirals* in another context when accused by Frenicle and Brûlart of having posed impossible problems in number theory (Fermat to Mersenne, VIII.1653(?), FO.II.260): "J'ai pourtant à leur représenter que tout ce qui parait impossible d'abord ne l'est pas pourtant, et qu'il y a beaucoup de problèmes desquels, comme a dit autrefois Archimède," οὐκ εὐμέθοδα τῷ πρώτῳ φανέντα χρόνῳ τὴν ἐξεργασίαν λαμβάνοντι. Cf. Archimedes *Opera omnia*, ed. Heiberg (Leipzig, 1880–1881), Vol. II, p. 3, 11–13. In discussing Archimedes' treatise below, we use Heiberg's edition, as well as the English paraphrase of Heath, *The Works of Archimedes*, pp. 151–188.

plain in other cases, Archimedes only proved a result of which he had prior knowledge. He did not furnish the analysis that had led to the solution. Yet, to Fermat, the line of proof itself suggested an heuristic method, which not only made it possible to treat an infinity of spirals analogous to Archimedes', but which also could be easily transformed for application to curves other than spirals.

Archimedes began his proof with the construction of a figure circumscribed about the spiral and one inscribed in it, such that the two figures consisted of similar sectors of circles and differed in area by less than any assigned value. The purpose of the construction was merely to show that it is generally possible to fulfill these conditions, in order to be able to apply the famous "method of exhaustion" to the main proposition concerning the area of the spiral.[20] Since, however, the construction provided Fermat with his analytic method of quadrature, it will be of value to follow Archimedes' reasoning and then see what Fermat made of it.[21]

Let O be the center of a circle of radius R, which serves as referent circle for the spiral

$$\rho = \frac{R}{2\pi} \vartheta.$$

Divide the circle into N equal sectors, and through the points at which each of the N radii intersect the spiral draw arcs of circles concentric to the referent circle, letting each arc extend over the neighboring sector on either side of the point. In one sector the arc will lie outside the spiral; in the other, inside it. Consider now the i-th sector of the referent circle counted from A. It contains a sector of the spiral lying between a circumscribed circular sector and an inscribed one. The radius of the circumscribed sector is to that of the referent circle as $i:N$, which is also the ratio of the subtending arcs. Hence, the area of the circumscribed sector is to the area of the sector of the referent circle as $i^2:N^2$. By similar reasoning, the area of the inscribed sector is to that of the sector of the referent circle as $(i - 1)^2:N^2$. It follows immediately from the construction that the area of the i-th inscribed sector is equal to that of the $(i - 1)$th circumscribed sector. Hence the total difference between the circumscribed step-figure and the inscribed step-figure is equal to the area of the Nth sector of the referent circle. That sector, however, is to the area of the whole referent circle as $1:N$ and may, by suitable choice of N, be made as small as

---

[20] Eudoxus of Cnidus (390–337 B.C.) is traditionally credited with the invention of the method of exhaustion, though Archimedes was undoubtedly its most skilled practitioner. For the best secondary account of the method, one of the most frequently misunderstood ancient mathematical techniques, see Oskar Becker's series, "Eudoxos-Studien I-IV," *Quellen und Studien zur Geschichte der Mathematik, Astronomie und Physik*, Vols. II, III (Berlin, 1933–1936).

[21] We paraphrase Prop. 21 of *On Spirals:* "Given an area bounded by any arc of a spiral and the lines joining the extremities of the arc to the origin, it is possible to circumscribe about the area one figure, and to inscribe in it another figure, each consisting of similar sectors of circles, and such that the circumscribed figure exceeds the inscribed by less than any assigned area."

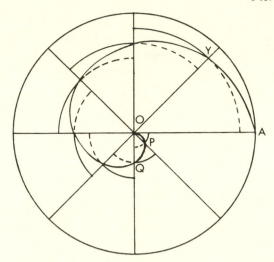

one might wish. *A fortiori*, the difference between the area of either step-figure and that of the spiral can be made less than any preassigned value.

The "method of exhaustion" now calls for a double *reductio ad absurdum* argument. Suppose, in the first case, that the area of the spiral, A_s, is not equal to one-third of the area A_c of the referent circle, but rather $A_s < \frac{1}{3}A_c$. By the above construction, one can circumscribe about the spiral a step-figure A_f such that

$$A_f - A_s < \frac{1}{3}A_c - A_s,$$

or

$$A_f < \frac{1}{3}A_c.$$

One need only choose N large enough to make A_c/N less than $\frac{1}{3}A_c - A_s$. But the area of this circumscribed step-figure will be to that of the referent circle as the sum of the squares of the radii of the spiral to the sum of the squares of the radii of the referent circle, i.e.

$$\frac{A_f}{A_c} = \frac{1}{N^3} \sum_{i=1}^{N} i^2.$$

By Proposition 10 of *On Spirals*, however,[22]

[22] Heath, p. 162: "If $A_1, A_2, A_3, \ldots, A_n$ be n lines forming an ascending arithmetical progression in which the common difference is equal to A_1, the least term, then $(n + 1) A_n^2 + A_1(A_1 + A_2 \cdots + A_n) = 3(A_1^2 + A_2^2 + \cdots + A_n^2)$. *Cor* .1. It follows from this proposition that

$$n \cdot A_n^2 < 3(A_1^2 + A_2^2 + \cdots + A_n^2)$$

$$\sum_{i=1}^{N} i^2 > \frac{1}{3} N^3,$$

whence $A_f > \frac{1}{3}A_c$, which contradicts the assumption just made. To prove that A_s cannot be greater than $\frac{1}{3}A_c$, one need only employ an inscribed step-figure $A_{f'}$ and show by a similar argument that $A_{f'}$ would have to be both greater and less than $\frac{1}{3}A_c$. In this second case one would have

$$\frac{1}{3} < \frac{A_{f'}}{A_c} = \frac{1}{N^3} \sum_{i=1}^{N} (i - 1)^2 = \frac{1}{N^3} \sum_{i=1}^{N-1} i^2.$$

But, from the corollary to Proposition 10,

$$\sum_{i=1}^{N-1} i^2 < \frac{1}{3} N^3,$$

whence emerges the second contradiction.

In the construction of circumscribed and inscribed step-figures and in Archimedes' elegant use of inequalities involved in the sum of the squares of the integers, Fermat discerned a direct heuristic path to the result. For, working directly from the areas of the circumscribed and inscribed figures, he had, on the one hand,

$$\frac{1}{N^3} \sum_{i=1}^{N} i^2 > \frac{A_s}{A_c} > \frac{1}{N^3} \sum_{i=1}^{N-1} i^2,$$

which held for all values of N, however large. From Proposition 10, on the other hand, he had

$$\frac{1}{N^3} \sum_{i=1}^{N} i^2 > \frac{1}{3} > \frac{1}{N^3} \sum_{i=1}^{N-1} i^2,$$

which also held for all values of N, however large. To equate A_s and $\frac{1}{3}A_c$ directly required only a short step of the imagination, albeit a step perhaps more in keeping with Fermat's algebraic bent than with the classical tradition: if two quantities A_1 and A_2 are always simultaneously less than some quantity $L_1(N)$ and greater than another quantity $L_2(N)$ for all values of N, and the

---

and also that

$$n \cdot A_n^2 > 3(A_1^2 + A_2^2 + \cdots + A_{n-1}^2)."$$

Heath summarizes the proofs on pp. 107–109. Clearly, the substitution of the integer i for the segment A_i yields the desired inequality in the text above.

difference between L_1 and L_2 can, by suitable choice of N, be made less than any preassigned value, then $A_1 = A_2$.[23] Such a lemma, employed heuristically, need never be stated explicitly; the classical pattern of the "method of exhaustion" could always be used to justify the equating of the two bounded quantities. The usefulness of this direct path to quadrature depended rather on one's ability to establish the numerical middle value between the two sums that held for all N. To see this last problem clearly, and to see Fermat's adjustment of Archimedes' method in use, we need only turn to Fermat's memoir on the quadrature of the "Galilean spiral," sent to Paris during the summer of 1636.

Fermat had been working with spirals for some time before he encountered one in *Galileo's Dialogue*. In a letter of 3 June 1636 he described to Mersenne the spiral $\rho^2 = a\vartheta$, which he (in all probability mistakenly) wished to identify as the "marvelous" curve of Menelaus, mentioned by Pappus.[24] He announced that the area of the first turn of this spiral was equal to one half of that of its referent circle, that the area of the second turn was twice that of the first, and that the areas of all further turns were each equal to that of the second. This quadratic spiral, in which the radius increases quadratically, leads quite naturally to the spiral in which the difference between a radius of the referent circle and the corresponding radius of the spiral grows by squares. As Fermat immediately recognized, such a spiral represents qualitatively the path of the

---

[23] No ancient author ever attempted to translate the method of proof by exhaustion into such a lemma stated once for all. Archimedes, for example, who employed the method frequently and to greatest advantage, laboriously supplied the complete double *reductio ad absurdum* at each occurrence. Newton was apparently the first to state the method explicitly as a lemma in his *Principia*, Book I, Section I, Prop. I. (I wish to thank my former student, Philip Kitcher, for calling my attention to the import of Newton's lemma.)

[24] FO.II.13: "J'en dresserai un Traité exprès, où je vous ferai voir de nouvelles hélices aussi admirables qu'on en puisse imaginer; pour vous en donner l'avant-goût, en voici une, qui est peut-être cette ligne que Ménélaüs appelle *admirable* dans le Pappus. Esto helix AMB in circulo

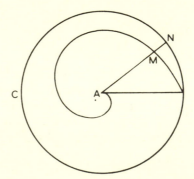

CNB, cujus ea sit proprietas ut, ductâ qualibet rectâ, verbi gratia AMN, tota circuli circumferentia sit ad ejusdem circumferentiae portionem NCB ut AB quadratum ad quadratum AM." The editors of FO identify the reference to Pappus as *Mathematical Collection*, IV, 36 (ed. Hultsch, p. 270, 1.26) and call Fermat's conjecture here "peu probable."

falling body discussed in the Second Day of the *Dialogue*.[25] Although Fermat, through Mersenne, pointed out this fact to Galileo, who had said the path would be a semicircle, Fermat's own interest in the curve was strictly mathematical.[26] A memoir discovered by DeWaard and published in the *Supplement* to *Oeuvres de Fermat* contains the details of Fermat's quadrature of the Galilean spiral.[27]

> When Galileo (a man of great genius) dubiously asserted in his *Dialogues* that, under the supposition of a diurnal motion of the earth, a naturally falling body would describe a semicircle, we had occasion to inquire more closely into the truth. Now, having already demonstrated the error of this opinion, we give the true curve, which, unless I am mistaken, is the same as that which, according to Pappus, Menelaus called "marvelous."[28]
>
> Let *ABKC* be a circle on the earth, which we suppose here to be the equator, let the origin of motion of some heavy body be *A*, and let it fall to the center *E* in a time which is to the time of 24 hours as arc *AKC* to the whole circumference of the circle (we posit only a diurnal motion of the earth, and not an annual motion). I say that the falling body will describe not a semicircle, but some spiral, such as that which is denoted in the diagram by the points *A, D*, 21, 18, 15, 12, etc. Under the assumption of

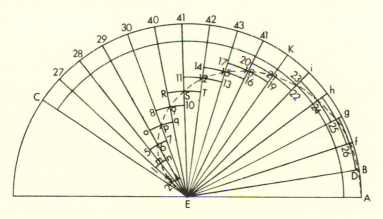

[25] Cf. Galileo, *Dialogue Concerning the Two Chief World Systems*, trans. Stillman Drake (Berkeley, 1962), p. 165ff.

[26] See the resulting correspondence between Carcavi and Galileo and between Diodati and Galileo, published by De Waard in FO.*Suppl*.46–71. Mersenne's personal copy of his *Harmonie universelle* (photorepr., Paris, 1963) contains the following pertinent marginal comment in his hand to Prop. IV of Book II (p. 96): "Il [Galileo] a du depuis confessé par lettre qu'il n'avoit parlé de ce demicercle que par caprice et galanterie, et non à bon escient, et qu'il avoit [word illegible] l'hélice."

[27] The full text, together with an introduction justifying its attribution to Fermat, was published by De Waard in FO.*Suppl*.1–19.

[28] Cf. above, n. 26. This remark suffices by itself to identify the author of the passage; Fermat was alone in this notion concerning Menelaus' "marvelous curve."

the ratio of the descent of heavy bodies noted by Galileo, the spiral has this property: any straight line *EK* having been drawn from the center and cutting the spiral at point 21 and the circumference of the circle at *K*, and straight lines *EA* and *EC* having been drawn, line *EK* is to line *K21* in the duplicate ratio of circumference *AKC* to circumference *AK*. Another notable property is that the area of the spiral contained by *AD21E* and line *EA* is to sector *AKCE* as the number 8 to the number 15.

Written in modern notation, the curve Fermat describes here is

$$\frac{R}{R - \rho} = \left(\frac{\alpha}{\vartheta} \right)^2,$$

where $R = EA$ and $\alpha = \measuredangle AEC$. Fermat devotes the first part of the memoir to demonstrating that the curve in fact satisfies the conditions set by Galileo's analysis of falling bodies in the *Dialogue*. That is, he shows that, if the whole sector *AEC* is divided into N equal sectors, the segments $BD, f26, g25$, etc., of the radii between the spiral and the referent circle increase as the square numbers from 1 to N^2, starting from point *A*. More interesting for the present, however, is Fermat's analytic demonstration of the second property he cites, i.e. that the area of the spiral is equal to 8/15 of that of sector *AEC*.

One need look no further than Fermat's diagram to see the path his analysis will take; the figure comes almost straight out of *On Spirals*. Fermat works directly from the diagram in taking a concrete number, 16, of subsectors and computing the areas of the circumscribed and inscribed figures. Since, however, he asserts clearly that the argument itself holds for any number of divisions—indeed, the demonstration rests on that foundation—one does no historical violence to his memoir by generalizing the argument from the start.

By the property of the spiral first demonstrated, as the length of the arc from *A* increases according to the series of natural numbers from 1 to *N*, the corresponding segments of the radii measured from the circumference increase as the squares of those numbers. If, then, one numbers the subsectors from 1 to *N*, starting at *A*, the segment between the circle and the spiral at the end of the *i*th sector is to that at the end of the $(i - 1)$th as $i^2:(i - 1)^2$. Hence, in the *i*th subsector, the radius of the inscribed arc is to the radius of the circle (both now measured from the center) as $N^2 - i^2:N^2$ and the radius of the circumscribed arc to the radius of the circle as $N^2 - (i - 1)^2:N^2$. Now, since the areas of similar sectors of circles are to each other as the squares of the radii of the circles, the area of the inscribed sector is to that of the sector of the referent circle as $(N^2 - i^2)^2:N^4$, and for the circumscribed sector the ratio is $(N^2 - (i - 1)^2)^2:N^4$. For any value of *N*, the ratio of the area of the *i*th sector of the spiral to the area of the *i*th sector of the referent circle lies between these two limits, whence, for the sum of the *N* segments, one has

$$\frac{1}{N \cdot N^4} \sum_{i=1}^{N} (N^2 - (i - 1)^2)^2 > \frac{A_s}{A_c} > \frac{1}{N \cdot N^4} \sum_{i=1}^{N} (N^2 - i^2)^2,$$

which holds for any and all values of N. The problem facing Fermat at this stage of his analysis should be clear; he requires the numerical ratio p/q such that

$$\frac{1}{N \cdot N^4} \sum_{i=1}^{N} (N^2 - (i - 1)^2)^2 > \frac{p}{q} > \frac{1}{N \cdot N^4} \sum_{i=1}^{N} (N^2 - i^2)^2,$$

for any and all values of N. As Fermat knew, his problem ultimately reduces to that of finding summation formulas for $\sum_{1}^{N} i^2$ and $\sum_{1}^{N} i^4$, since

$$\sum_{i=1}^{N} (N^2 - i^2)^2 = N^5 - 2N^2 \sum_{i=1}^{N} i^2 + \sum_{i=1}^{N} i^4.$$

Fermat had the formulas. Before completing the above analysis and continuing the development of the method of quadrature derived from it, we should look at his derivations.

B.   *From Polygonal Numbers to Series Summations*

In the realm of summation formulas for the powers of the integers, Fermat's Archimedean model, Proposition 10 of *On Spirals*, could offer little inspiration. Almost impossible to follow because of its geometric style, the model was powerless before higher powers because of the dimensional restrictions of classical geometry. Instead, Fermat found his inspiration in Bachet's appendix to Diophantus' treatise *On Polygonal Numbers*.[29] To have done so was fully in keeping with his fundamental belief in the unity of mathematics through algebra.

Polygonal (and polyhedral) numbers date from another time at which the realms of geometry and arithmetic were united.[30] The view of a number as a collection of units led (according to tradition) the Pythagoreans to arrange the units of a number—represented, say, by N small stones—in geometrical patterns. Numbers were then classified in groups according to their common visual patterns. Any number N could, of course, represent a line. The square numbers N^2, however, could be arranged to form a square of side N. The

---

[29] Cf. FO.II.84. Bachet's *Appendix ad librum de numeris polygonis* appeared as part of his edition of Diophantus' *Arithmetica* (Paris, 1621).

[30] Perhaps the best account of the Pythagorean doctrine of figurate numbers is that of Paul-Henri Michel, *De Pythagore à Euclide* (Paris, 1950), pp. 295–328.

numbers 1,3,6,10, . . . , i.e. the successive sums of the integers, could be arranged in the form of equilateral triangles, the sum of the first N integers in an equilateral triangle of side N. Analogously, the numbers 1,4,10,20, . . . , i.e. the sums of the triangular numbers, formed a tetrahedron, and so on for most of the regular plane and solid figures.

The visual display of such numbers presented to the eye many of their basic generational properties. For example, the figure of the square numbers leads immediately to the formula $(N + 1)^2 = N^2 + 2N + 1$. What about the

```
  .              . . . . .
  . .            . . . . .
  . . .          . . . . .
  . . . .        . . . . .
  . . . . .      . . . . .
```

formula for the triangular numbers? Consider the square of side $N + 1$, containing $(N + 1)^2$ units. Removing the main diagonal of length $N + 1$, one sees clearly that the remainder consists of two triangular numbers of side N. Hence, each triangular number of side N, i.e. the sum of the integers from 1 to N, is equal to

$$\frac{(N + 1)^2 - (N + 1)}{2} , \text{ or } \frac{N(N + 1)}{2} .$$

If, then, in the series 1,2,3, . . . , N, one calls N "the last side," one has, as Fermat puts it in the language of polygonal numbers: "The last side multiplied by the next greater [side] makes twice the triangle."

Now, though Bachet could give this rule for the triangular numbers, and indeed for polygonal numbers in general, he offered in his commentary no rule for any of the polyhedral numbers (e.g. the series 1,4,10,20, . . .) or numbers of higher dimension. Fermat had discovered the general rule; how is not important for the moment. As he wrote to Mersenne, and later to Roberval:[31]

> The last side multiplied by the next greater makes twice the triangle. The last side multiplied by the triangle of the next greater side makes three times the pyramid. The last side multiplied by the pyramid of the next greater side

[31] Fermat to Roberval, 4.XI.1636, FO.II.84–85; cf. Observation 46 (on Diophantus' *De multangulis numeris*, IX), FO.I.341 (discussed below, Chap. VI, p. 336), and Fermat to Mersenne, IX or X.1636, FO.II.70. The precise date of the letter to Mersenne is in doubt. Jean Itard, who would like to date it 1638, argues his case and reviews the arguments of the editors of FO in "Sur la date à attribuer à une lettre de Pierre Fermat," *Rev. d'hist d. sci.* 2 (1948), 95–98. Cornelis de Waard (*Correspondance de Mersenne* VII (1962), 272) supplies additional arguments for dating it to the beginning of June 1638; cf. Jean Itard's review of the first edition of this book, *Historia Mathematica* 1(1974), 470–5; at 473.

makes four times the triangulotriangle. And so on by the same progression in infinitum.

"All these propositions," he added in the letter to Roberval, "however pretty in themselves, have aided me in the quadrature that I'm pleased you value." How had they aided? A proposition that Fermat again cites in both letters gives the answer. He writes: "If you multiply four times the greatest number increased by two by the square of the triangle of numbers, and from the product you subtract the sum of the squares of the individual numbers, five times the sum of the fourth powers will result." That is,

$$5 \sum_{i=1}^{N} i^4 = (4N + 2)\frac{N^2(N + 1)^2}{4} - \sum_{i=1}^{N} i^2.$$

One concrete result of Fermat's propositions on polyhedral numbers, then, had been the summation formula for the fourth powers of the integers from 1 to N. But this was not the only result. The propositions had enabled him, he claimed, to obtain the summation formula for any power and thereby to prove the lemma Roberval had recognized at the heart of Fermat's work on quadrature, the lemma Fermat doubted Roberval could prove rigorously.

It does no violence to Fermat's patterns of thought here to take advantage of the economy of modern notation, although doing so of course masks the genius that the lack of such notation demanded of Fermat. Nonetheless, his method was algebraic, even if his language was not.[32] The propositions for polyhedral numbers given above may be expressed in a single proposition schema:

$$\sum_{i=1}^{N} \frac{i(i + 1)(i + 2)\cdots(i + k - 1)}{k!} = \frac{N(N + 1)(N + 2)\cdots(N + k)}{(k + 1)!}.$$

Mathematical induction provides an immediate proof.[33] Rewriting the numerator of the summand in the form[34]

$$i^k + a_1 i^{k-1} + a_2 i^{k-2} + \cdots + a_{k-1} i$$

[32] Compare, for example, Fermat's reasoning in the letter to Mersenne with the derivations of Archimedes cited above, n. 22.

[33] There is no direct evidence in his works that Fermat knew of the technique of proof by induction, though his method of infinite descent (see below, Chap. VI, §V) may be described as "reverse induction." Neither, however, does Fermat provide the slightest hint as to his derivation or his proof of this theorem schema, and the best one can do is to verify its validity. For a detailed analysis of Fermat's result and its relation to figurate numbers, see A.W.F. Edwards, *Pascal's Arithmetical Triangle* (London, 1987), Chap. 1, esp. 14–15.

[34] The a_i here are simply the elementary symmetric functions of the expression in the numerator of the summand.

one can then express the left-hand side of the proposition schema in the form

$$\frac{1}{k!} \left[\sum_{i=1}^{N} i^k + a_1 \sum_{i=1}^{N} i^{k-1} + \cdots + a_{k-1} \sum_{i=1}^{N} i \right].$$

Transposition of terms then leads to a recursive formula for the sum of the kth powers of the first N integers:

$$\sum_{i=1}^{N} i^k = \frac{N(N + 1)(N + 2) \cdots (N + k)}{k + 1} - \left[a_1 \sum_{i=1}^{N} i^{k-1} + \cdots + a_{k-1} \sum_{i=1}^{N} i \right].$$

One could in principle use the corresponding recursive formulas for the powers from $k - 1$ to 1 to derive a general formula containing only N and k, but in practice the algebraic computations, even with modern notation, quickly get out of hand. It would seem far more likely that Fermat worked in the other direction, using the formula to compute the sum of the kth powers on the basis of the known sums for the powers from $k - 1$ to 1.

Application of the general scheme in this way for $k = 4$ (the summation formulas for $k = 1,2,3$ had been known since Antiquity[35]), would have given Fermat

$$\sum_{i=1}^{N} i^4 = \frac{N(N + 1)(2N + 1)(3N^2 + 3N - 1)}{30}.$$

This is not the form in which Fermat states the result in his letters to Mersenne and Roberval, and one may be tempted at first to question whether he took another path to the result. However, it seems unlikely that Fermat would have stated the formula merely as a polynomial in N. As such, it does not lend itself to easy expression in terms of a last side. Far more elegant is his own recursive formulation of the result in terms of the "last side," the square of the triangle of that side (= the sum of the cubes), and the sum of the squares of the individual terms. His formulation has both economy and the correct classical tone, which may serve as reasons for his having juggled the direct formula given above.

Whatever the form in which he gave or got his results, Fermat's success in the realm of summation of the powers of the integers supplied him with more than enough material for his research on quadrature. Each of the finished formulas made it evident that, for integer k,

---

[35] Fermat could not have known of the summation of the fourth powers by Arabic mathematicians; cf. A. P. Yushkevich, *Istoria matematiki v srednie veka* (Moscow, 1961; German trans. *Mathematik im Mittelalter*, Leipzig, 1964), 278. The works of Ibn al-Haitham and Thabit ibn Qurra in which that result appears were not known in Europe until the late nineteenth century.

$$\frac{1}{N^{k+1}} \sum_{i=1}^{N} i^k > \frac{1}{k+1} > \frac{1}{N^{k+1}} \sum_{i=1}^{N-1} i^k,$$

for all N, and that the difference between the boundary sums is $1/N$ and hence, by a choice of suitably large N, may be made as small as one might wish. Moreover, his successful determination of the sums of powers of the integers made possible the determination of the more complicated sums resulting from such analyses as that of the Galilean spiral.

C.   From Spirals to Conoids

Fermat's analysis of the quadrature of the Galilean spiral had placed the ratio of the area of the spiral to that of its referent circle (or sector) between the limits

$$\frac{1}{N^5} \sum_{i=1}^{N} (N^2 - (i - 1)^2)^2 \quad \text{and} \quad \frac{1}{N^5} \sum_{i=1}^{N} (N^2 - i^2)^2$$

for any N however great. On the basis of the summation formulas he had worked out, Fermat could show that

$$\frac{1}{N^5} \sum_{i=1}^{N} (N^2 - (i - 1)^2)^2 > \frac{8}{15} > \frac{1}{N^5} \sum_{i=1}^{N} (N^2 - i^2)^2$$

for all N. In both cases the difference between the two bounding sums is $1/N$ and hence can be made less than any assigned magnitude. From the "lemma of containment," then, which could in any case be verified by a classical proof by exhaustion, the area of the spiral is to that of the referent sector as $8:15$. It is a sign of the new approach Fermat was taking that he does not carry out the classical proof.

The generality of Fermat's technique of series summation enabled him to treat an infinite class of spirals, indeed two infinite classes. All curves of the forms

$$\rho = (a\vartheta)^k \quad \text{and} \quad \frac{R}{R - \rho} = \left(\frac{\alpha}{\vartheta}\right)^k$$

lay within the power of his modified Archimedean method of quadrature. Fermat apparently soon recognized, however, that his new method was not restricted to spirals. An immediate, almost obvious, transformation of the method to a rectangular framework opened the way to the quadrature of another infinite family of curves, the "higher parabolas."

The transformation of the method itself suggests its possible range of applicability. For let AB be a curve referred to rectangular axes, such that its vertex is at A and such that point B corresponds to a given point x_0, y_0 on the curve.

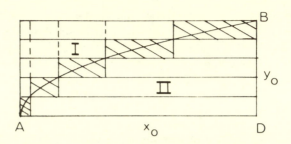

Divide the given length y_0 into N equal parts by drawing N lines of length x_0 parallel to axis AD. Let the segments of these lines falling to the left of the curve be denoted by $x_1, x_2, \ldots, x_0$. Consider first the area I. In the i-th strip, the circumscribed segment has the area $\frac{y_0}{N} x_i$; the inscribed segment, the area $\frac{y_0}{N} x_{i-1}$. Hence, the total area I lies between the limits

$$\frac{y_0}{N} \sum_{i=1}^{N} x_i \quad \text{and} \quad \frac{y_0}{N} \sum_{i=1}^{N-1} x_i$$

for all N. From the diagram alone, however, it is evident that the sum of the differences in area between the circumscribed and inscribed segments is the topmost strip $\frac{y_0}{N} x_0$. This last quantity is therefore the difference between the two boundaries given above, and, since x_0 and y_0 are given quantities independent of N, this difference may clearly be made less than any assigned magnitude by taking N sufficiently large.

Obviously, then, a vast number of curves could be squared by this method, the only restriction being one's ability to determine from the series the numerical value of the quantity contained between the bounding sums. Fermat knew what series he could evaluate. He could establish the middle value for the series of integers to any power. That is, for any curve for which the x_i increased according to the series i^k, Fermat could show that

$$\frac{1}{N^{k+1}} \sum_{i=1}^{N} i^k > \frac{1}{k+1} > \frac{1}{N^{k+1}} \sum_{i=1}^{N-1} i^k,$$

for all N. Similarly, on the model of the Galilean spiral, he could determine the middle value for series of x_i of the form $N^k - i^k$ or even $(N^k - i^k)^m$ (for

234

integer *m*). What curves had these properties? They were none other than the curves of the form $y^k = px$, i.e. an infinite subclass of the "higher parabolas" about which Fermat had so much to say in his letters at the end of 1636. He had so much to say, yet it could be said so succinctly. From the above inequality, it followed immediately that, for any curve $y^k = px$,

$$\text{Area I} = \frac{1}{k+1} x_0 y_0$$

and hence

$$\text{Area II} = \frac{k}{k+1} x_0 y_0.$$

Thus Fermat could indeed square the ordinary parabola "autrement qu'Archimède." For, where Archimedes had carried out his quadrature of the parabola by inscription and circumscription of triangles, Fermat had taken the model of the quadrature of the spiral and extended it for systematic application to curves referred to orthogonal axes. His reading of Apollonius had led to the *Isagoge* and a system for introducing into mathematics an infinity of new curves. Now his reading of Archimedes had led to the quadrature of at least one infinite subclass of those curves. Needless to say, the method of tangents directly assured that all the new curves had tangents.

Indeed, Fermat's Archimedean technique had two important corollaries that served to extend the range of its applicability. The "new conoid" of which Fermat spoke in his letter to Roberval, i.e. the solid generated by the rotation of the curve $y^2 = px$ about an ordinate, represented a whole class of such figures subject to cubature by Fermat's technique. For the technique could determine the volume of any solid generated by rotation about an ordinate or about the axis of any curve of the form $y^k = px$ and of certain other curves. Take first Fermat's concrete example. Fermat never offered Roberval the analysis of the cubature of the new conoid, but, as the quotation that opens this section indicates, that analysis followed the model of the Galilean spiral.

Let *AB* be the parabola $y^2 = px$, of which *A* is the vertex and *B* a given point. Setting $DB = y_0$, $AD = x_0$, divide up the plane section *DAB* into inscribed and circumscribed segments in the manner shown above. The vol-

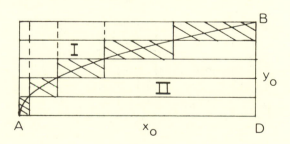

ume of the new conoid will clearly lie between the volumes of the two step-cones generated by the rotation of these segments (area II) about fixed axis *DB*. The ratio of the volume of the circumscribed step-cone to that of the inscribed cone *ABD* will be

$$3 \sum_{i=1}^{N} (N^2 - (i - 1)^2)^2 : BD \cdot AD^2,$$

and similarly that of the inscribed step-cone to that of cone *ABD* as

$$3 \sum_{i=1}^{N} (N^2 - i^2)^2 : BD \cdot AD^2.$$

Whence it follows that

$$3 \sum_{i=1}^{N} (N^2 - (i - 1)^2)^2 > \frac{\text{conoid}}{\text{cone}} > 3 \sum_{i=1}^{N} (N^2 - i^2)^2.$$

One need pursue the problem no further. Except for the constant factor 3, the problem of the volume of the new conoid is precisely the same as the problem of the area of the Galilean spiral. Clearly, the cubature of a conoid generated in the same way by any of the curves $y^k = px$ required only the ability to handle

$$\sum_{i=1}^{N} (N^k - i^k)^2,$$

a problem that lay fully within Fermat's grasp. Similarly, the cubature of any segment of the conoid cut off by a plane parallel to the base required only the comparison of partial sums, again a problem for which Fermat had the necessary tools.

The cubature of conoids generated by rotation of the curves about the axis, however, required a second corollary of Fermat's basic method. This corollary not only extended the applicability of the method, it also led Fermat to recog-

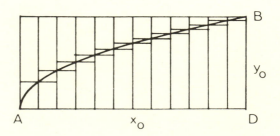

nize the limits of the method. Take the example of the ordinary Archimedean paraboloid. The construction of step-cones within and about the paraboloid necessitates the division into N equal segments of the axis rather than the ordinate. That is, divide AD into segments of length x_0/N Since $y^2 = px$, the radius of the i-th cylindrical segment will be

$$\left(\frac{i}{N} px_0 \right)^{1/2},$$

or

$$\left(\frac{i}{N} \right)^{1/2} y_0.$$

The volume of the paraboloid lies therefore between the limits

$$\frac{\pi x_0 y_0^2}{N} \sum_{i=1}^{N} \frac{i}{N} \quad \text{and} \quad \frac{\pi x_0 y_0^2}{N} \sum_{i=1}^{N-1} \frac{i}{N}.$$

But, it follows from previous results that

$$\frac{1}{N^2} \sum_{i=1}^{N} i > \frac{1}{2} > \frac{1}{N^2} \sum_{i=1}^{N-1} i,$$

whence the volume of the paraboloid is one-half that of the cylinder with base radius y_0 and altitude x_0, or three-halves that of the cone with the same base and altitude. Thus the cubature of the ordinary Archimedean paraboloid follows from a direct application of Fermat's Archimedean method of quadrature.

Pursuing the same model for the paraboloid generated by the curve $y^4 = px$, however, leads to a minor difficulty. The bounding limits are

$$\frac{\pi x_0 y_0^2}{N} \sum_{i=1}^{N} \left(\frac{i}{N} \right)^{1/2} \quad \text{and} \quad \frac{\pi x_0 y_0^2}{N} \sum_{i=1}^{N-1} \left(\frac{i}{N} \right)^{1/2}.$$

The difficulty is at first minor because, although Fermat had no summation formula for the square roots of the integers, he did have something almost as good and wholly sufficient to his immediate needs. For consider again the analysis of the cubature of the ordinary paraboloid. The plane section to be rotated, i.e. the semiparabola, was divided into circumscribed and inscribed segments of area

$$\frac{x_0 y_0}{N} \left(\frac{i}{N} \right)^{1/2} \quad \text{and} \quad \frac{x_0 y_0}{N} \left(\frac{i-1}{N} \right)^{1/2},$$

respectively. Were one to sum those segments without rotating them, the area under the semiparabola would lie between the limits

$$\frac{x_0 y_0}{N} \sum_{i=1}^{N} \left(\frac{i}{N} \right)^{1/2} \text{ and } \frac{x_0 y_0}{N} \sum_{i=1}^{N-1} \left(\frac{i}{N} \right)^{1/2}$$

for all N. But Fermat already knew the middle value that must lie between those limits, for he knew the area of the semiparabola. It is $\frac{2}{3} x_0 y_0$. It followed, therefore, that

$$\frac{1}{N} \sum_{i=1}^{N} \left(\frac{i}{N} \right)^{1/2} > \frac{2}{3} > \frac{1}{N} \sum_{i=1}^{N-1} \left(\frac{i}{N} \right)^{1/2}.$$

Indeed, Fermat could go somewhat further. Since his method of quadrature extended to all curves of the form $y^k = px$, it followed that

$$\frac{1}{N} \sum_{i=1}^{N} \left(\frac{i}{N} \right)^{\frac{1}{k}} > \frac{k}{k+1} > \frac{1}{N} \sum_{i=1}^{N-1} \left(\frac{i}{N} \right)^{\frac{1}{k}}$$

for all N. Thus the earlier restriction that the power of y in the equation $y^k = px$ be an integer disappeared in part. It could also take on the values of the reciprocals of the integers. Hence, for example, the volume of the rotational solid $y^4 = px$ is two-thirds that of its circumscribed cylinder, or twice that of its inscribed cone. And, in general, Fermat could show that the volume of the solid generated by rotation of the curve $y^k = px$ about its axis is $k/(k + 2)$ that of the circumscribed cylinder, *provided that* k *is an even integer*. Were k an odd integer, determination of the volume of the rotational solid would require the determination of a middle value for

$$\frac{1}{N} \sum_{i=1}^{N} \left(\frac{i}{N} \right)^{\frac{2}{m+1}} \text{ and } \frac{1}{N} \sum_{i=1}^{N-1} \left(\frac{i}{N} \right)^{\frac{2}{2m+1}}$$

and Fermat had neither a direct nor an indirect method for determining that value, as he sadly admitted to Roberval in his letter of 16 December:[36]

> you will easily find the ratio [of solid parabolas to their cones] for those that follow alternately after the square (*post quadrata alternatim*), e.g. quadratoquadrates, cubocubes, etc., for which you give an example of the first. But in parabolas that are cubic, quadratocubic, and so on alternately *ad infinitum*, the method that we have used does not yield the ratio of the conoids to their cones. By our method [of centers of gravity], however, we

[36] FO.II.95.

have found the center of gravity of any conoid whatever. Hence, by your proposition, their ratio to their cones will be given.

The restrictions on Fermat's method, made clear by his analysis of the rotational solids $y^k = px$ for odd k, apparently pointed to the restrictions inherent in the method for problems of quadrature itself. For, clearly, the method could be applied only to those higher parabolas of the form $y^{km} = px^m$ (or, of course, $y^m = px^{km}$). It could not, however, be applied to the general curve $y^q = ax^p$, since Fermat had no method for determining the values of bounding sums of the form $\sum_{i=1}^{N} i^{p/q}$. He had no reason to assert that the inequality

$$\frac{1}{N^{k+1}} \sum_{i=1}^{N} i^k > \frac{1}{k+1} > \frac{1}{N^{k+1}} \sum_{i=1}^{N-1} i^k,$$

for all N also held for all fractional values of k. Indeed, one may doubt that the thought ever occurred to him. For, though Fermat did not lack daring as a mathematician, he lacked even the slightest hint of such an extended generalization or of how one might go about justifying it. More precisely, one should say Fermat never made the generalization explicit. As will be clear presently, it arises implicitly in his work on centers of gravity—there, however, under entirely different circumstances.

In fact, as Fermat's remarks in the letter to Roberval indicate, the method of centers of gravity represented for a time Fermat's faint ray of hope in overcoming the restrictions of his Archimedean method. His future research into quadrature, with the aim of finding a way past the restrictions of the early method, would in time meet with success. For the moment, however, he had to seek help from his method of centers of gravity which, he claimed, arrived at the correct results without requiring explicit knowledge of the area or volume of the figure. Since Roberval had claimed to be able to determine such volumes and areas on the basis of the known center of gravity, Fermat's method promised at least an indirect path to the quadrature of curves his direct method could not handle.

III.   THE METHOD OF CENTERS OF GRAVITY

A study of Fermat's technique for determining the centers of gravity of various plane and solid figures represents a carry-over from the previous chapter. For the technique derived from the method of maxima and minima and employed the basic assumptions of that method. Nonetheless, it properly belongs to Fermat's work on quadrature, because of the role Fermat saw for it in reaching the quadrature of some curves.

Fermat gave only one explicit example of the method in a memoir he sent to Roberval in the spring of 1638, though he referred to its use in connection with every plane figure and solid he ever investigated.[37] Roberval voiced some concern over the generality of the method, though not, as one might think, in relation to the use of maxima and minima. Roberval was bothered rather by the very step that Fermat himself found to be the method's strongest point, to wit, the step that seemed to eliminate the need for concrete knowledge of the volume or area of the figure under consideration. It will help to follow Fermat through the one detailed analysis he provided and then to focus on the questionable step.

In the memoir, Fermat attacks his problem with no introductory remarks. He proposes to determine the center of gravity of the Archimedean paraboloid. Let *CAV* be a parabola ($CI^2 = pAI$) which, rotated about axis *AI*, generates a paraboloid. Let *O* be the center of gravity of that paraboloid. By symmetry, *O* lies somewhere on the axis *AI*; let $AO = x$ denote the distance

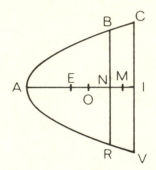

from vertex *A* to point *O*. Now imagine the paraboloid to be cut by a plane drawn parallel to the base *CV* at some arbitrary distance $NI = y$ from it. Let *E* denote the center of gravity of the upper segment *BAR* and let *M* denote that of the frustrum *CBRV*.

Now the two paraboloids *CAV* and *BAR* are similar solids, and Archimedes has demonstrated that similar paraboloids have similarly placed centers of gravity. Hence, if $AI = b$, then

$$\frac{b}{x} = \frac{b - y}{AE},$$

whence it follows that

$$EO = x - AE = \frac{xy}{b}.$$

[37] Text in FO.I.136–139. According to the editors of FO, it was dispatched to Mersenne for transmission to Roberval with Fermat's letter to Mersenne of 20.IV.1638. De Waard (FO.*Suppl*.xvi) gives the date II.1638 for its composition.

Now apply the law of the lever: if O is the center of gravity of the whole figure CAV, then it is the point at which the segments BAR and $CBRV$, acting at their centers of gravity, will balance. Hence

$$\frac{CBRV}{BAR} = \frac{EO}{OM}.$$

One now needs an expression for the left-hand ratio in terms of segments on the axis AI, i.e. in terms of x. By Proposition 26 of Archimedes' *On Conoids and Spheroids*,

$$\frac{CAV}{BAR} = \frac{AI^2}{AN^2} = \frac{b^2}{(b - y)^2}.$$

Hence,

$$\frac{EO}{OM} = \frac{CBRV}{BAR} = \frac{b^2 - (b - y)^2}{(b - y)^2}.$$

Now, since M is the center of gravity of the frustrum, it will always lie on AI between points N and I. That is, no matter what the value of y, the length OM will always be less than length OI. Though Fermat does not put it in so many words, he evidently sees in OI the same sort of quasi-extreme value he discerned in the ordinate BC in his analysis of the tangent problem.[38] Hence, treating OI as a maximum, he takes the step of adequating OM and OI; that is,

$$OI = b - x \approx OM = OE \cdot \frac{BAR}{CBRV}.$$

Substitution of the algebraic equivalents of the geometric quantities yields the adequality

$$b - x \approx \frac{b^2xy - 2bxy^2 + xy^3}{2b^2y - by^2},$$

and this adequality, treated according to the method of maxima and minima, gives the final result

$$x = \frac{2}{3} b.$$

Fermat's method has, then, five main steps, each resting on a lemma. The first lemma is that similar figures have similarly placed centers of gravity. Fermat makes no attempt to establish the similarity of the figures; he takes it to be immediately evident. For his analysis of the paraboloid, he could borrow the lemma directly from Archimedes. For solids generated by other parabolas, i.e. the "higher parabolas," or by other curves, he had no ancient authority. Nonetheless, he seems to have been sure the lemma held generally and that it could be proved, for example for hyperboloids of revolution. The second step,

[38] See above, Chap. IV, p. 167.

the balancing of the segments about the center of gravity of the whole, rested on the law of the lever and on the definition of the center of gravity, both of which counted as axiomatic truths. Jumping for a moment to the last two steps, one finds assumptions and techniques peculiarly Fermat's. Neither needs commentary, since the reasoning underlying the adequation of quantities and the method for transforming adequality into strict equality derived wholly from that of the method of maxima and minima discussed in Chapter IV. It is rather the third step on which Fermat and Roberval focused attention, as we should.

Fermat made much of the fact that he did not need actually to compute areas or volumes in order to find an expression for the ratio of the volumes of the two paraboloids in terms of segments of the axis. Indeed, he went farther to claim that application of his method never required such computation. In reacting to the memoir in a letter to Fermat on 1 June 1638, Roberval expressed his doubts on the matter.[39] More precisely, Roberval expressed doubt about Fermat's step. Fermat had written that the frustrum was to the segment of the paraboloid as $AI^2 - AN^2$ to AN^2. In trying, however, to apply the method to the ordinary parabola, Roberval had found he could not take this step, which, he recognized, rested on the specific properties of the paraboloid. That is, he could not write

$$\frac{\text{area } CBRV}{\text{area } BAR} = \frac{AI^2 - AN^2}{AN^2}.$$

Where equality failed, Roberval would look toward inequality, and he suggested that Fermat could save his method by replacing the "=" with "<."

As the editors of *Oeuvres de Fermat* note in a classic understatement, "Roberval had not understood all the resources of Fermat's method." Clearly he had understood the method about as well as he understood the method of maxima and minima or the method of tangents, to wit, not very well. Nonetheless, as was the case with the method of maxima and minima itself, a critic's lack of understanding had the historical virtue of drawing clarification from Fermat. On 15 June 1638, Fermat wrote to Mersenne:[40]

And on the subject of [centers of gravity] I will show that the center of gravity of a figure can be found without one's knowing its quadrature. For in the parabola, for example, I use no other means than that:

(1) In the segments cut off by the applicates, the centers of gravity are similarly placed;

(2) The segments of this sort have the same proportion to corresponding triangles of the same base and height, even if we do not know what that proportion is;

[39] FO.II.149–150.
[40] FO.*Suppl*.84–86.

(3) If the perimeter of any figure is bent in the same direction, the center of gravity of the figure lies within it;

(4) The parts of the figure are to one another reciprocally as the straight lines drawn from their centers to the center of the whole.

All these means are easily proved by the Archimedean way, without supposing the quadrature of the parabola. So that, having thus found the center of gravity of the parabola, we can consequently derive from it its quadrature, a technique of quadrature which must be added to the two of Archimedes and to that of the quadratures of the infinite parabolas which I have discussed several times with M. Roberval.

The important "means" here is the second. In modern terms it asserts that, for any curve $y^k = px$,

$$\frac{1}{y(a) \cdot a} \int_0^a ydx = \frac{1}{y(a + e) \cdot (a + e)} \int_0^{a+e} ydx.$$

Application of the method to hyperbolas of the form $x^p y^q = m$, besides requiring conditions on the values of p and q, rests on the analogous proposition

$$\frac{1}{y(a) \cdot a} \int_a^\infty ydx = \frac{1}{y(a + e) \cdot (a + e)} \int_{a+e}^\infty ydx.$$

Now Fermat is correct, but he is correct in spite of himself. Any rigorous justification of his lemma must ultimately involve the actual quadrature of the curve. Fermat's intuitive grasp of the lemma could have found support only in his successful quadrature of the parabolas $y^k = px$ for integral k (or for $k = 1/m$) and his cubature of certain of the corresponding rotational solids. Such concrete support must have seemed adequate basis for his generalization. Nonetheless, the possibility of circularity that he sought to deny in the last paragraph of the passage quoted above cannot be waved away so easily. His argument is not possibly circular; it is actually circular.

More interesting than the circularity, however, is the relation of Fermat's lemma to his quandary about fractional exponents. He could not square the curves $y^q = ax^p$ directly, nor did he feel justified in generalizing the value $1/(k + 1)$ for fractional k. The method of centers of gravity provided his path around the hurdle. Yet for Fermat to make the above statement (2) is precisely to assert, in modern terms, that

$$\lim_{N \to \infty} \frac{1}{N^{k+1}} \sum_{i=1}^N i^k = \frac{1}{k + 1}$$

for all k, fractional and integral. Geometric intuition had come to the aid of algebraic analysis, but Fermat could not recognize the gift.

For the families of curves to which Fermat applied his lemma (2) in the course of his research, i.e. the families $y^q = ax^p$ and $x^p y^q = a$ (p,q positive integers), it held. For other curves for which he sought the center of gravity, he could rely on concrete quadratures, either his own or those of others. Possession of the quadrature makes the lemma unnecessary, and the area or volume may be fed into Fermat's third step directly. The method of centers of gravity did not depend essentially on the lemma.

Fermat's method of determining centers of gravity, like the method of maxima and minima from which it had sprung, never underwent essential alteration or development. The method Fermat had devised in his earliest research remained a finished method. But his method of quadrature, derived from the inspiration of Archimedes' On *Spirals*, provided no such sense of perfection. Fermat knew its limitations, and in the years following 1638 he sought to devise new methods for squaring the new classes of curves his *Introduction* had suggested to him. Toward the mid-1640s he succeeded in establishing the necessary foundation for an attack on the general problem of quadrature.[41] From that foundation he moved to the imposing system of his major treatise on quadrature, composed sometime around 1658.

IV. THE TREATISE ON QUADRATURE (CA.1658)

Except for a clumsy attack on the quadrature of the cycloid,[42] Fermat remained almost completely silent during the twenty years from 1637 to 1657 about any efforts he might have been making in the realm of quadrature. His *Response to the Questions of Cavalieri*, written in 1643, shows, however, that he was actively engaged in such research during the period. In a letter written to Kenelm Digby in 1657 Fermat makes reference to material he had sent to Torricelli in 1646.[43] That material dealt with the quadrature of curves of the form $x^p y^q = k$, and such curves, as we shall presently see, lay beyond the power of the early method of quadrature. Moreover, the appearance of John Wallis' *Arithmetic of Infinites* in 1657 drew from Fermat not only criticism of Wallis' methods but also Fermat's claims to have long since achieved the same results, results that again lay beyond the power of the method of quadrature presented above.[44] Indeed, Wallis' work provoked Fermat to set down his

[41] The results exchanged between Fermat and Torricelli (see above, n. 5) in 1644–1646 involved curves beyond the reach of Fermat's early methods. Similarly, his *Response* to *the Questions of Cavalieri* (FO.I.195–198) addresses the general problem of squaring higher parabolas "in which the abscissas on the diameter are to one another as any powers of the applicates." The specific example he treats is the curve $y^4 = ax^3$, which again he could not have approached by his early method.

[42] See his letters to Mersenne of VII.1638 and 27.VII.1638 in FO.*Suppl*.87–95. It is this work on the cycloid that Descartes so harshly condemned in his letter to Mersenne of 23.VIII.1638; cf. above, Chap. IV, n. 101.

[43] See above, n. 5.

[44] Fermat to Digby, 20.IV.1657, FO.II.337–342.

results in a treatise of his own, and that treatise, completed within a year or so, could only have been the product of many years' research. The lengthy title alone indicates the extent to which Fermat had revised his thinking on the problem: *On the transformation and alteration of local equations for the purpose of variously comparing curvilinear figures among themselves or to rectilinear figures, to which is attached the use of geometric proportions in squaring an infinite number of parabolas and hyperbolas.*[45]

This *Treatise on Quadrature* (the short title we shall use hereafter) has two distinct parts. In the first, Fermat presents a new technique for the direct quadrature of any of the higher parabolas and hyperbolas. In the second, he establishes a new form of "reduction analysis" by which the quadrature of any curve given by its equation can be reduced either to a series of direct quadratures or to the quadrature of the circle. Of these two parts, the first represents a continuation of themes and problems evident in Fermat's early work on quadrature, while the second embodies a radical departure from those themes. It will help, therefore, to treat the two parts separately.

A.   *Adequality and the "Logarithmic Method"*

The method of direct quadrature contained in the *Treatise on Quadrature* overcomes the restrictions of Fermat's earlier method of quadrature by altering the manner in which the axis of the figure is divided into intervals. It also differs in the introduction of the concept of adequality, which was absent from the earlier method. Though Fermat left no explicit indication of the path along which he arrived at the new method, the nature of the difficulties posed by the old provide relatively clear hints. For one is struck not only by the applicability of the new method to all curves of the form $y^q = kx^p$, i.e. to all higher parabolas, but most particularly by the appearance of a new class of curves previously untreated by Fermat. The new method also yields the direct quadrature of all higher hyperbolas, i.e. all curves of the form $x^p y^q = k$, with the exception of the simple hyperbola $xy = k$. Although, in his previous work, Fermat had never referred to the quadrature of these hyperbolas, he had spoken of the curves. They were as old as the higher parabolas. One may assume with some certainty that Fermat had been interested in their quadrature, which, as the references to the correspondence with Torricelli indicate,

---

[45] *De aequationum localium transmutatione et emendatione ad multimodam curvilineorum inter se vel cum rectilineis comparationem, cui annectitur proportionis geometricae in quadrandis infinitis parabolis et hyperbolis usus.* The text in FO.I.255–285 is taken from the *Varia*, pp. 44–57. The treatise bears no explicit date, but 1658 or 1659 seems most likely. On p. 266 Fermat explains that the centers of gravity of the figures he is squaring may be found by a technique derived from this method of maxima and minima, which he had made known to mathematicians "more or less twenty years ago"; the method of centers of gravity was announced in 1638. Citation of the treatise on rectification (see below, §V) on p. 263 indicates that the two works were composed at about the same time, a point that has some importance for the argument to follow (see below, p. 262ff.).

Fermat had accomplished by 1646. If, however, one tries to apply the early Archimedean method of quadrature to them, one quickly stumbles on difficulties that quite likely led Fermat to devise his new method of quadrature.

Take, for example, the curve $x^2y = k$. Given the point x_0, y_0 on the curve, one must determine the area under the curve to the right of the ordinate y_0. In keeping with the earlier Archimedean method, one would first divide either

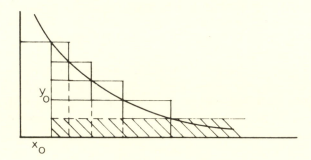

the ordinate (base) of the segment or its axis into N equal segments. The infinite length of the axis, however, would seem at first to limit the choice. One can only divide the ordinate y_0 into N segments of length y_0/N. But does such a division in fact satisfy the first requirement of the method? Does it allow one to circumscribe and inscribe step-figures that approach the desired area as closely as one wishes? By analogy with the division of the parabola, all the individual differences add up to the area of the bottom segment; that much is evident from the diagram. But can the bottom segment be made as small as one wishes? No matter how small the interval y_0/N, the length of that bottom segment is infinite. Division of the ordinate, therefore, might fail to meet even the first demand of the Archimedean method. One must turn again to the axis.

Geometrically, at least, it is evident that some sort of division of the x-axis into an infinite number of intervals would permit one to construct the step-figures such that the total difference were less than any preassigned magnitude. For, however many divisions are made, the sum of the differences will be equal to the first circumscribed segment, which has one finite side and one side that one should be able to make as small as one pleases. However, any such division of the x-axis requires the use of an infinite number of intervals, a condition that Fermat's Archimedean method could not accommodate. That method had begun with a finite number of intervals and then determined boundaries that held for any such finite number and could, upon making that number large enough, approach each other as closely as desired. But how does one divide an infinite length into N finite segments that may be made as small as one wishes, depending on the size of N?

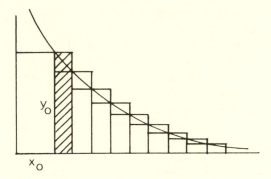

Fermat's solution to this dilemma opens the *Treatise on Quadrature*. It clearly shows the influence of the advanced notion of adequality already evident in the 1643 memoir on tangents. If division of the *x*-axis into equal segments will not work, division into adequal segments will turn the trick. But what sort of division? "Imagine," Fermat writes,[46]

> the terms of a geometric progression, extended *in infinitum*, of which the first is *AG*, the second *AH*, the third *AO*, etc., *in infinitum*, and let them, by approximation, approach each other as closely as is necessary in order that, by the Archimedean method, the rectilinear parallelogram *GE·GH is adequal*, as Diophantus says, or almost equal to the mixed quadrilateral figure *GHIE*; and further, such that the first of the rectilinear differences of the proportionals, *GH*, *HO*, *OM*, and so on, are almost equal to one another, in order that the Archimedean method of demonstration by *reductio ad absurdum*, by circumscription and inscription, can be easily set up. It suffices to mention this once and not worry about frequently teaching and repeating a technique already well known to any mathematician.

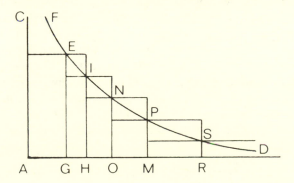

[46] FO.I.257.

Fermat required a subdivision of the *x*-axis that would satisfy the two requirements of the Archimedean method. He had to be able to circumscribe and
inscribe the whole area by rectangles erected on intervals of the axis; that is,
here, he had to fill an infinite axis with intervals. And he had to be able to
carry out the construction in such a way that the difference between the circumscribed and inscribed step-figures—and hence between either figure and
the hyperbolic segment—was less than any preassigned small value. The
second requirement called for a construction in which the sum of the individual differences was equal to the first circumscribed rectangle, which in turn
could be made indefinitely small by choosing a suitably small interval as base.
To make the sum of the differences equal to the first rectangle, however, one
must have equal intervals on the axis. The construction described above meets
the requirements through the medium of adequality. In modern terms, since
for suitably large values of *m* and *n*, *m* > *n*, the value of the ratio *m*/*n* can be
set as close to unity as one might wish, the infinite geometric series

$$x_0, \ \left(\frac{m}{n} \right) x_0, \ \left(\frac{m}{n} \right)^2 x_0, \ \ldots, \ \left(\frac{m}{n} \right)^i x_0, \ \ldots.$$

will yield, first of all, a series of lengths *AG*, *AH*, *AO*, measured from the
origin, that divides the *x*-axis into intervals and also "fills up" that axis. Moreover, since

$$\left(\frac{m}{n} \right)^i - \left(\frac{m}{n} \right)^{i-1} = \left(\frac{m}{n} \right)^{i-1} \left(\frac{m}{n} - 1 \right),$$

those intervals follow the same geometric progression as the segments, and
suitable values of *m* and *n* not only bring any two intervals as nearly equal to
one another as desired but also make it possible to set the length of any interval, in particular that of the first, less than any preassigned value. Thus Fermat has *GH* ≈ *HO* ≈ *OM* ≈ . . . and *GH* → 0. He later terms this method of
division, "our logarithmic method."[47]

The construction shows how the notion of adequality did more than serve as
a path around the problem of dividing an infinite axis. For Fermat not only
speaks of the intervals being adequal, he also speaks here for the first time of
the adequality of the first rectangle to the hyperbolic segment about which it is
circumscribed, and by implication of the adequality of the stepfigures to the
whole of the hyperbolic segment. The temptation to view a curvilinear area as
the limit of a sequence of rectilinear figures had always been present in the
ancient method of exhaustion, but careful thinkers in Antiquity had resisted
the temptation.[48] The notion of adequality, however, broke whatever final
resistance Fermat had to this temptation, for, following the construction of the

[47] FO.I.265. "[P]ropter nostram methodum logarithmicam." This represents Fermat's only reference to logarithms, and he does not explain his meaning further.
[48] Cf. above, n. 23

step figures, that notion further transforms the old method of quadrature. To see how, let us reconstruct how the quadrature of the hyperbola $x^2y = k$, which is Fermat's first example in the *Treatise on Quadrature*, would have progressed in the earlier method.

Having constructed the circumscribed and inscribed step-figures, one next faces the problem of determining their areas. The manner in which the axis has been divided in one sense simplifies that problem. For it is easy to show, as Fermat does in the *Treatise,* that any circumscribed or inscribed rectangle is to its immediate predecessor as $n{:}m$. Hence, if ΔA is the area of the first circumscribed rectangle, i.e. $\Delta A = \left(\dfrac{m}{n} - 1 \right) x_0 y_0$, the circumscribed rectangles form the series

$$\Delta A, \ \left(\frac{n}{m} \right) \Delta A, \ \left(\frac{n}{m} \right)^2 \Delta a, \ \dots .$$

Similarly, the inscribed rectangles form the same series, less the first two terms. But $m > n$; hence both series are decreasing geometric series. The summation of such series had been known since the late Middle Ages; Fermat gives the summation rule as a lemma at the beginning of the *Treatise*:[49]

> Given any geometric proportion, the terms of which decrease *in infinitum*, the difference of the terms constituting the progression is to the smaller term as the greatest term of the progression is to all the remaining terms taken *in infinitum*.

That is, in modern terms, if

$$S = a + \left(\frac{n}{m} \right) a + \left(\frac{n}{m} \right)^2 a + \dots ,$$

then

$$\frac{m - n}{n} = \frac{a}{S - a} .$$

Hence, if S_c and S_i denote the areas of the circumscribed and inscribed step-figures respectively,

$$S_c = \left(\frac{m}{n} \right) x_0 y_0, \text{ and } S_i = \left(\frac{n}{m} \right) x_0 y_0 .$$

The area of the hyperbolic segment from y_0 on out must, therefore, lie between the limits S_c and S_i for all values of m and n, $m > n$. It requires no sophisticated summation formulas to recognize that

---

[49] FO.I.255–256. Fermat offers no proof.

$$\left(\frac{m}{n} \right) x_0 y_0 \ > \ x_0 y_0 \ > \ \left(\frac{n}{m} \right) x_0 y_0$$

and hence that the area sought is $x_0 y_0$, or the rectangle $AG \cdot GE$.

That is how Fermat would have reasoned in 1636. Some twenty years later, however, he takes an essentially different line of analysis. Instead of setting upper and lower bounds to the area sought, he sums only the circumscribed rectangles, adequates, and moves to the limiting case. That is, by application of the summation lemma, he obtains

$$\frac{GH}{AG} = \frac{EG \cdot GH}{\text{remaining rectangles}}.$$

Those remaining rectangles constitute, he then argues, "by Archimedean adequation, . . . the figure contained by HI, the asymptote HR, and the curve IND extended indefinitely."[50] But the ratio GH/AG is equal to the ratio $(EG \cdot GH)/(EG \cdot AG)$, whence Fermat sets

$$EG \cdot AG \approx \text{figure } DHIR.$$

Therefore, rectangle $AG \cdot GE$ ($= x_0 y_0$) is:[51]

> adequal to the aforesaid figure [figure *DHIR*]. If you add to that figure the parallelogram $GE \cdot GH$, which because of the extremely small division disappears and goes to nothing, there remains the most true result, confirmable also by an Archimedean demonstration (albeit more prolix): in this sort of hyperbola, parallelogram AE *is* equal to the figure contained by base GE, asymptote GR, and curve ED extended *in infinitum*.

Even if one overlooks for the moment the paradoxical notion of "Archimedean adequation," a careful reading of Fermat's words in the *Treatise* reveals a curious double use of adequality. First, the sum of the circumscribed rectangles, less the first, is set adequal to the hyperbolic segment from HI. Then, the rectangle $AG \cdot GE$ is set adequal to that same segment, the difference from complete equality being the rectangle $GE \cdot GH$. Fermat removes the second adequality by observing that that difference rectangle is less than any finite value and hence zero, but he says nothing explicit about the removal of the first adequality.

It is too easy to follow previous commentators in assigning incoherencies in the *Treatise on Quadrature* to Fermat's failure to publish it and to sloppy editing on the part of his son, Clément-Samuel, in 1679. In the present instance, the difficulty just raised may, in fact, offer a clue to Fermat's path to his new method and to the reason for the introduction of the concept of ade-

[50] FO.I.258.
[51] FO.I.259.

quality into the summation procedures of that method. For assume for the moment that Fermat's first efforts with his new method took the more conventional path of summation of both limiting areas, circumscribed and inscribed, and that his reasoning at first proceeded as conjectured above. He may well have noticed a striking feature of the bounding inequalities, to wit, that the area of the hyperbolic segment is precisely equal to the sum of the circumscribed rectangles, less the first, no matter what the values of m and n. That is,

$$S_c = \left(\frac{m}{n} \right) x_0 y_0 = \left(\frac{m}{n} - 1 \right) x_0 y_0 + x_0 y_0,$$

no matter how the division is made. But the geometric picture, viewed by a man schooled in the Archimedean technique, clearly indicates that that sum should be less than the hyperbolic segment, indeed that it should fall short by something slightly less than the area of the first circumscribed rectangle. Another area that falls short of the hyperbolic segment by something slightly less than the area of the first circumscribed rectangle almost jumps out from the diagram: the segment measured from *HI*. Hence, it may have occurred to Fermat to adequate (indeed, perhaps to equate) the sum of the inscribed rectangles to the area of the hyperbolic segment from *HI*, and to do so without particular concern for explicitly transforming the adequality to equality. Since the sum of the inscribed rectangles falls short of that segment by the area of the second inscribed rectangle, the same argument applies *mutatis mutandis*, leading Fermat to focus on the circumscribed rectangles alone.

A new method of quadrature, then, began to emerge. The division of the axis into an infinite number of adequal segments led to infinite series of circumscribed and inscribed rectangles. But the very nature of the infinite series eliminated the need for upper and lower boundaries. They suggested rather the feasibility of obtaining one sum and then employing the concept of adequality to remove the difference between that sum and the area of the hyperbolic segment. Through the introduction of the concept of adequality into Fermat's work on quadrature, the search for a middle value between two finite sums was transformed into the search for the limiting value of one infinite sum. The implications of this transformation become clear in the second part of the *Treatise on Quadrature*; but, before turning to that second part, we should examine Fermat's solution to the long outstanding problem of the higher parabolas.

The essentially new method of direct quadrature, developed to meet the demands of the infinite axis of the higher hyperbolas, had immediate feedback on the old problem of higher parabolas with finite axes. For in the *Treatise on Quadrature* Fermat next presents a new quadrature of the parabola, which now employs an actually infinite subdivision of the axis. It is a method which removes the restrictions inherent in Fermat's earlier approach and which yields the quadrature of all curves of the form $y^q = kx^p$ for all positive integer values of p and q.

The new quadrature of the parabola follows that of the hyperbola in Fermat's treatise, and the reason is clear. The former clearly represents a corollary of the latter. For, although the infinite axis of the hyperbola obviously motivated the infinite subdivision of its axis, only the striking success and simplicity of the quadrature of the hyperbola could have motivated the infinite subdivision of the axis of the parabola. In the case of the parabola, in fact, the end dictates the means, since one can only subdivide axis BC infinitely by

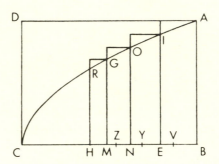

employing a convergent series of adequal segments. That is, lengths CB, CE, CN, CM, CH, ... now follow the convergent sequence (taking $CB = x_0$):

$$x_0, \left(\frac{n}{m} \right) x_0, \left(\frac{n}{m} \right)^2 x_0, \ldots ,$$

where, as before, $m > n$ are two integers that may be so chosen that the ratio n/m approaches unity as closely as one might wish.

In determining the resulting ratio of any circumscribed segment to its immediate predecessor, Fermat must take an additional step not found in the quadrature of the hyperbola. Its inclusion alters slightly the pattern of reasoning found in the preceding quadrature and serves to entrench the concept of adequality even more firmly in Fermat's new approach to the subject. The additional step springs, however, from hard necessity. To see why, follow for a moment the direct pattern of the hyperbola. If $BC = x_0$, $AB = y_0$, and the curve CA is the parabola $y^2 = x$ (for simplicity's sake, $k = 1$ here), then the rectangle on the i-th segment, $\left(\frac{n}{m} \right)^{i-1} \left(1 - \frac{n}{m} \right) x_0$, will be

$$\left(\frac{n}{m} \right)^{i-1} \left(1 - \frac{n}{m} \right) x_0 \cdot \left(\frac{n}{m} \right)^{(i-1)/2} y_0$$

or

$$\left(\frac{n}{m} \right)^{3(i-1)/2} \left(1 - \frac{n}{m} \right) x_0 y_0.$$

Hence, the rectangles do form a decreasing geometric series, but the proportional constant of that series is $(n/m)^{3/2}$. If the summation formula is directly applied, one must, to obtain an answer, be in a position to evaluate the limit of

$$\left(1 - \frac{n}{m}\right) \left(\frac{m^{3/2}}{m^{3/2} - n^{3/2}}\right)$$

as n/m approaches 1. That is, one must be in a position to apply series expansions for fractional exponents. Fermat stood powerless before such a demand. Of course, it is anachronistic to put his dilemma in such terms. Nonetheless, Fermat clearly knew that the direct pattern of the quadrature of the hyperbola would lead him, in the case of the parabola, into a mathematical situation he could not handle. Forseeing the problem in his analysis, he adds to the basic subdivision of the axis at points $B, E, N, M, H, \ldots$ a second subdivision at points $V, Y, Z, \ldots$. The latter points represent the mean proportionals between the segments in the interval of which they fall; i.e. CV is the mean proportional between CB and CE. It then follows directly that the ratio of rectangle $BA \cdot BE$ to rectangle $EN \cdot EI$ is BC/YC, and so on; i.e. that the proportional constant of the geometric series is YC/BC (instead of $(EC/BC)^{3/2}$). Fermat can now argue directly from his summation formula that

$$\frac{BA \cdot BE}{\text{remaining segments}} = \frac{BC - YC}{YC}.$$

By adequality,

$$\frac{BA \cdot BE}{\text{parabola}} \approx \frac{BC - YC}{BC} = \frac{AB(BC - YC)}{AB \cdot BC}.$$

Alternando,

$$\frac{AB \cdot BC}{\text{parabola}} \approx \frac{BY}{BE}.$$

Now the removal of the adequality depends on what value the ratio BY/BE obtains as the rectangle $BA \cdot BE$ goes to zero, i.e. as the intervals on the axis approach complete equality. Here Fermat gives his concept of adequality its own head: "But BY is to BE (because of the adequality and the extremely small sections, wherein the lines BV, VE, EY representing the differences of the proportionals are supposed almost equal to one another) as 3 to 2."[52] That is, just as the intervals of the first subdivision are adequal, so too are the intervals of the second subdivision, not only to each other but also to those of the first. Hence, in the limiting case, $BY/BE = 3/2$, which is, therefore, the ratio of the rectangle $x_0 y_0$ to the area of the parabolic segment.

Fermat completes the first section of his treatise by showing first that the quadrature of the parabola $y^2 = kx$ is directly transferrable to curves of the

[52] FO.I.263.

form $y^q = kx^p$. The trick lies in taking p mean proportionals in the second subdivision of the axis, and the general result (arrived at by extrapolation rather than strict proof) is that the parabolic segment is to the containing rectangle as $q/(p + q)$. Then Fermat generalizes the result for the higher hyperbolas; if $x^p y^q = k$ ($q \neq p$), then the area of the hyperbolic segment from some ordinate y_0 on out is to the rectangle $x_0 y_0$ as $q/(p - q)$.

Sometime before 1657, then, and most probably before 1646, Fermat had solved the basic problem inherent in his early Archimedean method of quadrature. Not only could he now directly square all higher parabolas, he could also carry out the quadrature of any higher hyperbola. His new method did far more than extend the range of his Archimedean method. It radically altered that method by introducing the concept and technique of adequality, a non-rigorous but powerful heuristic device that Archimedes would never have countenanced. Through adequality the method of exhaustion, originally conceived by the Greeks as a technique of proof, became in Fermat's hands an analytic problem-solving technique. Just how radically the concept of adequality and the direct quadratures based on it altered the traditional problem of quadrature becomes evident in the second part of Fermat's *Treatise on Quadrature*.

B. A Theory of Equations for Quadrature

If the opening pages of the *Treatise on Quadrature* treat a pair of problems that had occupied Fermat's attention for some time and that have obvious ties to his early research on quadrature, the remainder of the treatise is without precedent in Fermat's extant papers and letters. In this second part, Fermat grouped together all his mathematical forces—his analytic geometry, his method of maxima and minima, his method of tangents, and his direct quadrature of the higher parabolas and hyperbolas—to construct a brilliant "reduction analysis" for the quadrature of curves. The title of the treatise itself indicates that Fermat meant to emphasize this new analysis. At the same time, it points to the model Fermat had in mind: Viète's major treatise on the theory of equations bore the title *On the Emendation of Equations*. As Viète had investigated the transformation and alteration of determinate equations in order to analyze their properties and their solvability, so too Fermat intended to transform and alter the indeterminate equations of curves to establish their quadrability. If the model is evident, the motivation is not. Fermat is silent regarding the sources and development of the ideas he presents as finished work.

Having accomplished the direct quadrature of the higher parabolas and hyperbolas, Fermat anticipates the question of where his results might lead: "It is a wonder how much the task of quadrature benefits from what has been said above, for from it emerges the easy quadrature of an infinity of figures contained by curves, of which up to now nothing has entered the mind of either the ancient or the new mathematicians. We abbreviate the matter in several rules."[53] Characteristically, Fermat abbreviates where contemporaries and posterity cry for detail. His first example, which, in addition to the interpretive problems it poses, affords an excellent example of the complete change of style that separates the two parts of the *Treatise on Quadrature*:[54]

> Let there be a curve whose property gives the following equation: *Bq.* − *Aq.* equals *Eq.* (it will be immediately clear that this curve is a circle); it is certain that the unknown power *Eq.* can be reduced by application or parabolism to a side.
>
> For we can suppose *Eq.* to be equal to *B* · *U*, since we are free to equate an unknown quantity *U*, multiplied by the known *B*, to the square of *E*, also unknown.
>
> This posited, *Bq.* − *Aq.* will equal *B* · *U*; but the homogeneous quantity *B* · *U* can be composed of as many homogeneous quantities as there are in the correlative side of the equation; and the homogeneous quantities of this sort should be noted by the same signs. Let us suppose therefore *B* · *U* to be equal to *B* · *I* − *B* · *Y*. For in the manner of Viète we have always taken vowels for the unknown quantities. Therefore *Bq.* − *Aq.* equals *B* · *I* − *B* · *Y*.
>
> Let the individual members of one side equal the members of the other side; that is, let *Bq.* equal *B* · *I*; therefore it will be given that *I* equals *B*. Now set −*Aq.* equal to −*B* · *Y*, that is, *Aq.* equal to *B* · *Y*; the endpoint of the straight line Y will lie on a primary parabola. Therefore all [terms] in this case can be reduced to a square, and therefore, if you apply all *E* squares to the given straight line, a given and known rectilineal solid will result.

That is, translating into the Cartesian symbols we will be using hereafter, we have the following series of equations:

$$b^2 - x^2 = y^2$$

$$y^2 = bu = bv - bz$$

$$b^2 = bv, \text{ whence } b = v$$

$$x^2 = bz.$$

Compared to the above, the *Method* must rank as a model of lucidity. Yet, with study, Fermat's line of thought here becomes clear. Insight into his meaning essentially depends on understanding the sense of the phrase "apply all *E* squares to the given line." Understood in the classical sense, as, for example,

[53] FO.I.266–267.
[54] FO.I.267–268.

it is used in the first paragraph, the phrase would mean "divide the sum of all the E squares—whatever that would mean—by the line segment B." But then the result would have to be linear in dimension, and Fermat expressly says that the result of the operation is a solid, i.e. a three-dimensional figure. Clearly, we confront here an entirely new concept of "application."

The nature of that concept becomes unambiguously clear in the course of the treatise. Application of all y^2 to a given straight line means nothing less than the determination of the limit-sum of the parallelepipeds formed by the squares of the ordinates y multiplied by indefinitely small subdivisions of a segment of length b on the x-axis. In other words, in the case of an equation $y = f(x)$, one "applies all y to a given line b" by carrying out the quadrature of the curve corresponding to the equation over a segment b of the axis, and that quadrature is carried out according to the pattern of the first part of the treatise. One circumscribes the area with rectangles erected on intervals of the axis, sets the sum of these rectangles adequal to the area under the curve, and then removes the adequality by taking the limit of that sum as the intervals go to zero. The concept of the area of a curve as the limit-sum of the circumscribed rectangles, a concept made possible by adequality, takes effect now in the second part of the treatise. Fermat now speaks simply of "application of all y (or y^2, etc.) to a given line." The reason for the new language will be evident presently.

What, then, is the sense of Fermat's transformations above? Consider the following diagram, in which the segment b on the x-axis has been subdivided into small intervals.[55] Fermat requires the limit-sum of the solids formed by

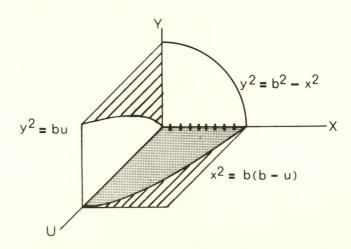

[55] Needless to say, no such three-dimensional diagram appears in Fermat's text. Indeed, what makes this particular section of the *Treatise on Quadrature* so difficult to follow and understand is precisely the total absence of any illustrative diagrams. Unless his sketches for this section alone are missing, Fermat would seem to be operating on wholly abstract, algebraic foundations.

256

the products of these intervals times the squares of their corresponding ordinates y. But that sum cannot be directly determined by the methods so far presented in the treatise because it does not correspond to any area in the diagram. If, however, we set $y^2 = bu$, we obtain an ordinary parabola. Visualizing the area beneath that parabola as the sum of all u, that is, as all u applied to the length b, we can apply direct methods of quadrature. Nonetheless, the product-sum required depends on the manner in which the x-axis has been divided, and that axis does not enter into the parabola. One needs a further transformation. What is the relation of the new variable u to the variable x? In paraphrase, Fermat's reasoning runs:

$$bu = b^2 - x^2, \text{ or}$$

$$x^2 = b(b - u).$$

Taking $b - u = z$, he obtains $x^2 = bz$. Now the sum of all z applied to the segment b on the x-axis of this last parabola is the area outside the parabolic segment shaded in the diagram. That area follows directly from the method of quadrature that opens the treatise, and it is equal to a determinate rectilinear area. Therefore that same area gives the sum of all $b - u$ applied to the same segment of the x-axis. But the sum of all the b so applied is simply b^2. Hence the sum of all the ordinates u applied to the segment b on the x-axis is equal to the area of the parabolic segment shaded in the diagram. Hence the sum of all bu applied to the same line is b times that area, i.e. a determinate rectilinear solid.

Fermat's next example should now be easier to follow. Let $x^3 + bx^2 = y^3$. What is the sum of all y^3 applied to a given segment b of the x-axis? Set first $y^3 = b^2u$ and then write b^2u as the sum of b^2v and b^2z. Set the left-hand side of the original equation equal term by term to this latter sum; i.e. $x^3 = b^2v$ and $bx^2 = b^2z$ (or $x^2 = bz$). The first of these two equations corresponds to one of Fermat's higher parabolas and thus is subject to the method of direct quadrature. So too is the second curve, an ordinary parabola. Hence the sum of all v applied to segment b, and the sum of all u applied to segment b, are determinate rectilinear areas. Multiplied by b^2, the sum of the areas yields a determinate rectilinear fourth-dimensional quantity (Fermat, using the language of the new algebraists, says "quadratoquadrate") on the one hand and the sum of all b^2u applied to the segment b on the other. But this last sum is the sum of all y^3 applied to the same segment b. So far Fermat has exploited only one part of his direct method of quadrature, i.e. the quadrature of the higher parabolas. His next two examples—we need consider only one—demand the other part. Let

$$\frac{b^6 + b^5x + x^6}{x^4} = y^2.$$

To transform this curve into directly quadrable segments, set $y^2 = bu = bw + bv + bz$. Term-by-term equation yields $b^5 = x^4w$, $b^4 = x^3v$, $x^2 = bz$. That is, we confront two higher hyperbolas and one ordinary parabola, each of which, by results thus far obtained, is directly quadrable and has a determinate rectilinear area. The sum of these areas, multiplied by b, yields a rectilinear solid equal to the sum of all y^2 applied to the segment b on the x-axis.

One sees now the power and novelty of Fermat's approach to quadrature. Expressed in modern terms, it sets forth in a single uniform stroke the integration of any algebraic polynomial

$$\sum_{i,j} \frac{a_i x^i}{b_j x^j}$$

over a given interval of the x-axis by reducing that integration to the quadrature of higher parabolas and hyperbolas. Nonetheless, Fermat still lies some distance from his goal. Quadrature, and not definite integration, is the subject of his investigations, and quadrature demands that, in the examples given above and in many others, the sum of all y applied to a given segment of the x-axis be found, and not the sum of some power of those y. Fermat knew full well, however, that that sum could not be determined (as a rectilinear area) for the very first of his examples. For the curve $y^2 = b^2 - x^2$, that sum of all y applied to b would represent the area of a quadrant of a circle, and Fermat knew better than to try to square the circle. Nevertheless, other curves were rectilinearly quadrable. The next and climactic phase of the *Treatise on Quadrature* is the development of a method for separating the quadrable curves from those of which the quadrature depends on the quadrature of the circle, and for generating families of quadrable curves.

This method derives from a lemma that Fermat states as if it were common knowledge and for which offers no explicit proof. He writes:[56]

Let . . . *ABDN* be any curve, of which *HN* is the base, *HA* the diameter, *CB* and *FD* applicates to the diameter, and *BG* and *DE* applicates to the base; and let the applicates constantly decrease from the base to the vertex, as here *HN is* greater than *FD* and *FD* greater than *CB*, etc.

The figure composed of the squares of *HN*, *FD*, *CB* applied to the line *AH* (i.e. the solid under *CB* squared times *CA* and under *FD* squared times *FC* and under *NH* squared times *HF*) is always equal to the figure under the rectangles *BG* times *GH*, *DE* times *EH*, taken twice and applied to base *HN* (i.e. to the solid under twice *BG* times *GH* times *GH* and under twice *DE* times *EH* times *EG*) etc., both *in infinitum*.

[56] FO.I.271–273. The introduction to this unprecedented theorem is limited to the remark: "Sed ut ulterius progredi et opus tetragonismicum promovere nihil vetat."

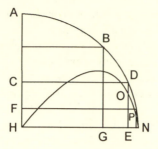

With the same ease, one carries out the reduction of homogeneous [terms applied] to the diameter to homogeneous [terms applied] to the base, in the remaining powers *in infinitum*.

It is all too simple to recognize from hindsight the meaning of this passage. In modern terms, it asserts that, if the curve $y = f(x)$ monotonically decreases over the interval $(0, b)$, where $f(0) = d$ and $f(b) = 0$, then

$$\int_0^b y^2\,dx = 2\int_0^d xy\,dy.$$

Indeed, one finds in the last paragraph of this passage a statement equivalent to the more general transformation

$$\int_0^b y^n\,dx = n\int_0^d xy^{n-1}\,dy.$$

This at least is the operational meaning of Fermat's lemma. Its conceptual meaning for Fermat poses problems, as does the question of how he derived it.

Fermat gave neither proof nor derivation of the proposition. Had he come upon it himself, or did he have it from a source readily recognized by and available to the reader? Only one candidate seems likely: Pascal's *Traité des trilignes rectangles, & de leurs onglets*, published as part of the *Lettre de A. Dettonville à M. Carcavy* in 1658.[57] Fermat told Carcavi the following February that he had received a copy but had not yet been able "to apply myself

[57] *LETTRE DE A. DETTONVILLE A MONSIEUR DE CARCAVY, EN LUY ENVOYANT Vne Methode generale pour trouver les Centres de grauité de toutes sortes de grandeurs. Vn Traitté des Trilignes & leurs Onglets. Vn Traitté des Sinus du quart de Cercle. Vn Traitté des Arcs de Cercle. Vn Traitté des Solides circulaires. Et enfin Vn Traitté general de la Roulette,* Contenant *La solution de tous les Problems touchant LA ROULETTE qu'il auoit proposez publiquement au mois de Iuin 1658*. Paris, [no publisher], 1658; The reprinting by Dawsons of Pall Mall (London, 1966) also includes several other letters published in 1658–1659, all of which were joined in a single volume, *LETTRES DE A. DETTONVILLE CONTENANT Quelque-vnes de ses Inventions de Geometrie,* Paris: Chez Guillaume Desprez, 1659.

seriously to reading it"; a year later, he insisted that he had already achieved his results on cycloids before seeing Pascal's work.[58]

In the *Traité*, Pascal worked from a general configuration that differs from Fermat's in potentially significant ways. Let *ABC* be a three-sided figure bounded by two perpendicular lines and an undefined curve. Draw an indefinite number of ordinates *DF* and *EG*, dividing the axis and the base into segments of equal size.[59] On the other side of the axis *AB*, let *AK* be another, "adjoint" curve bounded by *CA* extended and *BK* parallel to it. Extend the ordinates *FD* to meet it at points *O* and parallel to them draw the *counter-ordinates GRI*, meeting *AB* at *R* and *AK* at *I*. Now imagine *ABC* as the base of

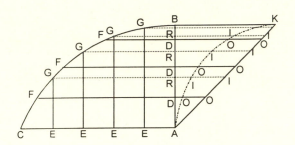

a quasi-cylinder, and, erecting the figure *ABK* at right angles to *ABC*, imagine the line *AC* moving parallel to itself along the boundary *AK* to cut off a portion of that cylinder.[60] The section points on the axis and base will thereby determine two sets of planes having as their sides the ordinates to the two curves. "Now it is evident [*visible*]," says Pascal,

---

[58] Fermat to Carcavi, 16.II.59, FO.II.430: "J'ai reçu le Traité de M. Pascal depuis deux jours, et n'ai pu encore m'appliquer sérieusement à le lire; j'en ai pourtant conçu une grande opinion, aussi bien que tout ce qui part de cet illustre." It is not clear which treatise Fermat was referring to in particular, though it was most likely the *Traitté general de la Roulette*, since that is what he referred to a year later, lest Carcavi think Fermat had borrowed anything on that subject from Pascal. Cf. Fermat to Carcavi, ⟨IV.1660⟩, FO.II.446: "Je pourrais ajouter le théorème général pour tous ces cas, c'est-à-dire pour l'invention des paraboles égales aux roulettes accourcies et pour l'invention de l'agrégé des droites et des circulaires égales aux allongées. Mais ce sera pour une autre fois. Ma méthode générale ne dépend que du chiffre que je vous envoyai l'année passée, avant que j'eusse vu le Livre de M. Dettonville."

[59] Pascal established the configuration in his "Méthode générale pour les centres de gravité de toutes sortes de lignes, de surfaces, et de solides," which forms the body of the letter to Carcavi. There he addressed the details of the division of the axes into segments (*Lettre*, 18): "Soient divisées en un nombre indefini de parties égales, tant AB aux points D, que AC aux points E, et encore la courbe même BC aux points L; et chacune des parties de AB, soit égale à chacune des parties de AC, et encore à chacune des parties de la courbe BC. (Car il ne faut pas craindre l'incommensurabilité, puisqu'en ôtant d'une de deux grandeurs incommensurables une quantité moindre qu'aucune donnée, on les rend commensurables.)"

[60] This is Gregory of St. Vincent's *ductus plani in planum*; see his *Opus geometricum quadraturae circuli et sectionum coni decem libris comprehensum* (Antwerp, 1647), Book VII.

that the sum of the sections made by each of these orders of planes are equal each to the solid, and consequently to one another, (since the indefinite portions *AE*, *EE*, etc. of the base are equal both to one another and to the equal and indefinite portions *AD*, *DD*, etc. of the axis); that is, the sum of all the rectangles *FD* times *DO* is equal to the sum of all the portions *RIA*.[61]

That is, the rectangles $FD \cdot DO$ are infinitesimally thick cross-sections of the solid sliced along the axis *AB*, and the figures *ARI* correspond to the cross-sections of the solid determined by the slices *GE* along the axis *AC*. If one accepts the infinitesimalist reasoning by which the sums each constitute the same solid, then they are plainly equal for any curves *BC*, *AK*, provided that they are single-valued over the interval.

A second lemma specifies the adjoint curves of particular interest to Pascal here, namely the triangle and family of parabolas $RI = AR^n$. The lemma is none other than the result Fermat had derived decades earlier, namely that the area *ARI* under the curve is equal to $\dfrac{1}{n + 1} AR^{n+1}$. From this by now well-known result and the general lemma, Pascal then derives a series of propositions setting out the "ratios between the ordinates to the axis and the ordinates to the base of any right-angled trilinear figure whatever." The second reads:

> The sum of the squares of the ordinates to the base is twice that of the rectangles comprised by each ordinate to the axis and by its distance from the base. That is, the sum of all the *EG* squared is twice the sum of all the rectangles *FD* times *DA*.

The proposition follows by taking *ABK* as an isosceles right triangle, thus setting each *DO* and *RI* equal to the corresponding *AD* and *AR* respectively. Then the sum of the rectangles $FD \cdot DO (= FD \cdot DA)$ is equal to the sum of the triangles *ARI*, or the sum of the products $\frac{1}{2}AR \cdot RI = \frac{1}{2}AR^2 = \frac{1}{2}GE^2$. To show in Proposition III that the sum of the EG^3 is three times the sum of the products $FD \cdot DA^2$, Pascal takes *ABK* as the simple parabola, $RI = AR^2$ (whence also $DO = AD^2$).

Clearly, Fermat's lemma is identical in substance with the first section of Pascal's *Traité*, which supplies the derivation and proof that Fermat omits. However, Fermat's configuration differs from Pascal's in several ways. Where Pascal differentiates between the segments on the base and on the axis, Fermat says nothing about segments on either axis. Unlike Pascal, Fermat gives no hint of an auxiliary curve to mediate between the sets of products. Moreover, the language differs. Pascal speaks of sums of products in a line; Fermat of products applied to a line. Do those differences signify? Do they point to an independent result, reached by a path left unmarked? Certainly, there is noth-

[61] *Traité*, p. 2.

ing in Pascal's version that Fermat could not have devised on his own. Fermat's previous experience with quadrature and cubature had already required him to divide up the area by rectangles based on each of the two main axes—for example, in the cubature of the two conoids generated by the ordinary parabola. We have already discussed how a change in the axis to be divided into intervals enabled him to extend the original method of summing series to cover the reciprocals of the integers as powers.[62] The quadrature of the "higher parabolas" was an early achievement in which he took particular pride—indeed one might say pride of ownership—especially when using it to counter the claims of Cavalieri and Torricelli.

But, by the same token, Fermat had not until now used the language of indivisibles and infinitesimals, adding up "all the lines." He claimed to have achieved the Italians' results by his own methods, not to have devised their methods. The same held true of his exchanges with Roberval. So he had not spoken to these people in the terms of their methods, which is why the language takes the reader of Fermat's correspondence and other works by surprise. Up to the time of the composition of the *Treatise on Quadrature*, it had not expressed Fermat's way of thinking about the subject. Suddenly, the language and the thinking changed. Given the timing, the reason seems to lie at hand.

Pascal's *Lettres* may well have arrived as Fermat was composing his treatise, stimulated initially by the exchange with Wallis and Brouncker and, perhaps, given added motivation by Lalouvère's encouragement to publish his work on rectification (see below). Having arrived at the quadratures of the general parabola and hyperbola, he may have been alert to ways of extending them to polynomials and may have grasped immediately the implications of Pascal's propositions, especially given the role of the higher parabolas in them. Fermat had already thought out the question of dividing the axes into segments and of the convergence of the segments in the limiting case, so he would have looked past Pascal's concern to separate the two sets of ordinates.

Whether or not the lemma itself was Fermat's own, he used it to build something quite original for the time, taking the methods of indivisibles and infinitesimals along new algebraic paths. To follow him, it will help to establish a symbolic notation less verbose than Fermat's and yet less anachronistic than the modern symbols of the integral calculus. The notation must, however, be in keeping with the innovative thinking that underlies Fermat's words. In one sense, that thinking makes the modern notation less alien to seventeenth-century mathematics than it might otherwise be. From what has gone before, Fermat clearly visualizes the "application of all y^n to the diameter" as the limit-sum of the products $y^n\Delta x$, where the intervals Δx go to zero. Nonetheless, the modern notation suggests the inverse relationship of differ-

---

[62] See above, p. 237ff.

entiation and integration and for that reason presents a real danger of anachronism. For present purposes it will help to borrow from Leibniz' original notation in this area.[63] We will write "$Omn_d \, y^n$" for "all y^n applied to the segment d of the diameter," or "$Omn_x \, y^n$" when the axis of reference is defined, but not the segment itself.

Expressing the above results in the new notation, we have from Fermat's earlier studies that, for curves monotonically decreasing over the first quadrant,

$$Omn_x \, y \; = \; Omn_y \, x.$$

That is, no matter on which axis the rectangles are based, or to which axis the ordinates are applied, the area under the curve is the same. The lemma now extends that transformation of axes for powers of y, that is, again for montonically decreasing curves,

$$Omn_x \, y^2 \; = \; Omn_y \, 2xy,$$

or, even more generally,

$$Omn_x \, y^n \; = \; Omn_y \, ny^{n-1}x.$$

Fermat's first example shows clearly where this lemma can lead.[64] Given the circle $b^2 - x^2 = y^2$, he has already shown that $Omn_b \, y^2$ is equal to a determinate rectilinear solid. Since the circle in the first quadrant satisfies the conditions noted above—i.e. it decreases monotonically—it follows that

$$Omn_b \, y^2 \; = \; Omn_b \, 2xy,$$

where the segment b is now taken on the y-axis. Hence $Omn_b \, xy$ is also a determinate rectilinear solid. If now we set $xy = bu$, we clearly define a new curve $F(u,y)$, shown in the diagram as *HOPN*. By substituting bu/y for x in the original equation, we obtain

$$b^2y^2 - y^4 \; = \; b^2u^2.$$

In addition, we have the following chain of transformations:

$$Omn_b \, y^2 \; = \; Omn_b \, 2xy \; = \; 2 \, Omn_b \, xy \; = \; 2 \, Omn_b \, bu \; = \; 2b \, Omn_b \, u.$$

But $Omn_b \, u$ is the sum of all u applied to the segment b on the y-axis, which in turn is precisely the area under the curve $F(u,y)$. Hence, not only is that derived curve quadrable; its quadrature follows directly from the sum of all y^2 applied to b in the original circle.

Another example points to the flexibility of Fermat's new technique of reduction as well as to the manner in which the *Treatise on Quadrature* unifies

---

[63] Cf. J. M. Child, *The Early Mathematical Manuscripts of Leibniz* (Chicago, 1920), Chap. IV, *passim*.
[64] FO.I.273.

the whole of Fermat's achievement in the realm of algebraic analysis.[65] The curve $bx^2 - x^3 = y^3$ permits the direct evaluation of $Omn_b\ y^3$. But one cannot immediately apply the lemma, because that curve does not uniformly decrease over the interval $(0,\ b)$ on the x-axis. Rather, the ordinates increase to a maximum and then decrease to zero. To overcome this minor hurdle, Fermat employs his method of maxima and minima to determine the value of x for which y attains a maximum and the value of that maximum. Let $y_m = \max y = y(x_m)$. Using the transformation techniques of the *Introduc-*

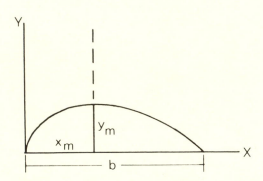

tion, Fermat shifts the origin of the curve from 0 to x_m and then applies his lemma to establish that

$$Omn_b\ y^3 = Omn_{x_m}(x_m - x)y^2 + Omn_{b-x_m}(x - x_m)y^2.$$

By means of this transformation he is then able to show that the direct evaluation of $Omn_b\ y^3$ leads to the rectilinear quadrature of the curve $b^5u^2y^2 - y^9 = b^6u^3$.

Fermat's long experience with curves leads him in one example to pass over possible difficulties in carrying out his transformation of bases.[66] From the possibility of determining $Omn_b\ y^3$ directly from the equation

$$\frac{b^5x - b^6}{x^3} = y^3,$$

he employs the transformation lemma to get $Omn_y\ 3xy^2$ and the transformation $b^2u = xy^2$ to arrive at the derived quadrature of the curve

$$buy = u^3 + y^3$$

[65] FO.I.274–275. "Et est generalis, ad omnes omnino casus extendenda in infinitum, methodus. Notandum porro et accurate advertendum in translationibus curvarum, quarum applicatae ad diametrum versus basim decrescunt, aliam omnino viam analystis ineundam, a praecedenti diversam." (p. 274)

[66] FO.I.275–276.

which he calls "Schooten's curve" in the treatise, but which he must have recognized as the curve to which Descartes had challenged him to determine the tangent some twenty years earlier.[67] The difficulties he ignores in his paper concern the asymptotic relation between the original curve and the y-axis; the $3xy^2$ are applied to an infinite segment on the y-axis. Yet the transformations are valid, as Fermat was in a position to see. That is, in the original equation, let $b^2z = y^3$, whence

$$zx^3 = b^3x - b^4.$$

The curve $F(x, z)$ clearly contains a finite area between the curve, the x-axis,

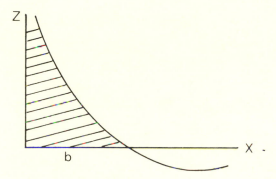

and the z-axis. Hence, on the basis of his previous work, Fermat may write

$$Omn_b\, z = Omn_z\, x.$$

Indeed, he may also write:

$$Omn_b\, y^3 = Omn_b\, b^2z = b^2Omn_b\, z = b^2Omn_z\, x = Omn_y 3xy^2.$$

These few examples suffice to show that the quadrature of any algebraic curve directly determines the quadrature of a whole family of curves.[68] The stricture that the original curve decrease monotonically over the first quadrant is, as Fermat shows both explicitly and implicitly, less restrictive than might at first appear. For the *Introduction* supplies the means to transform by change of axes any monotonic curve into a monotonically decreasing curve, and the method of maxima and minima provides the means for dividing any curve into

---

[67] FO.I.276: " . . . quae est curva Schotenii, cujus constructionem tradit in Sectione 25 *Miscellanearum*, pag. 493 [of van Schooten's Latin edition of Descartes' *Geometrie*, Leiden, 1649]." Did Fermat really forget the curve to which Descartes had challenged him to find the tangent (see above, Chap. IV, p. 181), the curve that Descartes' followers were already calling the *folium Cartesii*? Or had Clerselier's revival of the controversy opened old wounds, leading Fermat deliberately to slight the author of the curve by attributing it to someone else?

[68] Note that both Leibniz, in his *Analysis tetragonistica ex centrobarycis* (1675, Child; p. 65ff.), and Newton, in his *Quadrature of Curves* (1704; D. T. Whiteside, *The Mathematical Works of Isaac Newton*, New York, 1964, p. 141ff.), follow this same approach to quadrature.

monotonic segments. The usefulness of the method in this regard depends only on the ability of the mathematician to employ these techniques and to see the proper transformations.

But what if one is simply given an algebraic curve and asked whether it can be squared? Here, too, Fermat's reduction analysis points the way to an answer. Consider, for example, the curve $b^3 = x^2y + b^2y$.[69] Is $Omn_x\,y$ equal to some rectilinear figure? First set $by = u^2$, whence

$$Omn_x y = \frac{1}{b}\,Omn_x u^2,$$

and, by the main lemma,

$$Omn_x\,u^2 = Omn_u\,2xu.$$

Now set $bv = xu$, whence

$$Omn_u\,xu = Omn_u\,bv.$$

If, therefore, the original curve were quadrable, so too would be the curve that results from the two transformations of variable. But what is that derived curve? Substituting u^2/b for y and bv/u for x, Fermat gets $b^2 = u^2 + v^2$, which is the equation of a circle. Therefore, $Omn_u\,v$ is equal to no rectilinear figure, and neither is $Omn_x\,y$.

In the *Treatise on Quadrature*, then, one may find the full inventory of the analytic techniques Fermat developed in the course of his career. One may also see how several of those techniques, but adequality in particular, exercised a dynamic of their own, leading Fermat on the one hand to take steps his classical forebears would never have countenanced and yet on the other hand, by their origins in Fermat's early classical work, enabling him to believe that he was remaining in the classical tradition. Not only the stark algebraic style of the second part of the treatise belies Fermat's oath of allegiance to the *via ordinaria, legitima et Archimedea* made in his letter to Digby. Adequation, limit-sums, application of all y to a given line: at all these concepts Archimedes would have shuddered; certainly, he would never have published them. That Fermat could not or did not see this fact, that he did not fully appreciate the almost revolutionary character of his own advances, serves to show how Greek mathematics exercised a formal influence long after its content had become obsolescent.

In his other major treatise of the late 1650s, however, Fermat did consciously reject that influence and tradition. With the exception of Archimedes, ancient mathematicians had accepted Aristotle's dictum that curved lines and straight lines could not be compared. In Fermat's own day, Descartes had reasserted this taboo. Fermat believed otherwise, and his *Treatise*

[69] FO.I.279.

on Rectification, written simultaneously with, or perhaps even prior to, the *Treatise on Quadrature*, aimed to show that infinitely many curved lines were equal in length to straight lines.

V. THE TREATISE ON RECTIFICATION (1660)

In 1660 the Toulousan Jesuit Antoine de Lalouvère published as an appendix to his own *The Geometry of the Ancients Advanced in Seven Books on the Cycloid* the only treatise of Fermat's to appear in print in his lifetime. Entitled *Geometrical Dissertation on the Comparison of Curved Lines with Straight Lines* (hereafter, *Treatise on Rectification*), the treatise contains a complete treatment of the problem of determining the rectilinear length of any segment of a given curve.[70] The treatise takes the student of Fermat's mathematics somewhat by surprise, since it appears without any hint of its development in Fermat's papers. Moreover, its starkly classical geometrical style tends to mask the underlying analytical methods familiar from Fermat's other work.

From all appearances, the treatise represents a single inventive effort by Fermat, most probably in reaction to his reading of Pascal's *Lettres de A. de Dettonville* and to Wren's successful rectification of the cycloid in the same year.[71] Fermat first received a copy of Pascal's *Lettres* on 14 February 1659, and within a year his own treatise on rectification was in print.[72] That treatise opens with a reference to Wren's result and to the controversy that surrounded all such efforts at rectification. Mathematical purists, in particular the followers of Descartes, denied the possibility of comparing curved and straight lines. Wren's success and Pascal's proof of the equality of the length of the Archimedean spiral to that of the parabola had been possible, they argued, only because the definitions of the curves themselves presupposed rectification.[73] That is, in the case of the cycloid, the generation of the curve presupposed that an arc of a circle could be measured along a straight line. Fermat agreed with these purists on the specific matter of the cycloid, but he refused to grant the general proposition that rectification of a curve was possible only when it had already been assumed in the definition of that curve. In his trea-

---

[70] *De linearum curvarum cum lineis rectis comparatione dissertatio geometrica.* The text in FO.I.211–254 is taken from the *Varia*, pp. 89–109.

[71] See Tannery's notes in FO.I.211–212.

[72] Fermat to Carcavi, 16.II.1659, FO.II.430. Fermat addressed two critiques of the work to Carcavi, one in September 1659 (FO.II.441–444) and one in February 1660 (FO .II.445–446).

[73] Pascal, *Lettres* (*Oeuvres*, edn. of 1779, V, p. 413): "A quoi M. de Sluze ajouta cette belle remarque dans sa réponse du mois de septembre dernier, qu'on devoit encore admirer sur cela l'ordre de la nature, qui ne permet point qu'on ne trouve une droite égale à une courbe, qu'apres qu'on a déjà supposé l'égalité d'une droite à une courbe." (Quoted by Tannery, FO.I.211, n. 3). Cf. Fermat's introductory remarks to the *Treatise on Rectification* (p. 212): " . . . ii quippe hanc esse legem et ordinem naturae pronuntiant ut non sinat inveniri rectam curvae aequalem, quin prius supposita fuerit alia recta alteri curvae aequalis."

tise, he proposed to rectify segments of curves that in no way presupposed in their definitions the equality of curved and straight lines.[74]

The very first sentence of the treatise shows, on the one hand, that Fermat had not previously given much thought to his subject and was unaware of the work of others and, on the other, how his basic belief in the possibility of rectification had been part of his mathematics for some time. "Never, so far as I know, have mathematicians compared straight lines to purely geometrical curved lines." In fact, Hendrick van Heurat had done just that in an appendix to the 1659 Latin edition of Descartes' *Geometry*.[75] That Fermat had not seen this text, or even heard of it, may indicate the increasing isolation into which he was slipping toward the end of the 1650s. More interesting than Fermat's failure to keep up with the literature is his choice of the Latin word for "compare"—*adaequarunt*. Clearly used here in its classical, non-technical sense, it nonetheless recalls immediately the one memoir in which Fermat had occasion to compare explicitly an arc with a straight line. In his paper on the tangent to the cycloid, sent to Roberval in 1643 and discussed at the end of Chapter IV, Fermat employed the notion of adequality to set an arc of a circle equal to the tangent segment subtending it. That use of adequality, combined with the techniques of quadrature set down in the *Treatise on Quadrature* (themselves strongly dependent on the notion of adequality), forms the analytic core of Fermat's *Treatise on* Rectifcation. In a sense, the latter required no long genesis; much of it follows almost as a corollary from Fermat's other work.

Indeed, the simplicity of Fermat's solution to the problem of rectification lends to the *Treatise* an almost Baroque air. for the treatise hides that simplicity beneath an overwhelmingly complex superstructure of geometric proofs and constructions. Archimedes and Apollonius dictate the style; synthesis replaces analysis; and Viète's algebraic notation disappears. Fermat's reasons for choosing the classical geometrical style of exposition are far from clear. Perhaps he was responding to the distaste for analytic exposition that still prevailed in some mathematical circles. Perhaps he wished to fit his style to that of the work to which the treatise was appended. Perhaps too he was simply indulging his old, well-documented preference for hiding his methods. Whatever the reasons, the *Treatise on Rectification* will bewilder the reader who mistakes the geometrical style for geometrical thinking. For, at heart, Fermat was an algebraist, and the treatise is basically as algebraic in its con-

---

[74] FO.1.212: " . . . nos enim curvam vere geometricam, et ad cujus constructionem nulla talis alterius curvae cum recta aequalitas praecessisse supponatur, rectae datae aequalem esse demonstrabimus et paucis, quantum fieri potuerit, totum negotium absolvemus."

[75] Henrici van Heuraet, *Epistola de transmutatione curvarum linearum in rectis* [sent to Franz van Schooten, 13 January 1659], in *Geometria a Renato Des Cartes anno 1637 Gallice edita; postea autem una cum notis Florimondi de Beavne . . . in latinam linguam versa, & commentariis illustrata, opera atque studio Francisci a Schooten . . .* Second Edition (Amsterdam: Elzevier, 1659–1661), 517–20.

cepts as the obviously algebraic *Treatise on Quadrature*. Hence, in what follows, we will try to strip Fermat's propositions of their geometrical veneer, most of it merely formalistic, and reveal the underlying analytical structure.

Only the subject itself is without precedent in Fermat's research. Its treatment displays many long-standing themes of his mathematics. For example, the first two propositions establish the modified Archimedean model of circumscription and inscription. In the first proposition, Fermat shows that, if *AHMG* is any uniformly convex or concave curve, *IHK* the tangent to that curve at point *H,* and *BI* and *DK* ordinates drawn to the tangent and cutting the curve at points *R* and *M* respectively, then $HI <$ $\overset{\frown}{RH}$ and $HK >$ $\overset{\frown}{HM}$. The

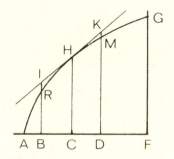

proof, which rests on a postulate borrowed from Archimedes' *Sphere and Cylinder* (Book 1, Postulate 2), is relatively straightforward and requires no comment here.

The purpose of this proposition emerges clearly from the introductory remarks to Proposition II.[76] Fermat intends to revise the Archimedean procedure, used in both the *Measurement of the Circle* and the *Sphere and Cylinder*, of approaching the length of the curved segment by circumscribed and inscribed rectilinear segments. At the same time, as Proposition II shows, he intends to achieve the same basic result, to wit, that it is possible to construct a series of segments less in length than the curved line and a series greater in length, such that the difference between these lengths, and hence between either of them and the curved length, is less than any preassigned value. Using Proposition I, Fermat can show that circumscribed segments alone will satisfy his needs, as circumscribed rectangles alone had done in the *Treatise on Quadrature*.

In the following two diagrams, which represent the same curve, drawn twice to avoid confusion, let the points *A, B, C, D, E, F,* and *G* divide the axis

---

[76] FO.I.213–214: "Ad dimensionem linearum curvarum non utimur inscriptis et circumscriptis more Archimedeo, sed circumscriptis tantum ex portionibus tangentium compositis, duas enim series tangentium exhibemus quarum una major est curva, altera minor: demonstrationem autem multo faciliorem et elegantiorem per circumscriptas solas evadere analystae experientur."

of the curve into equal intervals and erect at each point the ordinates *BP*, *CT*, *DY*, *EN*, *FO*, *GH* respectively. In the first case, draw through each of the

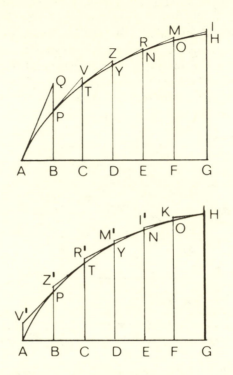

points *A*, *P*, *T*, *Y*, *N*, *O* the tangent segments *AQ*, *PV*, *TZ*, *YR*, *NM*, *OI*. By Proposition I, each of these segments is greater than the arc it subtends, and hence their sum is greater than arc *AH*. In the second case, draw the tangent segrnents *PV′*, *TZ′*, *YR′*, *NM′*, *OI′*, *HK*. Here, by Proposition I again, each segment is less than the arc it subtends, and hence the sum of the segments is less than $\widehat{AH}$.

Fermat now has two series of rectilinear segments, one greater in sum than the arc *AH*, one less in sum than the arc *AH*. By what quantity do the two sums differ? A corollary to Proposition I provides the answer. Since the intervals on the axis are equal to one another, *PV* = *PV′*, *TZ* = *TZ′*, and so on. Hence the sum of the greater segments minus the sum of the smaller segrnents is precisely equal to *AQ* − *HK*. Fermat now sets *HK* = *FG*, a step which, though he adds no comment, is made possible by his already established transformation procedures; that is, Fermat knows he can always carry out a transformation of axes which will make tangent *HK* parallel to the axis, whence *HK* = *FG*. But *FG* = *AB*, whence *AQ* − *HK* = *AQ* − *AB*.

Now, by the method of tangents, it is always possible to determine the ratio

AQ/AB for any given curve, therefore, that ratio is given. To carry out a construction, then, which will yield the two series of circumscribed segments differing by less than a given amount, one must merely solve the problem: given that $x/y = m$ and $x - y < k$, find x and y. And this problem is a simple variant of Proposition I.4 of Diophantus' *Arithmetic*.[77] Hence, in the first two propositions of the *Treatise on Rectification*, one finds two continuing themes of Fermat's research: the Archimedean model of circumscription and the method of tangents.

Proposition III forms the first half of a double keystone to Fermat's method of rectification, and it too shows the direct influence of Fermat's earlier work. To understand this proposition and the equally important one that follows, it will help to have some overview of the general direction Fermat's analysis is taking. To one familiar with the conceptual framework of the *Treatise on Quadrature*, it will be clear that Fermat views the tangent segments as adequal to the arcs subtending them, and hence their sum as adequal to the whole arc. The task is to find the limiting sum for which adequality becomes complete equality. A mere look at the diagrams shows that the adequality of the tangent segments is dependent on the manner in which the axis of the curve has been subdivided. Hence a study of Fermat's approach to quadrature justifies rephrasing his task in the following manner: what is the limit sum of the tangent segments ΔS as the intervals Δy on the axis go to zero? In Propositions III and IV, Fermat strives toward and achieves two goals: first, to find the relation between any ΔS and its corresponding Δy and, second, to find the limit sum of the ΔS via this relation to the limit sum of the Δy.

Fermat's formal path to these goals is, as was noted above, tortuously geometrical and synthetic. But the following analysis, using algebraic symbols, is completely in keeping both with the text and with Fermat's concepts and methods. It should, indeed, represent Fermat's own analysis of the problems. Fermat takes as an example the curve $y^3 = kx^2$, i.e. one of the higher parab-

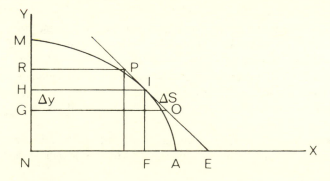

[77] Diophantus, trans. Heath, p. 132: "To find two numbers in a given ratio and such that their difference is also given."

271

olas.[78] In the diagram, let *AIM* be that curve, where *A* is the vertex, *AN* the *x*-axis, *MN* the *y*-axis, and where the segment *AN* (and hence *MN*) is given by the conditions of the problem. Through any point *I* on the curve draw the tangent *IE*, which intersects the *x*-axis at point *E*. From the method of tangents it follows that the subtangent *EF* is three-halves of the abscissa *AF*, or $2AE = AF$. Let *HG* be a small interval on the base *MN* corresponding to the segment *IO*. What is the ratio of *IO* to *HG*?

Because we will want to express *IO* in terms of segments on the axes, we start with the ratio IO^2/HG^2. From similar triangles, it is immediately clear that

$$\frac{IO^2}{HG^2} = \frac{IE^2}{IF^2}.$$

Further, $IE^2 = IF^2 + EF^2 = IF^2 + (3/2AF)^2$. *IF* and *AF* are, however, related to one another by the defining property of the curve; i.e. we have

$$IF^3 = kAF^2,$$

or

$$\frac{9IF^3}{4k} = \left(\frac{3}{2}AF\right)^2.$$

Substitution of this relation into the expression for IE^2 and from that into the proportion given above yields

$$\frac{IO^2}{HG^2} = \frac{(4/9)k + IF}{(4/9)k}.$$

But $4/9k$ is a constant quantity, and *IF* is the ordinate of point *I*. Hence, for any segment $IO = \Delta S$, we have

$$\frac{\Delta S^2}{\Delta y^2} = \frac{(4/9)k + y}{(4/9)k},$$

where Δy is the interval on the *y*-axis corresponding to ΔS, and *y* the ordinate of the point of tangency. This ratio holds for any one of the greater tangent segments. From the diagram, however, it is clear that IP^2/HR^2 has the same value, whence the ratio also holds for the smaller tangent segments. The limit sum in either case will be the same. A synthetic proof of the equality of the ratios discussed above is the substance of Proposition III of the *Treatise on Rectification*.

On the basis of his previous work on quadrature, the problem now facing Fermat almost dictated its own solution. Viewing the length of the curve as

[78] Fermat performs the direct quadrature of this curve as an example of the general efficacy of his new methods in the *Treatise on Quadrature* (FO.I.263ff.). He states explicitly that the curve appears in the *Treatise on Rectification*.

the limit sum of the ΔS as the Δy go to zero, he had a relation linking the squares of the two intervals, that is, he had

$$\Delta S^2 \approx \frac{(4/9)k + y}{(4/9)k} \Delta y^2,$$

But as such, the relation did not fit his needs. Fermat required a relation between ΔS and Δy themselves and hence had to be able to express the right-hand side as a perfect square. The technique of transformation of variable so fruitfully employed in the *Treatise on Quadrature* supplied not only the means to meet his needs but also the "reduction analysis" that would reduce rectification to quadrature. For let

$$u^2 = \frac{4}{9}k \left(y + \frac{4}{9}k \right).$$

This equation determines a parabola on the *y*-axis, with latus rectum equal to $(4/9)k$ and vertex at a point corresponding to $-(4/9)k$ on that axis. Substitution of this new variable into the above relation yields

$$\left(\frac{4}{9}\right)^2 k^2 \Delta S^2 \approx u^2 \Delta y^2,$$

or

$$\left(\frac{4}{9}\right) k \Delta S \approx u \Delta y.$$

Hence the limit sum of the ΔS, multiplied by the constant $(4/9)k$, will be equal to the limit sum of the products $u\Delta y$, that is

$$(4/9)kS = Omn_y\, u$$

But, clearly, the right-hand side of this equation is nothing other than the area of a segment of the parabola determined by the relation linking *u* and *y*. Hence, in the following diagram, in which the curve above the *y*-axis is the higher parabola under investigation, and the curve under that axis is the parabola given above, the length of any segment of the upper curve measured from vertex *A* to ordinate *MN*, multiplied by $AB = (4/9)k$, is equal to the area *KQM* of the parabola below the axis.[79]

The last result constitutes the essence of Proposition IV of the *Treatise on Rectification*. What one finds in the treatise itself is the synthetic geometric derivation of the result, clothed in the full regalia of the classical proof by double *reductio ad absurdum*. For all its finery, however, the proposition betrays to the critical eye its status as *parvenu*. It cannot hide the non- classical method of quadrature on which it is based, nor the non-classical means em-

---

[79] Note here how, in keeping with the geometrical style of the treatise, Fermat maintains the homogeneity of the two sides of the final equation; an area is equal to an area.

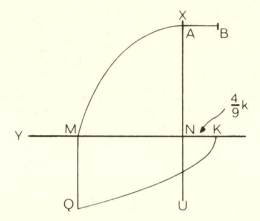

ployed to reduce the rectification to a problem of quadrature. The actual construction of a line equal in length to the given arc, with which Fermat means to crown Proposition IV, only touches up the disguise.

Insofar as Fermat was seeking to set forth a method of rectification, he had completely realized his goal with the proof of Proposition IV. He had provided the tools with which, given any algebraic curve, one could rectify any segment of it, provided of course that one could carry out the necessary quadrature. For, given any curve $y = f(x)$, one would start from the relation

$$\frac{\Delta S^2}{\Delta y^2} = \frac{z^2 + y^2}{y^2},$$

where z is the subtangent of the curve corresponding to ordinate y. The method of tangents directly provides an expression for z in terms of x, which we may write here in modern notation as $z = f(x)/f'(x)$. Substituting and simplifying, one then has

$$\frac{\Delta S^2}{\Delta y^2} = \frac{z^2 + f'(x)^2}{f'(x)^2}.$$

Hence, if one sets

$$u^2(y) = \frac{1 + f'(x)^2}{f'(x)^2} = x'(y)^2 + 1,$$

the determination of S reduces to the determination of $Omn_y u$ for the given length of the ordinate.

Yet, Fermat's treatise does not end with Proposition IV; rather, it continues for some pages. Fermat was, after all, polemicizing on the matter of the possibility itself of rectification. He needed more space to drive his point home. He needed in particular an example which would directly show that an infinite number of curves could be rectified. (The general method only reduces the problem to quadrature, and not all quadratures are possible.) He had, moreover, among his old papers on the method of tangents just the sort of example he needed.

When, in 1638, Fermat had been trying to cement his victory over Descartes by flooding his Parisian correspondents with new and difficult tangent problems, he had aimed those problems at the weak point in Descartes' method. He had deliberately chosen curves termed "mechanical" by Descartes, and thereby excluded from the *Geometry*, in order to show that his own method had far greater applicability than Descartes'. Since Descartes had in particular excluded the possibility of rectification of curved lines, Fermat bore down on curves defined by rectification, even though at that time he lacked any means of directly constructing those curves. The cycloid, for example, was such a curve; we have already seen how Fermat determined its tangent.[80] But Fermat had, in other contexts, gone beyond the cycloid. First he had rephrased its originally physical definition—the path of a fixed point on a circle as that circle is rolled along a straight line—to read:[81]

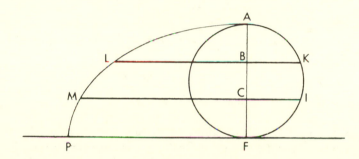

The principal property of the curve [described by a point on a rolling circle], as it is easy to demonstrate, is that, if you take a point such as *L* in it, from which you draw *LBK* perpendicular to *AF*, the line *LB* is equal to the line *BK* plus the [arc] segment *AK* of the semicircle; in the same way the line *MC* is equal to the line *CI* plus the [arc] segment *IA* of the semicircle; and similarly for the other [ordinates] up to the line *PF*, which is equal to the whole semicircle.

Then Fermat had defined other curves for which the ordinate *LB* was of the form *BK* + *n* · *AK* for integer *n*; these curves, of course, completely lost contact with physical situations. They also opened the way to other similar families.

In a letter of 22 October 1638 to Mersenne, Fermat offered the following example of the efficacy of his method of tangents:[82]

[80] See above, Chap. IV, p. 212ff.
[81] Fermat to Mersenne, VII.1638, FO.*Suppl*.89.
[82] FO.II.172.

Let *EDAG* be a parabola of which *AG is* the axis and *A* the vertex. Let *ABF* be another curve with the same axis and vertex such that the applicate *BC* is equal to the [arc] segment *DA* of the parabola and the applicate *FG is* equal to the [arc] segment *EA* of the parabola, and so on *ad infinitum*. To find the tangent to this new curve at point *B*.

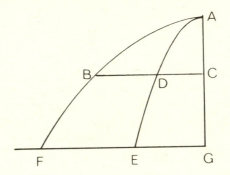

Draw applicate *BDC*. Let *O* be the focus of the parabola. Construct the proportion

$$OA + AC{:}AC = BC^2{:}CN^2.$$

The line *BN* will be tangent to the curve *FBA*.

Although Fermat never employed the term, we may call curve *ABF* the *rectificatrix* of curve *EDAG*. As early as 1638, Fermat had not only been thinking in terms of such curves, but, in the case at least where the base curve was a parabola, he had managed to derive the tangent to such curves.

These early efforts achieve full generality in the later *Treatise on Rectification*. In Proposition V, the *propositio generalis*, Fermat defines an infinite family of curves based on the parabola $y^3 = kx^2$, each of which is the *rectificatrix* (measured from a fixed point *C* on ordinate *BC*) of the one preceding it in order; in Fermat's words:[83]

Let . . . *CMA* be our parabolic curve, *AB* its altitude, *CB* its semibase, and from this curve let an infinite number of others be formed in such a way that, any perpendiculars *DMNL*, *EKIH* having been drawn to the base and cutting the curve in points *M*, *K*, the new curve *CNIG* formed from it is of the sort that the straight line *DN* is always equal to the portion *CM* of the

[83] FO.I.226–227. "Si haec non sufficiant ad obtinendum a geometris ut nostra haec curva parabolica inter admiranda Geometriae collocetur, illud fortasse ab ipsis quae mox sequentur impetrabunt. Quid enim mirabilius quam ex una hac curva derivari et formari alias numero infinitas, non solum ab ipsa, sed inter se, specie differentes, quae tamen singulae rectis datis aequales esse demonstrentur?"

prior curve corresponding to it. Similarly, straight line *EI* is equal to the portion *CMK* of the prior curve, and so one for all other perpendiculars. This new curve *CNIG* will be different in species from the former.

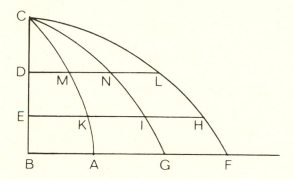

In the same way let there be formed from this curve a third curve *CLHF*, in which the straight lines *DL*, *EH* are always equal to the portions *CN* and *CNI* of the second curve. And, by the same reasoning, let a fourth be formed from the third, a fifth from the fourth, a sixth from the fifth, proceeding in this order *in infinitum*.

The purpose of introducing the curves is to show that every one of them is rectifiable. Since the definition of the base curve does not presuppose any rectification, none of the succeeding rectificatrices assumes any rectification other than that of its predecessors in the family and, ultimately, that of the base parabola.

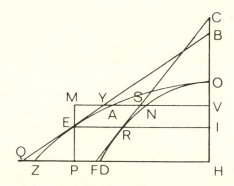

In Proposition VI, Fermat then extends his earlier work on tangents to rectificatrices, showing first that, if curve *ORD* is any member of the family and curve *OEZ* its rectificatrix, then *RC* and *EB* are the tangents to each curve respectively at the same ordinate if and only if $RC/CI = IE/IB$. On the basis of this result, Fermat then shows that, if

$$u^2 = \frac{(4/9)k + y}{(4/9)k}$$

is the auxiliary curve for the base parabola, then

$$u_2^2 = \frac{2(4/9)k + y}{(4/9)k}$$

is the auxiliary curve for its rectificatrix, the second member of the family. In fact, he goes on to generalize, the auxiliary curve for the nth member of the family is

$$u_n^2 = \frac{n(4/9)k + y}{(4/9)k} .$$

Hence, as Fermat next demonstrates in Proposition VII, the rectification of any member of the family of rectificatrices based on the curve $y^3 = kx^2$ reduces to the quadrature of an ordinary parabola.

Despite Fermat's own restriction of the results above to the specific family he has defined, both the results and their proofs are fully general. That is, the relation between the tangents to the same ordinate of any curve and its rectificatrix is the same as that established by Fermat in Proposition VI; curve *ORD* need not be restricted to any family. Moreover, using now the modern notation introduced on p. 274, one can see that, if

$$u^2 = x'(y) + 1$$

is the auxiliary curve of the base curve of any family of rectificatrices, then

$$u_n^2 = x'(y) + n$$

is the auxiliary curve of the nth member of the family. Hence, the auxiliary curves of any family are themselves all of the same type, differing only in the parameters. The rectification of any rectificatrix depends only on the rectification of the base curve.

With the rectification of this infinite family of curves, and himself aware most probably that the specific results are in fact quite general, Fermat ends the body of his *Treatise on Rectifcation* with a triumphant reminder of his other achievements in the realm of analysis:[84]

> We will add nothing concerning the solids [of revolution] generated by the said infinite [family of] curves, nor of their curved surfaces, nor of the centers of gravity, either of these lines or of the said solids or of the curved surfaces, since the general methods in this matter, made common knowl-

---

[84] FO.I.236–237.

edge by the greatest and most outstanding mathematicians, do not allow these things to be ignored, *once the specific property of the given curve is known*; although we do believe that it would not be without use for anyone to add his own industry to the task.

The phrase we have emphasized by italics is the crucial one, for it points to the theme raised at the beginning of the present chapter. The power and unity of Fermat's various analytic methods rest on the fact that one can determine all the properties noted above—and, of course, the tangent and the quadrature— on the basis of a single piece of information: the equation of the curve. For all his triumph, Fermat could still not lay down his pen. He closed the *Treatise on Rectification* with the statement of another problem, the solution of which he provided in an appendix to the treatise. The problem itself sets up a new family of rectificatrices to the base curve $y^3 = kx^2$, this time measured from the vertex on the x-axis. The complexity of the solution in the appendix derives from the difficulty of setting up the expression for the ratio $\Delta S^2/\Delta x^2$ needed to determine the auxiliary curve for the base curve. Nonetheless, the proof follows basically the same pattern as those of the rectifications referred to the y-axis in the treatise itself, and the details need not concern us here. More interesting as a final example of Fermat's method of quadrature and rectification is the second of a series of propositions published separately in the same volume, again as an appendix to Lalouvère's treatise.[85] In that proposition, repeated (with a minor mistake) in a letter to Carcavi in 1660,[86] Fermat determines the surface of revolution of the new conoid he spoke about in his earliest letters to Roberval.

Only the result is given, yet the derivation is relatively transparent, given what has gone before. Let AB be an ordinary parabola drawn on axis AC, with

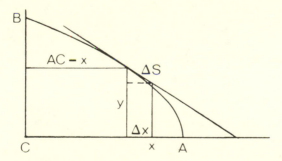

latus rectum AE; in algebraic symbols, $y^2 = px$, where $p = AE$. Let the semiparabola AB be rotated about ordinate BC; Fermat seeks the resulting

[85] Published in FO.I.199–210 under the title *Propositiones ad Laloveram*, the memoir appeared originally as the first part of the second appendix to Lalouvère's treatise on the cycloid.
[86] FO.II.446–447.

curved surface. Taking any tangent segment ΔS and referring it to the x-axis, we have

$$\frac{\Delta S^2}{\Delta x^2} = \frac{(2x)^2 + y^2}{(2x)^2} = \frac{x + (p/4)}{x},$$

since $2x$ is the subtangent of any tangent drawn to the point with abscissa x. Hence, as a preliminary result,

$$S = Omn_x\, u,$$

where

$$u^2 = \frac{x + (p/4)}{x}.$$

For the surface of revolution, however, we need, first of all, all products $(AC - x)\Delta S$. That is, we view the generation of the surface as the revolution of tangent segment ΔS about ordinate BC at radius $AC - x$. Fermat (and we) will take account of the necessary constant 2 by a subsidiary construction. The surface of revolution, lacking this constant, will be

$$Omn_x\,(AC - x)u,$$

or

$$AC\, Omn_x\, u - Omn_x\, xu.$$

Now the first term is merely $AC \cdot S$; but what about the second? Here Fermat's method of quadrature comes into play. A simple transformation of variable turns the trick. Let $ux = v$, or $u = v/x$. Then, on the one hand,

$$u^2 = \frac{v^2}{x^2} = \frac{x + (p/4)}{x},$$

i.e.

$$v^2 = x^2 + \frac{p}{4}\, x,$$

or

$$v^2 = \left(x + \frac{p}{8}\right)^2 - \left(\frac{p}{8}\right)^2.$$

On the other hand,

$$Omn_x\, xu = Omn_x\, v.$$

Hence, in accordance with Fermat's solution, we subtract from the product $AC \cdot S$ the area contained by the segment AC of the axis and the hyperbola of which the transverse axis and latus rectum are both one-quarter of the latus rectum of the original parabola. The product $AC \cdot S$ may be determined either directly from the hyperbola (x,u) given above or indirectly from the hyperbola

$$\left(\frac{p}{2}\right)^2 v^2 = y^2 + \left(\frac{p}{2}\right)^2,$$

by taking $(p/2)S = Omn_y\, v$, as Fermat does in the preceding proposition *Ad Laloveram* as well as in the letter to Carcavi.

VI. FERMAT AND THE CALCULUS

An interesting question momentarily passed through Fermat's mind as he concluded his "Méthode expliquée" for Descartes in 1638. Having shown how to determine the tangent to any given curve, he mused:[87]

> Finally one could seek the converse of the proposition and, being given the property of the tangent, seek the curve that this property fits. The problems concerning burning glasses proposed by M. Descartes reduce to this problem. But that merits a discourse in itself and, if he is agreeable, we will confer on it when he wishes.

As Fermat's method of quadrature assumes algebraic shape, one looks expectantly for a sign of recognition by him that it contains the answer to that question. One waits in vain. The thought never seems to occur to him. The method of tangents serves as a tool for the method of quadrature, but nowhere in his writings does Fermat even hint that the two methods might be inversely related. No other aspect of his work so clearly denies him the right to the honor accorded to Newton and Leibniz for the invention of the calculus.

Why did Fermat not see the inverse relationship that, in retrospect at least, would appear to have lain before his eyes? The answer, of course, rests in the phrase, "in retrospect." Fermat did not see the relationship, because he could not see it. And he could not see it, because it did not in fact lie in the questions he was asking and the problems he was trying to solve. What was the purpose of the method of tangents? Obviously, to determine the tangent to any point on a given curve. But how did one determine the tangent?

One *constructed* the tangent by linking the given point on the curve with the endpoint of the subtangent measured on the axis from the foot of the ordinate. Hence, the method of tangents aimed at, and succeeded in, establishing the length of the subtangent. In Fermat's work, that method never called attention to the angle contained between axis and tangent.[88] The only derivative to be found in the method of tangents is the one the modern reader puts into it.

---

[87] FO.II.162.

[88] With one exception, to wit, Fermat's method for determining points of inflection (see above, Chap. IV, n. 113). Consideration of the angle of the tangent with the axis appears only in that single instance, however, and concludes what was Fermat's last work on the method of tangents.

Similarly, only in retrospect does Fermat's method of maxima and minima set the derivative of the polynomial equal to zero. In Fermat's mind, one establishes the relations among the coefficients of the polynomial such that the equation formed by setting the polynomial equal to zero has a repeated root. To find the differential calculus in the material of Chapter IV is to read that material with questions quite different from those in Fermat's mind.

So too one can find the integral calculus in Fermat's work on quadrature only by putting it there in retrospect. Quadrature, as Fermat understood it, involved finding a rectangle equal in area to a figure contained by curved lines. In the last analysis, it too required a construction. And when Fermat was looking for an area, even in the case where the ordinate and abscissa bounding the area were arbitrary and hence variable, he was not likely to see a curve in the answer. Yet, the path to the calculus lay precisely in that direction, as the work of Newton shows. The germ of the integral calculus, indeed of the calculus itself, rested in seeing that quadrature with a variable abscissa generated another curve.

Hence, one cannot say with any degree of fairness or objectivity that Fermat's work in analysis of curves was even heading in the direction of the calculus. For it was not pointed toward the concept that underlies the calculus and its fundamental theorem. For all the suggestive and promising hints in his specific results, the manner in which Fermat understood and set up his problems precluded his arriving at the notion of the quadrature of a curve as a function of the abscissa, or of the tangent as such a function. For all the personal enmity that lay behind the observation, Descartes was probably quite right when he said to van Schooten that Fermat was a brilliant problem-solver but basically inept at conceiving systematic questions. The *Treatise on Quadrature* contains a general method for resolving problems of quadrature. To construe it as a treatise on the integral calculus is to betray the spirit in which it was conceived and written. In it, Fermat sought neither curves nor functions; he sought areas, and he found them. In all the papers and memoirs that constitute the historical material of Chapters IV and V, Fermat found the answers to the problems he had posed. He did not invent the calculus.

CHAPTER VI

Between Traditions

I know that the algebra of this country is not suited to solving these problems, or at least no one here has yet found the way to apply it to them. That is what leads me to believe that you have recently devised for yourself some sort of special analysis for probing into the most hidden secrets of numbers, or that you have found some way of using for that purpose the algebra you have customarily employed in other contexts.
Frenicle to Fermat

So let arithmetic reclaim the doctrine of whole numbers as a patrimony all its own.
Fermat

I. INTRODUCTION

Fermat's work in number theory displays most clearly the Janus-like quality of his mathematics first discernible in the mature method of quadrature. In number theory especially, one sees the paradox of Fermat's mathematical career: in seeking to renew and continue old, classical traditions, he shattered them to lay the foundations of a new, modern tradition. In the area of quadrature the source of the paradox lay partly in the very nature of Viète's analytic program which, by applying the symbolic theory of equations to ancient problems, subjected those problems to an alien form of treatment. There, Fermat shared his ambiguous approach to mathematics with others of his contemporaries. In number theory, however, the paradox was peculiarly his own. Here, he did more than realize the promise of the theory of equations in analyzing and enhancing the solution techniques of his classical source, in this case the *Arithmetic* of Diophantus of Alexandria. He eventually broke with Viète's vision of algebra as the language and method that united the diverse realms of mathematics and assigned to the integers a realm of their own, in which continuous magnitude and its methods had no place.

If, in his method of quadrature, Fermat appealed to a tradition still very much alive in the seventeenth century, in number theory he looked back to an ancient tradition to which most of his contemporaries no longer responded. Fermat sought to renew arithmetic as Plato had understood it, as the doctrine of whole numbers and their properties.[1] Despite his long-standing fascination with Diophantus' *Arithmetic*, in the end he disavowed much of Diophantus' work because it aimed at rational, rather than integral, solutions. Following a pattern set by his work in other fields, Fermat's search for number-theoretical methods stricter than those of Diophantus led finally to a number theory that had little in common with classical arithmetic and that was fully appreciated only by mathematicians of later centuries. With his eye turned to the past, Fermat created modern number theory.

He did so alone. That is what accentuates the paradox and makes it peculiarly Fermat's. His methods of maxima and minima and of quadrature, his algebraic approach to geometry, all evoked the interest and reinforcement of Fermat's forward-looking contemporaries. His work on number theory fell largely on deaf ears. Standing outside the mathematical communities of Paris, Leiden, and London, Fermat could exercise influence only when he pursued their agenda. Taking another direction, he lacked the strength either to renew an old tradition or to begin a new one. His efforts to do so only reinforced his isolation and frustrated his correspondents into silence. The situation involved a change of character. The same Fermat who so meekly confronted a raging Descartes in 1638 became, as a result of his frustration over lack of response to his proudest efforts in number theory, an aggressive, even belligerent combatant in his correspondence with Frenicle and Brûlart in the early 1640s and in his dispute with John Wallis in 1657–1658.

Fermat had not, however, completely changed his habits by 1657. If he proved more willing to engage in open combat, he continued to keep his arsenal secret. The dispute with Descartes brought posterity the dividend of a look into the genesis of the method of maxima and minima. To fight Descartes, Fermat had to show his weapons. But his disputes with Frenicle, Wallis, and others seldom progressed farther than assertion and counterassertion of theorems or got beyond a basic disagreement over what number theory was or should be, or whether it was worth doing at all. As a result, the record of the dispute with Wallis, i.e. the correspondence that passed between the participants and was published by Wallis as the *Commercium epistolicum* in 1658, offers little or no insight into Fermat's number-theoretical methods themselves. In the very field in which Fermat tried to create a following, to take a lead, he most successfully guarded his methods from outside eyes. A prophet unwilling to share his vision, he could win no disciples.

Fermat's secretiveness about his number theory makes the historian's task particularly difficult. In no other aspect of Fermat's career are the results so striking and the hints at the underlying methods so meager and disappointing.

---

[1] Plato, *Republic* (trans. F. M. Cornford, Oxford-New York, 1956), p. 245 (525D-E): "Good mathematicians, as of course you know, scornfully reject any attempt to cut up the unit itself into parts: if you try to break it up small, they will multiply it up again, taking good care that the unit shall never lose its oneness and appear as a multitude of parts."

It was the results—the theorems and conjectures—and not the methods that drew the attention of men such as Euler, Gauss, and Kummer. Wallis said of his contemporaries' efforts to regain the analysis of the ancients, that they found it easier to create their own methods than to recreate the original.[2] The same holds true for number theorists of the eighteenth and nineteenth centuries. In the end they laid down their own paths to Fermat's results, following him as far as his few hints pointed but then striking out on their own. It is difficult to locate the point of separation and hence to know with confidence how much of their thinking one may ascribe to Fermat in the absence of direct evidence from him. Hence, although, with proper care, modern calculus has helped to reconstruct Fermat's line of thought in previous chapters, modern number theory offers aid of a more ambiguous sort here. The development of the calculus followed an essentially continuous path from Fermat to the present, and Fermat was one of several people setting out on that path at the time, talking to one another as they went. Number theory, by contrast, made a jump from Fermat to Euler and thence to the early nineteenth century, and jumping back is an inherently uncertain process. "May we credit Fermat, in 1640, with a degree of abstract thinking that Euler reached only towards the end of his long career?" asks André Weil, who then immediately answers "Probably not."[3]

The situation is not entirely hopeless.[4] If Fermat carefully guarded the precise details of his methods, he did occasionally describe some of them in vague outline. Moreover, he spoke in his correspondence more about his re-

[2] Wallis to Digby, 21.X.1657, *Commercium epistolicum* (Oxford, 1658), XVI, p. 43: "Sed nec Archimedes solus, verum & veterum plaerique [!] omnes, Analyticen suam, (quam habuisse, extra dubium est,) eousque celarunt posteros, ut Recentioribus facilius jam fuerit novam suo marte comminisci (quod praesenti praeteritoque seculo factum est) quam indagasse veterem." ["But not only Archimedes, but just about all the ancients so hid from posterity their method of analysis (which they doubtlessly possessed) that it was easier for more recent mathematicians to invent a new analysis of their own (which has been done in the present and the last century) than to seek out the old."]

[3] André Weil, *Number Theory: An Approach Through History from Hammurapi to Legendre* (Boston: Birkhäuser, 1984), 65. "Il est toujours suprêmement imprudent," Weil notes in another discussion of Fermat, "de prétendre reconstituer après coup une démonstration perdue. Même lorsqu'il s'agit de son propre travail, un mathématicien y échoue le plus souvent; une fois le but atteint, on est tout surpris, au bout de quelques années, de retrouver dans ses vieux papiers la trace du chemin détourné, tortueux, illogique par lequel on y était parvenu." ("Fermat et l'équation de Pell," in *ΠΡΙΣΜΑΤΑ: Naturwissenschaftsgeschichtliche Studien: Festschrift für Willy Hartner*, ed. Y. Maeyama and W. G. Saltzer (Wiesbaden: Steiner, 1977), 441–448; at 443).

[4] As J. E. Hofmann has shown in three seminal articles on aspects of Fermat's number theory: "Neues über Fermats zahlentheoretische Herausforderungen von 1657," *Abhdlg. d. Preuß. Akad. d. Wiss.*, Jhrg. 1943, *Math.-naturw. Kl.*, Nr.9, Berlin, 1944; "Studien zur Zahlentheorie Fermats (Über die Gleichung $x^2 = py^2 - 1$), *ibid.*, Jhrg. 1944, *Math.-naturw. Kl.*, Nr.7, Berlin, 1944; and "Über zahlentheoretische Methoden Fermats und Eulers, ihre Zusammenhänge und ihre Bedeutung," *Arch. Hist. Ex. Sci.*, 1(1961), pp. 122–159. Hofmann has also dealt with individual problems treated by Fermat in several other articles. His pathbreaking work was indispensable for the first edition of this chapter. Weil's work on Fermat's number theory, cited above in the previous note, has provided new insights and guided revisions.

sults in number theory than about any other topic. He also left to posterity his often arcane *Observations on Diophantus*, which he had scribbled into the margins of his copy of the *Arithmetic*.[5] Finally, in one rare departure from his customary reticence, he revealed in a series of letters addressed in the mid-1650s to Father Jacques de Billy his improved method for solving the "double equation" so frequently employed by Diophantus. All of this material tends to form patterns of development which, if they do not explain the genesis of the results, at least indicate their motivation. That is, if the sources do not suffice to determine how Fermat arrived at his theorems, or precisely how he proved them, they do indicate what purpose the theorems were meant to serve and the context in which they were derived. One can gain an overview of Fermat's number theory and some sense of the role it played in his overall career as a mathematician. But the sense of deep insight afforded by previous chapters seems unattainable. Number theory is a game of inspiration, the possible paths to a given result are legion, and Fermat did not leave enough signposts.[6] He purposely failed to do so: that is the crux of the historical problem.

As noted above, Fermat in a sense created modern number theory by isolating it and its methods from the algebra of continuous magnitude. The twin concepts of divisibility and primality peculiar to the domain of integers form perhaps the most important central theme of his research. They in turn derive from the first complex of problems Fermat investigated, problems involving the sum of the *aliquot parts* (proper divisors) of a number. His first achievement in number theory, one that dates back to the period in Bordeaux, consisted of a general method for finding integers that are given submultiples of the sums of their aliquot parts. As will become clear below, the general problem places emphasis on determining whether a number is prime and, if not, on finding its divisors. Fermat's search for methods applicable to this general problem culminated in a theorem that today, in slightly different form, bears his name: (Fermat's Theorem) if p is prime and a is any number prime to p, then $a^{p-1} \equiv 1 \pmod{p}$.

The second category of problems to which Fermat devoted a major portion

---

[5] Originally published in 1670 as part of Samuel de Fermat's reedition of Claude Gaspar Bachet de Meziriac's 1621 edition of Diophantus, Fermat's *Observations on Diophantus* were excerpted from that work and published separately in FO.I.291–342. The title is taken from part of the title of the 1670 edition (". . . cum observationibus Domini Petri de Fermat . . . ") and is merely used to denote the unified body of individual comments collected in FO. Fermat certainly did not intend them to constitute a formal treatise. Translations of the *Observations on Diophantus* exist in several European languages, but unfortunately not in English.

[6] "Thus we are left to reconstruct, as best we can," observes Weil (*Number Theory*, 114–115), "some of Fermat's most interesting proofs by descent, and again, as on former occasions, to seek Euler's and Lagrange's help. In doing so one should not forget that it is in the nature of such proofs to admit a number of possible variants, one as good as the other; examples for this abound in the writings of Euler; an explanation of this phenomenon is to be found only in the modern arithmetical theory of curves of genus 1."

of his research, the genesis of Pythagorean number triples satisfying various conditions, also by its very nature laid particular emphasis on integers, even though it was in part approachable by rational algebraic methods. Primes again played a central role, but divisibility gave way to partition or decomposition as the main concern. For example, the solution to the question of the number of triples in which a given number can appear as hypotenuse rests on the well-known prime factorization theorem and the following original result of Fermat's: if p is prime and of the form $4k + 1$, then p is uniquely the sum of two squares. If p is of the form $4k - 1$, then it cannot be decomposed into two squares. Other, similar results concerning the decomposition of primes of various forms into sums of squares led Fermat to claim a proof of Diophantus' conjecture that every number can be expressed as the sum of at most four squares. They also led to preliminary results in the direction of what became with Gauss the theory of quadratic forms. Most importantly, however, various problems concerning Pythagorean triples steered Fermat toward the study of the general indeterminate equation $x^2 - py^2 = \pm 1$ for nonsquare p and integral x and y, and beyond that to preliminary results concerning the equation $x^2 - py^2 = q$. While the first equation was accessible through the theory of equations, analysis of the second depended on his theorems about the decomposition of primes into sums of squares.

The problem of decomposition into squares, originally part of the overall problem of rational right triangles, led in turn to the problem of decomposition in general. The various challenges that Fermat issued to Wallis and others in 1657–1658 included, for example, the "four cube problem": to divide the sum (difference) of two given cubes into the sum (difference) of two other cubes. Here Fermat dropped his insistence on integral solutions. For the deceptively similar problem of finding two cubes such that their sum is a cube, however, Fermat recognized that a solution in rational numbers involved a solution in integers or, conversely, that a proof of the impossibility of a solution in integers implied the impossibility of one in rational numbers. He had such a proof, he claimed, not only for cubes but for any power greater than 2 (Fermat's Last Theorem).

Fermat's close study of Diophantus' *Arithmetic* formed the original matrix of the results cited above, though the results themselves led number theory out of that matrix. Most of the marginalia that constitute the *Observations on Diophantus* conform, at least outwardly, to the traditional Diophantine pattern of rational solutions of indeterminate equations. Of the marginalia within the tradition, almost all represent Fermat's success in extending Diophantus' method of "single" and "double" equations to achieve infinite families of solutions. This success, communicated to Billy by letter and recast by him in the form of a treatise, *New Invention of the Analytic Doctrine,* prefixed to the Diophantus edition of 1670, enabled Fermat to perplex Frenicle and Brûlart with traditional problems that had traditional solutions, or at least solutions that modified tradition in non-radical ways. By contrast, the problems that Fermat meant to confound Wallis and others, including Frenicle, in 1657–1658 neither represented traditional problems nor had traditional solutions. They properly belonged to a new matrix that Fermat meant to establish. Within that new matrix, proofs of impossibility of solution based on the division and decomposition properties of the integers played a central role, and, in denying the usefulness or value of such "negative propositions," Wallis rejected the new matrix.

Failure to attract the interest of such men as Wallis led Fermat to outline in somewhat greater detail the sort of new arithmetic he had in mind. The outline, in the form of a memoir, "Rélation des nouvelles découvertes en la science des nombres," addressed to Carcavi for transmission to Huygens in 1659, affords a glimpse into Fermat's new methods, especially into the method of "infinite descent," but only a glimpse. It also represents Fermat's swan song as a number theorist. His last word on the subject, it caps his career in number theory as well as in mathematics as a whole. So too it caps this last chapter of Fermat's scientific biography.

But the "Relation" lies at the end of a long series of investigations, which is rich in detail so far only barely outlined here. Its origins, like those of Fermat's mature works in other fields, go back to that amazingly fruitful period of initial research in Bordeaux.

II. NUMBERS, PERFECT AND NOT SO PERFECT

As indicated in Chapter 1, Fermat's reputation as a mathematician preceded his first letter to Mersenne and the Parisian mathematical community. A passage in Mersenne's *Universal Harmony* (Paris, 1636) indicates one of the bases for that reputation:[7]

> Now, if I wished to speak of the men of high birth or quality who have thrived so well in this area of mathematics [music theory] that no one can teach them anything, I would repeat the name of him to whom the book on the organ is dedicated [Étienne Pascal], and I would add Monsieur Fermat, Councillor of the Parlement of Toulouse, to whom I owe the observation he made concerning the two numbers 17296 and 18416, the aliquot parts of each of which constitute the other, as do those of the two numbers 220 and 284; also concerning the number 672, which is the subdouble of its aliquot parts, as is the number 120. He knows the infallible rules and the analysis for finding an infinity of others of this sort.

Among the achievements, then, that attracted Mersenne's attention was Fermat's demonstrated prowess in solving problems concerning the aliquot parts,

---

[7] Préface, p. [9] The passage is quoted in FO.II.20–21.

or proper divisors, of numbers. The subject forms a recurrent theme of Fermat's early letters to Paris.

It was a subject deeply rooted in classical Greek tradition, and revival of interest in it in the late sixteenth and early seventeenth centuries fitted in with the overall revival of classical themes, of which the music theory of Mersenne's *Universal Harmony* is itself an excellent example. Ancient Greek mathematicians had been especially interested in the relationship between a number and the sum of its aliquot parts.[8] To an age for which numbers still had mystical properties, "perfect" numbers held special meaning. They were the numbers that were equal to the sum of their aliquot parts, e.g. $6 = 3 + 2 + 1$. The problem of determining perfect numbers occupied the Pythagoreans and their successors, and its solution is recorded in Euclid's *Elements* (IX,36): if $2^{n+1} - 1$ is prime, then $2^n(2^{n+1} - 1)$ is a perfect number. The solution is, of course, incomplete, since it contains no generating function for perfect numbers but rather leaves it to the mathematician to determine for various n whether $2^{n+1} - 1$ is prime. Other than the famous, but laborious, "sieve of Eratosthenes," the ancients offered posterity no effective means of testing the primality of a number or of finding its divisors.[9] Although, until the time of Fermat, several new perfect numbers had been determined, the procedures and theoretical foundations had remained the same.[10] Even less progress had been made in the realm of "friendly" numbers, that is, pairs of numbers of which each is the sum of the aliquot parts of the other. Until Fermat, no one had been able to add to the original pair discovered by the Greeks, 220 and 284.[11]

Although the last type of number mentioned by Mersenne obviously grows out of the same sorts of classical concerns, actual investigation of it was new to the seventeenth century. Ancient texts had divided numbers into three classes, depending on their relationship to the sum of their aliquot parts: numbers that were greater than the sum of their aliquot parts (e.g. all prime numbers, since the number itself did not constitute an aliquot part) were "abundant," those equal to it were "perfect," and those less than it "deficient." For reasons that are not clear, Mersenne and others first concentrated attention on one particular subclass of the "deficient" numbers, to wit, those of which the sum of their aliquot parts is an integral multiple. For example, the sum of the aliquot parts of 672 is 1344, or 2×672. It was to this class of problems in particular that Fermat claimed to have found a fully general and effective

---

[8] See, for example, the account of P.-H. Michel, *De Pythagore à Euclide* (Paris, 1950), pp. 329–364.

[9] The "sieve" constitutes nothing more than a systematic procedure for testing the primality of an odd number N by successively testing its divisibility by all primes from 3 to $\sqrt{N}$; cf. Heath, *History of Greek Mathematics*, I, 100.

[10] Cf. Leonard E. Dickson, *History of the Theory of Numbers* (3 vols., repr. New York, 1952), Vol. I, pp. 3–12.

[11] *Ibid.*, 38–40.

method of solution. "By my method," he wrote to Mersenne on 26 December 1638, "I can solve all problems concerning aliquot parts, but the length of the computations dissuades me, as does the determination of prime numbers, to which all these problems reduce."[12]

To follow Fermat's results in greater detail, it will help to introduce an abbreviatory symbol $s(n)$ to denote the sum of the aliquot parts of a number n, that is, the sum of all proper divisors of n including 1. One can then express the problems just discussed in the following form: the determination of "perfect" numbers corresponds to the solution of $s(x) = x$; "friendly" numbers are the solution pairs x, y of the equations $s(x) = y$ and $s(y) = x$, or of the equation $s(x) + x = s(y) + y$; and x will be the nth submultiple of the sum of its aliquot parts if it satisfies the equation $s(x) = nx$.

It is a mark both of the slow development of number theory prior to the 1630s and of the difficulties of gaining insight into Fermat's methods that the first documented derivation of explicit formulas for $s(n)$ is a memoir written in 1638 by Descartes.[13] By that time, Fermat had already announced his most striking results on the subject of aliquot parts, and one can hardly imagine his having achieved them without possessing such formulas. Yet, no trace of them exists in his extant works. In the absence of direct evidence from Fermat, it may help to follow Descartes' derivation in order to gain at least some sense of style of number theory at the time.

"In solving problems concerning the aliquot parts of numbers," Descartes wrote, "we imagine the numbers to be composed either of mutually prime numbers or of numbers that result from the repeated multiplication of some prime number by itself, or of a combination of the two." One begins, that is, with the expression of numbers as products of (unique) prime factors. A prime number itself has but one aliquot part, to wit, 1. A number of the form p^n, where p is prime, has n aliquot parts: $1, p, p^2, \ldots, p^{n-1}$. Their sum, as Descartes immediately recognized from his knowledge of the sum of any geometric progression, can be expressed in the closed form $(p^n - 1)/(p - 1)$.[14] To determine the sum of the aliquot parts of a number composed of several factors, Descartes operated recursively, beginning with the case of a prime number multiplied by another number of which the sum of the aliquot parts is known. In Descartes' notation, let x be the prime number, a the other number, and b the sum of a's aliquot parts. The aliquot parts of ax consist, then, of those of a taken alone, those of a each multiplied by x, a itself, and x itself.[15]

---

[12] FO.II.176–177.

[13] Dickson, pp. 52–53. Descartes, *De partibus aliquotis numerorum*, AT.X.300–302. In a letter to Frenicle dated 9.I.39, Descartes said that a year earlier he had not known about the sum of a̅ the aliquot parts of a number. In fact, a letter to Mersenne dated 31.III.38 is the first evidence of Descartes' knowledge of the above results (see AT.X.300, n.).

[14] Note that Fermat deals explicitly with the sum of a geometric progression in the *Treatise on Quadrature*; cf. above, Chap. V, p. 249.

Their sum is $(x + 1)b + a$, or $bx + a + b$; in the notation introduced above, $s(ax) = x \cdot s(a) + s(a) + a$. By a similar line of reasoning, consideration of the product of two distinct prime numbers raised to powers, $a^n c^o$, leads to the recursive formula $s(a^n c^o) = s(a^n) \cdot s(c^o) + a^n \cdot s(c^o) + c^o \cdot s(a^n)$ and thence to Descartes' explicit formula:

$$s(a^n c^o) = \frac{aa^n c^o + a^n cc^o - cc^o - aa^n - a^n c^o + 1}{ac - a - c + 1}$$

In the case, then, where a prime power x^n is multiplied by a number a, "which is prime with respect to the other, even though not absolutely prime, and of which the [sum of the] aliquot parts is given," the above result leads directly to the recursion formula $s(ax^n) = s(a) \cdot s(x^n) + a \cdot s(x^n) + x^n \cdot s(a)$.

Descartes summed up his results with the general recursion formula: "If we have two mutually prime numbers and their aliquot parts, we also have the aliquot parts of their product: e.g., if one is a, and its aliquot parts b, the other c and its aliquot parts d, the aliquot parts of ac will be $ad + bc + bd$." Clearly, this statement is equivalent to the formula $s(ac) = s(a) \cdot s(c) + a \cdot s(c) + c \cdot s(a)$ derived in stages above. Moreover, it suffices to determine the sum of the aliquot parts of any number. For, in the fully general case of the number $N = p_1^a p_2^b \ldots p_n^r$, one first computes the sum of the aliquot parts of each of the prime powers, taken separately and then applies the recursive formula to compute that of the product of the first two, followed by the product of the first two times the third, of the first three times the fourth, and so on.[16]

As noted above, Fermat must have possessed some similar means for rapidly computing the sums of aliquot parts of numbers, even though his papers offer no direct evidence to that effect. Rather, they supply only solutions to specific problems and then proceed to discuss the general problem fundamental to all perfect and multiply perfect numbers, the determination of primality. To see the issue involved, let us consider the derivation of Euclid's rule for perfect numbers; we shall paraphrase using modern notation.

As a first guess at the nature of a solution N of the equation $s(x) = x$, let us assume that N is of the form $2^n p$, where p is prime. By the formulas set forth above we have $s(2^n p) = s(2^n)s(p) + 2^n s(p) + p \cdot s(2^n)$, or $2^n - 1 + 2^n + p(2^n - 1) = 2^{n+1} - 1 + p(2^n - 1)$. From the initial equation we have, then,

$$2^{n+1} - 1 + p(2^n - 1) = 2^n p$$

[15] Since 1 is counted among the aliquot parts of a, the aliquot part x is accounted for in the product bx.

[16] A more direct route (non-recursive) would correspond to the formula:

$$\left(\frac{p_1^{a+1} - 1}{p_1 - 1} \right) \left(\frac{p_2^{b+1} - 1}{p_2 - 1} \right) \ldots \left(\frac{p_n^{r+1} - 1}{p_n - 1} \right) - N.$$

$$p = 2^{n+1} - 1.$$

Hence, $N = 2^n p$ will be a perfect number if p is a prime number of the form $2^{n+1} - 1$, and the problem of determining perfect numbers reduces to that of testing the primality of $2^{n+1} - 1$ for various values of n or, even more specifically, to that of determining the precise values of n for which it is prime.[17] The results that Fermat transmitted to Mersenne in the early years of their correspondence further emphasized the centrality of the problem of testing the primality of numbers of a certain form at the same time that they increased the difficulty of any solution of that problem. For example, in addition to the specific pair of "friendly" numbers that Mersenne cited in the preface to his *Universal Harmony*, Fermat also provided a general solution to the problem of "friendly" numbers that Mersenne included in the second part of the *Universal Harmony*; if $3^2 \cdot 2^{2n-1} - 1$, $3 \cdot 2^n - 1$ and $3 \cdot 2^{n-1} - 1$ are prime, then $2^n(3^2 \cdot 2^{2n-1} - 1)$ and $2^n(3 \cdot 2^n - 1)(3 \cdot 2^{n-1} - 1)$ are "friendly" numbers.[18] The crux of the solution lies, of course, in finding values of n for which all three factors are prime. Similarly, the general solution of the equation $s(x) = 2x$—the only case of multiply perfect numbers for which Fermat offered a general solution, though he reported several specific solutions for other values of the numerical constant, including 9/4—also rested on the determination of prime values: if

$$\frac{2^n - 1}{2^{n-3} + 1} \text{ is prime, then } 2^{n-1} \cdot 3 \cdot \left(\frac{2^n - 1}{2^{n-3} + 1} \right)$$

is equal to one half of the sum of its aliquot parts.[19] But for what values of n is the expression on the left-hand side prime? It was to such questions of

---

[17] Owing to the prominence given this problem by Mersenne, the numbers $2^p - 1$ are now generally called "Mersenne numbers" and primes of that form "Mersenne primes." Many of the results credited to Mersenne on the subject stem in fact from Fermat. Compare, for example, Dickson, *History of the Theory of Numbers*, p. 13, with the discussion below, p. 294 ff. Given the original assumption on which the derivation of the solution for perfect numbers rests, the question immediately arises whether there exist any perfect numbers of a form other than $2^n p$. It can be shown that all even perfect numbers are of this one form. No one, however, has yet been able to discover an odd perfect number or to prove that none exists. Cf. G. H. Hardy and E. M. Wright, *Introduction to the Theory of Numbers*, pp. 239–240.

[18] *Seconde partie de l'Harmonie universelle* (Paris, 1637), "Nouvelles observations physiques et mathématiques," Obs. XIII, repr. in FO.II.21–22. Fermat, whom Mersenne does not explicitly cite here, operates from an array of number series. The general symbolic solution given above translates the members of the series and the procedures by which they are manipulated.

[19] *Ibid*. Again, Fermat works from an array. None of Fermat's extant writings offers the slightest hint of how he derived these general formulas and other specific solutions of aliquot part problems; cf., for example, Fermat to Mersenne, 20.II.1639, FO.II.179, where he gives $s(2016) = \left(\frac{9}{4} \right) \cdot 2016$, or Fermat to Mersenne, 7.IV.1643, FO.II.255, where he identifies a solution of enormous size for the equation $s(x) = 5x$. "Pour ce qui est des nombres et de leurs parties aliquotes," he wrote to Roberval on 16.XII.1636 (FO.II.92), "j'ai trouvé une méthode

primality, rather than to the derivation of further solutions, that Fermat directed his research in the late 1630s.

The results that Fermat communicated to Mersenne in the first four years of their correspondence eventually attracted the attention of Bernard Frenicle de Bessy, a Parisian number theorist with a gift for manipulating large numbers and a sixth sense for their properties.[20] Frenicle knew from experience the immense amount of labor involved in testing the primality of large numbers by the traditional means and was intrigued by Fermat's claim to have a method by which he could solve all problems involving aliquot parts. Curiosity led to correspondence, which began in 1640 and continued, with some interruptions, until the late 1650s. If relations between Fermat and Frenicle were often strained, ultimately to the breaking point,[21] Frenicle nonetheless proved to be Fermat's most interested and sympathetic correspondent in the realm of number theory. Though in no way Fermat's equal as a mathematician, Frenicle alone among his contemporaries could challenge Fermat in number theory, and Frenicle's challenges had the distinction at least of drawing out of Fermat some of his carefully guarded secrets. Their correspondence is perhaps the most valuable source of information about Fermat's number theory.

Although Fermat freely admitted to Mersenne and others that he could devise no more effective method for testing the primality of an arbitrary number than the traditional sieve,[22] his rapid solution of Frenicle's first challenge suggested that he possessed a more efficient method for certain classes of numbers. In March 1640, Frenicle, writing via Mersenne, challenged Fermat to find a perfect number of twenty digits "or the next one following."[23] By the last phrase, Frenicle evidently meant the first perfect number of 20 or more digits and was testing how quickly Fermat could check the primality of 2^{n-1} for $n \geq 37$, where the sieve quickly becomes unmanageable.[24] Fermat took

générale pour soudre toutes les questions par algèbre, de quoi j'ai fait dessein d'écrire un petit traité." Regrettably, the "petit traité," like others promised on other subjects, never was written, and Fermat's secret died with him.

[20] On Frenicle (1605–1675), see the article in *Dictionary of Scientific Biography*, IV, and J. E. Hofmann, "Neues über Fermats . . . ." The latter discusses in detail Frenicle's major work, the solution of Fermat's first two challenge problems of 1657 (see below §IV).

[21] See below, p. 307ff.

[22] Fermat to Mersenne, 26.XII.1638, FO.II.176–177: "Pour les nombres, je peux trouver par ma méthode toutes les questions de parties aliquotes, mais la longueur des opérations me rebute età la recherche des nombres premiers, à laquelle toutes ces questions aboutissent. Sur lequel sujet je ne sais point de méthode que la vulgaire. . . ."

[23] FO.II.185. The letter deals mostly with magic squares, a subject to which Fermat could apparently contribute very little, if anything. See, for example, Frenicle's disparaging estimation of Fermat's results in this area; Frenicle to Mersenne, ⟨III.1640⟩, FO.II.182–185.

[24] See Colin R. Fletcher, "Fermat's Theorem," *Historia Mathematica* 16 (1989), 149–53. In correcting both my account in the first edition and Weil's in *Number Theory*, Fletcher points out that "*le prochainement suivant*" evidently points to the masculine *nombre* rather than the feminine *lettre*. As to the significance of 20, he observes that $2^{31} - 1$ is the prime base of a perfect number of 19 digits. The next one of 20 or more digits would correspond to the base $2^{37} - 1$, which is composite, as are all other candidates up $2^{61}-1$, which is itself a prime of 19 digits.

him to mean a perfect number of exactly 20 or 21 digits and responded almost immediately, promising more than a simple answer: "I have several shortcuts for finding perfect numbers, and I can say in advance that there is none of 20 digits, nor any of 21 digits. . . ."[25] Hence, Fermat added, the widely held belief that each decimal interval (from 10^n to 10^{n+1}) contains one perfect number is incorrect; neither the twentieth nor the twenty-first *dizaine* holds a perfect number. Clearly more fascinating than this statement, however, which told Frenicle something he already knew, was Fermat's reference to his "several shortcuts" (*plusieurs abrégés*). Fermat's next letter to Mersenne seems to indicate that Frenicle urged Mersenne to probe further into the matter.

In that letter, probably written in June 1640, Fermat lifted the curtain slightly on his methods.[26] "Here are three propositions," he wrote,"on which I hope to raise a great structure." The core of the problem of perfect numbers lay in the series of numbers of the form $2^n - 1$, which Fermat called the "radicals" of the perfect numbers, since each prime number in that series determined a perfect number. Fermat's three propositions reduced the primality of any radical to the primality of its index n. First, if n is composite (i.e. not prime), then $2^n - 1$ is composite. Second, if n is prime, then $2n$ divides the number $(2^n - 1) - 1$, i.e. n divides $2^{n-1} - 1$ (symbolically, $n | 2^{n-1} - 1$). Third, if n is prime, then the only possible divisors of $2^n - 1$ are those of the form $k(2n) + 1$. Other than to give a numerical example for each proposition, Fermat said nothing of his path to the propositions themselves or of how they might be proved. They had been found "not without difficulty," and they could be called "the foundations for the determination of perfect numbers." The third proposition, for example, had enabled Fermat to determine the composite nature of the radical $2^{37} - 1$, which is divisible by 223, or $3(2 \cdot 37) + 1$. Though he did not say so explicitly in the letter, that radical represents the only possible perfect number of twenty or twenty one digits, since the index n of the radical of all other candidates of that size is composite, whence the radicals themselves are composite.[27] This shortcut to the elimination of $2^{37} - 1$ as the radical of a perfect number was what enabled Fermat to provide Frenicle with such a quick reply to his challenge.

The three propositions were clearly only a taste of more to follow. "From these shortcuts," he wrote, "I already see a great number of others emerging, and for me it is like seeing a great light."[28] Fermat was just being coy. He had already seen the light, and his three propositions were merely faint rays of it.

[25] Fermat to Mersenne, ⟨V?1640⟩, FO.II.194.

[26] FO.II.195–199; cf. Fermat to Roberval, VIII.1640, FO.II.202: " . . . bien que la démonstration en soit assez cachée, ce que je vous prie d'essayer, puisque vous les avez vues."

[27] Since any perfect number is of the form $2^n(2^{n+1} - 1) = 2^{2n+1} - 2^n$, the least exponent $2n + 1$ would have to range between 63 and 73 to produce a perfect number of twenty or twenty-one digits. Therefore $31 \leq n \leq 36$, or $32 \leq n + 1 \leq 37$. Of the numbers from 32 to 37, only the last is prime and hence a candidate to yield a prime radical.

[28] FO.II.199: *mi par di veder un gran lume.* Neither Fermat nor the editors of FO identify the source of this Italian phrase.

As noted above, his methods for determining "friendly" and multiply perfect numbers placed emphasis on the rapid and efficient determination of the primality or of the divisors of numbers of the general form $a^n - 1$ and $a^n + 1$, and the radicals of the perfect numbers constituted only a small subclass of those numbers. Of the three propositions, the two concerning prime n were merely corollaries of a far more general theorem that Fermat announced directly to Frenicle on 18 October 1640 and that he must have already had at hand much earlier; it was "the foundation on which I base demonstrations of everything to do with geometric progressions." The theorem still bears his name, though it is now stated in quite different language and serves correspondingly different purposes. Frenicle first learned it from Fermat in the following form:[29]

> Without exception, every prime number measures one of the powers -1 of any progression whatever, and the exponent of the said power is a submultiple of the given prime number -1. Also, after one has found the first power that satisfies the problem, all those of which the exponents are multiples of the exponent of the first will similarly satisfy the problem.

That is, given a prime p and a sequence of numbers of the form $a^t - 1$ (t a positive integer) based on any integer a, then p divides some least member of the sequence, say $a^T - 1$, and $T|p - 1$; moreover, p also divides all members $a^{kT} - 1$ ($k = 1, 2, \ldots$) of the same series.

"This proposition," Fermat continued, "is generally true for all series and for all prime numbers; I would send you a demonstration of it, if I did not fear going on for too long." In this early instance, as in many later ones, he never did set down his proof or suggest how he had come upon the theorem in the first place. That makes it difficult to gauge how much significance to attach to the manner in which the general proposition goes beyond the particular instances he had shown Mersenne. For, recast in the Gaussian language of moduli and residues, it conveys what is called "Fermat's Theorem" in a form suggestive of later proofs. The usual form today is: if p is prime, and a is any integer prime to p, then $a^{p-1} \equiv 1 \pmod{p}$, or $a^p \equiv a \pmod{p}$. Fermat's version states that there is a (least) integer t such that $a^t \equiv 1 \pmod{p}$ and $t|p - 1$; moreover, $a^{kt} \equiv 1 \pmod{p}$ for all integer k, and hence, in all cases, $a^{p-1} \equiv 1 \pmod{p}$.[30] That way of stating it lies close to the line of proof followed since Euler, namely that the ascending powers of a starting from 1 form a series of residues modulo p, of which there are only $p - 1$ values.[31] Hence, the series must repeat over an interval $\leq p - 1$; that is, for any r

---

[29] FO.II.209.

[30] Fermat's own statement of the theorem fails to include the condition that a and p be mutually prime (in symbols to be used below, $(a,p) = 1$). If, in fact, $p|a$, then trivially $a^p - a \equiv 0 \pmod{p}$, but $a^{p-1} \not\equiv 1 \pmod{p}$.

[31] Weil, *Number Theory*, 57.

there is some $k \leq p - 1$ such that $a^{r+k} \equiv a^r \pmod{p}$, or $a^k \equiv 1 \pmod{p}$. If k is the least value for which that is true, then $\{1, a, a^2, \ldots, a^{k-1}\}$ is a set of k different residues modulo p, and, for any other residue b, the set $\{b, ba, ba^2, \ldots, ba^{k-1}\}$ is either the same as $\{1, a, a^2, \ldots, a^{k-1}\}$ or disjoint from it. Similarly, the sets so generated by any two residues b and c are either the same or disjoint, and hence the $p - 1$ residues must divide evenly into sets of k, i.e. $k|p - 1$.

Now, while the first part of demonstration along these lines—repetition within at most $p - 1$ terms—can be readily couched in Fermat's language of progressions and their divisors ($a^{r+k} = a^r + mp$, etc.), the second—that a cycle $< p - 1$ partitions the residues into disjoint sets of equal size—requires a more complicated and problematic translation, as Euler's own path to it makes evident.[32] More importantly, perhaps, Euler took that path in pursuit of problems different from those that had stimulated Fermat's investigation, at least up to the time when he stated the theorem. For Fermat, the theorem provided the theoretical framework for a series of shortcut procedures for determining the division properties of a very special class of numbers, the numbers that produced in turn perfect, friendly, and multiply perfect numbers; that is, numbers that offered answers to very old questions. Precisely what implications Fermat drew from the theorem and hence what role the theorem played in other, later investigations is a question in itself. That the theorem ultimately pointed toward radically new and different questions and their answers merely underlines the basic paradox of Fermat's mathematics as a whole.

In the absence of any direct evidence, one can make conjectures about how Fermat found and proved his theorem, conjectures that take account of the quality and content of the means available to him. Leibniz's later derivation is suggestive and worth presenting in this regard. In his *Nova algebrae promotio* (1695?), which remained in manuscript until its publication by Gerhardt in 1863, Leibniz gives in essence the following reasoning:[33] let $a, b, \ldots, r$ be any integers and let p be prime. Since $(a + b + \ldots + r)^p$ *is* equal to $a^p + b^p + \ldots + r^p$ plus a series of terms each containing p as a factor, the expression $(a + b + \ldots + r)^p - (a^p + b^p + \ldots + r^p)$ is clearly divisible by p. If $a = b = \ldots = r = 1$, then any integer n may be expressed as the sum of n such terms. Moreover, then $a^p + b^p + \ldots + r^p = 1^p + 1^p + \ldots + 1^p = n$. Therefore, p divides in particular the expression $n^p - n$. If, in addition, $(n, p) = 1$, then it also holds that $n^{p-1} - 1$ is divisible by p. There is no step here that Fermat could not easily

---

[32] For Euler's development of the theorem, see Weil, *Number Theory*, Chap. III, §VI.

[33] *Leibnizens mathematische Schriften*, ed. Gerhardt, Vol. VII, p. 180. Professor Helmuth Gericke, Munich, called this passage to my attention and noted that the context of the proof suggests that Leibniz was aiming at a different theorem altogether; to wit, that if p is prime then p divides all binomial coefficients $\binom{p}{k}$.

have taken, and the proof has the advantage of being both immediately apparent for the case of $2^p = (1 + 1)^p$ and readily generalizable from it. Moreover, viewed as a heuristic path, the proof suggests why Fermat did not include in his statement the condition that a and p be mutually prime. That is not necessary for $a^p \equiv a$ (mod p), but only for moving to $a^{p-1} \equiv 1$ (mod p). However, Leibniz' approach proves Fermat's theorem for $p - 1$, rather than for divisors of $p - 1$, and it does not suggest a proof corresponding to that form. One can imagine from Fermat's example of $3^3 \equiv 1$ (mod 13) that he discovered empirically that in some cases $a^t - 1 = mp$ for $t < p - 1$ and satisfied himself algebraically that then $a^{nt} - 1 = m'p$ for all n. Knowing that in all cases $a^{p-1} - 1 = kp$, he may then have reasoned along the lines that, if t is the first such exponent in the progression, then $p - 1 = nt + r$, whence $a^{p-1} = kp = a^{nt+r} - 1 = a^{nt}a^r - 1 = (a^{nt} - 1)a^r + a^r - 1 = m'pa^r + a^r - 1$; that is, $p|a^r - 1$. But $0 \le r < t$, and t is the smallest power > 0 with residue of 1; hence $r = 0$ and $p - 1 = nt$.

Fermat himself spoke of the theorem only in the one letter to Frenicle quoted above, and at a time when he and Frenicle had begun to shift their attention to Pythagorean number triples and the related problem of decomposition of numbers into sums of squares. Neither Frenicle nor anyone else at the time pressed Fermat for further information about the theorem, and Fermat never explicitly returned to the subject after 1640. Deprived of any derivation or proof, we can understand its role in Fermat's number theory only by noting the particular corollaries he drew from it and the reason for his choice of those corollaries.

Another look at the three "foundations for determining perfect numbers" reveals two of the corollaries. Indeed, the second proposition represents little more than a restatement of Fermat's Theorem for $a = 2$. That is, if n is prime, then $n|2^{n-1} - 1$, and hence (for $n > 2$) $2n|2^n - 2$. The third proposition requires, however, somewhat more involved reasoning. Suppose n is prime and $2^n - 1$ is divisible by some odd prime p (if it is divisible by any number, it is divisible by some prime). By his major theorem, Fermat knew that then $n|p - 1$, that is, $p - 1 = k'n$ for some integer k'. But, since p is odd, $p - 1$ is even and hence divisible by 2. Therefore, $2|k'n$. But n is prime, whence $2|k'$, or $k' = 2k$. Therefore, $p - 1 = 2kn$, or $p = 2kn + 1$.

On the basis of his three propositions, two of them derived from his major theorem and one from an algebraic identity,[34] Fermat could rapidly test the possibility that a given radical was prime. To start with, its exponent had to be prime. That prime exponent led in turn to a more limited range of possible divisors than merely all primes less than the square root of the radical; e.g.

---

[34] The first proposition regarding composite n follows from the identity:

$$\frac{a^{pq} - 1}{a^p - 1} = a^{p(q-1)} - a^{p(q-2)} + \ldots + 1.$$

$2^{37} - 1$ would be divisible only by primes of the form $74k + 1$. The determination of prime radicals for the perfect numbers was not, however, the only purpose of Fermat's theorem. Other numbers of a similar form emerged from the investigation of sums of aliquot parts. Although, for example, the simple algebraic identity

$$\frac{a^n - 1}{a - 1} = a^{n-1} + a^{n-2} + \ldots + 1 \; (a > 2)$$

told him that no number of the form $a^n - 1$ could be prime for $a > 2$, his theorem enabled him to limit the number of its possible prime divisors. It also provided him with the means to attack numbers of the form $a^n + 1$.

A letter written to Mersenne on Christmas Day 1640 suggests that Fermat possessed an elementary theorem concerning numbers of that form: $a^n + 1$ can be prime only if a is even and n is of the form 2^m. For, if a is odd, a^n is also odd, whence $a^n + 1$ is even and therefore divisible by 2. If n contains an odd factor k, i.e. $n = kp$, then $a^p + 1 | a^{kp} + 1$.[35] The theorem offers only the necessary conditions for the primality of $a^n + 1$, and a conjecture by Fermat in his letter to Mersenne seems to have been aimed at supplying the sufficient condition. He wrote:[36]

> In this regard, I would like to be clear as to whether one of my ideas is true, to wit, that in a progression [based on] an even number, say 6, all the powers $+ 1$ of the progression which have for their exponent 1, 2, 4, 8, 16, etc., are prime numbers if they are not measured by one of these: 3, 5, 17, 257, etc. If this proposition is true, it would be very helpful.

Fermat's sufficient condition, then, was that $a^{2m} + 1$ not be divisible by any number of the form $2^{2n} + 1$; all numbers of the latter form were, he was convinced, prime numbers.[37]

But just as Fermat's interest in numbers of the form $a^n - 1$ did not end with the impossibility of their being prime but rather extended to the determination of their prime factors, so too he sought to employ his major theorem to limit the number of possible prime factors of $a^n + 1$. In the letter to Frenicle, he drew the following conclusions, both evidently corollaries to the theorem: for a given prime p, let t be the least exponent for which $p | a^t - 1$. Then, if t is even, i.e. $t = 2s$, $p | a^s + 1$; if t is odd, then p divides no number of the form $a^n + 1$.[38]

---

[35] The result follows from the algebraic identity

$$\frac{a^{kp} + 1}{a^p + 1} = a^{(k-1)p} - a^{(k-2)p} + a^{(k-3)p} - \ldots + 1 \; (k \text{ odd}).$$

[36] FO.II.213.

[37] See below, p. 301.

[38] FO.II.209–210: "Mais il n'est pas vrai que tout nombre premier mesure une puissance $+ 1$ en toute sorte de progressions: car, si la première puissance $- 1$, qui est mesurée par le dit nombre premier, a pour exposant un nombre impair, en ce cas il n'y a aucune puis-

Neither Frenicle nor anyone else received anything more than the assertion of these two propositions from Fermat. His proof remained his secret. Nonetheless, the first conclusion follows directly from the assumptions in so obvious a manner that one can be reasonably sure that Fermat argued thus: let $t = 2s$ be the least exponent such that a given prime p divides some number $a^t - 1$. Then p divides $a^{2s} - 1 = (a^s - 1)(a^s + 1)$. But p cannot divide $a^s - 1$, since $s < 2s$. Therefore, $p|a^s + 1$. The path to the second conclusion is not obvious, but it does follow from the main theorem by an argument close to the one suggested above as Fermat's route to the theorem itself. Suppose that for given odd prime p and base a the least exponent t such that p divides $a^t - 1$ is odd and, further, that p divides some number $a^k + 1$. Then p divides $(a^k + 1)(a^k - 1) = a^{2k} - 1$ and, by the main theorem, $t|2k$.[39] Since t is odd, $t|k$, and hence, again by the main theorem, p divides $a^k - 1$. But if p divides both $a^k + 1$ and $a^k - 1$, then p divides their difference, i.e. $p|2$. But p is an odd prime and therefore > 2. The initial assumption thus leads to a false result, and hence p divides no member of the series $a^k + 1$.

The two corollaries to Fermat's major theorem made possible the elimination of certain primes from among the divisors of numbers of the form $a^n + 1$. For a given prime p, one had merely to find the least number of the form $a^t - 1$ divisible by p, and, by the main theorem, only the divisors of $p - 1$ (including $p - 1$ itself) had to be tested. Indeed, to eliminate p, Fermat had only to test the odd divisors of $p - 1$. In his letter to Frenicle, however, Fermat was seeking a method more efficient than the direct testing of one prime after another. He sought a theorem that would determine the whole class of eliminable primes. To his obvious regret, he could only pose the problem to Frenicle:[40]

> The whole difficulty consists in finding the prime numbers that do not measure any power + 1 in a given progression; for that serves, for example, for finding which of the prime numbers measure the radicals of perfect numbers and for thousands of other things, as, for example, it shows that the 37th power − 1 in the progression based on 2 is measured by 223. In a word, one must determine which prime numbers are the ones that measure their first power − 1 in such a way that the exponent of the said power is an odd number. I think this very difficult and await greater enlightenment from you. Please expand that place in your letter where you say that after having found that the divisor must be a multiple + 1 of the exponent,

---

sance + 1 dans toute la progression qui soit mesurée par le dit nombre premier. . . . Que si la première puissance − 1 qui est mesuré par le nombre premier donné a pour exposant un nombre pair, en ce cas la puissance + 1 qui a pour exposant la moitié dudit premier exposant sera mesurée par le nombre premier donné."

[39] See above, p. 295.
[40] FO.II.210.

there are also rules for finding the values of the said multiples +1 of the exponent.

Fermat's own attempts at solution of the problem had got no farther than a proposition "that I esteem highly, even though it does not reveal everything I seek." In summary it asserts that, for a given a, any prime p that divides an expression of the form $N^2 - a^{2k+1}$ for some integer N and is in addition of the form $4m - 1$ satisfies the conditions of the problem. That is, the least t for which $p|a^t - 1$ will be odd, and hence p will divide no number of the form $a^n + 1$.[41]

As usual, Fermat simply stated the proposition by means of several examples and a generalization based on them. Here again he offered no proof nor derivation, and here again the implications one can draw by hindsight sharpen the question of the conceptual structure behind his claims. For, if a prime p divides $N^2 - a^{2k+1}$, i.e. $N^2 \equiv a^{2k+1} \pmod p$, then there is a number n such that $n^2 \equiv a \pmod p$. If p is of the form $4m - 1$, then by Fermat's theorem $n^{4m-2} = (n^2)^{2m-1} \equiv a^{2m-1} \equiv 1 \pmod p$. That is, p divides $a^{2m-1} - 1$ in the series of powers of a. Couched in these terms, however, the derivation involves more than that immediate result. It establishes the basic criterion for quadratic residues. May we attribute it to Fermat? On the one hand, beyond the generality of the result for any base a, his insistence on the odd powers of that base suggests a focus on squared terms. On the other, moving from $N^2 \equiv a^{2k+1} \pmod p$ to $n^2 \equiv a \pmod p$ involves extending Fermat's theorem in ways for which there is no direct evidence in Fermat's work and which lead to the concepts of cyclic arithmetic that gave cause for hesitation in reconstructing the path to his main theorem above.[42]

---

[41] FO.II.210–211: "Voici un mienne proposition (que peut-être vous aurez aussi trouvée) que j'estime beaucoup, bien qu'elle ne découvre pas tout ce que je cherche, que sans doute j'achèverai d'apprendre de vous: En la progression double, si d'un nombre quarré, généralement parlant, vous ôtez 2 ou 8 ou 32 etc., les nombres premiers moindres de l'unité qu'un multiple du quaternaire, qui mesureront le reste, feront l'effet requis. . . . . En la progression triple, si d'un nombre quarré *ut supra* vous ôtez 3 ou 27 ou 243 etc., les nombres premiers moindre de l'unité qu'un multiple du quaternaire, qui mesureront le reste, feront l'effet requis. . . . En la progression quadruple, il faut ôter 4 ou 64 ou 1024 etc. à l'infini en toutes progressions, en procédant de meme façon." Cf. also Fermat to Mersenne, 15.VI.1641, FO.II.220: "En la progression de 3, tous les nombres premiers, qui sont différents par l'unité d'un multiple de 12, mesurent seulement les puissances -1. Tels sont: 11, 13, 23, 37, etc. En la même progression, les nombres premiers, qui sont différents par 5 d'un multiple de 12, mesurent les puissances $+1$. Tels sont: 5, 17, 19, etc. En la progression de 5, tous les nombres premiers, qui finissent par 1 ou 9, mesurent seulement des puissances -1. Tels sont: 11, 19 etc. Ceux qui finissent par 3 ou par 7 mesurent des puissances $+1$. Tels sont: 7, 13, 17, etc. Vous aurez une autre fois la règle générale en toute sorte de progressions." Unfortunately, the last promise is another Fermat would not or could not keep.

[42] That is, since $a^p \equiv a \pmod p$, $a^{p-(2k+1)}a^{2k+1} \equiv a^p \equiv a \equiv N^2 a^{(p-1)-2k} \equiv n^2$, where $n = Na^{1/2(p-1)-k}$. By Fermat's theorem, one may, of course, take an odd multiple of p to assure that the power of the last factor is both positive and even. If Fermat came this far, then a short step

Here too, as on other occasions, Frenicle did not press Fermat for a proof or derivation. Indeed, if Frenicle ever provided the information Fermat had requested, no record of it remains in the sources. As the correspondence of the two men shifted to another subject in number theory, they left that of aliquot parts *in media res*. Of all the material discussed above, only one proposition maintained a continuing hold on Fermat's attention: his conjecture that $2^{2^n} + 1$ is prime for all n. His first announcement of it in a letter of August 1640 to Frenicle characterizes all subsequent mention of it save the last:[43]

> But here is what I admire most of all: it is that I am just about convinced that all progressive numbers augmented by unity, of which the exponents are numbers of the double progression, are prime numbers, such as
>
> $$3, 5, 17, 257, 65537, 4\ 294\ 967\ 297$$
>
> and the following of twenty digits
>
> $$18\ 446\ 744\ 073\ 709\ 551\ 617, \text{ etc.}$$
>
> I do not have an exact proof of it, but I have excluded such a large quantity of divisors by infallible demonstrations, and my thoughts rest on such clear insights, that I can hardly be mistaken.

In the years that followed, Fermat's conviction of the validity of his conjecture grew, while a proof continued to elude him. The primality of all numbers of the form $2^{2^n} + 1$ underlay the climax toward which he steered Part III of the *Tripartite Dissertation*, and it figured among the first propositions with which he tried to engage Blaise Pascal's interest in number theory in 1654.[44] At each mention of the conjecture, Fermat bemoaned his inability to find a proof, and his tone of growing exasperation suggests that he was continually trying to do so. In view of Euler's disproof of the conjecture by counterexample in 1732 ($2^{2^5} + 1$ is divisible by 641), Fermat's quandary is understandable in retrospect.[45] So much more surprising, then, is his claim in the "Relation" to Carcavi in 1659 to have found the long-elusive demonstration.[46] Set

---

would have led him to the notion of division by multiplicative inverse, $a^r a^{(p-1)-r} \equiv 1 \pmod{p}$, which is essential for treating quadratic forms by way of quadratic residues. That he may not have taken the step would explain why he apparently failed to see that that proposition that no prime $p = 3k - 1$ can be expressed in the form $a^2 + 3b^2$ involves no more than showing that -1 is not a quadratic residue modulo 3; see below, §V.

[43] FO.II.206: " . . . et j'ai de si grandes lumières, qui établissent ma pensée, que j'aurais peine à me dédire."

[44] Cf. above, Chap. III, p. 140ff. and Fermat to Pascal, 29.VIII.1654, FO.II.309–310, where Fermat says: "C'est une propriété de la vérité de laquelle je vous réponds. La demonstration en est tres malaisée et je vous avoue que je n'ai pu encore la trouver pleinement; je ne vous la proposerais pas pour la chercher, si j'en etais venu à bout."

[45] Euler, "Observationes de theoremate quodam Fermatiano, aliisque ad numeros primos spectantibus," *Comm. Acad. Sci Imp. Petrop.*, VI (1733), pp. 103–107.

[46] See below, p. 356ff.

forth merely as a bald claim, however, it offers no further insight into Fermat's thinking on the subject of his conjecture in particular and aliquot parts in general. It does, however, elucidate simultaneous claims for proofs of other theorems not yet discussed, and hence it is best left for the moment.

The letter of 18 October 1640 to Frenicle represents the culmination of Fermat's research on aliquot parts and on the complex of problems that had motivated his major theorem on the divisibility of numbers of the form $a^n + 1$. By the spring of 1641, the two men had refocused their attention on a different set of problems, those concerning rational right triangles, and the subject of aliquot parts disappeared from Fermat's correspondence, at least in any creative sense.

III. TRIANGLES AND SQUARES

Like the subject of aliquot parts, perfect numbers, etc., the topic of rational right triangles belonged to the traditional complex of problems associated with "arithmetic" in the seventeenth century. And, just as Fermat's treatment of the former subject led to the foundation of an entirely new sort of arithmetic, the sort inherent in his theorem, so too his approach to rational right triangles eventually reshaped the subject itself. Here his achievement was twofold: on the one hand, he extended the effectiveness of traditional solution techniques and was thereby able to solve problems his predecessors and contemporaries thought impossible; on the other hand, he used rational right triangles and their techniques to attack a body of problems apparently unrelated to them and in doing so created a body of number theory that has outlived its source. Starting from triangles, Fermat arrived at theorems concerning the decomposition of numbers into sums of squares and at his first method for generating complete solutions to the so-called "Pell Equation."

Chronologically, as well as conceptually, the subject of rational right triangles marks the transitional stage in Fermat's revision of number theory. It appears in his earliest correspondence and becomes the focus of his exchange with Frenicle from 1641 on. There it largely conforms to the classical tradition from whence it comes, that is, to Book VI of Diophantus' *Arithmetic*. But it leads in the 1640s to increasing concern with decomposition into sums of squares and culminates in 1657 with the second "challenge" (the complete solution of the "Pell Equation"). What number theory Fermat does after 1657 belongs unmistakably to the new tradition of which he is the founder. As is Fermat himself, so too his study of rational right triangles is caught between traditions.

A. Triangles and Diophantus

"Rational right triangles" are triples of rational numbers satisfying the indeterminate equation $x^2 + y^2 = z^2$. They are apparently as old as mathematics itself. Babylonian clay tablets contain lists of such (integral) triples evidently generated by the same formation rule employed later by the Pythagoreans and set down in Euclid's *Elements* (X, 28, Lemma 1): take any two different numbers p and q; then $(p^2 + q^2, p^2 - q^2, 2pq)$ is a right triangle. A constellation of challenging problems emerges when one seeks to add conditions to the basic solution; for example, that the "hypotenuse" (i.e. $p^2 + q^2$) be itself a square, that the sum of the two "sides" be a square, and so on. Book VI of Diophantus' *Arithmetic* forms the *locus classicus* for such problems and for the techniques used to solve them. The same techniques provided the foundation for Fermat's treatment of rational right triangles. Basically there were two classical methods of solution: a suitable algebraic identity and the technique of "single" and "double" equations.

The method of algebraic identities requires little or no explanation. The solution of the basic problem, i.e. the indeterminate equation $x^2 + y^2 = z^2$, itself rests on the identity

$$(p^2 + q^2)^2 = (p^2 - q^2)^2 + (2pq)^2$$

Except in terms of extending the usefulness of a given identity through transformation into equivalent forms, there is little in the use of identities themselves to warrant the title of "method." Often they spring from the solution of problems by other methods and then serve as a general solution algorithm. Otherwise, the mathematician already knows a particular identity and recognizes its applicability to a given problem. Nonetheless, though it would serve no purpose here to list the many identities Fermat employed in his research, both their variety and their complexity display the side-benefits of his experience in the theory of equations with its frequent transformations to equivalent forms.

By contrast, Diophantus' method of "single" and "double" equations fully warrants the title of "method," though characteristically Diophantus never sets it out in general terms, but simply uses it. Described in general terms—with the use of anachronistic symbolism—the method determines for an indeterminate algebraic equation, $y^m = f^n(x)$ (where the exponent denotes the degree of the expression), a rational pair x,y of solution values ("single" equations) or for a pair of algebraic equations, $y^2 = f^2(x)$, $z^2 = g^2(x)$, a rational triple x, y, z of solution values ("double" equations). For example, in *Arithmetic* II,8, Diophantus seeks to split a given square number into two (rational) squares. Given m^2 (Diophantus uses the specific value 16), he sets one square equal to x^2, whence $m^2 - x^2$ must also be a square. Diophantus requires, then, a rational solution pair for the indeterminate equation $y^2 = m^2 - x^2$. To obtain it, he sets $y = kx - m$ (again taking a specific value, 2, for k). Then,

$$k^2x^2 - 2mkx + m^2 = m^2 - x^2,$$

or

$$(k^2 + 1)x = 2mk$$

or

$$x = \frac{2mk}{k^2 + 1}.$$

Though Diophantus is content with the particular numerical solution, $x = 16/5, y = 12/5$, that emerges from his specific example, one may note that the symbolic solution just derived leads directly to an identity that provides a general solution algorithm for infinitely many rational solutions to that specific example, to wit,

$$m^2 = \left[\frac{2mk}{k^2 + 1} \right]^2 + \left[\frac{m(k^2 - 1)}{k^2 + 1} \right]^2$$

Throughout the *Arithmetic*, the solution technique for single equations remains the same: given a quadratic expression in x to be set equal to a square (for the method to work, either the coefficient of x^2 or the constant term must be a square), equate the expression to the square of a binomial $mx + n$, where m, n, and the sign are so chosen as to eliminate the square term of the original expression and assure a positive solution. The elimination of the square term linearizes the problem, either directly, or indirectly by subsequent cancellation of the extraneous root $x = 0$. For example, in *Arithmetic* IV,20, Diophantus solves $y^2 = 9x^2 + 24x + 13$ by setting $y = 3x - 4$; i.e. $9x^2 + 24x + 13 = (3x - 4)^2 = 9x^2 - 24x + 16$. He chooses the coefficient 3 to eliminate the term $9x^2$, he takes the minus sign to keep the coefficient of x on the left-hand side positive, and he uses 4 because he requires a number of which the square is greater than 13 (so that the constant on the right-hand side remains positive). Although Diophantus remarks that 4 is an arbitrary value (any number $> \sqrt{13}$ would suffice), he does not move beyond the single solution provided by the choice of 4. His failure to do so is characteristic of the *Arithmetic*'s overwhelming orientation toward problem-solving (one answer to a problem suffices), though it probably also reflects the lack of a symbolism sophisticated enough to capture a general solution algorithm in visual format. Fermat had the symbolism, but, more importantly, he was seeking precisely those general solution algorithms rather than specific solutions.

Diophantus employs the method of "double" equations to greatest effect in Book VI of the *Arithmetic*, the section dealing exclusively with rational right triangles. Problem VI,6 illustrates how "double" equations arise, as well as the level of algebraic sophistication Fermat found in his main source. Diophantus proposes to find a right triangle such that the area added to one of the

304

smaller sides makes a given number; here 7 is the given number. He begins with the (possibly false) assumption that the desired triangle is similar to the 3, 4, 5 triangle; i.e. let $3x$, $4x$, $5x$ be the sides of the triangle sought. The conditions of the problem assert that $6x^2 + 3x = 7$. Diophantus then looks at the characteristic of this quadratic equation. "In order that this might be solved, it would be necessary that the square of half the coefficient of x, plus the product of the coefficient of x^2 and the absolute term, should be a square."[47] However, $(3/2)^2 + 6 \cdot 7$ is not a square. Hence, Diophantus reasons, one must choose as base triangle one in which the square of half of one perpendicular, plus 7 times the area, is a square. Taking 1 and m as the sides of that triangle, Diophantus has $(1/2)^2 + 7(m/2) = y^2$ as one equation and, since the triangle is right, $m^2 + 1 = z^2$ as the other. The problem then reduces to the solution of the "double" equation

$$14m + 1 = u^2 \text{ (where } u = 2y)$$

$$m^2 + 1 = z^2.$$

Diophantus' solution of this "double" equation follows a standard pattern. Subtraction of the first equation from the second yields $m^2 - 14m = z^2 - u^2 = (z + u)(z - u)$. Factor the left-hand side into $m(m - 14)$ and set $m = z + u$ and $m - 14 = z - u$. Elimination of z from this simultaneous system yields $u = 7$, whence $14m + 1 = 49$, or $m = 24/7$. Therefore, the base triangle must be $(1, 24/7, 25/7)$, or $(7, 24, 25)$. Beginning the problem again with the triangle $(7x, 24x, 25x)$, Diophantus then obtains a rational value for x.

The technique employed by Diophantus works generally for any pair of equations $y^2 = f(x)$, $z^2 = g(x)$, where $f(x)$ and $g(x)$ are both quadratic polynomials, or both linear, or one quadratic and one linear, subject only to the condition that either (1) for two expressions of the same sort either the coefficients of the leading terms or the constant terms are both squares or stand in the ratio of squares (as 8:2), or (2) for a mixed pair the constant terms are both squares or stand in the ratio of a square to a square. The condition stems from the need to linearize the system; that is, given a pair of equations $a_1x^2 + b_1x + c_1 = y^2$, $a_2x^2 + b_2x + c_2 = z^2$, one must be able to eliminate by subtraction either both leading terms or both constant terms. Suppose, for example, $c_1 \neq c_2$. Then each equation must be multiplied by the factor necessary to raise the constant term to the least common multiple of c_1 and c_2. In order for the system to remain a "double" equation (and hence subject to Diophantus' method of solution), however, the factor must in each case be a square; that is, one must arrive at a system, $p^2a_1x^2 + p^2b_1x + p^2c_1 =$

---

[47] We accept here for convenience's sake Heath's anachronistic substitution of the modern x for Diophantus' ς.

$p^2y^2 = (py)^2$ and $q^2a_2x^2 + q^2b_2x + q^2c_2 = q^2z^2 = (qz)^2$, where $p^2c_1 = q^2c_2$. Hence, necessarily, $c_1:c_2 = q^2:p^2$; i.e. c_1 and c_2 are either both squares or stand in the ratio of squares.

Depending, then, on whether the leading terms or the constant terms are eliminated, subtraction of one equation from the other will yield an equation of the form $mx + n = u^2 - v^2$ or of the form $kx^2 + lx = u^2 - v^2$ In either case, the left-hand side must be factored to correspond to the factors $(u + v)$ and $(u - v)$ of the right-hand side. One keeps in mind that the square of half the sum of the factors on the left-hand side will be equated to the greater of the two original expressions, or the square of half their difference to the smaller, and that, on the model of "single" equations, the guarantee of a rational solution for x rests on the elimination of the quadratic term in that expression. Hence, in the case of a square constant c_1^2, for example, one factors the difference $kx^2 + lx$ into the product $k'x(k''x + 2c_1)$, where $k'k'' = k$ and $2c_1k' = l$. Then in the equation

$$a_1x^2 + b_1x + c_1^2 = \left[\left(\frac{k' + k''}{2} \right)x + c_1 \right]^2$$

the constant squares cancel, leaving an equation with one extraneous root, $x = 0$, and one rational root.

The need for rational roots dictates the method for solving "single" and "double" equations. The condition cited above is both necessary and sufficient for the method to work. Some equations that fail to satisfy the condition do have rational solutions. For example, the value $x = 1$ satisfies the "double" equation $6x^2 + 2x + 17 = y^2$, $3x^2 + 13 = z^2$, which fails to meet condition (2). Nonetheless, that equation cannot be solved using Diophantus' technique in any way that eliminates guesswork or accidental factorability of the difference. His procedures aim at rational solutions without guesswork and, under the conditions cited, they achieve them.

Even the compact account of "single" and "double" equations just given treats the technique more fully as a technique than does Diophantus in the *Arithmetic*; indeed, the account does little more than record symbolically the actual procedures that Diophantus follows when he encounters a "single" or "double" equation in his problems. Nonetheless, that those procedures can be stated in general form shows that they constituted for Diophantus a systematic problem-solving technique, as does the fact that Diophantus usually deliberately sets up his problems in such a way as to arrive at a "single" or "double" equation. He evidently considered the technique one of his most powerful mathematical tools. Systematic though it may be, the technique never becomes itself the subject of mathematical analysis by Diophantus; rather, it serves to solve problems. Hence, Diophantus never generalizes the technique for the purpose of examining how many solutions it might yield—one solution suffices for him—nor does he discuss what happens when the technique does not work. If it always works for Diophantus, it is because he does not treat problems for which it fails, or he treats them by some other means. As he does for all other methods of the *Arithmetic*, Diophantus merely sets an example for his readers to follow; he leaves it to them to extricate themselves if his example seems to lead nowhere.

B. *Diophantus and the Theory of Equations: Billy's (Fermat's)* Inventum novum

In preparing a commentary to Proposition VI,24 of the *Arithmetic,* Bachet sought in 1621 to present a complete theory of "double" equations.[48] His presentation, rife with distinctions into different cases, accomplished nothing more than the account just given. Bachet's rules conformed to Diophantus' example: they offered an arbitrary single rational solution for "single" equations and a definite single solution for "double" equations. But Bachet, like Viète before him, knew that Diophantus' techniques did not always fulfill their promise. In particular, the method of "double" equations occasionally led to a negative rational solution. Seeing no way of overcoming the negative root, Bachet followed Viète in declaring the problem unsolvable. An example here may prove instructive in several ways.

In 1643, seeking to induce Brûlart to join in the discussion of number theory being carried on with Frenicle, Fermat posed three problems to him: 1. to find a right triangle such that the hypotenuse is a square and the sum of the two perpendiculars, or indeed of all three sides, is also a square (cf. Observation 44 of the *Observations on Diophantus*); 2. to find four right triangles having the same area (cf. Observation 23); 3. to find a right triangle such that the area plus the square of the sum of the smaller sides is a square.[49] Two letters written by Fermat to Mersenne immediately after this challenge indicate that both Brûlart and Frenicle viewed the problems and the man who proposed them with anger and distrust. They accused Fermat of having posed impossible problems. Though Fermat admitted the extreme difficulty of the problems, he assured Mersenne, and through him his two correspondents, that the problems had solutions.[50] Eventually he supplied them; those to the first two problems are also found in the *Observations on Diophantus.*

---

[48] *Diophanti Alexandrini Arithmeticorum libri sex, et De numeris multangulis liber unus, Nunc primum graece et latine editi, atque absolutissimis commentariis illustrati Auctore Claudio Gaspare Bacheto, Meziriaco Sebusiano* (Paris, 1621), Commentary on Prop. VI, 24, pp. 435–440.

[49] Fermat to Brûlart, 31.V.1643, FO.II.258–260.

[50] Fermat to Mersenne, (VIII.1643), FO.II.260–262; p. 260: "Vous m'écrivez que la proposition de mes questions impossibles a faché et refroidi MM. de Saint-Martin et Frenicle, et que ç'a été le sujet qui m'a rompu leur communication. J'ai pourtant à leur représenter que tout ce qui paraît impossible d'abord ne l'est pas pourtant, et qu'il y a beaucoup de problèmes desquels, comme a dit autrefois Archimède, οὐκ εὐμέθοδα τῷ πρώτῳ φανέντα χρόνῳ τὴν ἐξεργασίαν λαμβάνοντι. Vous vous étonnerez bien davantage si je vous dis de plus que toutes les questions

What would have led Frenicle and Brûlart to think the problems impossible? One may fairly assume that, following the example of Diophantus, they would have attacked the problems in the same way as Fermat apparently did. That is, they would have sought to reduce them to "double" equations. Take the third example. Let the two smaller sides of the desired triangle be x and 1. Then the square of their sum, plus the area, set equal to a square would lead to the equation

$$x^2 + \frac{5}{2} x + 1 = y^2.$$

Moreover, since they are the sides of a right triangle, it is also necessary that

$$x^2 + 1 = z^2.$$

Hence, the solution of this "double" equation should yield a rational value for x and, from it, the desired triangle. But application of the Diophantine method, though it provides a rational solution, unfortunately provides a negative rational solution, and triangles do not have negative sides. It would seem likely that, following Bachet, Frenicle and Brûlart took the negative solution to attest to the unsolvability of the problem as posed. The first problem similarly leads to a negative result.

"I am led to believe," Frenicle had written in 1641 in another context, "that you have recently devised for yourself some sort of special analysis for probing the most hidden secrets of numbers."[51] In this case, Fermat's "special analysis" was less special than simply more consequential than Frenicle's. In considering Bachet's exegesis on the method of "double" equations, Fermat had discerned how to extend that method in order to determine any number of solutions for a "double" equation and, hence, how to find a positive solution where a negative one first appears. His discovery, perhaps first made in the early 1640s, reveals not only the secret of his total mastery over the field of rational right triangles but also the manner in which the model of the theory of equations influenced his thinking in the realm of number theory as it had in geometry.

Fermat's work on "double" equations is contained in capsule form in Observation 44 and in greater detail in the contents of letters he addressed to Père Jacques de Billy in the late 1650s. Billy abstracted these letters and published their technical content in the form of a treatise, the *New Invention of the*

---

que je leur ai proposées sont possibles, et que j'ai découvert leur solution. Ce n'est pas qu'elles ne soient très malaisées et que, pour les soudre, il ne faille faire quelque démarche au delà du Diophante et des Anciens et Modernes. Mais, comme toutes les inventions n'arrivent pas et ne se produisent pas en même temps, celle-ci est du nombre de celles dont la méthode n'est pas dans les Livres et que je puis attribuer au bonheur de ma recherche." Cf. also Fermat to Mersenne, 1.IX.1643, FO.II.262–264.

[51] Frenicle to Fermat, 2.VIII.1641, FO.II.226–232; p. 227.

Analytic Doctrine as part of the Diophantus edition prepared in 1670 by Samuel de Fermat.[52] Although Billy was responsible for the actual text of the *New Invention*, he fortunately let Fermat speak for himself with regard to the manner in which he had extended the ancient method and made it more effective. Billy quoted directly from Fermat's (now lost) *Appendix to Claude Gaspar Bachet's Dissertation on Double Equations in the Style of Diophantus*. "Here," Billy said, "are his very own words:"[53]

> That subtle and most learned analyst Bachet has successfully set forth quite a few modes and cases of the double equation in his commentary on question VI,2 of Diophantus, but he has not reaped the full harvest. For nothing prevents advancing and extending those questions that he confined to one or at most two solutions to an infinite [number of solutions]; indeed, it is clearly accomplished by a simple operation.
>
> Take the sixth mode, which Bachet rather prolixly explains on pages 439 and 440. All the cases enumerated by him admit by our method, which we will soon present, infinitely many solutions, which are derived from the first by degrees *ad infinitum* through iterated analysis.
>
> The method is this: seek the solution of the proposed question according to the common method, i.e. according to the method of Bachet or Diophantus; the value of the unknown number or root will emerge directly. Having done this, repeat the analysis, taking for the value of the new root to be sought one root plus the number of units in the prior root. The question will be reduced to a new double equation, in which the constant terms in each expression will be squares by virtue of the first solution. Hence, the difference of the equations will consist only of numbers and squares, which are neighboring species. Wherefore this new double equation is resolved following Diophantus and Bachet. By the same technique a third double equation is derived from the second, a fourth from the third, and so on *ad infinitum*.
>
> That neither Diophantus nor Bachet, nor for that matter even Viète, discerned this has been the greatest defect of analysis up to now. But this special technique of our invention proves itself in those questions in which the original analysis exhibits for the value of the unknown root a number bearing the minus sign, which therefore is understood to be less than nothing. Indeed, in this case our method has bearing not only on problems solved by double equations, but generally on all problems, as experience will make known.
>
> Therefore, one proceeds as follows: solve the proposed problem in accor-

---

[52] The letters themselves are no longer extant. A French translation of the *Inventum novum* is included in FO.III.325–398.

[53] Original Latin passage in FO.I.338–339 (taken from the *Inventum novum*). The last paragraph is contained *verbatim* in Observation 44.

dance with the common method. If, at the conclusion of the operation, the solution does not emerge, because the value of the number bears the minus sign and hence is understood to be less than nothing, I confidently assert that the mind should not despair (that, as Viète says, was the tediousness both of him himself and of the ancient analysts). Rather, let us try the problem again and set for the value of the unknown root 1*N*—the number that we found in the first operation to be equal, with the minus sign, to the unknown root. Without doubt, a new equation will emerge which represents the solution of the question in positive numbers.

The rest of the *New Invention* consists of Billy's explication of the details of Fermat's method or technique. Billy never treats the matter in general symbolic terms, but rather adduces a host of specific examples to support Fermat's claims and to attest to the effectiveness of the method. Insofar as Billy supplies the details of solution of several problems that Fermat cites either in the *Observations on Diophantus* or in his letters, the examples provide welcome information on Fermat's own problem-solving techniques.

In considering Fermat's extension of the method of "double" equations, one should bear in mind the claims made by both Fermat and Billy of the originality of Fermat's method. "Neither Diophantus nor Bachet, nor for that matter even Viète" had thought of such a method. Nor, apparently, had anyone else with whom Fermat exchanged ideas on number theory. The failure of Frenicle and Brûlart to solve the problems posed to them in the spring of 1643 would seem to indicate that they knew nothing of it. Moreover, Billy insists throughout the *Inventum novum* that the method is entirely and uniquely Fermat's.[54] Billy may not have been a great mathematician, but he was an active professor of mathematics at several Jesuit institutions and did keep abreast of the latest developments in mathematics.[55] If anyone else had seen what Fermat saw, Billy would have known about it.

The point of originality assumes such importance only because of the apparent triviality of Fermat's method when it is translated into general terms. It asserts nothing more than the following: let $x = m$ be a solution of the "double" equation $f(x) = y^2$, $g(x) = z^2$; i.e. $f(m) = s^2$, $g(m) = t^2$. Now transform the original system into a new one by substituting $u + m$ for x; i.e. $f(u + m) = F(u) = y^2$, $g(u + m) = G(u) = z^2$. Let $u = n$ satisfy the new system; i.e. $F(n) = p^2$, $G(n) = q^2$. Then $m + n$ will be another solution of the original "double" equation, since $F(n) = f(n + m) = p^2$ and $G(n) = g(n + m) = q^2$. Moreover, it will always be possible to solve the second "double" equation because, by virtue of the first solution and the fact that $f(x)$ and $g(x)$ are polynomials, the constant term in each of $F(u)$ and $G(u)$ will be a square, to wit, the constant in $F(u)$ will be s^2 and that in $G(u)$ will be

---

[54] See in particular Billy's prefatory remarks to the *Inventum novum*.
[55] See the article "Billy" in *Dictionary of Scientific Biography*, II, p. 131.

t^2, and Diophantus' "common" method of solution always works in that case. Hence, in the case of an initial negative solution, one derives a second solution greater than the first, and so on, until a positive solution is reached.

The anachronistic symbolism does not render Fermat's method qualitatively more immediate, nor is it the source of the apparent triviality of the method. The sophistication of his own theory of equations was more than equal to expressing the above argument in terms of general polynomials. Rather, Fermat's genius in comparison with his contemporaries lies precisely in the fact that he *used* the theory of equations to analyze Diophantus' method of "double" equations and thus could see what was an obvious corollary to that method. Fermat's "special sort of analysis" consisted of nothing more special than the theory of equations consequentially applied to an algebraic technique of solution. His extension of the method of "double" equations reflects those habits of thought so evident in his work in the realm of geometry. He was a problem-solver concerned less with special solutions than with methods of solution and questions of solvability. He studied those methods and investigated those questions by subjecting them to the analysis made possible by the theory of equations. To take such an approach to Diophantus' *Arithmetic* constituted, however, a break with the Diophantine tradition, whether that break was conscious or unconscious. Fermat's contemporaries had not made the break.

When did Fermat make it himself? Circumstances surrounding the method of "double" equations and the nature of the method itself suggest sometime in the early 1640s. Moreover, they point possibly to an interesting connection between the method of "double" equations and Fermat's final version of the method of maxima and minima. Writing to Carcavi sometime in 1644, Fermat announced the solution of a problem posed to him in 1641 by Frenicle, who had been unable to solve it: to find a triangle in which the square of the difference of the two perpendiculars exceeds twice the square of the smaller side by a square number.[56] Fermat noted that he too had had difficulty in solving the problem. Now that problem also appears in Observation 44, where it is grouped with others solved by Fermat's extension of the method of "double" equations; indeed, the first problem discussed in that context is the first of the problems posed to Frenicle and Brûlart in 1643. Hence, it would seem likely that the difficulties posed by Frenicle's problem were the same as those presented by the problems Fermat set out in return, to which by 1643 he had the solutions.

That is, in fact, the case. Taking the couple $(x + 1, 1)$ as the generator of the triangle sought, one arrives at the "single" equation

$$x^4 - 12x^2 - 16x - 4 = y^2,$$

[56] FO.II.265–266; cf. Frenicle to Fermat, 6.IX.1641, FO.II.232–242; p. 241.

which poses two sorts of difficulties, both discussed in detail in the *Inventum novum*.[57] First, in the case of "single" equations involving higher-order terms, how does one linearize the problem in order to guarantee a rational solution? Fermat gives the answer in Part III, "containing the procedure for obtaining solutions in indefinite number giving squared or cubed values to expressions wherein more than three terms of different degree occur." Given, for example, a general quartic equation (1) $ax^4 + bx^3 + cx^2 + dx + e = y^2$, one must choose a trinomial $rx^2 + sx + t = y$ such that three terms on either side mutually cancel, leaving only two terms in proximate powers of x. One can accomplish this only if either the lead term or the constant term, or both, are squares. For example, let $a = r^2$. One then aims at eliminating the first three terms of the expression, and the theory of equations points the way to that goal. For consider the expansion of $(rx^2 + sx + t)^2$, i.e.

$$r^2x^4 + 2rsx^3 + (2rt + s^2)x^2 + 2stx + t^2.$$

Since $a = r^2$, one sets $s = \dfrac{b}{2r}$ and $t = \dfrac{c - (b^2/4a)}{2r}$. The final reduced form of equation (1) will then contain only a term in x and a constant term, whence a rational solution emerges.[58]

As Frenicle's problem shows, however, the procedure for establishing the proper values of r, s, and t limits the possibility, present in the case of a quadratic "single" equation, of guaranteeing a positive solution. For, setting

$$x^4 - 12x^2 - 16x - 4 = (x^2 - 6)^2$$

(here choosing r, s, and t so as to eliminate the quartic and quadratic terms) yields the solution $x = -5/2$.[59] But that value, substituted into the generator $(x + 1, 1)$, produces a triangle with a negative side. Hence, the second difficulty arises: what does one do with a negative solution? Fermat's extension of "double" equations, as he himself notes, "has bearing not only on problems solved by double equations, but generally on all problems, as experience will make known." And here too it comes to the rescue of the analyst facing a negative, hence impossible, solution. Setting $x = u - 5/2$ and substituting it into the original quartic expression, Fermat derives a new "single" equation $u^4 - 10u^3 + \frac{51}{2}u^2 - \frac{37}{2}u + \frac{1}{16} = y^2$. Setting $y = u^2 - 5u - \frac{1}{4}$, he obtains the value $u = 21$, whence $x = \frac{37}{2}$, and the generator becomes $(\frac{39}{2}, 1)$ or $(39,2)$ to yield the triangle (1525, 1517, 156). The justification of the procedure is the same as that given earlier for "double" equations: if

[57] I.e. if the triangle has hypotenuse $x^2 + 2x + 2$ and sides $x^2 + 2x$ and $2x + 2$, then the conditions of the problem require that $(x^2 - 2)^2 - 2(2x + 2)^2$ be a square. The use of an indeterminate generator $(x + 1,1)$ is a trick taken from Diophantus, who uses it frequently.

[58] Compare this material with that of Part II of the *Tripartite Dissertation*, discussed above, Chap. III, p. 132ff.

[59] In this particuiar case, the absence of a cubic term in the left-hand expression dictates that $s = 0$.

312

$f(m) = r^2$ and $u = n$ is a solution of $F(u) = f(u + m) = y^2$, then $x = n + m$ is another solution of $f(x) = y^2$.

The difficulties, then, surrounding the problem that Frenicle posed to Fermat in 1641 and that Fermat admitted having had trouble solving are all removed by the method of derived solutions, a method Fermat had well in hand by the time he challenged Frenicle and Brûlart in the spring of 1643. Hence, it seems clear that Fermat carried out his equation-theoretical analysis of Diophantus' method of "double" equations sometime in the interval of little more than a year between 1641 and 1643. The timing gains added importance when Fermat's method of derived solutions is viewed in light of his demonstration of the method of maxima and minima sent to Brûlart in April 1643. For both results rest on the same basis in the theory of equations, to wit, that substitution of $x + m$ for x in any algebraic polynomial $P(x)$ results in an expression of the form $P(x) + P(m) + R(x,m)$.[60] The common basis of the two results gives insight, therefore, into the sorts of mathematical ideas on Fermat's mind during the early 1640s.[61] Indeed, Fermat himself draws the two methods together in a remark that accompanies his first assurance to Frenicle and Brûlart that the challenge problems of 1643 do have solutions. In his letter to Mersenne of August 1643, in which he responded to Frenicle's and Brûlart's accusations and offered a solution of the first problem, he concluded:[62]

> One should propose the other problems for solution to those who say (as M. de Carcavi has written me) that I found my method of maxima and minima by accident. For perhaps they will not think that I found these problems *à tâtons et par rencontre*. M. Hardy is one of them. I'd be obliged if you would soothe M. de St.-Martin [Brûlart] on my behalf. Perhaps for his sake

---

[60] The two uses of this general theorem differ greatly, of course, in the importance assigned the various elements of the expanded form. In the 1643 letter, Fermat is most interested in $R(x,m)$ expressed as a sum of terms in ascending order of the powers of m, and he ignores $P(m)$. For "single" and "double" equations, $R(x,m)$ is simply taken together with $P(x)$ to form a new polynomial expression which, when added to $P(m)$, constitutes a "single" or "double" expression to be set equal to a square; the fact that $P(m)$ is always a square guarantees that the new problem will have a solution. Hence, the method here lays emphasis on the nature of $P(m)$.

[61] Weil (*Number Theory*, 103–108) discusses Fermat's method of single and double equations in terms of the algebraic geometry of curves of genus 1: indeed, "it is still the foundation for the modern theory of such curves." (*ibid.*, p. 104). By the same token, he observes that "it is noteworthy that the simple idea of intersecting a cubic with the straight line through two previously known rational points does not seem to have occurred to Fermat, either in geometric or in algebraic garb" and he questions at another point whether Fermat was aware of a geometric interpretation, saying that "[f]or lack of evidence, this intriguing question must remain unanswered" (*ibid.*, p. 108). Nonetheless, he concludes, "[i]t will be noticed that in all of this there is no number theory; everything would be equally valid over any field, in particular the field of real numbers. In Fermat's eyes such work has therefore to be classified as 'geometry'; we would describe it as algebraic geometry, just as the work of Diophantus and that of Viète of which it is an extension; as to this, cf. Chap. I, §X, and Fermat's comments, quoted there [see below p. 338], from his challenge to the English mathematicians (*Fe*.II.335)."

[62] Fermat to Mersenne, ⟨VIII.1643⟩, FO.II.260–262; p. 261.

I will set down in writing my discoveries concerning Diophantus, where I have uncovered more than I could ever have promised myself. The method for solving the problems that I proposed to him is a sample of my work.

As shown in Chapter IV above, it was precisely the charges of operating by luck that provoked the letter to Brûlart on the method of maxima and minima.[63] The striking parallel between the proof contained in that letter and the method of derived solutions of "double" equations places their common origin in Fermat's continuing studies in the theory of equations during the early 1640s and lends further weight to the notion that Fermat's highly touted "method" consisted primarily of just that theory of equations.[64]

C. Primes and Squares

For all practical purposes, the methods of the *Inventum novum* severed Fermat's ties to the Diophantine tradition, and hence 1643 marks the watershed between traditions in number theory. Rational right triangles ceased to hold any interest for Fermat, since he now had the technique for solving any problem in that area, indeed for generating infinitely many solutions. The method of "single" and "double" equations represented Diophantus' most powerful and widely applicable algebraic method for rational arithmetic. Fermat had reduced it to a compact algorithm and thus moved many of the problems of the *Arithmetic* from the realm of challenge to that of tedium. He began in 1643, or even a bit earlier, to concentrate instead on questions in the *Arithmetic* that involved a purer form of arithmetic, one not subject to rational algebraic analysis. Having laid to rest an old tradition, he now sought to institute a new one, to wit, arithmetic as the doctrine of whole numbers, or number theory in the modern sense. His work on aliquot parts had shown the promise of such an arithmetic, but the problems studied there arose out of sources other than Diophantus. Now Fermat brought his vision to bear on the contents of the *Arithmetic*.

Indeed, he had already begun to do so. When in late 1640 he and Frenicle shifted the subject of their correspondence from aliquot parts and magic squares—Fermat could force from himself no real enthusiasm for the latter subject, nor did he contribute anything original to it—to the field of right triangles, Fermat stunned his correspondent with the complete solution to a

---

[63] As Dickson notes *(History of Theory of Numbers*, I, p. 34), Descartes also charged Fermat with operating *a posteriori* in the matter of aliquot parts. Unable to believe Fermat had the method he claimed to possess, Descartes accused him of designing formulas to fit already discovered results. Nothing in Fermat's papers suggests he knew of these charges, though his knowledge of them would explain the shift to number theory in countering Descartes' assertions of operating *à tâtons*. Cf. above, Chap. IV, §IV-V.

[64] It is, moreover, the close similarity between the method of "single" equations in the *Inventum novum* and the reduction of equations in Part II of the *Tripartite Dissertation* that suggests that the latter was an offshoot of the research of this same period 1641–1643.

problem arising out of Proposition III,19 of the *Arithmetic* and Bachet's commentary on it. Fermat had not only solved the immediate problem in all generality, but he had also linked it to several *diorismoi* in other propositions of Diophantus' work to form a totally unprecedented, unified theory of the decomposition of numbers into sums of squares. Proposition III,19 requires that one find four numbers such that, if any one of them is added to, or subtracted from, the square of their sum, the result is a square. Diophantus attacks the problem through a property of right triangles: if (a, b, c) is a right triangle with hypotenuse a, then $a^2 \pm 2bc$ [$= (b \pm c)^2$] is a square. This property enables him to reduce the original problem to that of finding four different triangles with a common hypotenuse. That is, he seeks four numbers $2b_ic_i$ such that $b_i^2 + c_i^2 = a^2$ and $\Sigma 2b_ic_i = a$. Then, by the property just stated, $a^2 \pm 2b_jc_j$ will be a square for $i = 1, 2, 3, 4$. The specific values of b_i and c_i are derived from the four triangles (a, d_i, f_i) with common hypotenuse a by setting $\Sigma 2d_ixf_ix = 2x^2\Sigma d_if_i = ax$ and solving for x; then $d_ix = b_i$ and $f_ix = c_i$.

To find the four triangles with a common hypotenuse, Diophantus begins by taking the two smallest triangles, (3,4,5) and (5,12,13), and generating something akin to their least common multiple; i.e. 5 times (5,12,13) = (25,60,65), and 13 times (3,4,5) = (39,52,65). He now has two triangles with a common hypotenuse 65. He then notes that $65 = 7^2 + 4^2 = 8^2 + 1^2$, adding cryptically "which is due to the fact that 65 is the product of 13 and 5, each of which numbers is the sum of two squares." Since $p^2 + q^2$ is the hypotenuse of a right triangle with sides $p^2 - q^2$, $2pq$, those two decompositions of 65 into the sum of two squares yield two further triangles with 65 as hypotenuse: (33,56,65) and (16,63,65).

Diophantus now has the four triangles he seeks and, as usual, he obtains them without a word of explanation. Again, predictably, it is Viète who provides the justification for Diophantus' procedure in the form of symbolic algebraic identities. He shows in Proposition 46 of the *Notae priores*[65] that, if $B^2 + D^2 = Z^2$ and $F^2 + G^2 = X^2$, then

$$(B^2 + D^2)(F^2 + G^2) = (BG \pm DF)^2 + (BF \mp DG)^2 = (ZX)^2$$

Elsewhere, he draws the other, more obvious corollary, to wit,[66]

$$(B^2 + D^2)(F^2 + G^2) = \left\{ \begin{array}{c} B^2X^2 + D^2X^2 \\ Z^2F^2 + Z^2G^2 \end{array} \right\} = (ZX)^2$$

Setting $B = 4$, $D = 3$ (whence $Z = 5$) and $F = 5$, $G = 12$ (whence $X = 13$), one immediately gets Diophantus' four triangles from Viète's iden-

---

[65] Viète, *Ad logisticen speciosam notae priores*, "Genesis triangulorum," *Opera mathematica* (ed. Schooten, Leiden, 1646), p. 33ff; p. 34.
[66] Cf. *ibid.*, Prop. XIIff.

tities. Indeed, the identities enable one systematically to determine an infinite number of quadruples of right triangles with a common hypotenuse.

In commenting on Proposition III,19, however, Bachet was interested less in the complete solution of Diophantus' particular problem via Viète's identities than in a more general question suggested by the problem.[67] Noting that any number, itself the sum of two squares in two different ways, will be the hypotenuse of four different right triangles (its square will be the sum of two squares in four different ways), he then asked the question to which Fermat later responded: find a number that is the sum of two squares in a given number of different ways. Though Bachet could offer some specific partial answers, the complete solution eluded him.

Fermat recorded the major portion of the complete solution in Observation 7, a copy of which he sent to Mersenne on Christmas Day 1640. He began with the fundamentals:[68]

A prime number, which exceeds a multiple of four by unity, is only once [i.e. in one way] the hypotenuse of a right triangle, its square twice, its cube three times, its quadratoquadrate four times, and so on *in infinitum*.

The same prime number and its square are composed once of two squares, its cube and quadratoquadrate twice, its quadratocube and cubocube three times, and so on *in infinitum*.

If a prime number composed of two squares is multiplied by another prime number also composed of two squares, the product is twice composed of two squares; if it is multiplied by the square of the same prime, the product is composed of two squares in three ways; if it is multiplied by the cube of the same prime, the product is composed of two squares in four ways, and so on *in infinitum*.

For the present discussion, the third paragraph of this passage is the most important. In attacking Bachet's problem, Fermat took the characteristic step of reducing the question of the various decompositions of a given number into two squares to that of the various decompositions of its prime factors. Leaving aside for the moment the result that opens the first paragraph, let us look first at Fermat's use of Viète's identities in achieving that reduction.[69]

Given that a prime p decomposes uniquely into the sum $a^2 + b^2$, the identities lead to the following decompositions, which Fermat takes in each case to be the only ones possible but does not explicitly prove to be so:

$$p^2 = (a^2 + b^2)^2 = (2ab)^2 + (a^2 - b^2)^2$$

---

[67] See FO.I.293, n. 1.

[68] FO.I.293–297; cf. Fermat to Mersenne, 25.XII.1640, FO.II.213–214.

[69] The presentation to follow derives from, but does not directly reproduce, Fermat's discussion of the subject in his letter to Frenicle (15.VI.1641), FO.II.221–223.

$$p^3 = (a^2 + b^2)\,[(2ab)^2 + (a^2 - b^2)^2]$$

$$= [a(a^2 - b^2) + b(2ab)]^2 + [a(2ab) - b(a^2 - b^2)]^2$$

$$= [a(a^2 - b^2) - b(2ab)]^2 + [a(2ab) + b(a^2 - b^2)]^2$$

$$p^4 = (a^2 + b^2)^2\,[(2ab)^2 + (a^2 - b^2)^2]$$

$$= (a^2 + b^2)^2\,(2ab)^2 + (a^2 + b^2)^2(a^2 - b^2)^2$$

$$= [(2ab)^2 + (a^2 - b^2)^2]^2$$

$$= [2(2ab)(a^2 - b^2)]^2 + [(2ab)^2 - (a^2 - b^2)^2]^2.$$

Hence, from the decomposition of p into two squares follows the decomposition of p^2 into two squares, by use of the algebraic identity that underlies the rule for generating right triangles. That same identity, combined with Viète's identities, produces the twofold decomposition of p^3 and p^4 into the sum of two squares. Further application of these identities yields the remaining parts of Fermat's assertion in the second paragraph above.

Granted the uniqueness of the above decompositions, the statement of the first paragraph follows directly. If p is the hypotenuse of a right triangle, then p must satisfy the equation $p^2 = x^2 + y^2$ for some integral x and y. By the above, there exists but one pair of solutions for x and y, namely $2ab$ and $a^2 - b^2$ Hence, p is the hypotenuse of only one triangle. If p^n serves as hypotenuse, the equation becomes $p^{2n} = x^2 + y^2$, and by the above there are n such pairs x, y. Hence p^n is the hypotenuse of n right triangles.

For the statement of the third paragraph, consider two primes p and q that each decomposes uniquely into the sum of two squares; i.e. let $p = a^2 + b^2$ and $q = c^2 + d^2$ Then, by Viète's identities,

$$(a^2 + b^2)(c^2 + d^2) = (ad \pm bc)^2 + (ac \mp bd)^2$$

Hence, the product pq splits into two squares in two different ways. Moreover,

$$pq^2 = (a^2 + b^2)(c^2 + d^2)^2 = \begin{cases} a^2(c^2 + d^2)^2 + b^2(c^2 + d^2)^2 \\ (a^2 + b^2)[(2cd)^2 + (c^2 - d^2)^2] \end{cases}$$

Since, following the model for the decomposition of pq, the expression $(a^2 + b^2)[(2cd)^2 + (c^2 - d^2)^2]$ splits two ways, the product pq^2 splits a total of three ways. Again, the remainder of Fermat's statement follows by iteration of the identities.

The solution of Bachet's problem reduces, therefore, to the determination of which primes uniquely split into the sum of two squares. For that Fermat turns to the theorem announced at the beginning of the observation: if p (prime) is of the form $4k + 1$ (i.e. $p \equiv 1 \pmod 4$), then p is uniquely the sum of two integral squares. The theorem is the result of a line of research essentially unrelated to the immediate problem. Before turning to it, therefore, let us take it as an established proposition and pursue Fermat's application of it to Bachet's problem.

Fermat divides the problem into three parts: 1. to determine in how many ways a given number is the hypotenuse of a right triangle; 2. to find a number that is the hypotenuse of a given number of triangles; and 3. to find a number that is composed of two squares in a given number of ways. The solution of the first subproblem leads to the solution of the second two as corollaries; the third subproblem is, of course, Bachet's problem. Fermat continues in his observation:

> From this it is easy to determine how many times a given number is the hypotenuse of a right triangle. Take all the primes exceeding by unity a multiple of four that measure the given number; e.g. 5,13,17. If powers of the said primes measure the said number, arrange them together with the rest in place of their roots; e.g. let them measure the given number [as follows]: 5 by the cube, 13 by the square, and 17 by the simple root. Take the exponents of all the divisors; i.e. the exponent of the number 5 is 3 owing to the cube, the exponent of the number 13 is 2 owing to the square, and of the number 17 is unity only. Order, then, as you wish, all said exponents; as, if you will, 3, 2, 1. Multiply the first by the second twice and, adding the product to the sum of the first and second, the result is 17. Then multiply 17 by the third twice and, adding the product to the sum of 17 and the third, the result is 52. Therefore, the given number will be the hypotenuse of 52 right triangles. The method is not dissimilar for any number of divisors and their powers. The remaining prime numbers [dividing the given number], which do not exceed a multiple of four by unity, add or detract nothing from the question, nor do their powers.

Fermat proceeds here by reducing the question of the number of ways in which a given number can be the hypotenuse of a right triangle to that of the number of ways in which its square can be split into two squares. The answer to the reduced question in turn depends on the number of prime divisors of the given number that are of the form $4k + 1$. For Fermat not only knows that such primes uniquely split into two squares and that their powers, as well as the products of two or more such primes, split in several ways, he also knows that primes of the form $4k - 1$ (or $4k + 3$) do not split into two squares. He announced this theorem to Roberval in 1640.[70] Hence, the number of ways in which the square of a given number can be split into two squares depends only on its prime factors of the form $4k + 1$.

Let a given number $N = N_1 p^a q^b \ldots s^c$, where $p, q, \ldots, s$ are all

[70] See below, p. 320ff.

primes of the form $4k + 1$, and N_1 is the product of all prime factors of N of the form $4k - 1$. Then $N^2 = N_1^2 p^{2a} q^{2b} \ldots s^{2c}$. Fermat now faces essentially a problem of combinatorics. Given the basic rules for the number of ways in which the powers of primes $4k + 1$ and their products split into two squares, what is the general formula for the number of decompositions for the product $p^{2a} q^{2b} \ldots s^{2c}$? With no hint of a derivation, Fermat gives his answer: $\frac{1}{2}[(2a + 1)(2b + 1) \ldots (2c + 1) - 1]$. Though given in a different form in the passage cited above,[71] this formula underlies the answer to the two corollaries that follow. For, to find a number that is the hypotenuse of a given number of triangles, Fermat sets the following verbal pattern on the basis of the given number 7:

> Double the given number 7; 14 results. Add unity; 15 results. Take all primes that measure 15; they are 3 and 5. Having subtracted unity from each, take half the remainder; 1 and 2 result. Look for as many different primes as there are here numbers, namely two, and multiply them by each other according to the exponents 1 and 2, i.e. the one by the square of the other. In this case, the question will be satisfied, provided that the primes that you take exceed a multiple of four by unity. From this it follows that one can easily find the smallest number that is an hypotenuse as many times as one wishes.

One can generalize this pattern immediately. Take any number n and set $\frac{1}{2}[(2x + 1)(2y + 1) \ldots (2z + 1) - 1] = n$; the number of unknowns $x, y, \ldots, z$ remains for the moment indeterminate. Simplifying the equation, one obtains

$$(2x + 1)(2y + 1) \ldots (2z + 1) = 2n + 1.$$

Since $2n + 1$ is an odd number, it will be the product of a variable number of odd factors. For uniformity's sake, Fermat arbitrarily fixes that number by factoring $2n + 1$ into its prime factors, all of which are odd and some of which may be repeated, i.e. $2n + 1 = ab \ldots . c$. The number, say m, of these prime factors will be the number of unknowns $x, y, \ldots, z$. Setting $2x + 1 = a$, $2y + 1 = b, \ldots, 2z + 1 = c$, and solving for the unknowns, one gets $x = \dfrac{a - 1}{2}$, $y = \dfrac{b - 1}{2}$, etc. Hence, if one takes m prime numbers of the form $4k + 1$, say $p, q, \ldots, s$, the product $N = p^x q^y \ldots s^z$ will be the hypotenuse of n different triangles. Clearly, if $p, q, \ldots, s$ are the first m such primes, i.e. 5,13,17, etc., the number N will be the least number that is the hypotenuse of n triangles, provided, of course, that $x, y, \ldots, z$ are arranged in descending order.

[71] That is, Fermat presents the formula as an algorithm, $2[2ab + a + b]c + [2ab + a + b + c] \cdots$

In the second corollary and final solution to Bachet's problem, to find a number that is composed of two squares in a given number of ways, Fermat works again from a general formula for the number of ways in which the product of a series of primes of the form $4k + 1$ splits into two squares. If $N = p^a q^b \ldots s^c$ then N may be represented as the sum of two squares in $\frac{1}{2}[(a + 1)(b + 1) \ldots (c + 1)]$ ways. The solution to the problem then follows precisely the same pattern as that of the first corollary, including the minimization of N.

Although the details of Fermat's solution to Bachet's question, which Frenicle had also posed to him in 1640, attest to Fermat's command of combinatorics (and hence of the technical facility that underlay his solutions of Pascal's problems in probability in 1654), they provide no insight into the origins of the pregnant theorem that makes the combinatorial superstructure possible. They do not, that is, explain how Fermat knew that any prime number of the form $4k + 1$ splits uniquely into two squares.

Fermat himself left no explicit derivation of that theorem, but he did leave traces of his path to it. To see these traces, it is necessary to separate heuristic from demonstration. That is, how Fermat found the theorem may have been quite different from how he managed to prove it. His letter to Carcavi of August 1659, the famous "Relation" to be discussed in greater detail below, suggests that the distinction corresponds to historical fact. In speaking of his method of infinite descent (see below, Section V), Fermat tells Carcavi: "For a long time I was unable to apply my method to affirmative propositions, because the trick and slant for arriving at them is much more difficult than that which I use for negative propositions. So that, when I had to demonstrate that every prime number that exceeds a multiple of four by unity is composed of two squares, I found myself in a pretty fix."[72]

How Fermat adjusted his method for that theorem is less important for the moment than his testimony that he already had the theorem without being able to prove it. Hence, in looking for the source of his knowledge, it will be of little help to try to reconstruct his proof of the theorem by infinite descent.

In the search for an alternative reconstruction, the order in which Fermat announced two allied results may provide the most important clue. Before announcing the contents of Observation 7 to Mersenne in December 1640, Fermat had already informed Roberval in August that he had cleared up a related difficulty in the *Arithmetic*:[73]

But here is what I have since discovered on the subject of Proposition 12 of the fifth book of Diophantus, wherein I have supplied what Bachet admits not having known and at the same time restored Diophantus' corrupted text. It would be too long to derive for you [in detail]. It suffices that I give you

[72] FO.II.432.
[73] FO.II.203–204.

my proposition and that I first remind you that I demonstrated earlier that a number less by unity than a multiple of four is neither a square nor composed of two squares, not in integers nor in fractions.[74] At that time, I rested there, even though there are many numbers greater by unity than a multiple of four that nevertheless are neither squares nor composed of two squares, as 21,33,37, etc. That is what caused Bachet to say of the proposed division of 21 into two squares: "but this, I believe, is impossible, since it is neither a square nor by its nature composed of two squares," where the words "I believe" clearly denote that he did not know the proof of this impossibility, which I have finally found and brought together generally in the following proposition:

> If a given number is divided by the greatest square that measures it, and the quotient is measured by a prime number less by unity than a multiple of four, the given number is neither a square nor composed of two squares, not in integers nor in fractions.

Example: let the given number be 84. The greatest square that measures it is 4, the quotient 21, which is measured by 3 or indeed by 7, [both of them] less by unity than a multiple of 4. I say that 84 is neither a square nor composed of two squares, neither in integers nor in fractions.

Let the given be 77. The greatest square than measures it is unity; the quotient 77, which is here the same as the given number, is measured by 11 or by 7, [both] less by unity than a multiple of the quaternary. I say that 77 is neither a square nor composed of two squares, neither in integers nor in fractions.

I assure you frankly that I have found nothing in [the theory of] numbers that has pleased me as much as the demonstration of this proposition, and I would be pleased if you would make the effort to find it, even if only to learn whether I think more highly of my finding than it is worth.

The problem in Diophantus' *Arithmetic* to which Fermat is referring here comes as no stranger to someone studying Fermat's mathematics. It is (in the modern edition) Proposition V,9, the problem from which Fermat borrowed the term "adequality."[75] There, to split unity into two parts, x and $1 - x$, such that if the same number n is added to each part the result is a square, Diophantus reduces the problem to one of splitting the number $2n + 1$ into

---

[74] There is no trace of this earlier statement in Fermat's extant writings, though on 22.V.1638 Mersenne informed Descartes that Fermat had a proof of the theorem, as well as one of the theorem that any number of the form $8k - 1$ is composed of four squares only (cf. Fermat to Mersenne (IX or X.1636), FO.II.66, where the theorem is stated and a proof promised but not given); cf. FO.II.203, n. 3. One must assume Fermat told Mersenne of the result in some letter now lost.

[75] See above, Chap. IV, p. 163ff.

two squares differing by less than 1. The analysis leads to the requirement that $2n + 1$ be decomposable into two integral squares. Hence, Diophantus adds a *diorismos* to the statement of the problem, a *diorismos* which became garbled with the passage of time and before which Bachet stood helpless, though not without words. Fermat's restoration of the *diorismos* rests on the theorem stated above: n must be a number such that if $2n + 1$ is divided by its largest square factor the quotient contains no prime factors of the form $4k - 1$. Why? Because no number of the form $4k - 1$ can be decomposed into two squares, either integral or fractional.

The last theorem, cited by Fermat himself in the letter quoted above and also in a letter from Mersenne to Descartes in 1638, in which the theorem and a proof of it are credited to Fermat, has a fairly direct proof in line with Fermat's habits of thought. Certainly no number of the form $4k - 1$ can be a square, since all squares are either even (and, if so, divisible by 4) or odd; if odd, they are squares of odd numbers, and $(2k \pm 1)^2 = 4k' + 1$. Suppose, however, some number of the form $4k - 1 = x^2 + y^2$ for a pair of integers x, y. Since $4k - 1$ is odd, one of x^2 or y^2 must be odd, but not both. Let it be y^2. Then $4k = x^2 + (y^2 + 1)$, or $k = \left(\dfrac{x}{2} \right)^2 + \left(\dfrac{y^2 + 1}{4} \right)$. Now, since x^2 is even, so too is x, whence $\left(\dfrac{x}{2} \right)^2$ is an integer. Therefore $\dfrac{y^2 + 1}{4}$ must also be an integer; i.e. $y^2 + 1 = 4k'$ for some integer k'. But then $y^2 = 4k' - 1$, and it has already been shown that no square can be of that form. Hence, a contradiction arises, and the original assumption is disproved.

Once, therefore, one begins to consider decomposition problems in terms of remainders after division by 4, a proof of Fermat's central theorem concerning numbers of the form $4k - 1$ follows directly, at least for integral squares. The extension of the proof for the case of fractional decomposition is equally direct and leads immediately to Fermat's emendation of Diophantus. If $4k - 1 = \left(\dfrac{x}{z} \right)^2 + \left(\dfrac{y}{z} \right)^2$, then $z^2(4k - 1) = x^2 + y^2$. Now, clearly, if the fractions x/z and y/z are in reduced form, not all of x, y, z can be even. Suppose z is odd. Then z^2 is of the form $4k' + 1$ and, consequently, $z^2(4k - 1) = (4k' + 1)(4k - 1) = 4k'' - 1$, which cannot be decomposed into two integral squares, x^2 and y^2. Suppose z is even. Then z^2 is of the form $4k'$, and $z^2(4k - 1) = 4k'(4k - 1) = 4k''$. Hence, the sum $x^2 + y^2$ is even. But, since z is even, x and y cannot both be even; therefore, they are both odd. But, then, x^2 is of the form $4m + 1$ and y^2 of the form $4n + 1$, and their sum is of the form $4r + 2$. Yet, it has just been shown that their sum is of the form $4k''$. Therefore, the assumption of decomposition of

$4k - 1$ into two fractional squares also leads to a contradiction that invalidates it.

Not only does the above proof employ concepts fully at Fermat's command, it also has a feature that ties it directly to the breakthrough Fermat announced to Roberval in the letter quoted above, and hence a feature that lends it weight as a possible reconstruction of Fermat's own proof. Note the corollary to the last part of the proof: no number of the form $n^2(4k - 1)$ can be decomposed into two squares. That is precisely Fermat's "general proposition" in the letter to Roberval. To get the full content of Fermat's central theorem concerning the non-decomposability of numbers of the form $4k - 1$ into two squares, it remains only to show that no square-free product of primes of that form can be so decomposed. Again, a passage in the letter to Roberval reinforces the reconstructed derivation suggested above. First, it follows immediately from that derivation that no product of an odd number of primes of the form $4k - 1$ can be decomposed into two squares, since the product itself will again be of the form $4k - 1$. Consider, then, a product $(4k - 1)(4k' - 1)$ of two primes and suppose it could be written in the form $x^2 + y^2$ in integers. Then it would follow immediately that both $4k - 1$ and $4k' - 1$ are divisors of $x^2 + y^2$. Note then the statement that immediately follows the passage quoted above in the letter to Roberval: "Finally, I have demonstrated this proposition, which serves in finding prime numbers: If a number is composed of two mutually prime squares, I say that it cannot be divided by any prime number less by unity than a multiple of four."[76] Clearly, x^2 and y^2 are mutually prime, since any common factor, which would have to be a square, would also have to be one of $4k - 1$ or $4k' - 1$ by the prime factorization theorem. But, as has already been shown, neither prime can be a square.

Completion of the proof of the theorem concerning primes of the form $4k - 1$ rests, then, on the theorem just quoted from the letter to Roberval. How Fermat arrived at that theorem, however, is totally unclear. It may have emerged as a byproduct of a theorem that he announced to Mersenne in 1636. In a letter dated 2 September, Fermat reported that he was close to a proof of the proposition that no number composed of three squares can be decomposed into two squares, either integers or fractions.[77] It would follow directly from that theorem that, since both $4k - 1$ and $4k' - 1$ require at least three squares for their decomposition, their product would also require at least three and would be irreducible to two. It is also possible, however, that Fermat had at his disposal a more powerful theorem, with which he has been credited but

---

[76] FO.II.204.

[77] FO.II.58: "Or qu'un nombre, composé de trois quarrés seulement en nombres entiers, ne puisse jamais être divisé en deux quarrés, non pas même en fractions, personne ne l'a jamais encore demontré et c'est à quoi je travaille et crois que j'en viendrai à bout."

323

which does not appear explicitly in any of his works, to wit, that the divisors of any sum of two mutually prime squares must themselves be the sum of two squares.[78] In that case, both $4k - 1$ and $4k' - 1$ would have to be decomposable into two squares; but it has already been shown that such a decomposition is impossible.

The extant sources offer no basis for choosing between these alternatives, nor do they suggest the manner in which either of the alternative theorems might have been derived or proved. Fermat's reticence blocks the path to any further insight. Nonetheless, it has so far become clear that he knew that no square-free number that contains a prime factor of the form $4k - 1$ can be decomposed into two squares. Hence, he may well have reasoned, the fact that certain odd numbers, which must be of the form $4k + 1$, can be so decomposed means that it is uniquely the presence in those numbers of odd prime factors of the form $4k + 1$ that make the decomposition possible. That is, if a number can be decomposed into two squares, it is only because its non-square, odd prime factors are all of the form $4k + 1$. The next step was clear: Fermat set up as a theorem the assertion that all prime numbers of the form $4k + 1$ can be uniquely decomposed into two squares. If the decomposition were not unique, then, by the identities governing multiple decomposition, the number would have a factor other than itself or 1 and would therefore not be prime.[79] At that point, Fermat began the search for a proof of the theorem, a proof that by his own testimony eluded him for some time. The theorem became a staple of his efforts to enlist others in pursuit of number theory, but he gave no hint of a proof until describing the method of infinite descent to Huygens in 1659.

D. *Pell's Equation: A Case of Mistaken Identity*

The report by Mersenne to Descartes in 1638 that Fermat had proved the theorem concerning the decomposition of primes of the form $4k - 1$, Fermat's own letter to Roberval extending the range of that theorem, and his letter to Frenicle in 1641 containing the theorem about primes of the form $4k + 1$ and its implications date the shift in Fermat's thinking about the nature of arithmetic. At the same time that he came upon his method of derived solutions for "single" and "double" equations, a method that all but

---

[78] That is, any divisor of a number of the form $a^2 + b^2$ must itself be of that form. Fermat's lemma concerning the divisibility of the quadratic form $a^2 + 2b^2$ would suggest he knew the theorem, and in the "Liste des principales inventions numériques de Fermat" (FO.IV.231–237), A. Aubry credited Fermat with it (p. 232). But Aubry did not, as he did with every other theorem, cite an FO reference for it. In fact, it appears nowhere in FO.

[79] Fermat nowhere explicitly justifies his assertion of the uniqueness of the decomposition. It is most likely, however, that he was simply relying on the uniqueness and biconditionality of Viète's identities. That is, given two different pairs of squares having the same sum, Fermat probably assumed he could reverse Viète's generating procedures to find two pairs whose product yielded the twofold decomposition.

completed the traditional doctrine of rational arithmetic, he was making a breakthrough to a new arithmetic of integers. Having complained to Mersenne in the letter of 2 September 1636 about the lack of "principles" for arriving at such theorems as the irreducibility of a sum of three squares, he had now established some of those principles and had begun to redeem the promise made in that same letter: "If I can extend at this point the limits of arithmetic, you cannot believe the marvelous propositions we will derive from it." Unfortunately, Fermat's correspondence breaks off early in 1644 for a period of ten years. Nonetheless, the problems Fermat was engaged in just prior to the break and the results that he announced upon resumption of correspondence, beginning with the exchange with Pascal, indicate the substance of his research in the interim and reveal the flow of marvelous theorems that followed his extension of the limits of arithmetic.

Fermat had included in his letter to Mersenne and Frenicle two problems that ostensibly pertained to the doctrine of rational right triangles but in fact had a far wider range of applicability: first, given a number N, to determine the number of different triangles of which it is one of the smaller sides; second, given a number N, to determine the number of triangles of which it is the sum of the two smaller sides.[80] The solution of the first problem is contained in a fragment of a letter sent to Mersenne or Frenicle sometime in 1643.[81] There Fermat announced:

> Every non-square odd number differs from a square by a square, or is the difference of two squares, as many times [i.e. in as many different ways] as it is composed of two numbers. If the squares are mutually prime, the numbers composing them will also be prime. But, if the squares have a common divisor, the number in question will also be divisible by the same common divisor, and the numbers composing the squares will be divisible by the root of that common divisor.

As proof, Fermat offered only an example, but it suffices to convey his reasoning. Let $N = n_1 n_2$ be an odd number. Then both factors are also odd, and

$$\left[\frac{n_1 + n_2}{2} \right]^2 - \left[\frac{n_1 - n_2}{2} \right]^2 = n_1 n_2 = N,$$

where both squares are integers. As Fermat noted, the proof itself shows how one determines the squares in question. Moreover, since the requirement that

---

[80] Cf. FO.II.216 (given a number, to determine how many times it is the difference of two numbers, the product of which is a square) and FO.II.223, 226. In a right triangle ($p^2 + q^2$, $p^2 - q^2$, $2pq$), one of the smaller sides is always the difference of two squares. Hence, the number of different triangles for which a given number may serve as one of the smaller sides is equal to the number of different ways in which the given number can be expressed as the difference of two squares.

[81] FO.II.256–258; on the source and status of the fragment, see FO.II.256, n. 2.

N be odd is due to the necessity that the sum and difference of its factors be even, "one can say just about the same thing for evenly even numbers, except for 4, with some slight modification." That is, the theorem also holds for N of the form $2n_1 \cdot 2n_2$, where $n_1 + n_2 > 2$.

Clearly, the number of different pairs of squares of which N represents the difference depends on the number of different pairs of odd factors (or pairs of even factors) of which N is composed. An earlier letter of the same year, containing the numerical solution to the same problem in slightly different guise, illustrates the same use of combinatorics to determine that number of different pairs of factors as in Fermat's treatment of primes of the form $4k + 1$.[82] The number 1,803,601,800 is the difference of two squares in 243 different ways. Fermat does not say why, but the above theorem gives the answer. $1,803,601,800 = 2^3 3^2 5^2 7^2 11^2 13^2$. Since each of the squared odd primes has three possible factors, including 1, the product of odd primes can be written as the product of two odd numbers in $3^5 = 243$ ways. Since both factors must be even, i.e. only those products of the form $2n_1 \cdot 2n_2$ allow the division into squares, 2^3 can be factored in only one way, $2 \cdot 2^2$ Hence, the total number of ways of factoring the number remains 243.

The fragment of the letter containing this theorem also draws explicitly the wider implications of Fermat's result. He notes, "That posited, let a number, e.g. 2,027,651,281, be given me and let it be asked whether it is prime or composite, and, in the latter case, of what numbers it is composed." The specific example given lends itself directly to symbolic generalization:[83] Given N, to find a factorization $n_1 \cdot n_2$. If such a factorization exists, then the factors yield an integral solution of the equation $x^2 - y^2 = N$. Rewriting that equation in the form $x^2 - N = y^2$, one begins the search for factors by determining the integer m such that $m^2 < N < (m + 1)^2$. Is $(m + 1)^2 - N = y^2$ for some integer y? Often the last digits of the remainder will show immediately that it is not; "there remains 46619, which is not a square, because no square ends in 19." Add then to the left-hand side $2m + 3$ to yield $(m + 2)^2 - N$. Is this a square? If not, add $2m + 5$ to make the left-hand side $(m + 3)^2 - N$. If for some k, $(m + k)^2 - N = n^2$, then $N = (m + k + n)(m + k - n)$. Clearly, since $N = N \cdot 1$, the process cannot go on indefinitely; it will eventually lead to some value for k such that $(m + k)^2 - N = (m + k - 1)^2$. If, however, it reaches that point without discovering a square remainder, then N has no other factors than N and 1, in which case it is prime. In any case, the total number of test values for k will not exceed exceed $\dfrac{N - 2m + 1}{2}$.[84] Hence the factorization technique also

[82] Fermat to Mersenne ⟨27.1.1643⟩, FO.II.250.

[83] Cf. the description of this technique given by Oystein Ore, *Number Theory and Its History* (New York, 1948), pp. 154–158.

[84] Suppose N is prime and hence can be factored only in the trivial form $N \cdot 1$. Then $m + k - n = 1$, whence $n = m + k - 1$. Since, in that case, $m + k + n = N$, one has $2m + 2k - 1 = N$, or $k = (N - 2m + 1)/2$.

has value as a test for primality. That value as compared with the traditional "sieve" lies not in a decrease in the number of prime factors to be tested, but in the fact that it does not require that one know all the primes less than the square root of the given number. Moreover, the ease with which many remainders can be excluded as possible squares simply by examining the final two digits makes the method less cumbersome than direct testing of prime factors.

The second problem with origins in the doctrine of rational right triangles but implications that extend beyond it was first posed by Fermat to Frenicle in that same important letter of 1641. Modeling the problem on what had gone before, Fermat asked Frenicle, "Being given a number, find how many times it can be the sum of the two small sides of a right triangle."[85] An analysis of the problem suggests what Fermat had in mind. If (x, y) is the generator of the triangle sought, then the sum of the two smaller sides is $x^2 - y^2 + 2xy = 2x^2 - (x - y)^2 = (x + y)^2 - 2y^2$. The problem therefore requires the solution of the equation $2u^2 - v^2 = \pm N$ in integers, and subsequent developments indicate that Fermat expected of Frenicle a general examination of the solvability of that equation.

As usual, however, Frenicle merely solved the immediate problem without pursuing it further, if indeed he even saw its implications. In a letter to Fermat dated 2 August 1641, Frenicle noted first that every prime number of the form $8k + 1$ was the sum of the two smaller sides of a right triangle, and conversely. Playing then on Fermat's solution of the problem of hypotenuses, he added, "On this basis, one must do the same thing with these numbers that one would do with prime numbers of the form $4k + 1$ to find what is required by the problem if hypotenuses were required instead of the sum of the two smaller sides. It would be superfluous to derive this at any length; I am speaking to one who knows (*intelligenti loquor*)."[86] Frenicle should have written, "I am speaking to one who knows better," for he had again missed the point. Some years later, in the midst of the controversy of 1657–1658, Fermat set down on paper what he had been seeking from Frenicle.[87]

To be sure, Frenicle was correct in suggesting that, if two primes p_1 and p_2 can each be expressed as the difference of a square and the double of a square, then their product can also be expressed in the same form, but in two different ways. But the result was trivial in light of the fact that, for a prime p of the form $8k + 1$, the equation $2u^2 - v^2 = \pm p$ has an infinite number of solution pairs u, v. Given any particular solution, one can determine the least solution by a reduction process based on the identity

---

[85] FO.II.226.

[86] FO.II.231.

[87] "Pour servir de supplément à l'Écrit Latin de Monsieur Frenicle. Par M. F⟨ermat⟩." This appendix to some editions of Frenicle's *Solutio duorum problematum* . . . (see below, n. 113) was first brought to light by J. E. Hofmann in his "Neues über Fermats zahlentheoretische Herausforderungen von 1657" (see above, n. 3), pp. 41–44. It represents the most important addition to Fermat's extant works since DeWaard's publication of the 1922 *Supplement* to FO.

$$2(3u - 2v)^2 - (4u - 3v)^2 = 2u^2 - v^2$$

and, given the least solution, one can generate an infinite family of solutions from it on the basis of the identity

$$2(3u + 2v)^2 - (4u + 3v)^2 = 2u^2 - v^2.$$

In retrospect, it seems clear that Fermat was not looking for another exercise in combinatorics. He was probing whether Frenicle could solve equations of the form $x^2 - py^2 = q$ in integers for given p and q or could show which cases of the equation were unsolvable. That general problem, and the particular case of $x^2 - py^2 = \pm 1$, apparently had come to occupy the forefront of Fermat's attention by late 1643. In that year, for example, he spoke in his correspondence of only two sorts of problems, those involving the method of "double" equations and those involving solutions of cases of the equation $x^2 - py^2 = q$. Of the latter, he specifically considered the solutions of $x^2 - py^2 = q$, of $2u^2 - v^2 = \pm 1$, and of $2u^2 - v^2 = 7$. If the second of these found both its origin and its complete solution in the classical Greek doctrine of "side and diagonal numbers," its extension to a whole class of equations and the attempt to treat that class by methods designed to produce only integral solutions was Fermat's doing alone.

Yet, it is hard to know for certain just what methods Fermat had devised by then or ten years later when, resuming his scientific correspondence, he alluded to the proposition in letters to Pascal. Several years later, having set the particular equations of the form $u^2 - Nx^2 = \pm 1$ for non-square N as a challenge problem, he accepted the solutions by Wallis and Brouncker but criticized their authors for not having provided a "démonstration générale" that the method implicit in their solutions would always work if the equation had a solution and indeed set the conditions for its having one. Fermat claimed to have that demonstration, but he left no record of it, other than to tell Huygens in 1659 that he proved it by the method of infinite descent, "appliquée d'une manière toute particulière."[88] In "Fermat et l'équation de Pell," Weil offers a reconstruction suggested by Bachet's integral solution of the linear equation $Ax - By = C$ and by Euler's description of "eine sehr sinnreiche Methode" that he attributed to "ein gelehrter Engländer namens *Pell*."[89]

The method of continued fractions is based on a sequence of substitutions $u = nx + y, x = my + z, \ldots$, continued until the remainder is 0. The method of descent aims that sequence toward a repetition of the same expression for a smaller set of values. On the assumption that a solution exists, let u be of the form $nx + y$ for integer x, y, n, where n is the largest integer in

[88] Fermat to Carcavi, pour Huygens, VIII.1659, FO.II.433; see below, §5.

[89] André Weil, "Fermat et l'équation de Pell," in *ΠΡΙΣΜΑΤΑ: Naturwissenschaftsgeschichtliche Studien: Festschrift für Willy Hartner*, ed. Y. Maeyama and W. G. Saltzer (Wiesbaden: Steiner, 1977), 441–448. For Euler's account, see his *Vollständige Anleitung zur Algebra*, Part II, §II, Chap. 7, Par. 98 (p. 426 in Hofmann ed.).

$\sqrt{N}$. Substitution leads to the equation $(n^2 - N)x^2 + 2nxy + y^2 = \pm 1$, or, expressed in general form, $F(x,y) = Ax^2 - 2Bxy - Cy^2 \pm 1 = 0$, where $A = (N - n^2), B = n, C = 1$ are all positive and $B^2 + AC = N$. To preserve that form for the next iteration, one wants to express $x = \dfrac{By + \sqrt{N}y^2 \mp A}{A}$ in the form $my + z$ for positive integers y, z, m. The form of x suggests taking as the value of m the greatest integer in $\xi = \dfrac{B + \sqrt{N}}{A}$, the positive root of the equation $f(t) = At^2 - 2Bt - C = 0$; that is, $m < \xi < m + 1$. Since $A < 2B + C$, m must be ≥ 1. Substituting for x and regrouping terms leads to an equation of the form $A'y^2 - 2B'yz - C'z^2 \mp 1 = 0$, where $A' = -(Am^2 - 2Bm - C), B' = Am - B$, and $C' = A$. Since $m < \dfrac{B + \sqrt{N}}{A}$, $Am^2 - 2Bm - C < 0$, and $A' > 0$, and, since $\dfrac{2B}{A} < \dfrac{B + \sqrt{N}}{A} < m + 1 \leq 2m$, $Am - B = B' > 0$. Also $B'^2 + A'C' = B^2 + AC = N$. Once again the integer $y = \dfrac{B'z + \sqrt{N}z^2 \pm A'}{A'}$ must be of the form $kz + t$ for integer z, t, k, where k is the greatest integer in $\dfrac{B' + \sqrt{N}}{A'}$, and substitution will again yield an equation of the form $A''z^2 - 2B''zw - C''w^2 \pm 1 = 0$, the coefficients of which conform to the now familiar relations and the solution of which is smaller than that of its predecessor.

To confirm the last point, rewrite the expression $Ax^2 - 2Bxy - Cy^2 \pm 1 = 0$ in the form $f\left(\dfrac{x}{y}\right) \pm \dfrac{1}{y^2} = 0$, where $f(t) = At^2 - 2Bt - C$. In each step of the descent, $m = [\xi]$, where $f(\xi) = 0$, whence $m < \xi < m + 1$ and $f(m) < 0 < f(m + 1)$. For $y > 1$, $m < x/y < m + 1$. On the assumption that the initial equation $u^2 - Nx^2 = \pm 1$ has an integral solution u, x, then, the sequence of substitutions $x = my + z$ must eventually lead to the least solution in which $y = 1$ and either $z = 0$ or $z = 1$ (in which case the remainder will be 0 on the next iteration). That is, for $y = 1$, either $f(x) + 1 = 0$ or $f(x) - 1 = 0$, depending on the initial sign of the 1 and the number of iterations. In the first case, $f(x) < 0$ and $f(x + 1) = A(x + 1) - 1 + (C - 1)/x > 0$. But $f(m) < 0 < f(m + 1)$, and x and m are both integers. Therefore $x = m$. In the second case, $f(x) = 1$, and $f(x - 1) < 0$, whence $m = x - 1$.

A solution emerges most readily when a derived equation is of the form $y^2 - 2B'yz - C'z^2 - 1 = 0$, i.e. where $A' = 1$, and the sign of the constant term is $-$. To see that the sequence of transformations described above will eventually reach such a point, note first that the condition $N = B^2 + AC$ is satisfied by only a finite number of values and hence that the successive forms $F = (A,B,C)$ must eventually repeat. A closer look at the transformation reveals that it can be reversed by the same substitutions. That is, given the values A',B',C', one has $A = C'$, $B = C'm - B'$, $C = -(C'm^2 - 2B'm - A')$. Hence, one can associate with each $F = (A,B,C)$ a form $G = (C,B,A)$ which, owing to the conditions on the zeros of $f(t)$, will also be normal. Indeed, each of the derived forms will be binormal, allowing derivation in either direction, and thus the sequence of forms G will be the reverse of that of forms F. Suppose, now, that some form F repeats over a period p, that is (using subscript notation, where $F_{i+1} = F'_i$ and F_1 is the first derived from the original equation) $F_{i+1+p} = F_{i+1}$. Then $G_{i+1+p} = G_{i+1}$. But the ith derived form of G_{i+1} is G_1, and that of G_{i+1+p} is G_{1+p}, whence $G_{1+p} = G_1$ and, consequently, $F_{1+p} = F_1$. But then $C_{1+p} = C_1 = 1$, whence $A_p = 1$, and F_p will have the solution $(1,0)$, provided that the sign of the constant 1 is negative. If p is odd, the sign will change from one period to the next; if even, the solution will depend on the sign in the original equation.[90]

Because of this cyclical character, the "descent" becomes a means of generating a potentially infinite series of solutions by working down to the base solution and then building back up through the cycle, repeating it as desired. For example, let $u^2 - 5x^2 = 1$. Then $u = 2x + y$, and $x^2 - 4xy - y^2 + 1 = 0$. Let $m = [2 + \sqrt{5}] = 4$, and $x = 4y + z$. Then $y^2 - 4yz - z^2 - 1 = 0$. Similarly, $y = 4z + t$ yields $z^2 - 4zt - t^2 + 1 = 0$, and so on; evidently m will always be 4, and succeeding equations will differ only by the alternating sign of 1. The second derived equation has the trivial solution $y = 1$, $z = 0$, whence $x = 4$ and $u = 9$. But one may also repeat the substitution $x = 4y + z$ twice, generating pairs 17,4 and 72,17, of which the latter generates the solution $x = 72$, $u = 161$. Repeating it twice again will generate another solution, and so on. In the case of $u^2 - 13x^2 = 1$, $u = 3x + y$, $4x^2 - 6xy - y^2 + 1 = 0$, and $m = [(3 + \sqrt{13})/4] = 1$. For brevity, consider simply the coefficients A, $-2B$, $-C$ of the expressions and the sign of 1, e.g. $(4,6,1,+1)$. Then the sequence continues $(3,2,4,-1)$, $(3,4,3,+1)$, $(4,2,3,-1)$, $(1,6,4,+1)$, $(4,6,1,-1)$, $(3,2,4,+1)$, $(3,4,3,-1)$, $(4,2,3,+1)$, and $(1,6,4,-1)$. The last of these, i.e. the equation $v^2 - 6vw - 4w^2 - 1 = 0$ has the trivial solution

---

[90] In a outline of this reconstruction in *Number Theory*, Weil himself points to the interpretive difficulties posed by the use of subscripts, which seem essential to the argument, especially when applied to a further reconstruction of Fermat's proof that primes of the form $4k + 1$ decompose uniquely into two squares: "It does not seem unreasonable to assume that this may have been, in general outline, what Fermat had in mind when he spoke of his proofs in his letters to Pascal and Huygens; how much of it he could have made explicit must remain a moot question. Perhaps one of his most serious handicaps was the lack of subscript notation; this was introduced later by Leibniz, in still imperfect form, and its use did not become general until rather late in the next century" (*Number Theory*, 99). Historiographically, one might turn the situation around and ask whether the need for subscripts is a handicap for the historian trying to understand Fermat's thinking.

$v = 1$, $w = 0$. Building back up from these, one arrives at the solution $u = 649$, $x = 180$, and one may repeat the cycle for additional solutions.

In proposing the above as the elements of a reconstruction of Fermat's general and demonstrated solution of "Pell's equation," Weil posits that, like the English, Fermat built on Bachet's solution of the linear case and developed a method substantially the same as theirs. "On that assumption, we are going to show that a careful examination of that solution in particular cases such as those that have been described above can well inspire in a mathematician of Fermat's caliber a simple principle of demonstration."[91] To Weil's argument, proposed "à titre purement hypothétique" and based on the mathematical structure of the problem, one may perhaps add some evidence from Fermat's own work, beginning with the assertion that led Frenicle to surmise that Fermat had "a particular sort of analysis for probing the most hidden secrets of numbers." There was nothing "particular" about it, except, perhaps, for the way Fermat applied it, and then, perhaps, for the way he extended it.

In the letter of 15 June 1641 in which Fermat showed Frenicle how to determine the number of ways a number may be the hypotenuse of a right triangle, he extended the technique to finding how many ways it could be one of the smaller sides of a triangle. Depending on the particular division into factors, it could be the smallest or the middle side of the triangle. To the obvious question of the precise ratio of the factors that decided that outcome, Fermat responded that "it is impossible to determine it exactly in integers, because the algebraic equation produces irrational numbers, even though one approach it ever more closely *in infinitum* by integers."[92] The irrational ratio, he said, is 1 to $1 + \sqrt{2}$. He said no more, but it is clear what he had in mind.[93] The difference of the sides of a right triangle is $(p^2 - q^2)/2 - pq$,

---

[91] Weil, "Fermat et l'équation de Pell," 443: "C'est donc à titre purement hypothétique qu'on va indiquer un schéma de démonstration qui aurait pu, croyons nous, être celui de Fermat [Weil here cites Euler's *Vollständige Anleitung*]. En premier lieu, nous admettrons que la méthode de solution de Fermat ne différait pas substantiellement de la 'méthode anglaise.' Comme il est remarqué plus haut, celle-ci semble directement inspirée par celle de Bachet, que Fermat devait bien connaître. Qu'on passe tout naturellement de celle-ci à celle-là semble démontré, non seulement par le caractère du problème, mais par la rapidité avec laquelle les Anglais obtinrent leur solution.

"Cela admis, nous allons faire voir qu'un examen attentif de cette solution, dans des cas particuliers tels que celui qui a été décrit plus haut, pouvait bien inspirer à un mathématicien de la force de Fermat un principe simple de démonstration."

[92] FO.II.223.

[93] Or perhaps only by hindsight, since Frenicle wrote in reply that he knew the ratio was irrational, but not how to generate an infinite sequence of values approaching it, nor, evidently, how to analyze the question algebraically. "Je sais que l'Algèbre de ce pays-ci n'est pas propre pour soudre ces questions, ou pour le moins on n'a pas encore ici trouvé la manière de l'y appliquer: c'est ce qui me fait croire que vous vous êtes fabriqué depuis peu quelque espèce d'Analyse particulière pour fouiller dans les secrets les plus cachés des nombres, ou que vous avez trouvé quelque adresse pour vous servir à cet effet de celle que vous aviez accoutumé d'employer à d'autres usages. Si la démonstration de cette limitation étoit courte, vous m'obligeriez beaucoup de me l'envoyer: car, si elle est trop longue, je ne voudrois pas que vous vous détournassiez de vos études à cette occasion." Frenicle to Fermat, 2.VIII.1641, FO.II.227.

where p and q are the factors of the number. Whether pq is the least or the middle side depends on whether that difference is greater or less than 0 for the given values, and the dividing point is obviously the solution of $p^2 - 2pq - q^2 = 0$, or, expressed as a ratio, $(p/q)^2 - 2(p/q) - 1 = 0$. In 1641, then, Fermat was looking at an equation of precisely the form $f(p/q) = 0$ discussed above and was talking about approaching its solution.

Moreover, two letters of 1643 show Fermat exploring iterative solutions or analyses of equations: first, the fragment discussed above, setting out a technique for factoring large numbers based on the solution of the equation $x^2 - y^2 = N$, and, second, the letter to Brûlart explaining the method of maxima and minima and discussed above in Chapter IV. In the latter, Fermat had clearly been probing the nature of the coefficients of algebraic polynomials expanded about their arguments. In particular, taking E as both increment and decrement of the value A in $BA^2 - A^3 = M$, he developed $B(A \pm E)^2 - (A \pm E)^3$ as $(BA^2 - A^3) \pm (2BA - 3A^2)E + (B - 3A)E^2 \pm E^3$, and used the coefficient of the second term to explain how his method of maxima and minima reached "the same equation" for both an increment and a decrement and the sign to explain why the expression was set equal to 0. It is not far from this line of reasoning to expand $Ax^2 - 2Bxy - Cy^2 \pm 1 = 0$ around $x = my + z$ to get $-(Am^2 - 2Bm - C)y^2 - 2(Am - B)yz - Az^2 \mp 1 = 0$ and to then ask about the conditions under which it might be the "same equation," albeit with smaller solutions. Fermat devised his new account of the method of maxima and minima at about the same time that he began to press Frenicle to push beyond specific solutions to the structure of problems such as "Pell's equation" and the decomposition of primes of the form $4k + 1$.

If Fermat had indeed done what Weil conjectures, whether by the mid-1640s or by the late 1650s, he did not record it anywhere. As a result, history has ignored his achievement in two ways. It has named the equations after John Pell, an English mathematician who did no more than copy them from Fermat's letters to Digby in 1657 and 1658. Moreover, it has called the body of techniques aimed at eliciting purely integral solutions to algebraic equations "Diophantine analysis." Ironically, it was precisely because Fermat, the real author of the equations, had broken away from the Diophantine tradition of rational solutions that these techniques were first developed.

IV.   RECLAIMING THE PATRIMONY: THE CHALLENGES OF 1657

The ten-year hiatus in Fermat's correspondence with other mathematicians was broken with Blaise Pascal's request for aid in solving a problem in probability. After dealing directly with the questions Pascal had posed, Fermat then tried to interest him in number theory.[94] The results that Fermat chose from

---

[94] Cf. Fermat to Pascal, 29.VIII.1654, FO.II.309–310 (the prime number conjecture) and Fermat to Pascal, 25.IX.1654, FO.II.312–313.

the myriad of his earlier correspondence document conclusively the change his thinking had undergone regarding the nature of arithmetic and the sorts of problems he had been investigating in the interim. "I hope to send you on St. Martin's Day a sketch of everything worthwhile that I have discovered concerning numbers. Permit me to be concise and to express myself in a way suitable only to a man who understands everything when it is half said." Most important, Fermat had proved the theorem that constitutes Observation 18: every number is either a triangular number or composed of two or three triangular numbers; a square or composed of two, three, or four squares; a pentagonal number or composed of two, three, four, or five pentagonal numbers; and so on indefinitely. He did not disclose to Pascal how this marvelous proposition might be demonstrated. Indeed, writing for himself in the *Observations on Diophantus*, he did no more than promise: "Its demonstration, however, which is derived from many varied and most abstruse mysteries of the numbers, cannot be placed here. For we have decided to direct a treatise and a whole book to this task and to further arithmetic in this direction beyond the Ancients and the known limits in a marvelous way."[95]

Perhaps to provide a hint for the case of squares, Fermat did add in the letter to Pascal that it was necessary first to prove that every prime number of the form $4k + 1$ is composed of two squares, and he set Pascal a problem he had posed to Frenicle thirteen years earlier: given such a prime, find "by a general rule" the two squares composing it. Fermat then continued by noting that every prime number of the form $3k + 1$ is composed of a square and the triple of a square, and every prime number of the form $8k + 1$ or $8k + 3$ can be expressed as the sum of a square and the double of a square. He topped off these propositions with the claim to be able to prove that no integral right triangle can have a square area (i.e. that half the product of the two smaller sides cannot be a square number). "That will be followed by the finding of many propositions that Bachet admits not having known and that are missing in Diophantus. I am persuaded that, once you have become familiar with my manner of demonstrating propositions of this sort, it will seem pretty to you and will give you a chance to carry out many new proofs. For, as you know, it is necessary that many pass by in order that knowledge grow (*multi pertranseant ut augeatur scientia*)." Fermat dangled before Pascal's eyes the promise of an arithmetic more subtle and powerful than anything the Ancients (or, indeed, the moderns) possessed.

It would seem clear from what followed in the next four years that the particular results Fermat first announced to Pascal applied to more than the theorem concerning the number of polygonal numbers of which any number was composed. In Observation 34 (to *Arithmetic* VI,3), the main method of which is the use of a "single" equation, Fermat noted, almost in passing, that the equation $5x^2 - 1 = y^2$ has a solution because 5 is composed of two

[95] Observation 18, FO.I.305.

squares.[96] At the conclusion of the observation, he then showed that the solution of the problem in Diophantus—to find a right triangle such that its area added to a given number equals a square—always reduced to the problem of solving the equation $px^2 + 1 = y^2$, "which is generally quite easy, since 1 is a square."

The decomposition of prime numbers of the form $4k + 1$ into two squares is not the only theorem concerning the quadratic forms of numbers that has application to an indeterminate equation. As J. E. Hofmann has shown, the various theorems of this sort that Fermat mentioned in his letter to Pascal and in a later letter to Digby are precisely the theorems involved in any complete solution of the general equation $x^2 - py^2 = \pm q$.[97] Hence, it seems quite likely that Fermat had by this time arrived at such a solution (or at least a major portion of it) and was probing Pascal for the opportunity to announce it in such a way as to gain for it the admiration it deserved and thereby to initiate a new tradition in arithmetic.

Pascal dashed Fermat's expectations with the curt letter that ended their short mathematical collaboration. Having little enough interest left for mathematics as a whole, he could summon none for number theory. "Look elsewhere," he wrote,[98]

> for someone to follow you in your numerical researches, of which you have done me the honor of sending the statements. As far as I am concerned, I confess to you that they go right past me; I am capable only of admiring them and of begging you very humbly to take the first opportunity to complete them. All our gentlemen saw them last Saturday and admired them with all their hearts. One cannot easily bear waiting for things so pretty and so desirable. Please think about them, then, and be assured that I am, etc.

Pascal's genteel way of putting things could not assuage Fermat's disappointment in the rejection of his new arithmetic. Fermat was not looking for an encouraging pat on the head and the advice to get his work together and publish it. His reference to Bacon's motto should have made that clear to Pascal. He sought dialogue and exchange of ideas; he sought to establish a community of mathematicians all working together on a common body of problems, with a faint atmosphere of competition and mutual challenge to spur them on.[99] Pascal's rejection blocked that search for the moment.

By 1654 Fermat did not feel he could turn again directly to his two earlier

[96] FO.I.328–329.

[97] See his "Studien zur Zahlentheorie Fermats" cited above, n. 3.

[98] Pascal to Fermat, 27.X.1654, F0.II.314

[99] Cf. Roberval to Fermat, 4.VIII.1640, FO.II.202: "J'oubliais presque à vous dire que les nombres, dont vous avez déjà découvert des propriétés admirables, contiennent de grands mystères; mais, pour les mieux découvrir, il faudrait être plusieurs ensemble, d'accord et sans jalousie, et desquels le génie fût naturellement porté à cette spéculation, ce qui est très difficile à rencontrer." As it turned out, Roberval spoke with the gift of prophecy.

correspondents, Frenicle and Brûlart. The exchanges of 1641–1643 had generated bad feeling. The Parisians had felt that Fermat was consciously playing the righteous Archimedes by posing problems that had no solution in order to expose his rivals.[100] Fermat suspected that they, in turn, were siding with those who accused him of relying on clever guesswork. The bad feeling had not been overcome before correspondence broke off. Indeed, just before the break the men had been speaking to one another only through Mersenne, and that one common friend had died in 1648. Fermat had something of real importance to show people. He was not, however, ready to break a long-standing habit of not composing finished treatises for publication. He needed an audience with which he could interact. He could find none, and his frustration grew, perhaps to the point of desperation.

How else can one explain the actions that followed? In 1656, Fermat received a visit from Kenelm Digby, who brought with him a copy of John Wallis' recently published *Arithmetic of Infinites*. Hofmann would have it that the realization that others had published and claimed credit for results in quadrature that Fermat had long since established spurred him to take some radical action that would reassert his superiority in mathematics.[101] Certainly, Fermat did criticize Wallis' work, contesting in particular the completeness, generality, and rigor of his summation of series. But, in that respect, the *Arithmetic of Infinites* could only have comforted Fermat's mathematical ego. Wallis had gone no farther in his method of quadrature than had Fermat by the mid-1630s, and, as Chapter V has shown, Fermat had by the mid-1650s

---

[100] Since Fermat has already quoted once from Archimedes' introductory letter to his treatise *On Spirals* and will make explicit reference to it again during the controversy of 1657–1658, it would help to quote the opening paragraph here in full. It may offer insight into Fermat's behavior in number theory, especially given the importance of Archimedes' treatise in his work on quadrature. "Of most of the theorems which I sent to Conon, and of which you ask me from time to time to send you the proofs, the demonstrations are already before you in the books brought to you by Heracleides; and some more are also contained in that which I now send you. Do not be surprised at my taking a considerable time before publishing these proofs. This has been owing to my desire to communicate them first to persons engaged in mathematical studies and anxious to investigate them. In fact, how many theorems in geometry which have seemed at first impracticable are in time successfully worked out! [This is the line Fermat quoted to Frenicle and Brûlart in 1643; cf. above, n. 45] Now Conon died before he had sufficient time to investigate the theorems referred to; otherwise he would have discovered and made manifest all these things, and would have enriched geometry by many other discoveries besides. For I know well that it was no common ability that he brought to bear on mathematics, and that his industry was extraordinary. But, though many years have elapsed since Conon's death, I do not find that any one of the problems has been stirred by a single person. I wish now to put them in review one by one, particularly as it happens that there are two included among them which are impossible of realization and which may serve as a warning how those who claim to discover everything but produce no proofs of the same may be confuted as having actually pretended to discover the impossible." (Trans. T. L. Heath, *The Works of Archimedes*. Cambridge, 1897, p. 151.) One could, with very little adjustment, make this passage the *leitmotif* of Fermat's own career. Indeed, Lalouvère echoed Fermat's sense of being a new Conon in the Preamble to Appendix II of the *Veterum geometria promota* (Toulouse, 1660); see FO.I.199, n. 1.

[101] See Hofmann's "Neues über Fermats . . . Herausforderungen," where he carefully sets out the full details of the controversy in its earliest stages.

(probably even earlier) achieved a far more powerful and sophisticated theory of quadrature. Even if the challenge of Wallis' work did spur the composition of Fermat's own *Treatise on Quadrature*, Fermat took little or no action to publicize his efforts. Rather, the direct action he took had nothing to do with quadrature, though it was not entirely unrelated to the issue.

The very fact that Wallis did approach his subject in the same way that Fermat had at first, i.e. through summation of series of powers of integers, provoked a positive rather than a negative response from Fermat. To see why, one need only read Observation 46:[102]

> We append here without demonstration a most pretty and marvelous theorem which we have found:
>
> In the series of natural numbers beginning with unity, any number multiplied by the next greater makes twice the triangle of that number; multiplied by the next greater triangle, makes three times its pyramid; multiplied by the next greater pyramid, make four times its triangulotriangle, and so on *in infinitum* by a uniform and general method.
>
> I do not think there can be a prettier or more general theorem in numbers, but the margin neither has room nor allows its demonstration to be inserted.

The theorem, of course, formed the basis for Fermat's early method of quadrature, since, as shown in Chapter V, it led to a recursive formula for summing any series of powers of the integers. But here in the *Observations on Diophantus* it ranks as the most beautiful theorem of all in number theory, without reference to quadrature. In his frustration in 1656, Fermat could draw only one conclusion from Wallis' similar use of integral power series; Wallis must also be interested in number theory.

Wallis, therefore, represented for Fermat a renewed opportunity to open the discussion of number theory Pascal had so coldly refused. Where the plea for cooperation had failed, perhaps the challenge of competition would succeed. If the French would not respond to a countryman, perhaps they would react to a foreigner. On 3 January 1657, Fermat dispatched from Castres a pair of challenge problems. They were addressed to a Claude Martin Laurendière in Paris for transmission to mathematicians everywhere. Whatever title Fermat may have given them, the sentiments probably expressed in the accompanying letter (now lost) led Willem Boreel, the Dutch Ambassador to France, to head the problems "Two Mathematical Problems Posed as Insoluble to French, English, Dutch, and all Mathematicians of Europe by Monsieur de Fermat, Councillor of the King in the Parlement of Toulouse" when forwarding them to the mathematicians at the University of Leiden. In that copy, the problems read:[103]

[102] FO.I.341.
[103] FO.II.332–333.

First Problem: To find a cube which, added to all its aliquot parts, makes a square.

For example, the number 343 is a cube of side 7. All of its aliquot parts are 1,7,49, which, added to 343, make the number 400, which is a square of side 20. Sought is another cube of the same nature.

Second Problem: Also sought is a square number which, added to all its aliquot parts, makes a cube number.

The copy of the challenge sent to England simply bore the title "A Challenge from M. Fermat for D. Wallis, with the hearty commendations of the messenger, Thomas White."[104] That version began, "If it please them, let the following numerical problem be proposed to Wallis and the other English mathematicians." After stating the problems in the same wording as above, it concluded, "We await the solutions which, if England or Belgian and Celtic Gaul cannot give them, Narbonian Gaul will give and will offer and speak as a pledge of growing friendship to Mr. Digby."

The reference to Belgian and Celtic Gaul shows clearly, as the editors of Fermat's *Oeuvres* point out, that Fermat intended Frenicle to receive the problems, although the preface to Frenicle's solution (at the time unknown to the editors) does not support their conjecture that he received them in a personal letter from Fermat.[105] Fermat and Frenicle were not on cordial terms in 1657, and so Frenicle learned of the problems as they passed through Paris on their way to Leiden and London.

The challenge failed to bring the results Fermat had optimistically expected. He apparently thought that they would lead other mathematicians to the same subject to which they had led him and which had formed the core of his research over the previous dozen years. To take the first problem as an example, suppose the desired cube is x^3, where x is prime. Its aliquot parts are 1, x, and x^2. The problem, therefore, calls for the solution of $1 + x + x^2 + x^3 = y^2$ in integers. The left-hand side may be written as a product $(1 + x)(1 + x^2)$, each of the factors of which is even; however, they have only the factor 2 in common and are otherwise mutually prime. Hence, y^2 is of the form $2u^2 \cdot 2v^2$, where u and v are mutually prime squares; i.e. $1 + x = 2u^2$ and $1 + x^2 = 2v^2$. Any number x, therefore, that satisfies the first of Fermat's challenge problems must be such that the pair of equations $2u^2 - 1 = x$ and $2v^2 - 1 = x^2$ has integral solutions u and v.[106]

Fermat was prepared to show, as he later did, not only that the general form of these equations, i.e. $2u^2 - w^2 = p$, has solutions if and only if p is a

---

[104] FO.II.333.

[105] FO.II.332, n. 1.

[106] So, at least, Hudde interpreted the problem (cf. Hofmann, "Neues," p. 16), and Fermat's later solution would seem to bear him out.

prime of the form $8k \pm 1$ or contains prime factors of that form, but that the special cases to which the first challenge problem reduced have solutions only for the primes 1 and 7 and no others.[107] For he could show first that if u_0, w_0 is a solution of $2u^2 - w^2 = p$ ($p \equiv \pm 1 \pmod 8$), then $u_1 = 2u_0^2 \pm 2u_0w_0 + w_0^2$, $w_1 = 2u_0^2 \pm 4u_0w_0 + w_0^2$ are solutions of $2u^2 - w^2 = p^2$. If, moreover, u_0, w_0 is the least solution of the first equation, u_1, w_1 for the minus sign is the least solution of the second. Therefore, suppose $2u_0^2 - 1 = p$. Then $u_1 = 2u_0^2 - 2u_0 + 1$, $w_1 = 2u_0^2 - 4u_0 + 1$ is a solution of $2u^2 - w^2 = p^2$ But, clearly, u_0, 1 is the least solution of the first equation; hence, so too is u_1, w_1 the least solution of the second. By assumption, however, $w_1 = 1 = 2u_0^2 - 4u_0 + 1$. Therefore, $u_0 = 2$, and 2,1 is the only solution of the first equation. Hence, $p = 7$. Trivially, $u_0 = u_1 = 1$ gives a solution of both equations (which reduce to the same single equation) for $p = 1$.

Though Fermat nowhere explicitly discusses the second problem of the first challenge, it seems clear that he meant it to lead to the same theoretical end: to the study of one case of the general indeterminate equation $x^2 - py^2 = \pm q$. Although this was perhaps characteristic for Fermat, it was, to put it mildly, a strange way to proceed. Acting now as a theoretician, he could not shake loose from the problem-solver's habit of challenging others through specific problems. But, besides giving no hint of the theoretical aim of his challenge, Fermat was now committing the very transgression of which Brûlart and Frenicle had wrongly accused him in 1643. He was proposing specific problems which, as stated, had no solution. Perhaps because he soon recognized this, and also because he sensed (correctly, as it turned out), that the two problems did not suffice as signposts pointing in the proper direction, Fermat issued a second challenge dealing directly with the theoretical point at issue. That second challenge, written in February 1657, also stated in unmistakable terms Fermat's vision of a new tradition in arithmetic, which he intended to initiate. Since the challenge also exemplifies the paradox referred to at the beginning of this chapter, i.e. the development of an entirely new sort of mathematics under the guise of restoring an even more ancient one, it warrants quotation in full. "There is hardly anyone who proposes," Fermat began,[108]

> hardly anyone who understands, purely arithmetical problems. Is that not because up until now arithmetic has been treated more geometrically than arithmetically? Surely many volumes of both ancient and more recent writers indicate this; even Diophantus indicates it. Even though he departed from geometry somewhat more than the others by restricting his analysis to rational numbers only, Viète's *Zetetics*, in which Diophantus' method is

[107] See the paper cited above, n. 87.
[108] FO.II.334–335.

extended to continuous quantity and hence to geometry, is more than enough proof that this part [of arithmetic] is not entirely devoid of geometry.

So let arithmetic redeem the doctrine of whole numbers as a patrimony all its own. Let students of arithmetic take pains to further or to renew that doctrine, which was lightly adumbrated by Euclid in nothing less than the *Elements* but which was not sufficiently filled out by those who followed (unless, perhaps, it lies hidden in those books of Diophantus that the injustice of time has carried away).[109]

In order to light the way, we propose to those students of arithmetic the following theorem or problem, either to be demonstrated or to be constructed. And, if they solve it, they will think questions of this sort not to be inferior to more celebrated questions from geometry, either in subtlety, or in difficulty, or in demonstrative reasoning:

Given any non-square number, there are given infinitely many squares which, multiplied by the given number and added to unity, make a square. Example: 3 is given, a non-square number; 3 multiplied by the square 1 and added to unity makes 4, which is a square. Again, the same 3, multiplied by the square 16 and added to unity, makes 49, which is a square. And, in place of 1 and 16, one can find infinitely many squares with the same property; we seek, however, the *general* canon, given *any* non-square numbers. What, for example, is the square which, multiplied by 149, or 109, or 433, etc., and added to unity, makes a square?[110]

Fermat had mastered the solution of such equations in rationals by his emendation of Diophantus' methods. There was no longer any trick to it. The far greater challenge lay in resolving it directly and solely for integers, and success promised an open vista of beautiful mathematics made even more beautiful by its supposed pedigree.

Like that of many a noble house of Fermat's day, the pedigree was fanciful, a product of myopia and wishful thinking conditioned by the analytic tradition imbued in Fermat. True to that tradition, Fermat saw Viète's translation of the *Arithmetic* into the algebraic language of the "analytic art" as evidence for the original presence of an algebra of continuous magnitude in that work. In fact, it had not been present. Diophantus' unknowns never denoted continuous geometrical magnitudes; his "algebra" dealt solely with rational numbers. Heir to his own Greek tradition, Diophantus knew precisely the difference between continuous and discrete magnitude and the need for methods peculiar to each.[111] It was Viète, not Diophantus or any other Greek writer, who made

---

[109] Cf. Fermat to Roberval, 16.IX.1636, FO.II.63: "Vous ne sauriez croire combien la science du dixième Livre d'Euclide est défectueuse: je veux dire que cette connaissance n'a pas encore fait de grands progrès et qu'elle est pourtant de grandissime usage."

[110] Note that the last three non-square numbers given are all primes of the form $4k + 1$.

[111] Cf. Aristotle, *Posterior Analytics*, I, 7.

algebra the "analytic art," the system of formal reasoning that united the realms of the continuous and the discrete. It was Fermat, not Apollonius or Archimedes, who translated geometry into the theory of equations. And it was Fermat, not Diophantus, who used that same theory of equations to unleash the full power of the method of "single" and "double" equations. If in Fermat's day geometry obtruded on arithmetic, it did so because Viète and his followers, foremost among them Fermat, had used the "analytic art" to blur the distinction between the two, a distinction that is the very hallmark of classical Greek mathematics. And now, Fermat would reassert the distinction; he would "redeem the patrimony."[112] But, again like many a noble house of the day, the "patrimony" was in fact newly gained wealth, the result of recent conquests. It is the irony both of the passage just cited and of Fermat's career as a whole that the ancients had never dreamed of the mathematics that Fermat had in mind.

Unfortunately, neither did Fermat's contemporaries, perhaps because, less imbued with the analytic tradition, they saw more clearly the original nature of Greek mathematics. If Fermat had visions of himself holding the light for the *paides arithmetices* (the original Greek phrase translated above as "students of arithmetic") to reveal new vistas, they quickly and rudely disabused him of those visions.[113] Fermat probably expected no more from Frenicle than he in fact received, to wit, four solutions in large numbers for the first problem, along with the promise of bigger and better ones to come.[114] From his earlier contact with Frenicle, Fermat knew his erstwhile correspondent was simply incapable of approaching mathematics in terms of conditions of solvability and general techniques of solution. Frenicle would end his days computing large numbers, but he would add nothing to number theory. Fermat clearly did expect more from Wallis, but there too he was to be disappointed.

The first challenge did not reach Wallis until early March 1657, having gone by way of Paris and London (through the hands of Viscount Brouncker) before arriving in Oxford. In view of the fact that Fermat had an international reputation—Digby, after all, had traveled to Toulouse just to see him—and

---

[112] So, for example, in Observation 42 he writes (FO.I.333–334): "An autem alius in integris quadratus, praeter ipsum 25, inveniatur qui adsumpto binario cubum faciat, id sane difficilis primo obtuto videtur disquisitionis. Certissima tamen demonstratione probare possum nullum alium quadratum, praeter 25, in integris adjecto binario facere cubum. In fractis ex methodo Bacheti suppetunt infiniti, sed doctrinam de numeris integris, quae sane pulcherrima et subtilissima est, nec Bachetus, nec alius quivis cujus scripta ad me pervenerint, hactenus calluit."

[113] Fermat clearly takes the image from Bacon's motto of progress in *De dignitate et augmentis scientiarum*: "traditio lampadis ad filios" (see below, p. 351).

[114] Cf. Fermat to Mersenne, I.IV.1640, FO.II.187: "Pour Monsieur de Frenicle, ses inventions en Arithmétique me ravissent et je vous declare ingénûment que j'admire ce génie qui, sans aide d'Algèbre, pousse si avant dans la connaissance des nombres entiers, et ce que j'y trouve de plus excellent consiste en la vitesse de ses opérations, de quoi font foi les nombres aliquotaires qu'il manie avec tant d'aisance."

Wallis had just begun his career, Wallis' reaction to the two problems seems unwarrantedly curt:[115]

> The question [he wrote to Brouncker] is just about of the same sort as the problems ordinarily posed concerning the numbers called "perfect," "deficient," or "abundant." These problems, and others of the same sort, cannot at all or cannot completely be reduced to a general equation embracing all cases. Whatever the details of the matter, it finds me too absorbed by numerous occupations for me to be able to devote my attention to it immediately. But I can make at the moment this response: the number 1 in and of itself satisfies both demands.

Wallis' "numerous occupations" had, of course, led him to mistake the form of the first challenge for its substance. The problem bore only outwardly a resemblance to the traditional problems concerning sums of aliquot parts. Though he may have been right in asserting that the traditional problem allowed no general method—Fermat, of course, would have disputed the matter—the first challenge problem was not traditional, had little to do with sums of aliquot parts, and (as has been shown) did allow a fully general treatment.[116]

Fermat did not see Wallis' letter. It got no farther than Paris, where Digby showed it to Frenicle, who in turn then engaged Wallis in a continuing debate over whether 1 could even be considered a number, much less a cube—an interesting debate for its own sake, perhaps, but totally irrelevant to the issues posed by the challenge (1 is simply a trivial solution). Fermat did receive Brouncker's solutions, however, and his reply to them barely concealed his disappointment. Brouncker had said of the first problem that any number of the form $343/a^6$, where a is any integer, will serve as a solution. His solution to the challenge of February, a solution in which Wallis concurred, similarly aimed at fractional solutions. Brouncker took any number n (rational) and any square r^2 (also rational) and then argued that $\dfrac{4r^2}{(n - r^2)^2}$ constituted a square number which, multiplied by n and added to 1, produced a square, to wit $\dfrac{(r^2 + n)^2}{(r^2 - n)^2}$. To complicate the issue, Brouncker addressed his reply to Fermat in English, which Fermat, for all his linguistic talent, could not read. He had the letter translated by a young English student in Toulouse, but the translator knew no mathematics.

[115] *Commercium epistolicum*, #II; FO.III.404. Wallis adds a counterchallenge: solve $s(x^2) = s(y^2)$, $x \neq y$.

[116] Wallis evidently knew nothing about Fermat's correspondence of the mid-1630s on the subject of aliquot parts, nor apparently had he read the works of Mersenne in which some of the results in that correspondence were published.

Dimly at first, but then clearly, Fermat saw what Brouncker had done. His first reaction, in a letter to Digby dated 6 June 1657, offered his tentative reaction: even if he could not follow the details from the poor translation, it was evident that Brouncker was treating the problem too lightly to have understood it.[117] Certainly that was the case if Brouncker thought that fractional solutions would suffice. Indeed, in the case of the challenge of February, a solution in fractions lay within the means of "any beginner in arithmetic" (*quilibet de trivio arithmetico*). Fermat wrote to Digby again in August to reaffirm his reaction:[118]

> With regard to the numerical problems, I venture to say to you, with respect and without diminishing in the slightest the high opinion I have of your nation, that the two letters of Milord Brouncker, however obscure and badly translated, contain no solution at all. It is not that I mean thereby to renew the jousts and ancient tilting of lances which the English once carried out against the French. Rather, to continue the metaphor, I venture to maintain—and to you, Monsieur, who excel in both crafts, more justly than to any other—that accident and luck often intrude into scientific battles as much as in others, and that in any case we can say that "no field can bear every crop."[119]

"No field can bear every crop." Fermat possessed the rare talent for making an insult out of words of consolation. Wallis would receive those words with the same enthusiasm that Descartes showed upon learning from Fermat that the law of refraction was not quite right but that, after all, the problem was difficult and would require extensive cooperative effort. Fermat meant it as consolation, however, for he felt he had made his point. Number theory was difficult and worth pursuing. He now moved to secure it further. If Brouncker and Wallis preferred to work with rational solutions, Fermat would like them to solve the following problem:[120]

> Diophantus proposed the problem of splitting a given square number into two squares and also of splitting a given number composed of two squares into two other squares. But neither he nor Viète tried to raise the problem to the case of cubes. Why should we hesitate or postpone clarifying, therefore, a famous proposition left open for more recent analysts?
>
> So be it proposed to split a cube number into two cubes and, further, to split a given number composed of two cubes into two other rational cubes. We ask what England and Holland think of this matter.

[117] FO.II.341-342.
[118] Fermat to Digby, 15.VIII.1657, FO.II.343–344.
[119] "[N]on omnis fert omnia tellus"; according to FO.II.344, n. 1, the allusion is to Virgil, *Eclogues*, IV, 39: "omnis feret omnia tellus."
[120] FO.II.346; cf. observations 2 and 3, FO.I.291–292.

Fermat told Digby to include France in the person of Frenicle when passing this problem on. In fact, he added, he had some surprises for Frenicle. He had already asserted that the equation $x^2 + 2 = y^3$ had a unique solution 5,3 in integers, which Frenicle "could hardly believe and found too daring and too general." Now Fermat was ready to assert that the equation $x^2 + 4 = y^3$ had only two solutions, 2,2 and 11,5, and no others. "I don't know," he added in a feeler toward a possible new audience, "what you English will say of these negative propositions or if they will find them too daring. I await their solution and that of M. Frenicle, who has not responded to a long letter that M. Boreel gave him on my behalf. I am surprised about that, for I replied in detail to all his doubts and posed to him some questions of my own, to which I await the answer." Had Fermat sent the letter as it stood, events might have taken a different course. To the implied reproach of his consolation to Brouncker, however, he appended a lengthy critique of Wallis' *Arithmetica infinitorum.*[121] He had already written to Digby on the latter subject, pointing out to him and to Wallis that he, Fermat, had accomplished the same results some ten years earlier. The critique that went out now attacked the reasoning and style of Wallis' work directly. The criticism was summed up in the phrase that serves as *leitmotif* for Chapter V.

Fermat did not know that Frenicle had already stoked the fire that was breaking out. For Frenicle had also rejected Brouncker's solution of the challenge of February on the same grounds as Fermat, namely, the inadmissibility of fractional solutions. Moreover, Frenicle claimed the laurels for having solved the first challenge, and in doing so did not hesitate to emphasize the nationalistic overtones Fermat had introduced into the growing controversy.[122]

History repeated itself. Fermat started the fight but was then forced to the sidelines as others less talented than he carried it on. In 1657–1658 Frenicle and Digby played the roles Roberval and Pascal had exercised twenty years earlier. The debate between Frenicle and Wallis, for example, over whether 1 was a number or could be considered a cube echoed the caviling of Roberval, Pascal, and Descartes over whether the tangent to a curve could be considered the maximum distance from a given point on the axis to the curve. Together, Frenicle and Digby pressed Wallis and Brouncker for integral solutions of the two problems of the first challenge. For his part, Wallis assumed the role of the petulant Descartes, at first stoutly defending his solutions but arguing at the same time that the whole effort was a waste of time, and then slowly coming around to his opponents' position.

[121] Fermat to Digby, 15.VIII.1657, "Remarques sur l'Arithmétique des infinis du S. J. Wallis," FO.II.347–353.

[122] Frenicle, *Solutio duorum problematum circa numeros cubos, quae tanquam insolubilia universis Europae mathematicis a clarissimo viro D. Fermat sunt proposita et a D. Cl. M. Laurenderium . . . transmissa, a D. B[ernardo]. F[reniclio]. D[e]. B[essy]. inventa* . . . (Paris, 1657); cf. Gino Loria, "La réponse de Frenicle au premier défi de Fermat," *Bull. des sc. math.*, sér. 2, Vol. 54(1930), pp. 245–254.

The greater part of the resulting exchange of letters, released to the public by Wallis in 1658 as the *Commercium epistolicum*, sheds little light on Fermat's career.[123] It contains mostly the letters Wallis sent to Digby and Brouncker; to the latter as part of their joint effort to arrive at a method for the solution of the equation $px^2 + 1 = y^2$, to the former for the purpose of conveying the results. Hence, the solution that ultimately emerged in print, a solution based on the method of continued fractions, lends direct insight only into the mathematics of Wallis and Brouncker. That solution may also have formed the basis of Fermat had in mind; though, if Weil's reconstruction is accurate, Fermat took it much farther. We cannot know for sure. Closed out of the dialogue almost as soon as it had begun, Fermat never did disclose his complete solution of the problem; posterity inherited from the controversy only Wallis' and Brouncker's method, which Euler would attribute to Pell.[124]

Nonetheless, several of the things Wallis said in the course of the exchange did have bearing on Fermat's hopes for "redeeming the patrimony." Upon hearing of Fermat's rejection of Brouncker's fractional solutions of the first challenge, Wallis first tried simply to dismiss the problems themselves: "Besides, I added that I thought problems of this nature [i.e. aliquot part problems], of which it is easy to conceive a large number in a short time, to demand more work than they offered in utility or difficulty."[125] Later, he enlarged on this theme. Again defending his immediate solution, 1, to both problems of the first challenge, he added, "If I do not give others, it is not because I think they do not exist, but rather it is because he asks for no more and because I do not think the matter to be of such consequence (for to what end?) that it deserves research in detail, much less that all of England, with France and Holland, whom he challenges all together, should devote themselves to this study."[126] After posing a counter-challenge, Wallis continued: "Fermat can, if he wishes, attack this problem or, if he prefers, let it go. But I do not attach to it such importance that I would judge him more able if he succeeded, or less so in the contrary case."

Fermat had misjudged his man. Wallis had no interest in number theory. His solutions to the various challenge problems show that, because they uni-

---

[123] For an account of the mathematical details, see Gustav Wertheim, "Pierre Fermats Streit mit John Wallis," *Abh. z. Gesch. d. Math., 9*(1899), pp. 555–576.

[124] J. E. Hofmann supplies an ingenious reconstruction in his "Studien zur Zahlentheorie Fermats" (cf. above, n. 4). It is, however, quite hypothetical, relying on the sparsest of hints, and involving a sophistication of technique that Fermat's extant writings would seem to belie. Weil offers a fitting caveat to conclude his discussion of the matter in *Number Theory* (p. 99): "Faced with a difficult or complicated proof, Fermat might well be content with a careful analysis of some typical numerical cases, convincing himself at the same time that the steps involved were of general validity. Actually, he never succeeded in writing down his proofs. Were they such as Fermat himself, in his criticism of Wallis (cf. §III), has insisted that they should be in order to provide 'the foundations for a new branch of science'? That what we shall never know."

[125] Wallis to Digby, 3/13.IX.1657, *Comm. epist.*, #VII; FO.III.414.

[126] Wallis to Digby, 21.XI/1.XII.1657, *Comm. epist.*, #XVI; FO.III.429. The letter includes Wallis' reply to Fermat's critique of the *Arithmetica inflnitorum*.

formly overlook the theoretical aspect Fermat meant the problems to contain and to reveal upon analysis. Wallis steadfastly refused to go beyond his immediate solution to the first challenge and to explore the matter as a general problem, irrespective of specific solutions. His solution of the "four cube problem" simply listed twenty-two pairs of equal sums of cubes. Even the final solution of the equation $px^2 + 1 = y^2$ by continued fractions, though it did represent a general solution procedure, was nonetheless presented in a series of specific examples, most of them taken from Fermat's statement of the problem. Hence, Wallis frequently resorted to the technique of setting up tables of test values, a method Fermat would not countenance because it could not be expressed in general form and did not produce the sorts of theorems he was looking for.

Wallis' overwhelming sense that number theory consisted essentially of wearying computations closed his mind to the promises Fermat was making about the new arithmetic. His reaction to the "negative" propositions of which Fermat spoke in his letter to Digby amply illustrates Wallis' attitude toward number theory as a whole:[127]

> I say the same thing for his recent negative propositions. . . . I am not particularly worried whether they are true or not, since I do not see what great consequence can depend on their being so. Hence, I will not apply myself to investigating them. In any case, I do not see why he displays them as something of a surprising boldness that should stupefy either M. Frenicle or the English; for such negative conditions are very common and are familiar to us. His offer nothing better or stronger than if I were to say:
>
> There is (in integers) no cubocube (I mean sixth power), or even square, which added to 62 makes a square. Or, besides 4, there is no square which added to 12 makes a square. Or besides 16, there is no quartic which added to 9 makes a square. Or, there are no integer cubes of which the difference is 20, or aside from 27 and 8, of which the difference is 19. Or, there are no integer quartics of which the difference is 100 or (to say it all at once) any other even number not divisible by 16.
>
> It is easy to think of innumerable negative conditions of the sort.

Hence, not only would Wallis not take up the problem of splitting a cube into two cubes, but Fermat's later assertion that he could prove the problem to be unsolvable would fail to evoke Wallis' interest.

As E. T. Bell has pointed out, Wallis' negative propositions here are of an entirely different level of difficulty from Fermat's.[128] For example, to prove that $x^2 + 12 = y^2$ has only one integral solution, one need only express the equation in the form $y^2 - x^2 = 12 = (y + x)(y - x)$ and test all pos-

---

[127] *Ibid.*, FO.III.438.
[128] "Wallis on Fermat," *Scripta mathematica*, Vol. 15, pp. 162–163.

sible factorizations of 12. All of Wallis' propositions reduce to similarly facto-
rizable equations. By contrast, to show that the equation $x^2 + 2 = y^3$ has
only one solution, as Fermat asserted, one cannot rely on such direct factorization. Therein lies the difficulty and challenge of the problem. Wallis was in
no mood to see the difference.

When, then, upon receipt of the solutions to his second challenge by Wallis'
and Brouncker's method of continued fractions, Fermat tried to expand the
subject under consideration, he had already lost his audience and, with it, his
dream of instituting a new tradition in arithmetic. Fermat's letter to Digby of
June 1658 is poignant in its naive bliss. "I willingly recognize, indeed I rejoice," he began,[129]

> that the most illustrious men, Viscount Brouncker and John Wallis, have
> finally given legitimate solutions to the numerical problem I set forth. The
> most noble gentlemen did not wish even for a single moment to confess
> themselves unequal to the proposed problems. I should prefer that they had
> also straightaway acknowledged these problems as worthy of English ef-
> forts and that, once having arrived at their solutions, they had celebrated a
> triumph made more illustrious by the fact that the battle proved more ar-
> duous. They were of contrary opinion; that is to be attributed to the glory of
> their most illustrious and ingenious nation. However, in order that we here-
> after act frankly on both sides, the French will say that the English satisfied
> the proposed problems. But let the English say in turn that the problems
> were worthy of being proposed to them and let them not disdain in the
> future to examine and investigate more closely the nature of integers and
> also to foster this subject, in which they are esteemed for their subtlety and
> strength of mind.

Now, perhaps, Wallis and his English friends would be interested in some of
Fermat's findings with regard to Diophantus. Fermat announced with some
pride—bringing attention, in fact, to Descartes' failure to resolve the
problem—his proof of Diophantus' conjecture that any number can be expressed as the sum of at most four squares.[130] "And I can add many propositions that are not only celebrated but proved by the soundest demonstrations,"
he continued, citing the results he had offered earlier to Pascal.

But this was to be a dialogue. "We recall that Archimedes did not disdain to
put the final hand to Conon's true but unproved propositions and to confirm
their truth by those most subtle demonstrations. Why therefore should I not
expect similar help from my distinguished colleagues, a French Conon, as it

---

[129] FO.II.402–403.

[130] *Ibid:* "Id Renatus ipse Descartes incognitum sibi ingenue declarat in epistola quadam,
quam propediem edendam accepimus, imo et viam, qua huc perveniatur, difficillimam et ab-
strusissimam esse non diffitetur."

were, from English Archimedes?"[131] Could Wallis and his friends help with the proof to the following three propositions: every number of the form $2^{2^n} + 1$ is prime; the double of every prime of the form $8k + 1$ can be expressed as the sum of three squares; the product of any two primes of the form $20k + 3$ or $20k + 7$ can be expressed as the sum of a square and five times a square?

Fermat's letter reached Wallis in August 1658. By that time Wallis had prepared the *Commercium epistolicum* for publication and could do no more than include the letter as the last entry. He never replied to it. Fermat had again lost by winning. He had created the stir he intended with his first and second challenges. Letters had crossed the Channel, tempers had flared, but no one had changed his opinion of number theory. No one was prepared to assist Fermat in restoring the pure doctrine of whole numbers. As noted above, history has even compounded the futility of Fermat's effort. By a mistake of Euler's, the central bone of contention, the equation $y^2 - px^2 = 1$, has gone into the mathematical literature bearing the name of John Pell, whose only contribution was to write it down in his papers.

V. ONE FINAL ATTEMPT: THE "RELATION" TO HUYGENS (1659) AND THE METHOD OF INFINITE DESCENT

Launched with high hopes of awakening contemporaries to the beauty and subtlety of number theory, the challenges of 1657 encountered only lack of interest. Firm in the belief that England had carried the day against the French, Wallis could in 1658 publish the *Commercium epistolicum* and still not see what the real mathematical issues had been. He never ceased to wonder how a man of Fermat's standing could have staked his reputation on such patently simple and useless exercises in elementary algebra. Across the Channel, van Schooten had perceived, at least in outline, the underlying subtlety of the problems and the difficulty of the mathematics Fermat was pursuing, but he could not summon the enthusiasm or the energy to pursue it himself. Only Frenicle had both the interest and the understanding. Unfortunately, he lacked the mathematical talent to vie with Fermat. One mathematician had the talent, and his letters seemed to suggest that he also had the interest. Christiaan Huygens became Fermat's next and last target for proselytization.

Huygens had opened the correspondence in 1656.[132] His letters shone with warm respect for Fermat and pressed the aging mathematician to reveal what Huygens suspected to be a wealth of mathematical treasures. Huygens' solicitations should have reinforced the lesson Fermat could draw from the failure of his challenges. Faith in the mathematical insight of others to see clearly the implications of vaguely stated questions would not suffice to restore the an-

---

[131] FO.II.404. On Conon and Archimedes, see above, n. 100.
[132] See above, Chap. II, p. 67ff.

cient heritage of number theory. Fermat would have to take a more explicit route. He would have to expose something of his methods. But when he sat down in 1659 to take stock for Huygens of what he had accomplished and how he had accomplished it, his old habits retained their hold on him. Fermat again revealed no more than he had to reveal, and again it was not enough.

"Because the ordinary methods now in the books were insufficient for demonstrating such difficult propositions," Fermat began in his "Relation des nouvelles découvertes en la science des nombres" addressed to Carcavi but meant for Huygens, "I finally found a totally unique route for arriving at them."[133] He called the method "infinite descent" and had used it at first to prove the "negative propositions" Wallis thought such a waste of time. For example, no number of the form $3k - 1$ can be composed of a square and the triple of a square, or no right triangle has a square area.[134] The method, he continued, took the form of a *reductio ad absurdum*:

> If there were any integral right triangle that had an area equal to a square, there would be another triangle less than that one which would have the same property. If there were a second less than the first which had the same property, there would by similar reasoning be a third less than the second which would have the same property, and then a fourth, a fifth, etc., descending *ad infinitum*. Now it is the case that, given a number, there are not infinitely many numbers less than that one in descending order (I mean always to speak of integers). Whence one concludes that it is therefore impossible that there be any right triangle of which the area is a square. . . .
>
> I do not add the reasoning by which I infer that, if there were a right triangle of that nature, there would be another of the same nature less than the first, because the argument would be too long and because that is the whole mystery of my method. I will be content if the Pascals and the Robervals and so many other learned men search for it according to my indications.

The frustration of the previous two years had, then, taught Fermat nothing about himself or his colleagues. The man who so often cited Francis Bacon's title vignette of the *New Organon*, "Many will pass by and knowledge will increase," could still not bring himself to publish his methods, to share the wealth of his private mathematical storehouse. For this hint of how the method of infinite descent worked in fact showed nothing, and Fermat knew

---

[133] FO.II.431-436.

[134] As Weil points out (*Number Theory*, 75), it is strange that Fermat would have needed a proof by reduction that no prime of the form $3k - 1$ can be expressed in the form $a^2 + 3b^2$, since it follows straightforwardly from consideration of residues modulo 3. If it were the case, then $a^2 \equiv -1 \pmod 3$, which has no solution.

it. In his final effort to engage the interest of others, Fermat would not even disclose as much of the method as he had done in his *Observations on Diophantus*.

Before turning to that other description, however, let us read Fermat's "Relation" through to the end. The promise it contains (at least in retrospect), a promise that Fermat's stubborn secretiveness helped to smother, lends it a poignancy all its own. Fermat continued his description of the method by telling Huygens that its use in proving affirmative propositions had eluded him for some time. In particular he wished to prove that every prime number of the form $4k + 1$ is composed of two squares. "But finally oft-repeated meditation gave me the insight I lacked, and affirmative questions passed under the aegis of my method with the aid of some new principles it was necessary to add to it." Assume some prime of the form $4k + 1$ were not composed of two squares. Then there would exist another prime of the form $4k + 1$ smaller than the first, which also would not be composed of two squares. The reduction process from the first number to the second is generally applicable. Hence, if it is carried on long enough, it will lead to the smallest prime number of the form $4k + 1$, i.e. 5. And, by the reduction argument, 5 would not be composed of two squares. But $5 = 2^2 + 1^2$. Hence, by contradiction, the original assumption was false, and every prime number of the form $4k + 1$ is composed of two squares. What Fermat did not include in his "Relation" to Huygens, however, was the specific reduction process that turns this outline of a theoretically valid proof into an actual proof of the proposition at hand. How, that is, in detail does one generate from a given prime of the form $4k + 1$ assumed not to be composed of two squares another prime of the same form and less than the first with the same property? Fermat nowhere gives the answer, if indeed he had one.

Left to conjecture what he might have had in mind, historians may look to Euler, as Weil has, for a proof that lay within Fermat's conceptual reach.[135] Having established that a given prime p of the form $4k + 1$ divides some sum $a^2 + b^2$, one takes a' and b' as the residues $\leq 2k$ of a and b modulo p, respectively. Then $a'^2 + b'^2 < p^2/2$ and still contains p as a prime factor. The remaining prime factors are all $< p/2$ and are either 2 or of the form $4k + 1$. The reduction comes down to showing that, if those factors are all the sums of two squares, then so too is p. One route to that result comes by way of the fundamental identity $(a^2 + b^2)(c^2 + d^2) = (ad \pm bc)^2 + (ac \mp bd)^2$ to express p as the sum of two rational squares and then show (as Fermat admitted he could not in 1636) that the sum can be transformed into the sum of two integer squares. Euler took a more direct path by way of a lemma expressed by Weil in the form:

---

[135] Weil, *Number Theory*, pp. 67–69, to which the interested reader may turn for details of the argument to follow. Weil maintains (p. 67) that "we may with some verisimilitude attribute its substance to Fermat."

For any $N = a^2 + b^2$, let $q = x^2 + y^2$ be a prime divisor of N. Then N/q has a representation $u^2 + v^2$ such that the representation $N = a^2 + b^2$ is one of those derived by composition from it and from $q = x^2 + y^2$.

Both routes reduce the decomposition of p to that of smaller prime of the same form and hence ultimately to the smallest of that form, 5, which is the sum of two squares. But this is Euler's proof; as far as Fermat is concerned, we can at best say with Weil, "Except for minor details, Fermat's proof could have been much the same."[136] We cannot know for sure.

What else had Fermat accomplished by using his method of infinite descent? He listed his results for Carcavi and Huygens. With great difficulty, and some further adjustments of the method, he had proved Diophantus' conjecture that every number is a square or is composed of at most four squares. Again, Euler would have to supply what Fermat omitted, to wit, the details of a proof. "By infinite descent applied in a quite specific way," he had proved that the equation $py^2 = x^2 - 1$ for non-square p has an infinite number of integral solution pairs x, y. It was that proof Fermat had been seeking from Frenicle and Wallis, but again he went no farther than to claim he possessed it:

I grant that M. Frenicle has given various particular solutions, as has Mr. Wallis, but the general demonstration will be found by the descent duly and properly applied. I point that out to them so that they will add the demonstration and general construction of the theorem and of the problem to the special solutions they have given.

The method of infinite descent in fact lay behind all the challenges he had made in 1657: that no cube can be the sum of two cubes; that $x^2 + 4 = y^3$ has only two integral solutions; that $x^2 + 2 = y^3$ has only one; and that all square powers of 2 increased by one are prime. Finally, Fermat cited the proposition that had started it all: 1 and 7 are the only values of p for which the pair of equations $2x^2 - 1 = p$, $2y^2 - 1 = p^2$ has integral solutions x, y.

"After having run through these problems, most of them of diverse nature and requiring different means of proof, I passed over to the invention of general rules for solving the 'single' and 'double' equations of Diophantus." This was new material that Fermat's earlier correspondents had not seen. The two examples, the "single" equation $2x^2 + 7967 = y^2$ and the "double" equation $2x + 3 = y^2$, $2x + 5 = y^2$, do not meet the requirements placed on the "single" and "double" equations treated in the correspondence of the early 1640s, i.e. that the expressions contain square terms. "Bachet took pride in his commentaries on Diophantus in having found a rule for two particular

---

[136] *Ibid.*, p. 69. Note, however, that the lemma is precisely a theorem for which there is no explicit evidence in Fermat's works; see above, n. 78.

cases," but Fermat now had the general rule for solving all such problems or for proving that they had no solution. All this had made it possible to restore most of the defective propositions in the *Arithmetic*, not only the ones that Bachet had been unable to repair, but even some before which Diophantus himself had hesitated. Again, Fermat offered only claims.

Although Fermat had been unable to perfect his technique for determining primality, he had devised ways of limiting the number of divisors to be tested. "If M. Frenicle would send his thoughts on the matter, I think it would be a considerable help to learned men." (Who else but Fermat could in this context accuse another person of holding back valuable information?) Fermat had also been at work on restoring Diophantus' *On Polygonal Numbers*, but had not found a method that fully satisfied him.

Fermat did not intend the "Relation" as an apologia, but as propaganda. In the concluding paragraph he borrowed, perhaps unconsciously, perhaps with an eye toward the man he was addressing, a phrase from the concluding paragraph of Descartes' *Geometry*:[137]

> There in summary is an account of my thoughts on the subject of numbers. I wrote it only because I fear I shall lack the leisure to extend and to set down in detail all these demonstrations and methods. In any case, this indication will serve learned men in finding for themselves what I have not extended, particularly if MM. de Carcavi and Frenicle share with them some proofs by infinite descent that I sent them on the subject of several negative propositions. *And perhaps posterity will thank me for having shown it that the ancients did not know everything*,[138] and this relation will pass into the mind of those who come after me as a "passing of the torch to the next generation," as the great Chancellor of England says, following the sentiment and the device of whom I will add, "Many will pass by and knowledge will increase." (Emphasis added)

Only the phrasing derived from Descartes; the sentiment was peculiarly Fermat's. Descartes had hoped that posterity would appreciate the fact that, although capable of solving everything, he had left some crumbs in mathematics for it to learn by. Fermat claimed no more than to have devised a menu and shown the richness of the feast that awaited those with hearty appetites. He did so in the spirit of Francis Bacon, the prophet of science as a community effort. How paradoxical, then, that Fermat insisted, even in the "Relation," on

---

[137] It was, after all, Huygens to whom van Schooten related Descartes' remark, "Fermat est Gascon, moi non." Cf. van Schooten to Huygens, 19.IX.1658, FO.IV.122.

[138] FO.II.436: "Et peut-être la posterité me saura gré de lui avoir fait connaître que les Anciens n'ont pas tout su." Cf. Descartes, *Géométrie* (Leiden, 1637), p. 413: "Et j'espère que nos neveux me sauront gré, non seulement des choses que j'ai ici expliquées, mais aussi de celles que j'ai omises volontairement, afin de leur laisser le plaisir de les inventer."

keeping his recipes a guarded secret. It is the paradox of his career in number theory, indeed in mathematics itself.

VI. INFINITE DESCENT AND THE "LAST THEOREM"

Fermat lacked the volubility to fit a Baconian mold. Not only did he describe his methods to others in vague and often misleading terms, he also left his own notes in a cryptic form of mathematical shorthand that only he could follow. The several approaches made to Fermat, the last being that of Huygens, to edit and publish his work placed a heavier demand on Fermat than his encouragers thought. Fermat recorded the bulk of his mathematics in his mind and not on paper. A skilled and ingenious problem-solver, he wrote for himself no more than a sketch of his ideas, a sort of spur to the memory should he again have recourse to those ideas when facing some future problem. Time and again in the chapters above, we have had to flesh out the conceptual skeletons in Fermat's papers; indeed, we have often had to add a few bones. Having to do so has not only made the task of understanding Fermat's mathematics difficult, but it has also uncovered instances in which Fermat placed greater faith in the generality of his methods than they warranted; e.g. in his demonstrations of the method of maxima and minima, in his classification of curves, in his method of quadrature via centers of gravity. The state of Fermat's written records has repeatedly pointed to a man who lacked either time or will, or perhaps both, to work things out in detail, to polish them and make them complete. The method of infinite descent provides the final and best example.

The description of the method in Observation 45 would have offered Fermat's contemporaries little purchase on it, even if he had let it get beyond the margin in which it was scribbled. Understanding it and following out its hints required a way of doing mathematics that men such as Frenicle and Wallis had been unable to comprehend. The observation, the theorem that the area of a right triangle in numbers cannot be a square, begins with the same promise that permeates the "Relation":[139]

> We will append the demonstration of this theorem we have found, which demonstration we ourselves also uncovered, though not without hard and laborious thought. Indeed, this sort of demonstration will bring forth marvelous progress in arithmetic.

And then Fermat outlines the proof by infinite descent:

> If the area of a triangle were a square, there would be given two quadrato-quadrates of which the difference were a square. Whence it follows that two

[139] Observation 45, FO.I.340–341.

squares would be given, of which the sum and the difference would be squares. And thus a number composed of a square and the double of a square would be given equal to a square, under the condition that the squares composing it make a square. But if a square number is composed of a square and the double of another square, its root is similarly composed of a square and the double of a square, *as we can most easily demonstrate.* Whence one concludes that this root is the sum of the sides about the right angle of a right triangle, and that one of the squares composing it constitutes the base and the double square is equal to the perpendicular.

Hence, this right triangle is composed of two squares of which the sum and difference are squares. But these two squares will be proved to be smaller than the first squares initially posited, of which the sum and difference also made squares. Therefore, if two squares are given of which the sum and the difference are squares, there exists in integers the sum of two squares of the same nature, less than the former.

By the same argument there will be given in the prior manner another one less than this, and smaller numbers will be found indefinitely having the same property. Which is impossible, because, given any integer, one cannot give an infinite number of integers less than it.

Narrowness of space prevents inserting the whole demonstration explained in detail in the margin. By this argument we have recognized and confirmed by demonstration that no triangular number except unity is equal to a quadratoquadrate.

In citing the last corollary, Fermat omits another, similar theorem that would interest posterity even more intensely. To find that missing theorem, let us try to decipher Fermat's argument by infinite descent in detail.

If any triangle existed with a square area, one would exist such that its generator (p,q) contained mutually prime members. The area, then, would be $pq(p^2 - q^2)$. But $(p,q) = 1$ implies that also $(pq, p^2 - q^2) = 1$,[140] whence it follows that $pq = b^2$ and $p^2 - q^2 = c^2$ (i.e. if the product of mutually prime factors is a square, then each of the factors is a square). Again, $(p,q) = 1$ implies that $p = d^2$ and $q = f^2$, whence $d^4 - f^4 = c^2$ ("two quadratoquadrates of which the difference is a square"). From the last equation it follows that $(d^2 + f^2)(d^2 - f^2) = c^2$; and from $(d,f) = 1$, that $d^2 + f^2 = g^2$ and $d^2 - f^2 = h^2$ ("two squares would be given of which the sum and difference would be squares"). Thus, the initial assumption of a triangle having a square area leads to the conclusion that such a triangle would be generated by a pair of squares, $p = d^2$ and $q = f^2$, the sum and

[140] Suppose pq and $p^2 - q^2$ had a common factor $d \neq 1$. Then $p^2 + q^2$ would also have d as a factor. Let $p^2 + q^2 = md$ and $p^2 - q^2 = nd$ whence $2p^2 = (m + n)d$ and $2q^2 = (m - n)d$. From $(p, q) = 1$ it follows that both p and q cannot be even, whence neither can both pq and $p^2 - q^2$. Hence, $d \neq 2$. Therefore d divides both p^2 and q^2, whence d divides both p and q, which contradicts the assumption that $(p, q) = 1$.

difference of which are both squares. The area of the assumed triangle would therefore be $d^2f^2(d^2 + f^2)(d^2 - f^2)$. Now, since $d^2 = h^2 + f^2$, it follows that $g^2 = h^2 + 2f^2$ ("a number composed of a square and the double of a square would be given equal to a square, under the condition that the squares composing it make a square"). But that means that g is also of the form $k^2 + 2m^2$ ("if a square number is composed of a square and the double of another square, its root is similarly composed of a square and the double of a square, as we can most easily demonstrate").[141] On the one hand, then, $g^2 = (k^2 + 2m^2)^2 = (k^2 - 2m^2)^2 + 2(4k^2m^2) = h^2 + 2f^2$, whence we may set $f^2 = 4k^2m^2$. On the other hand, $g^2 = (k^2 + 2m^2)^2 = k^4 + 4m^4 + 4k^2m^2 = d^2 + f^2$. Since $f^2 = 4k^2m^2$, $d^2 = k^4 + 4m^4$. But $k^4 + 4m^4 = d^2$ means that k^2 and $2m^2$ are the sides of a right triangle of hypotenuse d ("this root is the sum of the sides about the right angle of a right triangle, and one of the squares composing it constitutes the base and the double square is equal to the perpendicular"). Clearly, though Fermat does not make the point explicitly, the new triangle has a square area, k^2m^2. And, hence, by a repetition of the initial argument, the new square area "is composed of two squares of which the sum and difference are squares"; that is, just as the original triangle was generated by a pair of squares, d^2, f^2, so too the derived triangle will be generated by some pair u^2, v^2. But, since d is the hypotenuse of the derived triangle, $d = u^4 + v^4$, and in the domain of integers, $u^4 + v^4 > u^2 + v^2$. Hence, $d > u^2 + v^2$, and, *a fortiori*, $d^2 + f^2 > u^2 + v^2$ ("if two squares are given of which the sum and difference are squares, there exists in the integers the sum of two squares of the same nature but less than the former sum"). By a repetition of the (fully general) argument, one can then generate another pair of squares, r^2 and s^2, such that $d^2 + f^2 > u^2 + v^2 > r^2 + s^2$ and another, and another, *ad infinitum*. But each of the sums is an integer, and in the domain of integers one cannot have an infinite, descending sequence. Hence, no right triangle can have a square area.

In his description of the method of infinite descent in Observation 45, Fermat does not call special attention to an immediate and important extension that he appears to have made elsewhere. For he asks in Observation 33, "But why not seek two quadratoquadrates of which the sum is a square?"[142] He then answers his own query, "Because this problem is impossible, as our method of demonstration can establish without a doubt." The reference is clear if we look at the proof just presented. The assumption of a triangle having a square area entails the assumption that there exists a triple of integers a, b, c such that $a^2 = 4b^4 + c^4$ (a is the hypotenuse of the triangle with a square area). But the assumption leads in its implications then to an infinite, descending sequence of triples, e.g. d, m, k, such that $d^2 = 4m^4 + k^4$ (where d is clearly less than a). Hence, no such triple exists.

---

[141] See below for a conjecture concerning the proof Fermat may have had in mind.
[142] FO.I.327.

Now there is another sort of triple very close in nature to the one just discussed, to wit, the triple a, b, c such that $a^2 = b^4 + c^4$. And, with only the slightest adjustments, the proof by infinite descent presented above can be adapted to disprove the existence of that triple, or (what is the same thing) the existence of a right triangle in which the two smaller sides are each squares. If such a triangle exists, then one exists such that its generator (p,q) contains mutually prime members. If, then, $a^2 = b^4 + c^4$, it follows that $b^2 = 2pq$ and $c^2 = p^2 - q^2$. Clearly, p and q must be of opposite parity and, if q were odd, c^2 would be congruent to -1 (mod 4). Hence, q is even, or $q = 2r$. Now $(p, q) = 1$ implies that $(p, r) = 1$ and, since $(b/2)^2 = pr$, p and r must both be squares, i.e. $p = d^2$ and $r = f^2$. Therefore, $c^2 = d^4 - 4f^4 = (d^2 + 2f^2)(d^2 - 2f^2)$. Since c is odd, and $(d,f) = 1$, $d^2 + 2f^2 = g^2$ and $d^2 - 2f^2 = h^2$. Therefore $g^2 = h^2 + 4f^2$, or $g^2 = h^2 + (2f)^2$. Since, however, g^2 is the sum of two squares, so too is g.[143] On the one hand, then, $g^2 = (k^2 + m^2)^2 = (k^2 - m^2)^2 + (2km)^2 = h^2 + (2f)^2$ whence $f = km$. On the other hand, $g^2 = (k^2 + m^2)^2 = k^4 + m^4 + 2k^2m^2 = d^2 + 2f^2$, whence $d^2 = k^4 + m^4$. But then k^2 and m^2 are the sides of a new triangle having the same property as the original one. The hypotenuse a of the original triangle, however, was equal to $d^4 + 4r^4$, whence $a > d^4$ and, a *fortiori*, $a > d$.[144] Clearly, then, d is only the first of an infinite, descending sequence of integers. Therefore, the original assumption cannot be valid. But, if there is no triple a, b, c of integers for which $a^2 = b^4 + c^4$, there is none for which $a^4 = b^4 + c^4$.

Is the method of infinite descent what Fermat had in mind when he spoke in Observation 2 (the famous "Last Theorem") of his "marvelous demonstration" that "one cannot split a cube into two cubes, nor a quadratoquadrate into two quadratoquadrates, nor in general any power beyond the square *in infinitum* into two powers of the same name"?[145] This would seem to be the case. In the "Relation" Fermat counts the impossibility of solving $x^3 + y^3 = z^3$ among the theorems proved by infinite descent. Indeed, whenever Fermat mentions the two cases of his "Last Theorem" (i.e. cubes and fourth powers), he does so in connection with the theorem concerning triangles with square areas. The same "narrowness of the margin" (*marginis exiguitas*) that prevented Fermat in Observation 2 from carrying out his "marvelous demonstration" also prevented him in Observation 45 from filling in all the details of the

---

[143] Note that this theorem is an immediate corollary of the theorem that all prime numbers of the form $4k + 1$, and only prime numbers of that form, are uniquely decomposable into the sum of two squares; cf. above, n. 72.

[144] Note the way in which the proof being carried out here is something of a "mirror image" of the one on which it is based. In a triangle with a square area, the hypotenuse is of the form $d^4 + f^4$ and its square of the form $4b^4 + c^4$. In the triangle now under discussion, the hypotenuse is of the form $d^4 + 4r^4$ and its square of the form $b^4 + c^4$.

[145] The first mention of the theorem for cubes and fourth powers goes back to the late 1630s; cf. Fermat to Mersenne ⟨IX or X.1636⟩ (the year is disputed by Itard; see above, Chap. V, n. 31), FO.II.65. See also Fermat to Mersenne ⟨IV?1640⟩, FO.II.195, where it is set as a challenge problem for Frenicle.

proof just presented, an immediate extension of which is the proof of the "Last Theorem" for fourth powers.[146] Fermat only mentioned his "Last Theorem" in full generality once, in Observation 2, but he cited the cases of cubes and fourth powers repeatedly in his correspondence. It would seem, therefore, most probable that the success in proving these two cases led him to assume that the method of infinite descent would work for all cases, and to make the assumption without carrying out the details. It would not have been the first time that he made a (possibly mistaken) conjecture because he had not attended to details. Indeed, in the "Relation" Fermat tells Carcavi that the method of infinite descent has led to the proof of a theorem he had been struggling with for years, to wit, that all square powers of 2 increased by 1 are prime.[147] He adds, apropos of the proof by infinite descent: "This last problem results from very subtle and very ingenious research and, even though it is conceived affirmatively, it is negative, since to say that a number is prime is to say that it cannot be divided by any number."

Clearly, the subtlety and ingenuity of Fermat's proof of his conjecture concerning prime numbers of the form $2^{2^n} + 1$ lay more in his faith in the applicability of the method of infinite descent than in having actually carried out the application in full. As Euler showed in 1732, $2^{2^5} + 1$ is divisible by 641.[148] And, strikingly, just as the prime number conjecture breaks down for $n = 5$, so too the demonstration by infinite descent of the "Last Theorem" becomes conceptually more sophisticated for $n \geq 5$. For $n \geq 23$, the method of infinite descent fails altogether.[149]

How far did Fermat pursue his theorem? We do not know, but again Euler offers some retrospective hints. He first encountered the general proposition in 1748 and in 1753 reported to Goldbach that

> In Fermat there is yet another very pretty theorem, the demonstration of which he says he has found. To wit, taking off from the Diophantine problem of finding two squares of which the sum is a square, he says that it is impossible to find two cubes of which the sum is a cube, and two fourth

---

[146] Cf. Observation 2, FO.I.291: "cujus rei demonstrationem mirabilem sane detexi. Hanc marginis exiguitas non caperet." Observation 45, FO.I.341: "Demonstrationem integram et fusius explicatam inserere margini vetat ipsius exiguitas."

[147] FO.II.433–434: "J'ai ensuite considéré certaines questions qui bien que negatives, ne restent pas de recevoir très grande difficulté, la méthode pour y pratiquer la *descente* étant tout à fait diverse des précédentes, comme il sera aisé d'éprouver. Telles sont les suivants: Il n'y a aucun cube divisible en deux cubes. Il n'y a qu'un seul quarré en entiers, qui augmenté du binaire, fasse un cube. Le dit quarré est 25. Il n'y a que deux quarrés en entiers, lesquels, augmentés de 4, fassent un cube. Les dits quarrés sont 4 et 121. Toutes les puissances quarrées de 2, augmentées de l'unité, sont nombres premiers. Cette dernière question est d'une très subtile et très ingénieuse recherche et, bien qu'elle soit concue affirmativement, elle est negative, puisque dire qu'un nombre est premier, c'est dire qu'il ne peut être divisé par aucun nombre."

[148] "Observationes de theoremate quodam Fermatiano, aliisque ad numeros primos spectantibus," *Comm. Acad. Sci. Imp. Petrop.*, 6(1733), pp. 103–107.

[149] Cf. Itard's introduction to Noguès (1966), p. iv.

powers of which the sum is a fourth power, and generally that this formula $a^n + b^n = c^n$ is always impossible when $n > 2$. I have now found a demonstration that $a^3 + b^3 \neq c^3$ and $a^4 + b^4 \neq c^4$, where $\neq$ means not possibly equal. But the demonstrations for these two cases are so different from one another that I see no possibility of deriving a general demonstration from them for $a^n + b^n \neq c^n$ if $n > 2$. However, one sees as through a net rather clearly that the larger n gets the more impossible the formula must be. In the meantime I have not yet even been able to prove that the sum of two fifth powers cannot be a fifth power. This proof depends to all appearances on a lucky insight alone, and, as long as one does not come upon it, all cogitation is in vain.[150]

Euler did not publish his demonstration at the time but included a revised version in his *Complete Introduction to Algebra* in 1770.[151] As reconstructed by Weil, the first version for $n = 3$ proceeds by infinite descent and bears structural similarities to Fermat's proof of the theorem concerning triangles having square areas and to the reconstructed proof of the "Last Theorem" for $n = 4$. Applying algebraic identities to quadratic forms $x^2 + 3y^2$, it contains no step alien in principle to Fermat's way of thinking. By contrast, the 1770 proof moves beyond quadratic forms to their foundation in complex numbers of the form $p + q\sqrt{-3}$, which did lie beyond Fermat's conceptual ambit.

Perhaps more important than Euler's proof itself, especially the first version, is his testimony that, while infinite descent provided a strategy for proving the theorem on a case-by-case basis, each case required its own special tactics. Alerted by him, one may find an intimation of the same thing in Fermat's "Relation" to Huygens. There Fermat based his description of infinite descent on the proof that no triangle can have a square area, of which the proof for $n = 4$ is a corollary. It constituted, as it were, the canonical application of the method. Only after discussing the problem of adapting the method to positive questions did he then cite his theorem for $n = 3$, including it among "certain questions which, albeit negative, do not cease to harbor

---

[150] Euler to Goldbach, 4.VIII.1753, in P.H. Fuss (ed.), *Correspondance mathématique et physique de quelques célèbres géomètres du XVIII ᵉᵐᵉ siècle* (St.-Petersburg: Impr. de l'Académie impériale des sciences, 1843), I, 618. Legendre first published a proof for $n = 5$ in "Sur quelques objets d'analyse indeterminée et particulièrement sur le théorème de Fermat," *Mem. Acad R. Sci.*, 6(1823), 1–60 (Dickson, *History of the Theory of Numbers*, II, 734). Cf. the proof in Legendre's *Théorie des nombres* (3rd ed. Paris, 1830), Vol. II, 361–368; Nogues, *Théorème de Fermat*, 72–76, provides an abbreviated version.

[151] *Vollständige Anleitung zur Algebra* (st. Petersburg, 1770), ed. H. Weber in *Leonhardi Euleri opera omnia*, ser. 1, Vol. I (Leipzig/Berlin, 1911), separate rev. ed. Joseph E. Hofmann (Stuttgart, 1959); Part II, §2, Par. 243. For Euler's theory of quadratic forms, which forms the crux of his proof and which (according to Hofmann, p. 564) lacks total rigor, see Part II, §2, Chap. 12. Nogués, *Théorème de Fermat*, pp. 61–62, presents an abbreviated version of Euler's proof. See also Weil, *Number Theory*, 114–116, 239–242.

great difficulty, since the method for applying descent to them is wholly (*tout à fait*) different from the preceding ones."[152]

Euler admitted defeat on the very next case, $n = 5$, and *a fortiori* despaired of finding a general pattern. Fermat said nothing more. Was that because he had found the trick and was loath to reveal it? Had he looked in vain and chosen to hide his failure? Or had he simply not looked far enough in the first place? In his correspondence Fermat referred only to his proofs for the cubic and quartic cases, leaving the general statement in the margin of his Diophantus. It seems quite likely that he rested his case with those two instances, confident he could adapt them to any exponent. One may doubt that he ever tackled $n = 5$.

Although the "Last Theorem" almost certainly was not Fermat's last theorem in number theory, it may well serve that purpose here. It sums up Fermat's work in that field. It is shrouded in mystery because Fermat could not or would not find the time to record his "proof" for posterity, or even for himself. The "proof" probably was no proof, because Fermat could not be bothered by detailed demonstration of theorems his superb mathematical intuition told him were true.[153] The theorem probably is true—no one to this day has doubted it—because that intuition seldom erred. And Fermat's contributions to number theory, unlike his work in other fields, never slipped into obscurity because the "Last Theorem," together with many others, has hitherto remained a seemingly elementary, intuitively true theorem lacking a proof. Fermat did not live to see his dream of a new tradition in arithmetic realized, but it was. It was, from the day Christian Goldbach first called Leonhard Euler's attention to Fermat's conjecture regarding primes.[154] It was, during the

---

[152] FO.II.433.

[153] Weil (*Number Theory*, 104) views the famous theorem in another light, speaking in terms of algebraic geometry: "As we have observed in Chap. I, §X, the most significant problems in Diophantus are concerned with curves of genus 0 or 1. With Fermat this turns into an almost exclusive concentration on such curves. Only on one ill-fated occasion did Fermat ever mention a curve of higher genus, and there can hardly remain any doubt that this was due to some misapprehension on his part, even though, by a curious twist of fate, his reputation in the eyes of the ignorant came to rest chiefly upon it. By this we refer of course to the incautious words '*et generaliter nullam in infinitum potestatem*' in his statement of 'Fermat's last theorem' as it came to be vulgarly called: 'No cube can be split into two cubes, nor any biquadrate into two biquadrates, *nor generally any power beyond the second* into two of the same kind' is what he wrote into the margin of an early section of his *Diophantus* (*Fe.*I.291, Obs. II), adding that he had discovered a truly remarkable proof for this 'which this margin is too narrow to hold.' How could he have guessed that he was writing for eternity? We know his proof for biquadrates (cf. above, §X); he may well have constructed a proof for cubes, similar to the one which Euler discovered in 1753 (cf. *infra*, §XVI); he frequently repeated those two statements (e.g. *Fe.*II.65,376,433), but never the more general one. For a brief moment, perhaps, and perhaps in his younger days (cf. above §III), he must have deluded himself into thinking that he had the principle of a general proof; what he had in mind on that day can never be known."

[154] Fuss, *Correspondance*, p. 10. Apropos the theorem mentioned at the end of Fermat's description of infinite descent, see Goldbach's "Demonstratio Theorematis Fermatiani, Nullum numerum triangularem praeter 1 esse quadrato-quadratum," *Supplementum actorum eruditorum*, VIII(1724), §XI, 483–484. The proof does not rely on infinite descent and is elegant in its simplicity.

twenty years in which E. E. Kummer devised a complete theory of the complex number field while trying to prove the "Last Theorem."[155] And it has been in this century with the development of modern algebraic geometry, culminating in what, as of this writing, appears to be a proof of the theorem by Andrew Wiles.

---

[155] Cf. Nogues, *Théorème de Fermat*, 95ff.

EPILOGUE

Fermat in Retrospect

The death of Fermat in 1665 marks the end of one of the most important transitional periods in the history of mathematics. It falls in the middle of what has come to be called Newton's *annus mirabilis*, those frenetic eighteen months during 1664 and 1665 when the young Cambridge student carried out his first penetrating experiments with the prism, gained his first insights into the workings of gravitational force, and invented his system of fluxional calculus.[1] In all three areas of research, Newton began where Fermat and his contemporaries had left off; he took for granted what they had toiled for years to achieve. The law of refraction, over which Fermat and Descartes had fought so bitterly, no longer represented an issue in itself.[2] By now almost universally accepted (with Fermat's help), it served as a foundation for Newton's theory of colors and his musings regarding a corpuscular theory of light. By 1672, the debate had shifted from the geometrical to the physical; only in its bitterness did it resemble the earlier dispute between Fermat and Descartes. So too Newton's first investigations of gravitational force rendered the geostatics in which Fermat had dabbled an amusing but distractingly over-mathematized dead-end in mechanics, a *reductio ad absurdum* of the Archimedean approach to nature.[3]

Most importantly, Newton's calculus of fluxions from the outset stood

---

[1] See *The Annus Mirabilis of Sir Isaac Newton, Tricentennial Celebration*, a special number of the *Texas Quarterly*, Vol. X, no. 3 (Autumn, 1967) and, for a comprehensive account of Newton's career with considerable attention to the early years, Richard S. Westfall's *Never at Rest: A Biography of Isaac Newton* (N.Y.: Cambridge University Press, 1981).

[2] See above, Chap. IV. §IV, and below, Appendix I, §II.

[3] See below, Appendix I, §I.

on the other side of the chasm Fermat had been unable to cross.[4] Newton began with the infinitesimal concepts and limit procedures it had taken Fermat a lifetime to discern even vaguely and with the inverse relationship between tangents and areas that Fermat had never recognized. If Fermat, with others, provided Newton with much of the technical apparatus necessary to make the calculus of fluxions work, it was Newton who first saw clearly the fundamental concepts that ultimately united that technical apparatus and made it a whole new approach to mathematical problem-solving. In mathematics at least, standing on the shoulders of giants did not so much mean seeing farther as it meant seeing more deeply into the nature of the techniques Newton had inherited.

Newton was not alone in his deeper perception. Only two years separated Leibniz' mathematical education in Paris in 1672 from his invention of the differential calculus in 1674.[5] In those two years he too made the leap in conceptualization that the sources from which he learned had been unable to make. He learned from Huygens, who had learned from Fermat (via van Schooten), but like Newton he began where they had left off.

That new beginning looms so large in the subsequent development of mathematics that it has come to represent an historical watershed dividing the ancient from the modern. The tributaries that conjoin to form the calculus flow through a "century of anticipation" (Boyer), deriving their historical importance only from the mainstream they ultimately feed. As both the metaphor and the preceding chapters suggest, however, those tributaries already lie on this side of the watershed, and one must pursue them to their source in order to uncover the origins of the final confluence. To drop the metaphor, Newton and Leibniz did not create the calculus in a fit of unprecedented genius. However novel their achievement, in a profound sense it merely realized the potential inherent in a new approach to mathematics of which they were not the originators but rather the inheritors and continuators. And one cannot fully grasp their achievement until one places it in the context of that

---

[4] Cf. Newton's earliest papers on the calculus in *The Mathematical Papers of Isaac Newton*, ed. D. T. Whiteside, Vol. I (Cambridge, 1967). For another perspective on the conceptual gap separating Fermat and Newton, see M. S. Mahoney, "Barrow's Mathematics: Between Ancients and Moderns," in *Before Newton: The Life and Times of Isaac Barrow*, ed. M. Feingold (Cambridge: Cambridge University Press, 1990), Chap. 3.

[5] More precisely, Leibniz developed his version of the calculus in a series of investigations stretching from 1673 to 1676. The crucial insight, however, into the inverse relationship between quadrature and the determination of the tangent took place during his research into Pascal's triangle from late 1673 to late 1674. See *The Early Mathematical Manuscripts of Leibniz*, trans. J. M. Child (Chicago/London, 1920), esp. Chap. III, "Historia et origo calculi differentialis." On Leibniz's extension of Fermat's methods to non-algebraic configurations, see M. S. Mahoney, "Infinitesimals and Transcendent Relations: The Mathematics of Motion in the Late Seventeenth Century," in *Reappraisals of the Scientific Revolution*, ed. D. C. Lindberg and R. S. Westman (Cambridge: Cambridge University Press, 1990), Chap. 12.

new approach, a context which transcends the calculus in its importance for the development of modern mathematics.[6]

As the chapters above, in particular Chapters I and II, have tried to show, the real watershed of mathematics in the seventeenth century is the analytic program of Viète, Fermat, and Descartes. These men were the leaders of a movement that fundamentally altered the practice of mathematics. They loosed it from the strict geometric model of classical Greek antiquity and reformulated it in terms of a new algebraic model. They gave new emphasis to what they felt centuries of awe and adulation of the Greek texts had obscured; to wit, that mathematics is concerned with solving problems. Like any art, it could not rest content with admiring past masterpieces, but rather had to analyze them to discover how they had been achieved and use them as a starting point for new work. And where past masters had obscured their techniques, present artists must devise their own. Only in that way could the art progress.

Algebraic analysis was the new technique of the seventeenth century. It was the technique of one of several competing schools of mathematics. It ultimately enabled its practitioners to dominate its competitors by incorporating them into the algebraic school, though the process of incorporation lasted well into the nineteenth century. At the beginning of the seventeenth century, the success of the new school was far from assured. It could hardly point to an illustrious past, which counted for much in a culture in which intellectual as well as social nobility depended on the right genealogy.[7] Even if Ramus, Viète, or Descartes found algebra among the forebears of classical Greek geometry, each man was forced to admit that it had been something of a black sheep; even Pappus refused to name it explicitly. Everyone knew it bore an Arabic name and had traveled in mercantile circles. No one before Ramus had seriously thought of admitting it into the intellectual nobility of the school or university. Moreover, if the Greeks admitted analysis as a member of the family, they nonetheless had tried to keep it in the background, barring it from any formal appearances. Only Pappus and a few scholiasts talked about it at any length. Hence, even to achieve equal status with purer, more traditional lines of mathematical research, algebraic analysis had to prove its worth by accomplishing what the others could not.

Fermat, more than any other mathematician of his day, vindicated algebraic analysis. Ramus could do no more than name names and identify algebra as analysis. Viète tried to bring algebra back into the fold by clothing it in the

---

[6] For a fuller analysis of this context, see M. S. Mahoney, "The Beginnings of Algebraic Thought in the Seventeenth Century," in *Descartes: Philosophy, Mathematics and Physics*, ed. S. Gaukroger (Sussex: The Harvester Press; Totowa, N.J.: Barnes and Noble Books, 1980), Chap. 5.

[7] On the incorporation of algebra into the academic culture of sixteenth-century Paris, see Giovanna C. Cifoletti, *Mathematics and Rhetoric: Peletier and Gosselin and the Making of the French Algebraic Tradition* (Ph.D. diss. Princeton, 1992).

finery of Greek neologisms and making it a formal science of mathematical reasoning. Though he thereby created the theory of equations, he could not himself extend the problem-solving power of algebra much beyond the state achieved by the sixteenth-century cossists. Moreover, by trying to teach algebra Greek, he tended to becloud the vital simplicity that was its main virtue. Descartes brought out that simplicity by keeping algebra in the vernacular and offered a glimpse of the strength of the new art in his *Geometry.* But Fermat really made it work.

How he did so has been set forth in close detail in the chapters above, and a summary here would run the gauntlet between repetition and oversimplification. What he did, however, does bear repeating in the context of the present discussion. His *Introduction to Plane and Solid Loci* reduced all earlier discussions of loci (i.e. of plane curves) to the theory of indeterminate equations in two unknowns. At the same time that it multiplied greatly the sheer number of curves known by and accessible to the mathematician, its *Appendix* and the later *Triparite Dissertation* ordered those curves according to an algebraic system that reduced the earlier classifications of Pappus to mere verbiage. Whatever the differences and disagreements in detail, the analytic geometries of Fermat and Descartes fundamentally altered the balance between the visual and the abstract in mathematics. The tangible picture of a curve gave way to its abstract characterization in an algebraic equation, a symbolic statement of its essential metric properties.

This shift from the visual to the abstract perforce changed traditional habits of mathematical thought and, like any radical change, required concrete justification.[8] Fermat, more than Descartes, provided the justification. The *Method of Maxima and Minima* applied the theory of equations to the equations of curves and thereby reduced to a simple algorithm what had formerly required problem-by-problem ingenuity. The turning points (and, later, points of inflection) of any and every curve now emerged from the application of a single, uniform method. So too the tangents (and, with them the asymptotes) to curves lost their mystery. The *Doctrine of Tangents* not only allowed the straightforward determination of the tangent to any point of a given curve, but it also made clear that the tangent lay implicit in the essential property of the curve, for it derived directly from the equation that expressed that property. The *Treatise on Quadrature* and the *Treatise on Rectification* brought out the same characteristic of the area under a curve and the length of its arc; these features, too, followed from algebraic analysis of the curve's defining equation.

In short, Fermat provided tangible evidence, as none of his contemporaries

---

[8] On the changing role of visualization in mathematics, see M. S. Mahoney, "Diagrams and Dynamics: Mathematical Perspectives on Edgerton's Thesis," in *Science and the Arts in the Renaissance,* ed. J. Shirley and F. D. Hoeniger (Cranbury, N.J.: Associated University Presses, 1985), Chap. 10.

could, that the problem-solving benefits of algebraic analysis far outweighed the loss of intuitive understanding that a concrete picture and tactile operations of geometric construction seemed to provide. In doing so, he spoke directly to a new generation of mathematicians for whom utility had become more important than esthetics. In a very real sense, Fermat presided over the death of the classical Greek tradition in mathematics. At the start of his career that tradition was, as Chapters I and II have shown, both alive and thriving. Even for Viète, it represented a vital repository of mathematical knowledge; it provided him with problems and methods of solution and hence exercised a directional influence on his work. Again, as Chapters III–V have shown, the Greek tradition constituted the starting point for Fermat's work in analysis. Over his career, however, Fermat moved so far beyond his original sources as to make them obsolete. If the Greek texts still provided Newton with the form of his *Principia*, if they offered an occasional theorem necessary to carry out its program, by 1665 those texts nonetheless belonged largely to the history of mathematics. Nothing shows this so clearly as the fact that the restitution of lost Greek texts had passed from the hands of men like Viète, Snel, Ghetaldi, and Fermat into those of second-rate antiquarians whom history has forgotten. Newton learned eagerly from Descartes, Fermat, Kinckhuysen, and other modern analysts; despite his reputation for preferring the methods of classical geometry, his own effort at restoring the ancient "treasury of analysis" begins with references to equations and ends with the algebraic quadrature of curves.[9] A new tradition had taken hold of mathematics by 1665, a tradition Fermat had largely established, a tradition that spoke to the needs and demands of the young mathematicians of the last third of the seventeenth century.

Only the existence of an eager audience explains how Fermat could have exercised his demonstrable influence without ever publishing his work. As the conspectus in Appendix II shows, many of his papers spread over Europe in copies made by those who wanted his work for easy reference. Surely other copies have simply not survived. Moreover, Fermat's influence seems to have transcended his writings. People clearly talked about him. The close similarities in style, direction, and program between Fermat's *Treatise on Quadrature*, on the one hand, and Newton's *Quadrature of Curves* and Leibniz' *On Quadrature by Means of Centers of Gravity*, on the other, make coincidence improbable; if the latter do not derive directly from the former—its unique existence in the 1679 *Varia* would seem to indicate they do not—they nonetheless follow it in spirit.

Fermat's failure to publish did not preclude his influence in the development of mathematics in his age and later. It meant, rather, that the influence would be severed from his name. It meant that Beaugrand would pass off the

---

[9] See *The Mathematical Papers of Isaac Newton*, ed. D. T. Whiteside, Vol. 7 (New York: Cambridge University Press, 1976), Part 2.

method of tangents as his own in a letter to Thomas Hobbes. It meant that Johann Hudde would embellish the method of maxima and minima and claim it as his own invention, despite the defense of Fermat's priority by Huygens. It mean ultimately that the fundamental algorithm of the *Method of Maxima and Minima* would go down in history as "De Sluse's Rule." And Fermat would be forgotten. Analytic geometry would be called "Cartesian," the second-derivative characteristic for extreme values would be anonymous, even the equation $px^2 - y^2 = \pm 1$ would acquire the title "Pell's Equation." Only number theory would remain Fermat's undisputed province; it would do so, ironically, because Fermat could interest none of his contemporaries in it.

Despite the immediate obscurity into which Fermat's name fell, in retrospect his determinative effect on the development of mathematics in the seventeenth century is clear. And it extended beyond the technical content and outward style of mathematics to touch the very way in which men came to think about the subject, and beyond that to the manner in which they practiced it. Algebraic analysis meant the replacement of geometric construction by the solution of algebraic equations, the replacement of synthetic proof by analytic derivation, the replacement of theorems by solution techniques. But, beyond that, it meant the replacement of the treatise by the memoir, the book by the article, and ultimately the "amateur" by the professional.

Algebraic analysis became, by the late seventeenth century, the single discipline that mathematics had lacked earlier and that could form the core of an emerging profession of mathematics. Oughtred, Hérigone, even Descartes and van Schooten, provided the textbooks for the discipline, and men like Billy brought it into the curriculum. It became the mathematics of the Academy of Sciences in Paris and the new Imperial Academy in St. Petersburg. It filled the pages of the *Journal des Savants*, the *Acta eruditorum*, the *Mémoires de l'Académie*, and the *Commentarii Academiae Imperialis Petropolitanae*. It formed the common language of Continental mathematics by the early decades of the eighteenth century and hence the language anyone who would be called a mathematician had to command. Algebraic analysis became, by the end of the seventeenth century, what Thomas S. Kuhn has called, in other contexts, a "paradigm for normal science."[10] The standard models were clear and universally accepted: Cartesian symbolism; analytic geometry; Leibniz' differential and integral symbolism; Newton's method of fluxions; the fundamental algorithm for differentiation; and the inverse relationship between differentiation and integration, to name just the outstanding features. And the problems were clear: the reduction of transcendental functions by infinite series; the summation, transformation, and convergence

---

[10] See his *Structure of Scientific Revolutions* (Chicago, 1962). Though Kuhn expressly exempts mathematics from his discussion, the development of algebraic analysis in the seventeenth century displays several of the features he thinks crucial to scientific revolutions. On the general issue, see Donald Gillies, ed., *Revolutions in Mathematics* (Oxford: Clarendon Press, 1992).

of series; the solution of ever more complex differential equations, both ordinary and partial; and so on. If mathematicians of the eighteenth century disagreed over many things, they agreed on much more, for they agreed fundamentally on how mathematics should be pursued. By realizing the dream of the analytic school of Viète, Fermat made that agreement possible.

Hence, the career of Fermat illustrates more than mathematics in transition; it reveals the mathematician in transition. With Leibniz, Fermat was one of the last great mathematicians to pursue the subject as a sideline to an essentially non-scientific career. He was also among the last to stand on the frontiers of research in all areas of mathematics in his day. The explosive growth of the subject, which he played such an important role in releasing, entailed the increasing specialization of mathematical research and competence and the concomitant necessity for formal training and continuing contact with a community of practitioners. In retrospect, Fermat's achievement meant the end of men like himself. It placed mathematics beyond the point where a provincial lawyer far removed from a center of scientific activity could, on his own and in his spare time, make fundamental advances, and thus it set the boundaries that would increasingly separate the professional from the amateur in mathematics.

APPENDIX I

Sidelights on a Mathematical Career

I have always believed that it is rather difficult to shake and destroy the principles of the sciences, for, being founded on the laborious experience of those who have sought them out, it seems that it is rather difficult to make them more precise. And it is still more futile to summon reason to the aid of the senses, since in its operations reason always supposes the senses to be exact and reliable.[1]

In keeping with the title of this book, the chapters above have focused on Fermat's mathematics. As Chapter II pointed out, however, Fermat did not limit his attention exclusively to mathematics narrowly defined. His introduction to mathematical circles in Paris resulted from his work in geostatics; he carried on a lively correspondence with Roberval on the subject during the first year of their acquaintance. If DeWaard is correct in his attribution to Fermat of two letters published in the *Supplement*, Fermat made known to Galileo his objections to the theory of motion presented in the *Two World Systems*.[2] A letter addressed to Gassendi in 1646 offers a welcome contemporary demonstration of Galileo's assertion in the *Two New Sciences* that the assumption that the velocity of a uniformly accelerating body is proportional to the distance through which it has moved entails instantaneous motion. As Chapter IV has outlined, Fermat and Descartes clashed bitterly over the latter's derivation of the laws of reflection and refraction. The dispute outlived Descartes as Fermat confronted Clerselier and Rohault in the late 1650s, and it culminated in Fermat's

---

[1] Fermat to Mersenne, 26.IV.1636, FO.II.3.

[2] The first of the memoirs (FO.*Suppl.* 15-19) attacks Galileo's assertion that the true path of a body falling from a tower on a rotating earth is a semicircle linking the top of the tower with the center of the earth (cf. *Dialogue Concerning the Two Chief World Systems*, trans. S. Drake, Berkeley, 1962; p. 165); Fermat insists that the path is a spiral in which the square of the radius is proportional to the angle. In the memoir itself, discussed in detail above, Chap. V, Fermat concentrates on the quadrature of any segment of the spiral. According to Drake (*Ibid.*, p. 480), Galileo was informed of Fermat's objection by Carcavi in 1637 but by that time had already determined the parabolic path presented in the *Two New Sciences*. Propositions III and IV of Book II of the *Second Part* of Mersenne's *Universal Harmony* indicate, however, that the spiral path was well known in 1636, when Mersenne was gathering material for the book. Mersenne's personal, annotated copy of the *Universal Harmony* (Paris, 1963) contains two interesting marginalia on p. 96. In regard to Proposition IV, "Montrer qu'il est impossible que les

own derivation of the same laws by means of the method of maxima and minima. Finally, as Chapter II noted, Fermat reestablished contact with Paris in 1654 after a ten-year lapse by exchanging with Blaise Pascal a series of original thoughts concerning the calculation of probabilities. For historians of science that exchange marks the beginning of the theory of probability.

For several reasons none of these activities has received close attention in the chapters above. They are, first of all, essentially irrelevant to the development of Fermat's mathematics. If his concern with geostatics, for example, suggests how to fill a gap in the demonstration of Proposition II,5 of the *Plane Loci*,[3] it offers no insight whatever into the central feature of that proposition, to wit, the shift from geometry to algebra. Fermat's letter to Gassendi may help to date more precisely the point at which he made the breakthrough that led to the basic solutions of the *Treatise on Quadrature*, but it tells us nothing about the mathematics of quadrature itself. The method of maxima and minima had lain complete for almost three decades before Fermat thought to apply it to optics. Nowhere do the letters exchanged with Pascal shed light on Fermat's work in number theory; at best, they fix certain of his results firmly in time.

Secondly, except perhaps for geostatics, none of these subjects held Fermat's attention for any length of time. The optical dispute with Descartes involved only two letters from Fermat and one from Descartes; lasting essentially from May to December 1637, the argument quickly became sidetracked on the issue of maxima and minima. The correspondence with Clerselier in the late 1650s shows clearly that Fermat had given no thought to the subject in the intervening twenty years. Fermat and Pascal discussed a single basic problem in probability for less than six months before Fermat tried to change the subject to number theory and Pascal withdrew from correspondence. The dispute with Roberval in 1636 over geostatics died a natural death from lack of interest on Fermat's part by the end of that year.

Indeed, whatever the contribution of Fermat's correspondents to the short life of these subjects in his thoughts, his own lack of interest was probably determinative. Both his initial letter to Mersenne and his first critique of Descartes' *Dioptrics* explain why Fermat largely ignored mathematical physics

---

corps pesants descendants jusqu'au centre de la terre décrivent le demicercle précédent, et donner la ligne par laquelle ils descendraient, si la terre tournait en 24 heures autour de son essieu," Mersenne noted later: "Voyez l'hélice de cette descente, décrite par M. Fermat" and "[Galilée] a du depuis confessé par lettre qu'il n'avait parlé de ce demicercle que par caprice, et galanterie, et non à bon escient, et qu'il avait [word illegible] l'hélice."

The second memoir (FO.*Suppl.*36–43) first criticizes Galileo's account of accelerated free fall on the grounds that a body cannot begin to move from complete rest, but must have some motion at the outset. Fermat's position rests on the strict dichotomy between rest and motion, a dichotomy that Galileo abandoned explicitly in the *Two New Sciences*. The rest of the memoir deals with the motion of bodies on inclined planes, which Fermat treats from the geostatical point of view; see below, § I.

[3] See above, Chap. V, § II.

unless forced into the subject by his friends. Simply put, he did not believe in it. He was either too poor or too good a philosopher to be swept up by the movement to mathematize nature. He saw no *a priori* reason to assume that natural phenomena obeyed exact mathematical laws of behavior, and he saw a good deal of *a posteriori* evidence to suggest they did not. The octave, he explained to Mersenne, may be determined by the exact ratio of 2:1, but it may also be that the human ear cannot distinguish between that precise interval and imprecise ones close to it.[4] Influenced perhaps by his reading of Bacon's *New Organon*, Fermat insisted that the only way to find the law of refraction lay in careful experimentation.[5] From an Aristotelian point of view, several of his objections to Descartes' approach to natural philosophy were well taken; they were Aristotle's arguments.[6] Perhaps only against that background can one appreciate Fermat's utter surprise at arriving in 1661 at the same law of refraction as Descartes. As a mathematician, Fermat had his goals clearly marked out, and they did not transcend the subject proper. Mathematics was not about nature.

In part for that very reason, Fermat did not earn in physics the fame that is deservedly his in mathematics. When forced to do physics at all, he did it like a mathematician, and in retrospect at least his physics suffered. From all evidence, he lacked the physical intuition of a Galileo, a Descartes, a Huygens, or a Newton. He insisted on exact reality where others sought insight from an ideal world; if he could abstract from the physical, he could not abstract from the real. Nothing reveals this so clearly as his work in geostatics, to be discussed immediately below. In that field, at least, Fermat could not loose himself from the classical Greek tradition and hence could take no part in the seventeenth-century revolution in mechanics.

But probability is not physics, and one may well wonder why, given the promising start of the correspondence with Pascal, Fermat did so little with

[4]Fermat to Mersenne, 26.IV.1636, FO.II.4: "... j'estime qu'il serait bien malaisé de trouver une proportion différente de la double qui fît l'octave plus exactement que celle-là. Je vous avoue bien qu'il y en a infinies, qui effectivement feront des accords différents et desquels néanmoins la différence ne sera pas comprise par l'ouïe la plus délicate qui puisse être; et de là on pourrait conclure que peut-être la vraie octave ne consiste pas précisément en la proportion double. Mais, puisque, en ce principe que les Anciens nous ont baillé, nous n'avons jusqu'à présent su découvrir d'erreur sensible, rendons-leur ce respect de le croire véritable, jusqu'à ce que le contraire nous ait apparu."

[5]Fermat to Mersenne, IX.1637, FO.II.111: "J'avais fait dessein de vous discourir ensuite de mes pensées sur ce sujet; mais, outre que je ne puis encore me satisfaire moi-même exactement, j'attendrai toutes les expériences que vous avez faites ou que vous ferez à ma prière, sur les diverses proportions des angles d'inclination et ceux de réfraction."

[6]For example, *ibid.*, pp. 108–109: "Je doute premièrement, et avec raison, ce me semble, si l'inclination au mouvement doit suivre les lois du mouvement même, puisqu'il y a autant de différence de l'un à l'autre que de la puissance à l'acte; outre qu'en ce sujet, il semble qu'il y a une particulière disconvenance, en ce que le mouvement d'une balle est plus ou moins violent, à mesure qu'elle est poussée par des forces différentes, là où la lumière pénètre en un instant les corps diaphanes et semble n'avoir rien de successif. Mais la Géométrie ne se mêle point d'approfondir davantage les matières de la Physique."

the subject and hence why it has here been relegated to an appendix. That Fermat in fact did little will become clear below. One may only guess why, but his involvement with number theory may provide the soundest, if paradoxical, conjecture. As Chapter VI has shown, Fermat was at the height of his creative period in number theory when Pascal approached him on the subject of probability. He had too much number theory to reveal, too much with which to astonish the world, to be receptive to the challenge of a new and essentially different subject. One senses at times the impatience with which Fermat listened to Pascal, waiting for the first opportunity to broach the subject of number theory. By the time he might have been ready to pursue probability further, van Schooten and Huygens had stolen a march on him. He had neither the leisure nor the will to push beyond what they had accomplished. Whatever the possible ties between probability and number theory, they were for Fermat remote enough to preclude any fruitful interaction in his own work.

To all these reasons for placing Fermat's work on geostatics, optics, and probability in an appendix, one may add still another. They require an understanding of the contemporary context that his mathematics does not. As the introduction to the present book points out, perhaps the most striking feature of Fermat's mathematical career is the extent to which it was self-contained; it does not so much fit into a context as establish one. That is why the main chapters have been able to focus so intently on Fermat's papers themselves; except for the analytic program of which the papers are the culmination, they reflect little or no input from outside influences. The same does not hold true of the three subjects under discussion. To the very extent to which Fermat's work in these areas represents his (at times grudging) response to the demands of others, they require a larger context than does his mathematics. To have worked that context into the main story would have cluttered an already complex setting; and, as has just been indicated, it would have done so with little compensatory benefit. Even the present appendix must rely on the reader's acquaintance with the larger picture of scientific thought in the early seventeenth century. A. I. Sabra's *Theories of Light from Descartes to Newton* is an example of the sort of wider context necessary to appreciate fully Fermat's dispute with Descartes over the laws of reflection and refraction. One cannot attempt here to emulate such a study (indeed, to emulate it in three separate areas); one can only try to offer some idea of what Fermat discussed and the terms in which he discussed it.

I. MECHANICS

In 1636 Jean de Beaugrand's *Geostatics* was published in Paris.[7] Even before its publication, the work provoked controversy, as its main conclusions be-

[7]*Joannis de Beaugrand Regi Franciae Domui Regnoque ac aerario sanctiori a consiliis secretisque Geostatice seu de vario pondere gravium secundum varia a terrae ⟨centro⟩ intervalla Dissertatio mathematica*; in citing this title, the eds. of FO (II.5, n. 1) note that the dedication to Richelieu is dated 20.IV.1636.

came known to the circle of mathematicians gathered about Mersenne. Roberval and Descartes, for example, who seldom agreed on anything, were united in condemning Beaugrand's results and in ridiculing his pretensions as a scientist.[8] Beaugrand badly needed support from some quarter,[9] preferably a non-partisan one. He apparently thought, then, of Fermat, whom he had taught in Bordeaux and who he knew shared his position in geostatics, without being party to the other activities that had cast a shadow over Beaugrand's reputation. On a journey to Italy in the previous year, Beaugrand had called Fermat's work to the attention of Cavalieri and others.[10] In Paris his reports of Fermat's talents undoubtedly received support from Carcavi, who had recently arrived from Toulouse to take up his new position as *bibliothécaire du Roi*. Sometime in March 1636, then, Mersenne wrote to Fermat, inviting him to join the discussion.

That discussion belongs to a largely forgotten chapter of the history of mechanics, as indeed does the subject of geostatics itself.[11] Fermat joined the fray at just the time that Galileo, Descartes, and their followers had begun to abandon geostatics as an approach to a mathematical understanding of the physical world; their new line of attack, which forms the basis of the new mechanics of the seventeenth century, relegated geostatics to obscurity, with the result that it is often difficult to perceive the precise issues, assumptions, and problem structure that underlay the hot debate of the mid-1630s.

As a mechanical tradition, geostatics had its roots in Archimedes' treatises *On Floating Bodies* and *On the Equilibrium of Planes* and in late medieval discussions of weight as a force causing bodies to move toward the center of the earth. The two sources had been brought together in the writings of sixteenth-century mechanicians such as Tartaglia, Benedetti, and Guid' Ubaldo dal Monte, and the problems they raised formed the basis for Galileo's early investigations in mechanics.[12] Two main questions emerged from these writings, the answers to which had both substantive and procedural repercussions on theoretical mechanics. First, in deriving the law of the lever, Archimedes

[8] There is no secondary account of the ongoing war between Beaugrand and Descartes during the 1630s, a war that extended beyond geostatics to encompass all of mathematics. Beaugrand had already been pamphleteering against Descartes before 1636, and in 1637 launched a vicious campaign against Descartes' *Discourse* and *Essays*, accusing him, among other things, of having plagiarized Viète in the *Geometry*. Those who sided with Descartes, for example Desargues, also felt Beaugrand's sting. For some orientation in the primary sources that contain the story, see Henry Nathan, "Beaugrand, Jean," *Dictionary of Scientific Biography*, I (New York, 1970), pp. 541–542; and René Taton, *L'Oeuvre mathématique de G. Desargues* (Paris, 1951), *passim*. See also FO.*Suppl*.33–34.

[9] He already had the support of Castelli and Cavalieri in Italy; see the sources cited in the previous note.

[10] See above, Chap. II, n. 58.

[11] For example, both Dijksterhuis' *Mechanization of the World Picture* and Dugas' *La mécanique au XVIIe siècle* effectively ignore the subject.

[12] See Stillman Drake and I. E. Drabkin, *Mechanics in Sixteenth-Century Italy* (Madison, Wisc., 1969).

had assumed that the lines of action of the suspended weights are perpendicular to the balance beam, i.e. parallel to one another. Yet, the diagrams alone in *On Floating Bodies*[13] emphasized the equally fundamental assumption that all heavy bodies tend toward the center of the earth (or, later, at least to some common center);[14] hence, in a real balance the lines of action of the suspended weights converge. Therefore, although Archimedes' theorem asserting the inverse proportionality of weight and distance from the fulcrum might serve as a useful approximation to the mathematical conditions governing small balances, it was clearly only an approximation, the accuracy of which depended on the size of the balance with respect to that of the earth. However consonant with experience, it left open the question of the true and absolute law of the lever independent of its size.

Second, Archimedes had also assumed that a given weight had the same effect on a balance no matter where that weight was placed on its line of action. That is, a weight acted in the same way whether placed directly below the balance arm or suspended from a cord several feet long. Two considerations restricted the validity of this assumption. First, it was intimately linked to the assumption of parallel lines of action, since in the case of convergent lines any lengthening of the suspension cord meant a shortening of the distance to the line of support perpendicular to the balance at the fulcrum. Second, it tacitly assumed that the weight of a body was invariant with respect to the distance of the body from the center of the earth. Few mechanicians of the late sixteenth and early seventeenth centuries were willing to grant this assumption of invariance, and even those who were willing agreed that it required proof. The question of the relationship between the weight of a body and its distance from the center constituted in itself a problem for investigation.

It took time for these problems to emerge clearly. For example, in arguing with those who maintained that all disturbed balances in equilibrium return of their own accord to a horizontal position, Guid' Ubaldo dal Monte in his *Book of Mechanics*[15] (the one work of the Italian school that Fermat cites, other than Galileo's *Two World Systems* and *Two New Sciences*) freely shifted back and forth between parallel and convergent lines of action, de-

[13]Cf. Proposition II of Book I (*The Works of Archimedes*, ed. Heath, p. 254): "The surface of any fluid at rest is the surface of a sphere whose center is the same as that of the earth."

[14]Even within the traditional Aristotelian cosmology, which placed the earth at rest in the center of the universe, there was some debate about whether the center toward which heavy bodies tended coincided with the center of magnitude of the terrestrial sphere. Copernicus' assertion of a moving earth placed an entirely new emphasis on the problem of the center toward which bodies tended and may well account for the great interest shown by the Italian mechanicians, many of whom became Copernicans during their careers.

[15]*Mechanicorum liber* (Pesaro, 1577), trans. into Italian by Filippo Pigafetta as *La Mechaniche* (Venice, 1581); Engl. trans. from the Italian by S. Drake in *Mechanics in Sixteenth-Century Italy*, pp. 241–328.

pending on the dialectical exigencies of the moment. By the mid-1630s, however, to judge from Fermat's work, the issues were somewhat clearer. If Fermat and his contemporaries argued no less vehemently over the questions that divided them than did the Italians of the sixteenth century, a quick glance at the writings that emerged shows that the later disputants had a much more precise notion of what they were arguing about.

Even before the appearance of Beaugrand's treatise, Mersenne had asked Fermat about its most controversial assertion, to wit, that the weight of a body varied directly as its distance from the center of heavy bodies.[16] Not having been in recent contact with Beaugrand and not knowing the exact nature of the debate going on in Paris, Fermat restricted his reply to a straightforward "geostatical proposition," which he sent to Mersenne in May 1636.[17] He began with a lemma, easily proved, he maintained, "by following in Archimedes' footsteps": let B be the center of the earth, AB the earth's

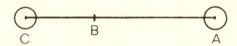

radius, BC a portion of the earth's radius, and W_A and W_C weights placed at A and C, respectively. If $AB : BC = W_A : W_C$, then the weights are in equilibrium. From this lemma he then derived his "rather marvelous proposition": let B again be the center of the earth, AB its radius, and BN some portion of that radius. Placing a weight W_N at N, one can imagine a "power" R_A acting

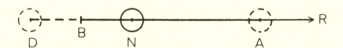

at A in the direction AR away from B. Then, if $AB : BN = W_N : R_A$, R_A will just sustain W_N in its position at N; if R_A is increased the slightest amount, it will lift W_N.[18] Hence, "the more closely a weight approaches the center of the earth, the less power is required to lift it."[19]

To prove his theorem, Fermat began with a system that was undoubtedly in

[16]*centre des corps pesants*; Fermat and his correspondents are usually careful at the outset to leave open the question raised above, n. 14, but then shift, as if by habit of thought, to *centre de la terre* as the point in question.

[17]FO.II.6–9, from *Varia*, 143–144.

[18]Most mechanicians of the late sixteenth and early seventeenth centuries distinguished between the force required to sustain a body (static force) and that required actually to lift it (dynamic force). Many, however, felt the difference was infinitesimal and hence for all practical purposes equal to zero.

[19]". . . quò proprius pondus accedit ad centrum terrae, minorem potentiam ad tollendum illud requiri."

equilibrium. Taking a segment $DB = BN$, he imagined a weight $W_D = W_N$ placed at D. Moreover, he imagined a weight W_A placed at A and exactly counterbalancing R_A, i.e. $W_A = R_A$. Clearly, with no resultant force acting at A, and equal weights placed at equal distances from the center B, the system W_D, W_N, W_A, R_A is in equilibrium about that center. By the preceding lemma, however, if $AB : BD = W_D : W_A$, then the system W_D, W_A is in equilibrium. Hence, if the latter system is removed, the system W_N, R_A remains in equilibrium. Moreover, $AB : BD = AB : BN$, and $W_D : W_A = W_N : R_A$. Therefore, if $AB : BN = W_N : R_A$, the system W_N, R_A is in equilibrium, and conversely.

Mersenne was sufficiently impressed by Fermat's geostatical proposition, which agreed in substance with Beaugrand's main result in the *Geostatics* (as Fermat correctly guessed it would), to include a French translation of it in the *Second Part* of the *Universal Harmony*, published in 1637.[20] A letter from Fermat to Mersenne, dated 24 June 1636, suggests, however, that for all its lucidity the demonstration did not entirely settle the question that had originally prompted Mersenne to invite Fermat's opinion. In corroborating Beaugrand's assertion, Fermat, too, seemed to imply that weight (taken as an intrinsic property of a body) was not constant or absolute, but varied with respect to position. The letter of 24 June 1636 is a reply to Mersenne's no longer extant critique of that implication. Fermat hastened to assure Mersenne that the implication did not in fact follow from the proposition and that Mersenne misunderstood it if he thought it did. "Every heavy body, wherever it is located in the world (except the center), always weighs the same when taken in itself and absolutely." Fermat added that he would have taken this as axiomatic if he had not seen it contested.[21] In fact, however, it was irrelevant to his geostatical proposition, for there Fermat was not taking weights "in themselves and absolutely," but as part of a balance system and hence in relation to others. Hence, he was not claiming that W_N in itself weighed more or less at N than at some other point on the radius, but merely that, if it were suspended from point A on a cord AN, the force at A necessary to support it would be less than if, say, the weight were placed at some point between N and A.[22]

---

[20] Book VIII, Prop. XVIII, p. 61ff. (reprinted in FO.II.10–11); the excerpt begins: ". . . Or, puisque Monsieur Fermat, Conseiller au Parlement de Tholose et trés-excellent Géomètre, m'a donné le raisonnement qu'il a fait sur les différentes pesanteurs des poids, suivant qu'ils approchent davantage du centre . . . , je veux faire part au public de ses pensées sur ce sujet."

[21] Apparently, he first saw it contested in Beaugrand's *Geostatics*, which he seems to have received at the beginning of June 1636. At the end of a letter to Mersenne dated 3.VI.1636, Fermat noted he had seen Beaugrand's work and that ". . . [je] me suis étonné d'abord d'avoir trouvé ma pensée différente de la sienne." Since he basically agreed with Beaugrand on the variation of weight with distance from the center of the earth, the "difference" referred to can only be in regard to absolute vs. relative weight.

[22] Fermat to Mersenne, 24.VI.1636, FO.II.20: "Soit le centre de la terre A, le grave E au point E, et le point N dans la superficie ou ailleurs, plus éloigné du centre que le point E. Je ne dis pas que le point [*sic*; should read *poids* or *grave*] E pèse moins étant

As would frequently be the case in the following years, Fermat found himself forced to go farther than he perhaps would have liked. Mersenne's misunderstanding of the thrust of his geostatical proposition required, he felt, that he take up the whole matter of geostatics, for the proposition fitted into the larger context of Archimedes' mishandling of the law of the lever. To the letter of 24 June 1636, Fermat appended his *New Theorems in Mechanics*, with the characteristic plea that it not be published until he had had a chance to refine it.[23] In the letter itself, he briefly set out the justification for a new approach to the subject. Archimedes, he explained, had mistakenly assumed that the center of gravity of any two equal weights hung from a balance arm lay at the midpoint between them. But, he argued, consider the balance arm *DAB* perpendicular to the earth's radius at *A*. It is true that equal weights

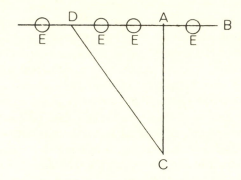

en *E* que s'il était en *N*, mais je dis que, si le point *E* est suspendu du point *N* par le filet *NE*, la force étant au point *N* le retiendra plus aisément que s'il était plus proche de la dite force, et ce, en la proportion que je vous ai assignée."

Fermat reiterated his critique of Beaugrand's position and his defense of his own in a letter to Mersenne dated (conjecturally) 15.VII.1636 (FO.II.27–30). If Beaugrand supposed that a heavy body *per se* becomes more or less heavy according to its distance from the center of the earth, "moi, je soutiens (en quoi je répondrai à votre seconde raison) qu'en soi il ne change point de poids, mais qu'il est tiré avec plus ou moins de force, ce qui est bien différente du reste." Though it would be precipitous to read into Fermat's words the distinction later made by Newton between mass and weight—especially since Fermat did not in his own work pursue the implications of the passage just cited—his remark nonetheless illustrates the context in which the need for such a distinction was becoming clear.

[23]*Nova in mechanicis theoremata Domini de Fermat*, FO.II.23–26, from *Varia*, 142–143. Cf. Fermat to Mersenne, 24.VI.1636, FO.II.18: "Or l'équivoque, sans doute, est venue de ce que je ne vous ai pas assez expliqué les nouvelles pensées que j'ai sur le sujet des Méchaniques et lesquelles vous verrez grossièrement crayonnées sur le papier que je vous envoie; c'est pourtant à la charge que vous m'obligerez de ne les communiquer à personne et que vous me donnerez le loisir pour en faire les démonstrations exactes ou plutôt pour les mettre au net, car elles sont déjà faites." A letter from Fermat to Roberval and Pascal, dated 23.VIII.1636, speaks of the mechanical propositions as having been sent to Carcavi (FO.II.50); unless Fermat was confused, two copies of the paper existed, whence, as the editors of FO note, it is impossible to tell which of them was included in the *Varia*.

placed at equal distances on either side of *A* counterbalanced, but not equal weights placed at equal distances on either side of *D*. In the latter case, the weights were in fact at unequal distances from the center of the earth and hence (by his geostatical proposition) had unequal inclinations toward that center. Only if all heavy bodies fell along parallel lines would Archimedes' assumption be correct. And here Fermat stated explicitly the credo of the geostatician: "It is not that there is any sensible difference in practice, but there is pleasure in seeking the finest and subtlest truths and in removing all ambiguities that might arise."

Fermat began the *New Theorems in Mechanics* on that theme. He had long suspected that Archimedes did not treat the foundations of mechanics with full accuracy, particularly when the Syracusan assumed that bodies fall along parallel paths. Fermat did not dispute that the assumption agreed with experience; due to the great distances from the center of the earth, the assumption of parallel lines of fall was no less valid than the assumption of parallel rays from the sun. "But, to those seeking the intimate and accurate truth, these assumptions do not suffice." One should study and analyze the nature of balances as they behave anywhere in the universe (and not just on the surface of the earth) or whatever their size, and hence one needs new foundations of mechanics.

Though Fermat did not say so explicitly, those new foundations themselves rested on one fundamental assumption: all heavy bodies move toward a single center. That assumption did lie implicit, however, in his initial distinction between two sorts of balances, those in which the motion of the endpoints is rectilinear and those in which it is circular. The ancients, he noted, had considered only the second sort, though he was sure that they knew of the first. In fact, the first balance Fermat had in mind had never entered the thoughts of the ancients (as it has since disappeared from our mechanical universe); it is peculiar to Fermat and to his age. It is, as Fermat later notes, the "great balance of the earth," called such in imitation of Gilbert's "great magnet of the earth."[24] It is the balance that forms the model for the lemma to his geostatical proposition; its center is the center of the earth, and weights placed on it incline along its arms toward that center.

Only during the ensuing debate with Roberval and Pascal did Fermat reveal the line of reasoning that lay behind his first type of balance. Though it is not set out in the *New Theorems*, it should be noted here, because it involves a circularity that vitiates Fermat's whole endeavor. Consider, Fermat argued in a letter to Roberval in August 1636,[25] two equal weights joined by a weightless line. Allow them to fall toward the center of the earth and ask where they will come to rest. Only one answer seemed reasonable to Fermat: they will

[24]*New Theorems*, FO.II.24: "Haec est prima propositio cujus respectu terra ipsa magnus vectis dici potest, ad imitationem Gilberti qui eam magnum magnetem vocat."
[25]FO.II.31-35; cf. especially p. 32.

stop when the midpoint of the line joining them coincides with the center of the earth. Symmetry alone dictates that answer. But Fermat had more than symmetry in mind. As he pointed out in response to Roberval's critique of his position, the midpoint represents the center of gravity of the two equal bodies, and no one would doubt that the center of gravity of every body unites with the center of the earth.

It is the notion of center of gravity that Fermat apparently used to move from the case of two equal bodies to that of two unequal ones, the case treated by Proposition I of the *New Theorems*. The notion that any system composed of two bodies linked by a weightless line will come to rest when its center of gravity coincides with the center of the earth, combined with the standard determination of the center of gravity of such a system, leads directly to the conclusion that the system will come to rest when the distances of the bodies from the center are inversely proportional to their weights. But the standard determination of the center of gravity derives in turn from Archimedes' law of the lever,[26] which is precisely what Fermat is trying to overturn! To prove Proposition I of the *New Theorems* directly (which Fermat never did explicitly in the extant papers) would have been to follow the paradigm of Archimedes' *Equilibrium of Planes*, i.e. to establish the case of two equal weights at equal distances by symmetry and then to treat two unequal weights by dividing them into a number of equal weights distributed equally along a line centered at the center of the earth. But that mode of proof would have involved Fermat (or perhaps did involve him) in two major difficulties. First, the dependence of weight on distance (which is Proposition II of the *New Theorems*) would make the notion of "equal weights" ambiguous at best; second, even if that notion were univocal, the proof would still require the assumption that any two of those equal weights placed at unequal distances from the center can be considered as acting through their center of gravity placed midway between them. But then the center of gravity would again be assumed as a well-defined notion prior to its definition. In short, either way Fermat's treatment of the first sort of balance lay tangled in a vicious circle with little indication of the way out of the tangle.

Indeed, the extent of the tangle emerges most clearly as a result of the third and final proposition of the *New Theorems*, Fermat's revised law of the lever for the second sort of balance, which he calls the "Archimedean balance." "But the inverse proportion between weight and distance, which we have demonstrated for the simple balance, does not obtain for the Archimedean balance, nor therefore do the sixth and seventh propositions of Archimedes [*Equilibrium of Planes* I, 6-7]." Fermat attempts to treat all three variations of the second balance (straight-beam, bent-beam, and beam parallel to the horizon) in one law: let DBC be a balance, of which the center B lies outside

---

[26] Indeed, Fermat's own technique for determining the centers of gravity of plane and solid bodies employed Archimedes' law; see above, Chap. V, §III.

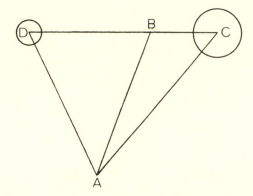

the center of the earth, and let weights W_C and W_D hang from points C and D, respectively. The system will balance at B if and only if $W_C:W_D = (DA:CA)$ ($\angle BAD:\angle BAC$).

Fermat promises, but does not provide, proof of his three theorems. Having stated them, he concludes by drawing two implications. First, the ancient definition of center of gravity (the definition found in Pappus' *Mathematical Collection*, Book VIII) can no longer be maintained. No body except the sphere contains within it a point such that, if suspended from that point, it remains in any position in which it is initially placed.[27] Rather, Fermat argues, we must speak of a point within a body such that, if that point were coincident with the center of the earth, the body would remain in its initial position; "only in that case does the center of gravity mean anything."[28] Second, Guid' Ubaldo dal Monte and his followers must abandon their notion that the arms of a balance can be in equilibrium in any position other than the horizontal.[29]

A letter from Fermat to Mersenne in July 1636 indicates that the *New Theorems* either satisfied Mersenne's questions or pushed beyond his command of the subject.[30] Whatever the case, Roberval and Pascal picked up the

[27]Fermat was by no means alone in maintaining this position. Simon Stevin, for example, noted in his *Art of Weighing* (Postulate V, Commentary; *Principal Works*, Vol. I, p. 113): "From this it also follows that among all the corporeal forms existing in nature, there is mathematically speaking, none but the sphere which remains at rest in any position given to it, when conceived to be suspended from its center of gravity, or which is divided by any plane through the center of gravity into parts of equal apparent weight, but because of the infinite variety of forms there will be an infinite number of different centers of gravity in them." Stevin did go on, however, to argue that the Archimedean approximation was more than sufficient for practical purposes.

[28]Roberval and Pascal refused to accept this counterfactual ideal definition of center of gravity on the grounds that it could in no way be tested experientially; cf. their letter to Fermat, 16.VIII.1636, FO.II.35–50, esp. pp. 36, 50.

[29]See Guid' Ubaldo's long and intricate account of the balance in his *Mechanicorum liber* (1577), trans. Drake in *Mechanics in Sixteenth-Century Italy*, pp. 259–298.

[30]Fermat to Mersenne, <15.VII.1636>, FO.II.27: "Puisque j'ai été assez heureux pour vous ôter l'opinion que vous aviez eue, que j'eusse suivi en ma *Proposition* le même

cudgel against Fermat's geostatical ideas; for the next five months the three men debated with seemingly endless subtlety the ramifications of the *New Theorems*. In the end, neither side managed to convince the other; having reached a dialectical impasse, the correspondents shifted to mathematical issues. A few examples should suffice to convey the flavor of this fruitless debate and to show why it ended from sheer exhaustion.

Roberval (writing for both himself and Pascal) began the attack:[31] he refused to accept Fermat's argument by symmetry that in the first balance equal weights would equilibrate at equal distances about the center of the earth. Exploring various causal mechanisms for the phenomenon of weight, he concluded that one had no way of deciding among them experimentally. Insofar as Fermat's "axiom" hung on what *would* take place at the center of the earth, surface-bound mortals had no basis for accepting or rejecting it; hence, it could not serve as an axiom. Anxious to overturn the foundations of Fermat's new system, Roberval and Pascal indulged in a sort of intellectual humility uncharacteristic of the age: "We believe," they wrote, "that it is hardly possible for the human mind to determine the proportions of the increase or decrease of weight according to varying distances from the center of the earth."[32]

From Fermat's geostatical proposition, they then turned to the new law of the lever. Concentrating on the case of the lever of which the arms are sectors of a circle about the center of the earth, they questioned the full generality of Fermat's proposition that the balance would be in equilibrium if and only if the weights suspended were inversely proportional to the central angles subtended by the balance arms.[33] Fermat had, they claimed, omitted to note that this could be true only if the total length of the balance was less than a 180° arc. Any balance greater than that would have one or both of its arms tending to lift it from its fulcrum rather than weighing down on it. The objection, they added, was not as trivial as it might sound at first. Any "Archimedean" demonstration of the proposition would involve the distribution of equal weights at equal distances along the arms as well as the assump-

---

raisonnement que M. de Beaugrand, j'espère qu'avec la même facilité je vous ôterai tous les autres scrupules." Except for an occasional general remark, the subject of geostatics then drops from Fermat's correspondence with Mersenne.

[31] Roberval's first critique of the *New Theorems*, sent sometime in late July or early August 1636, has been lost, but its contents can be reconstructed from Fermat's reply to it, FO.II.31–35. Its main elements were also reiterated in the letter from Roberval and Pascal to Fermat, 16.VIII.1636, FO.II.35–50.

[32] Roberval and Pascal to Fermat, 16.VIII.1636, FO.II.40.

[33] The balance of which the arms are arcs of a circle about the center of the earth was for Fermat the only balance that truly obeyed Archimedes' inverse proportionality of weight and arm-length. For such a balance, the lengths of the arms are as the central angles they subtend. For a rectilinear balance beam supported at a point outside the center of the earth, Fermat maintained that equilibrium was achieved when the weights were inversely proportional to the product of their distance to the center of the earth and the central angle subtended by the balance arm; cf. *New Theorems*, FO.II.25.

tion that each of those weights had the same effect on the balance when distributed as when it formed part of a single weight. But suppose two of the distributed weights were placed at opposite ends of a 180° balance; far from having any counteracting rotatory effect about the balance's fulcrum, they would form a balance of the first sort and have no effect whatever on the original balance. Moreover, if they were separated by more than 180°, they would combine to lift the balance from its fulcrum.

To cap their *reduction ad absurdum* of Fermat's new principle of the balance, Roberval and Pascal pointed to a situation of which Galileo was to make much in his own work.[34] Take, they said, a balance GIR, of which the

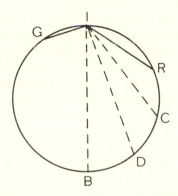

fulcrum is at I. Holding the position of G constant, consider the weight at R to occupy positions C, D, and B, the last position lying diametrically opposite the fulcrum. By Fermat's law, the lengthening of the balance arm, $IR < IC < ID < IB$, would imply that an ever greater weight would be necessary at G to maintain equilibrium; the greatest weight would be required for the balance GIB. But, placed at B, the weight would act directly along the radius BA, and "the least weight placed at G would cause the arm IB to move toward point D; and, however little the weight on arm IG causes the arm IB with its weight to move toward D (which, it seems, no one could deny), then, since both G and B leave the circumference, one can infer a clear absurdity against your position."

The attack of Roberval and Pascal evoked from Fermat two fervent defenses, one in a letter of 23 August and another in a letter of 16 September.[35]

---

[34] See, for example, Chap. 14 of Galileo's *On Motion* (1590), trans. Drabkin in *Galileo Galilei On Motion and On Mechanics* (ed. and trans. by I. E. Drabkin and S. Drake, Madison, 1960), p. 64ff, where Galileo uses the figure described by Roberval and Pascal to derive the law of the inclined plane.

[35] Fermat to Roberval and Pascal, 23.VIII.1636, FO.II.50–56; Fermat to Roberval, 16.IX.1636, FO.II.59–63.

Though filled with detailed analyses, the letters did not really speak to the arguments they were meant to refute. Fermat noted that he had expected objections to the new law of the lever for the second sort of balance, but added (rather weakly) that, right or wrong, he had at least succeeded in getting Roberval and Pascal to agree to the geostatical proposition (in fact, there is no indication they had agreed). The specific objections to the second balance Fermat tried to refute by granting his opponents' position: of course he had only meant to treat balances less than 180° in length. For those balances, he offered his rigorous demonstration. But the demonstration had nothing to offer, for both Roberval and Pascal already knew how to prove Fermat's law with the restriction in length. Rather than addressing the issues raised by them, Fermat was merely reviewing common ground. For the next three months, the debate made no headway.

Fermat may have realized the weakness of his rebuttal. Roberval and Pascal had seized on a contradiction in his theory to argue the superiority of Roberval's approach to mechanics in his *Treatise on Mechanics*, a work typical of the Italian Archimedean tradition.[36] In December 1636, Fermat tried to turn the tables by showing how Roberval's proposition regarding bent-beam balances contradicted his proposition regarding inclined planes.[37] Both propositions had become, as a result of the Italian tradition, widely accepted: in a bent-beam balance, the weights are inversely proportional to the perpendiculars drawn from the points of suspension to the line of support of the balance; on an inclined plane, the effective weight along the plane is as the sine of the angle of inclination measured from the horizontal.

Apparently overlooking the fact that both propositions depended on the assumption of parallel lines of action of weights (i.e. on overlooking the convergence of those lines at the center of the earth), Fermat tried to use convergence to make his point. Consider, he said, a plane *ANGF* parallel to the horizon and perpendicular to the radius of the earth at *N*. A sphere *B* placed at *N* can be moved by the slightest force, or, if untouched, will remain at rest at *B*. At any other point of the plane, say *G* or *F*, the sphere would not remain at rest, but would tend to move (i.e. descend) toward *N*, and hence some determinate force would be necessary to retain it or to move it in the opposite direction. Let *Z* be the force retaining the sphere at *G* by pulling along line *CZ* parallel to the plane. Connect points *C* and *G* with the center *H* of the earth, draw *CG* and draw *GI* perpendicular to *CH*. One can then analyze the forces acting on the sphere in terms of a bent-beam balance, of which the fulcrum is at *G*, the weight of the sphere acts through arm *GI* along *CH*, and

---

[36] *Traité de Mechanique. Des poids soustenus par des puissances sur les plans inclinez à l'Horizon. Des puissances qui soustiennent un poids suspendu a deux chordes. Par G. Pers. de Roberval* . . . , first published by Mersenne as an appendix to Book III of the *Harmonie universelle* (Paris, 1636).

[37] *Objecta a Domino de Fermat adversus propositionem mechanicam Domini de Roberval*, <XII.1636>, FO.II.87–89.

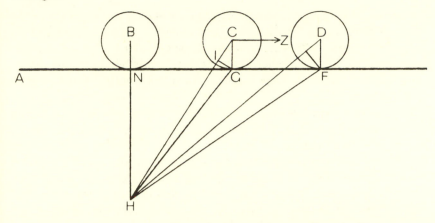

the restraining force Z acts through arm GC along CZ. By Roberval's law of the bent-beam balance, $GI:GC = Z$:weight of the sphere.

Now, if the sphere were placed at point F farther away from N than is G, the angle DFH would be more acute than angle CGH, and hence the arm through which the weight of the sphere acts would be greater. Therefore, to maintain equilibrium, the force restraining the sphere would have to increase commensurately. In short, the greater the distance of the sphere from point N, the greater the force necessary to hold it in place. But $ANGF$ is an inclined plane, and Roberval's law of the inclined plane asserts, first and foremost, the constancy of the restraining force anywhere on the plane. Moreover, since any inclined plane is tangent to the surface of the earth at some point when extended, the restraining force is not constant on any such plane, and Roberval's theorem is generally invalid.

Two more letters went off to Paris from Toulouse in December 1636,[38] but the debate had clearly run its course. The disagreement was fundamental. Roberval and Pascal insisted on the validity of Archimedes' assumptions, and Fermat insisted on the necessity of taking the actual converging lines of action of weights into account in any theory of mechanics.[39] Moreover, Roberval and Pascal were sympathetic to analysis by moving forces, while Fermat felt that "to establish the proportion of freely moving weights, one should not have recourse to the moving forces; on the contrary, the free weights should serve as the rule for all other violent motions."[40] Ultimately, the mechanicians who analyzed isolated systems under arbitrary forces and applied the results

[38] Fermat to Roberval, 7.XII.1636, FO.II.89–92, and 16.XII.1636, FO.II.92–99.

[39] Note, for example, that Fermat's correction of Galileo regarding the path of a ball falling from a tower on a moving earth, discussed above, n. 2, derives precisely from his insistence that the mathematical description must conform exactly to physical reality, in this case uniform motion about the center of the earth and accelerated motion toward that center.

[40] FO.II.90.

to nature confronted the geostatician who made the tendency of bodies toward a determined center a prerequisite to considerations of natural situations. Each side had its reasons, each side could point to tradition; neither side could convince the other. And there, in December 1636, the matter rested. In their future correspondence, Fermat and Roberval never returned to the subject.

Indeed, only once after 1636 did Fermat again express any opinion on mechanics. It is an opinion of some historical importance, for it sheds light on one of the most discussed passages in Galileo's *Two New Sciences*. The Italian scientist's long struggle with the kinematic foundations of accelerated motion is too well known to present in detail here.[41] He himself summarized his trials in the Third Day of the *Two New Sciences* when, having asserted that uniform acceleration must be taken to be the acquisition of speed over time (i.e., $v \propto t$), he admitted that, like Sagredo, he too had once maintained that it corresponded to the acquisition of speed over distance (i.e., $v \propto S$). But, he added, he had come to see that the second definition was "as false and impossible as that motion should be completed instantaneously."[42] Though he offered a "demonstration" in the form of a concrete example, his discussion is vague at best, and even today it is difficult to be sure what he had in mind.[43]

In a memoir addressed to Gassendi in 1646, Fermat provides a full-scale demonstration of Galileo's assertion.[44] Whether or not it coincides with Galileo's own, it retains enough value as a contemporary interpretation to warrant presentation here in some detail. Moreover, the details may help to confirm the thesis of Chapter V above that Fermat's breakthrough in the theory of quadrature occurred in the early 1640s. Fermat begins with an easily proved lemma: if any number of line segments measured from a common point are in continued proportion, the differences between them are in continued proportion in the same ratio. That is (we replace Fermat's geometric notation with algebraic), if $a:b = b:c = c:d = d:e = \dots$, then $(a - b):(b - c) = (b - c):(c - d) = (c - d):(d - e) = \dots$.

On that lemma, Fermat then bases his main proposition:

If one imagines a motion from point F to point A continually accelerated according to the ratio of the spaces traversed, and one sets out any number

---

[41] Essay 11 of Stillman Drake's *Galileo Studies* (Ann Arbor, 1970) is one of the most recent accounts.

[42] Galileo, *Dialogues Concerning Two New Sciences*, trans. H. Crew and A. de Salvio (New York, 1914), Third Day, pp. 167–168.

[43] See the reconstruction by I. B. Cohen, "Galileo's Rejection of the Possibility of Velocity Changing Uniformly with Respect to Distance," *Isis*, 47(1956), pp. 231–235.

[44] Fermat to Gassendi, <1646>, FO.II.267–276. The letter, first published in the *Oeuvres de Gassendi* (Lyon, 1658), Vol. VI, pp. 541–543, is apparently a response to Gassendi's 1646 treatise *De proportione qua gravia decidentia accelerantur Epistolae III* (*Oeuvres*, III, p. 564); cf. FO.II.267, n. 1.

of spaces in continued proportion, say *AF, BF, CF, DF, EF*, etc., the time in which the moving body traverses the space *ED* will be equal to the time in which the same moving body traverses the space *DC*; furthermore all the spaces *ED, DC, CB* will each be traversed in the same time.

Fermat begins by showing that *CB* and *BA* are traversed in the same time. The proof follows by double *reductio ad absurdum*: if the times are not equal, the time to traverse *BA* is greater or less than the time to traverse *CB*. Assume first that it is greater; let time over *BA*:time over *CB* = some line *Z:BF*, where *Z > BF*. Construct as many mean proportionals *RF, MF, NF*, ...between *AF* and *BF* as are necessary to make the smallest of them, say *NF*, less than *Z*. Clearly the construction must arrive at some such *NF*. Then, *AF:RF = RF:MF = MF:NF = NF:BF*. Moreover, since, by the initial lemma, *AF:BF = BF:CF = AB:BC*, one can insert between *BF* and *CF* the same number of continued proportionals in the same ratio, i.e. *BF:OF = OF:VF = VF:XF = XF:CF*.

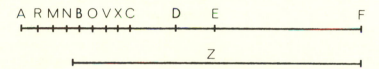

Compare now, says Fermat, the intervals *AR, RM, MN, NB* one-to-one with the intervals *BO, OV, VX, XC;* e.g., compare *AR* with *BO*. If the moving body were to traverse *AR* uniformly at the speed acquired at point *R*, and *BO* uniformly at the speed acquired at *B*, then, by Proposition *V* of Galileo's *On Uniform Motion*,[45]

$$\frac{\text{time to traverse } AR}{\text{time to traverse } BO} = \frac{AR}{BO} \times \frac{\text{speed at } B}{\text{speed at } R}.$$

But, *AR:BO = AF:BF* (by the lemma) and speed at *B*:speed at *R = BF:RF* (by the assumption of speed proportional to distance). Hence,

$$\frac{\text{time to traverse } AR}{\text{time to traverse } BO} = \frac{AF}{BF} \times \frac{BF}{RF} = \frac{AF}{RF}.$$

Mutatis mutandis, one arrives by the same line of reasoning at the results:

---

[45] Galileo, trans. Crew and De Salvio, p. 158: "If two particles are moved at a uniform rate, but with unequal speeds, through unequal distances, then the ratio of the time-intervals occupied will be the product of the ratio of the distances by the inverse ratio of the speeds."

$$\frac{\text{time to traverse } MR}{\text{time to traverse } OV} = \frac{RF}{MF}$$

$$\frac{\text{time to traverse } NM}{\text{time to traverse } VX} = \frac{MF}{NF}$$

$$\frac{\text{time to traverse } BN}{\text{time to traverse } XC} = \frac{NF}{BF} \ .$$

But $AF{:}RF = RF{:}MF = MF{:}NF = NF{:}BF$; that is, the body traverses any two corresponding intervals in the same ratio of times. Therefore,

$$\frac{\text{time of all motions through the whole of } BA}{\text{time of all motions through the whole of } CB} = \frac{AF}{RF} = \frac{NF}{BF} \ .$$

The "time of all motions through the whole of BA" represents the sum of the times required by a series of "fictitious" uniform motions over the intervals $RA, MR, NM, BN;$ in each interval, the time required for uniform acceleration over the interval is less than the time required for uniform motion at the speed acquired at the start of the interval (i.e. the time for uniform acceleration over RA is less than the time for uniform motion over RA at the speed acquired at R, and so on). Therefore, the time of accelerated motion over the whole interval BA is less than the time of all "fictitious" uniform motions over that interval. By contrast, since, for example, the time required for acceleration over the interval OB is greater than the time for uniform motion over that interval at the speed acquired at B (i.e. at the greatest speed acquired in the interval), the time of accelerated motion over CB is greater than the time of all "fictitious" uniform motions over OB, VO, XV, CX. Therefore,

$$\frac{\text{time of accelerated motion over } BA}{\text{time of accelerated motion over } CB} < \frac{NF}{BF} \ .$$

But the ratio on the lefthand side has already been posited to be $Z{:}BF$, where $Z > NF$. From the result just obtained, however, it follows that $Z < NF$, whence a contradiction arises.

By an entirely similar chain of reasoning one can show that the second possibility, i.e. that the time to traverse AB is less than the time to traverse CB, leads to a contradiction. Hence, the times are equal. "These things established, we reveal the thought of Galileo, or show the truth of his proposition, in a third proposition." Fermat imagines a body accelerating from rest at A toward point $H;$ "if possible, suppose that the velocity of the falling body is accelerated in proportion to the spaces traversed." Let the time over AH be one minute or any other determinate period of time. From H, let the body

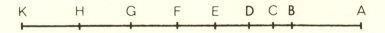

continue on to *K*. "I say that the motion from *H* to *K* takes place instantaneously." The proof is again by *reductio ad absurdum*.

Suppose, says Fermat, that the body requires some determinate time to traverse *HK*. Then there exists some multiple of that time, say 5, which exceeds the time required to traverse *AH*. Take, then, any number, greater than 5, of continuously proportional intervals, $AK:AH = AH:AG = AG:AF = AF:AE = AE:AD = AD:AC = AC:AB$. By the main proposition just proved, the body will require the same time to traverse each of the intervals *HK*, *HG*, *GF*, *FE*, *DE*, *DC*, *CB*; that is, the time required to traverse *HB* will be some multiple greater than 5 of the time required to traverse *HK*. But that means in turn that the body takes longer to traverse a portion of *AH* than it does to traverse the whole! The absurdity of this conclusion invalidates the original assumption that some determinate time is required to traverse *HK*; hence, it is traversed instantaneously.

Fermat's own disjunction, "we reveal the thought of Galileo, or we show the truth of his proposition," constitutes sufficient warning against treating the above demonstration as a reconstruction of Galileo's ideas. By the same token, however, there is nothing foreign to Galileo's thinking in the demonstration; it is, in more than mere time, entirely contemporaneous. It is also a piece of Fermat's science that stands alone. Nothing in Fermat's work prepares one to encounter it, nor did Fermat proceed farther with it. It does offer one little aid to dating developments in his mathematical thought. The demonstration rests on the clever use of continued proportion; so too does the new method of quadrature that forms the foundation of the *Treatise on Quadrature*. Fermat's testimony that the new method dates back to the early 1640s finds, then, corroboration in the dated memoir to Gassendi on uniform acceleration. That memoir was Fermat's last word on mechanics.

II. OPTICS

The closing of the debate over geostatics in December 1636 afforded Fermat only a short respite before he again became embroiled in a physical controversy. In May 1637 he undertook (at what he thought to be the request of Beaugrand and Mersenne) a critique of Descartes' *Dioptrics*, in particular of the derivations of the laws of reflection and refraction contained in it. He thereby unleashed the bitter dispute that Chapter IV above has examined in some detail, though not with specific reference to the optical issues involved. At the time, Descartes was concerned more with turning the issue back onto Fermat's mathematics than with answering Fermat's criticism of the *Dioptrics*, and Chapters IV through VI have shown at several points how effectively he succeeded. Hence, the importance of the original dispute in 1637 lies more in its effect on Fermat's reputation as a mathematician, and in the actions he took to counter that effect, than in the actual substance of his objections to Descartes' derivations.

Indeed, the technical issues did not become important until twenty years later when, in 1658, Claude Clerselier, acting through Kenelm Digby, approached Fermat with the request for copies of letters he wished to include in Volume III of his *Letters of Descartes*.[46] Intending to devote part of that volume to the controversy of 1637-1638, Clerselier had reason to believe that Fermat had written more than the two letters addressed to Mersenne for Descartes in May and December 1637;[47] unable to find any others beyond those two, he asked Fermat to supply copies. Having written no others, Fermat understandably misinterpreted Clerselier's request. Thinking Clerselier had been unable to get hold of the original two letters, Fermat composed in March 1658 a long letter directly to Clerselier, in which he restated his earlier objections and added new ones.[48] The restatement was not entirely accurate, as Clerselier, in possession of the original letters, knew full well.[49] Hence, it appeared to him that, by adducing new arguments against Descartes' derivations, Fermat was seeking to reopen the dispute. He felt his suspicions confirmed when, in response to his and Rohault's defense of Descartes,[50] Fermat continued his attack all the more stoutly. The result was a series of letters between Clerselier and Fermat that continued over the next four years.[51]

The independent development of Fermat's own thinking on optics injected into the renewed dispute an element that had been absent in 1637. His original critique of the *Dioptrics* had concentrated solely on the cogency of Descartes' derivations. He himself had had no commitments on the subject, except perhaps for a natural reticence to abandon the foundations of natural philosophy as he had been taught it. If he had doubted the sine law of refraction, it was because he had not known what the right law was and had not found Descartes' proof of the sine law convincing. In 1657, however, Fermat received and read Marin Cureau de la Chambre's *Light* and found in it a physical principle to which he could subscribe and on which he felt a truly convincing mathematical derivation of the laws of reflection and refraction could be based. Moreover, by the late 1650s, he had come, with other mem-

---

[46]Clerselier's letter is no longer extant; see FO.II.365, n. 2 and 3 for the complicated circumstances surrounding the letter and Fermat's response to it.

[47]*Ibid.*, n. 4.

[48]Actually, Fermat wrote two letters, dated 3.III.1658 (FO.II.365-367) and 10.III.1658 (FO.II.367-374). The first of these simply acknowledged Clerselier's request, noting that Carcavi was in possession of all versions of the method of maxima and minima, "with and without demonstration." The second was devoted to the *Dioptrics*. Both were sent together to Clerselier via Digby (FO.II.366).

[49]In his new critique Fermat did not emphasize, as he had in 1637, what he took to be Descartes' arbitrary choice of directional components (see the discussion below). Moreover, those earlier letters had attacked the derivation of the law of reflection as well as the law of refraction. In the letter of 10.III.1658, Fermat concentrated on the latter.

[50]Concerned more to refute Fermat's most recent objections, Clerselier left it to Rohault to reply in detail to the criticisms of 1637. See Clerselier to Fermat, 15.V.1636, FO.II.389.

[51]In all, eight letters were exchanged. Their Baroque *politesse*, especially in the letters of Clerselier, barely masks the anger and indignation in which they were written.

bers of the scientific community, to accept the sine law as correct. Hence, in the renewed debate with Clerselier, Fermat was to some extent arguing from a positive position of his own. Concentrating all the more heavily on the weakness of Descartes' derivations, he pressed his attack more strongly than he had in 1637. In addition, of course, he had a personal grudge to repay. Descartes' treatment of him after the controversy of 1637-1638, indeed his treatment by Cartesians in general, still rankled twenty years later and, though essentially unconnected with that treatment, signs of the impending failure of the project to restore number theory exacerbated his sense of having been ill treated. In 1658 Fermat was ready to do battle with Cartesian optics, as he had not been in 1637.

It is impossible within the compass of the present appendix to give a fully detailed account of the optical controversy of 1637 and 1658-1662. To do so would require first a long and technical account of Descartes' theory of light and of the derivations in the *Dioptrics* (that account is already available in A. I. Sabra's *Theories of Light from Descartes to Newton*, together with the main substance of Fermat's critique).[52] Moreover, the full account really says more about Descartes than about Fermat. Whatever Fermat's motivation to criticize Descartes, his actual criticism derived not so much from any positive position than from ambiguities and flaws in Descartes' presentation that anyone might have perceived, indeed that many others besides Fermat had perceived. To show this, and to convey at least some of the flavor of the controversy, a brief summary of Descartes' account of light and of his approach to the phenomena of reflection and refraction should suffice. It should, in addition, provide some of the elements of the contrast between Descartes' and Fermat's derivations, a contrast that exemplifies the latter's mathematical attitude toward physics.

A quirk of fate (to wit, the trial and condemnation of Galileo) dictated that Descartes' *Dioptrics* appeared without the cosmological treatise on which it was based. The short account of the nature of light that opens the *Dioptrics* could not replace the more extended and more carefully argued theory of *The World, or On Light*, which Descartes had withdrawn from publication upon hearing of Galileo's trial.[53] Moreover, the crucial steps in his derivations of the laws of reflection and refraction depended on the laws of motion presented in the suppressed treatise. Without the precise context of *The World*, the appearance of those laws in the *Dioptrics* seemed arbitrary at best; they

[52] See also M. S. Mahoney, "Descartes, René (Mathematics and Physics)," *Dictionary of Scientific Biography*, Vol. IV, pp. 58–60.

[53] *Le monde, ou Traité de la lumière* had already been announced in Paris and was due shortly to appear. It seems clear that Descartes had no fear that his cosmological theory, which embraced the Copernican system, would place him in personal danger; he was living in Holland, and the Inquisition had never established itself in France. He did, however, fear that publication of *Le monde* might bring censure from the Parisian theologians and hence censorship of his writings from the government, thus endangering his grand plan for the reform of philosophy.

were not set out properly until the publication of the *Principles of Philosophy* in 1644. Only against the background of *The World* or of the *Principles* can one fully understand and appreciate the thrust of Descartes' arguments in the *Dioptrics*. Only against that background can one understand Fermat's critique, for it focused precisely on the points that required the fuller context.

Light, for Descartes, was an impulse, or pressure, transmitted instantaneously and rectilinearly by the subtlest of the three sorts of corpuscles that filled the world. Descartes compared the transmission of light with the action of a blind man's cane. As the cane comes into contact with a solid object, it transmits to the blind man's hand an impulse which he feels and thereby "sees" the object. So too a luminous body emits an impulse that is transmitted by the ethereal particles to the surface of the eye and thence, after refinement and sorting of rays in the eye, along the optic nerve to the pineal gland. And just as the cane transmits the impulse rectilinearly and instantaneously, so too do the ethereal particles.

Descartes, then, understood the nature of light and its transmission in purely mechanical terms. Just as hard bodies collide with greater or less force, so too the light impulse transmitted by the pressing of the subtle particles on one another may vary in magnitude. That is, the transmission of light is the transmission of a "force" identical in nature to the force of collision of bodies, and the quantity of that "force" can vary.[54] Like any mechanical force, the force of light can be resolved into its directional components, or (in Descartes' terminology) its "determinations," in conformity with the parallelogram of motions and forces.[55] The resolved components are independent of one another, and one can be altered without effect on the other. All alteration, of course, is the result of direct impact.

The phenomena of reflection and refraction arise from the changes in the relation of the components of the force of light when the impulse meets (i.e. impacts on) an obstacle. In the case of reflection, the obstacle is completely immovable and incapable of transmitting the impulse in its original direction. Since the total amount of force must remain the same, however, the impulse simply changes direction with no loss of magnitude. To determine the precise nature of the change of direction, one must distinguish between two particular components of the incident force. The obstructing surface resists the transmission of that force only in the direction of the normal to that surface. Hence, if one resolves the determination of the force into orthogonal components, one perpendicular to the surface and one parallel to it, only the

[54] It followed in particular from Descartes' cosmology that the strength of the impulse varied directly as the density of the transmitting medium. Fermat found this implication contrary to both reason and experience and argued repeatedly against it.

[55] On the difficult concept of "determination" in Descartes' analysis, see Sabra, *Theories of Light*, pp. 116–135. This discussion reviews Fermat's main arguments against Descartes' derivation, since those arguments involved precisely the notion of "determination."

perpendicular component will encounter resistance and be altered. The parallel component will remain unaffected.

In the case of refraction, the impulse meets an interface demarcating a different medium. Since the force of light is directly proportional to the density of the medium through which it is transmitted, a change in both the magnitude and directional components of the impulse will take place at the interface. On the one hand, the magnitude of the force will be increased or decreased according as the new medium is more or less dense than the old. On the other hand, only one component of the determination will be altered, to wit, the component normal to the interface at the point of entry. The component tangent to the interface at that point will again remain unaffected. When combined with the change in magnitude of the impulse, the change in the ratio of the directional components will produce a new direction for the impulse toward or away from the normal, again according to the relative densities of the two media. The change occurs only at the interface; once in the new medium the impulse is propagated rectilinearly in the new direction.

The model of light just described is the real physical model of *The World*; at first glance, it seems to bear little resemblance to the model used in the *Dioptrics* to derive the laws of reflection and refraction. In fact, however, the latter model derives directly from the former. Just as, in *The World* and in *Principles*, Descartes analyzed transmission of mechanical force by means of the collision of bodies moving in a straight line, a situation that could not actually occur in his micro-universe, so too, in the *Dioptrics*, he translated his model of instantaneous transmission of an impulse of light into the model of a moving ball colliding with an obstruction. The force of the impulse became the velocity of the ball; constant force in the one model corresponded to constant velocity in the other.[56] As force could be resolved into its magnitude and its direction, and the latter into directional components, so too velocity could be resolved into speed and determination, and the latter into directional components. The media of transmission were replaced by the empty space of ideal geometric motion.

In the case, then, of reflection, Descartes imagined a ball moving uniformly along a straight path inclined toward an immovable and impenetrable surface. Since upon impact the ball can transmit none of its speed to the surface (to do so, it would have to move the surface),[57] that speed remains the same before and after impact. But since the ball cannot penetrate the surface, it must change direction. Consider, then, its directional components. As in the impulse model, the surface resists the motion of the ball only in the direction

---

[56] For Descartes, the force of impact corresponded to the quantity of motion (the product of the magnitude of a body and its speed); hence, constant force implied constant speed, and conversely.

[57] *Le monde*, Chap. VII (Alquié.I.354): "Je suppose pour seconde Règle: Que, quand un corps en pousse un autre, il ne saurait lui donner aucun mouvement, qu'il n'en perde en même temps autant du sien; ni lui en ôter, que le sien ne s'augmente d'autant."

normal to the surface. Hence, if one resolves the determination of the motion of the ball into two components, respectively perpendicular and parallel to the surface, the parallel component will be unaffected by the collision.[58] Since the total speed of the ball is also unchanged, the ball will clearly travel as far in the parallel direction after collision as it did before. Simple geometrical considerations show that, under these circumstances, the angle of incidence will be equal to the angle of reflection.

In the case of refraction, Descartes' heuristic model of the ball striking a surface preserves the uniqueness of action at the interface by imagining the ball to break through a sheet of silk or tissue. In breaking through, it will lose some of its speed; how much it loses will be a function of the breaking strength of the material. It will lose more than speed; since the silk surface resists motion only in the normal direction, the perpendicular component of the determination of the ball's motion will also be altered. Its parallel component will remain the same, however, and as a result the total determination will be changed. Since the speed of the ball has been diminished, the ball will take longer to travel the same distance after collision as before. Hence, the unchanged parallel component will have proportionately longer to act on the ball. Again, simple mathematical considerations yield the sine law of refraction.

Those mathematical considerations also show, however, that in losing speed the ball will veer away from the normal to the interface, whereas experience dictates that, upon entering a denser medium, light turns toward the normal. Here Descartes had to appeal to his physical model, in which an impulse entering a denser medium increases in force. To translate that aspect of the real model into the heuristic model, he imagined that, at the instant the ball breaks through the silk, it is struck again in the direction of the normal, thus receiving an increase in speed and in the perpendicular component of its determination. With that adjustment, the heuristic model conformed to observed phenomena.

The above account, which the reader should compare with Descartes' own presentation in the *Dioptrics*, has deliberately avoided the use of geometric diagrams in order to emphasize a feature of the derivation often obscured by such diagrams. It is an eminently *physical* argument; mathematics plays practically no role in it. To be persuaded by it, one must adhere to or accept temporarily all the main tenets of Cartesian cosmology and mechanics, among them: the impulse nature of light with the corollary that the force is proportional to density; the resolvability of force and motion into magnitude and determination, and the resolvability of determination into mutually independent components; the validity of each of the correspondence principles that

[58] Descartes here applies to the parallel component of the ball's determination his first rule (Alquié.I.351): "Que chaque partie de la matière, en particulier, continue toujours d'être en un même état, pendant que la rencontre des autres ne la contraint point de le changer."

underlie the translation from the real to the heuristic model; the laws of motion governing impact, including the law of inertia and the law of conservation of quantity of motion. And, in particular, one must be fully conscious of the underlying physical model. For example, although Descartes does appeal to that model in order to justify the striking of the ball at the instant of impact, he does not emphasize that in the heuristic model the medium is the same above and below the interface, i.e. that all motion is occurring in empty geometrical space, and that any and all changes in motion occur only at the interface at the instant of penetration.

It is, therefore, easy to see how someone unfamiliar with *The World* or, later, with *Principles* could misunderstand, misinterpret, or simply miss the point of Descartes' derivations in the *Dioptrics*. Just to take them as *mathematical* derivations is to misconstrue them; their central arguments all hinge on elements of Cartesian physics and mechanics. But it was precisely as mathematical derivations that Fermat attempted to understand them and, failing at that, to criticize them.[59] He apparently knew little or no Cartesian physics, and he certainly would have contested many of its postulates. At no time did his correspondents—Descartes, Clerselier, or Rohault—seem to realize Fermat's ignorance of the fundamental physical principles on which Descartes' derivations rested. They seem, rather, to have assumed that Fermat did not properly understand those principles.

As a result, the controversy of 1637 and 1658–1662 took place amid great confusion. Fermat, for his part, took the *Dioptrics* at face value; it purported to present mathematical derivations of the laws of reflection and refraction. Fermat denied that it in fact did. His opponents, for their part, failed to comprehend how much physics the derivations contained and how that physics might not be immediately clear to even a well-intentioned reader. Hence, the details of the resulting controversy are far more relevant to an understanding of Descartes and of the reception of his ideas than to an appreciation of Fermat's career. Fermat's role was largely restricted to that of an intelligent and mathematically learned critic responding to the often laconic and vague manner in which a highly complex and radically new system of physics was being presented to the public. In general, Fermat's feel for the weak points in Descartes' arguments was unerring; in responding to him, Clerselier and Rohault occasionally displayed their own lack of complete understanding of the philosophy they were espousing.[60]

[59] Cf. Sabra, *Theories of Light*, p. 128: "In order to understand correctly Fermat's arguments, it should be borne in mind that his attitude was always that of a mathematician, not of a physicist. If by his patient and sincere analysis he helped to expose certain physical difficulties, that was only by implication. His objections against Descartes' arguments, and those of his defenders, were always of a mathematical or logical character."

[60] Cf., for example, Clerselier to Fermat, 15.V.1658, FO.II.388, where Clerselier confuses determination with speed, or Rohault's attached critique (FO.II.392), where he too conflates the two quantities.

It would be disingenuous to cast Fermat entirely in the role of a disinterested critic during the debate of 1658–1662, though that was clearly his role in 1637. By the later period, he had adopted a stance on the behavior of light, and that stance was contrary to Descartes'. Moreover, he had found, he thought, a more reliable basis for a mathematical derivation of the laws of reflection and refraction. With that fact in mind, therefore, one must be cautious in interpreting his frequent claims in the correspondence with Clerselier that he would be more than pleased to find that Descartes was correct and that his criticisms were ill founded.[61] If twenty years of thought and reading about the subject had not altered his original view, nothing Clerselier or Rohault could say would do so. And yet, insofar as toward the end of the correspondence Fermat insisted more heavily on the flaws in the mathematical style of presentation than on the substance of the derivations, he may well have begun to sense the nature of the physical system lying behind them and the extent to which they might be cogently defended in terms of that system. By then, however, Fermat himself had a position to defend; he too had a derivation of the law of refraction.[62]

As noted above, the origins of that derivation date back to the summer of 1657, when Fermat's long-time acquaintance and intermediary to Chancellor Séguier, Marin Cureau de la Chambre, sent a copy of his newly published treatise simply titled *Light* and dedicated to Cardinal Mazarin.[63] Fermat wrote in August 1657 to thank Cureau for the gift and to reject Cureau's modest suggestion that Fermat set the work aside unread.[64] If only to undo the loss of a discourse on light that Fermat himself had addressed to Cureau some years earlier,[65] he felt he should respond in substance to Cureau's effort. Indeed, he insisted, "you and I are largely of the same mind, and I venture to assure you in advance that if you will permit me to link a little of my mathematics to your physics, we will achieve by our common effort a work that will immediately put M. Descartes and all his friends on the defensive."

The "physics" to which Fermat was referring reduced to a single physical principle, that "nature always acts along the shortest paths." The principle appealed to him both as a mathematician and as a thinker wary of much of the physical speculation going on about him. As a mathematician, he immedi-

[61]Fermat to Clerselier, 3.III.1658, FO.II.366: "Pour la question de Dioptrique, je vous proteste, sans nulle feintise, que je souhaite de m'être trompé; mais je ne saurais obtenir sur moi, en façon quelconque, que le raisonnement de M. Descartes soit une démonstration, et même qu'il en approche."

[62]What follows is little more than a précis of Fermat's derivation of the law of refraction and is based directly on the primary sources in FO. For a fuller account, see Sabra, *Theories of Light*, Chap. V.

[63]*La Lumière à Monseigneur l'Eminentissime Cardinal Mazarin par le sieur De La Chambre, conseiller du Roy en ses Conseils et son Médecin ordinaire* (Paris, 1657).

[64]Fermat to de la Chambre, VIII. 1657, FO.II.354–359.

[65]The discourse in question has disappeared without a trace. Strangely, this one remark to Cureau is the sole indication that Fermat ever wrote such a paper; he mentions it in no other letter.

ately perceived how the principle could be applied to the problems of reflection; more precisely, he thought he saw how it could be applied to refraction, since in *Light* Cureau had already used it to derive the law of reflection. That derivation consisted in showing that, given a plane surface *ABC* and two points, *D* and *E*, located above the plane, the shortest path from *D* to the plane and thence to *E* is that for which angle *DBA* is equal to angle *EBC*.[66]

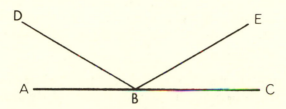

The law of refraction, Fermat reasoned, corresponded then to the solution of the following problem: given a plane interface *DB* and two points, *A* and *C*, lying on opposite sides, find the shortest path from *A* to *C*. "It would appear at first," he warned Cureau,

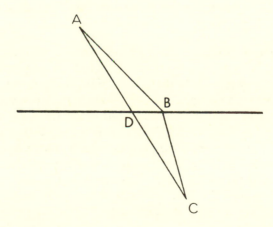

[66] To judge from Fermat's letter, Cureau's derivation had already been attacked on the grounds that, in the case of convex mirrors, the path of reflection was not the shortest path possible. Fermat offered the defense that reflection from a curved surface must be understood in terms of reflection from a plane mirror tangent to the curved surface at the point of reflection; with respect to the tangent plane, the path of reflection was indeed the shortest.

that this approach cannot succeed, and that you yourself raise an objection that seems invincible. For since, on page 315 of your book, the two lines *CB* and *BA* which contain the angle of incidence and that of refraction are longer than the straight line *ADC* which serves as base of the triangle *ABC*, the ray from *C* to *A*, which constitutes a shorter path than that of the two lines *CB* and *BA*, should, in the sense of our principle, be the only true route of nature, which however is contrary to experience. But one can easily rid oneself of that difficulty by supposing, with you and with all those who have treated this matter, that the resistance of the media is different, and that there is always a certain ratio or proportion between these two resistances, when the two media are of a certain consistency and are uniform in themselves.

Hence, Fermat went on to argue, the added factor of media of differing resistance altered the nature of the geometrical problem. The path to be minimized was not simply the sum of the lines *AB* and *BC*, but rather a sum involving multiples of those lines, the multiples being determined by the ratio of the resistances.

The way in which, both here and in later papers, Fermat constantly referred to a "path" as the quantity being minimized clearly shows the totally mathematical set of his mind in dealing with the law of refraction. Although his diagrams illustrate the paths of the rays in the two media, the quantity actually minimized in his analysis is not the total path, but the total time over that path, as becomes clear when he undertakes to introduce the factor of resistance into the problem:

If these two media were the same, the resistance to the passage of the ray along the line *CB* would be to the resistance to the passage of the ray along the line *BA* as the line *CB* to the line *BA*. For, the media being the same, the resistance to passage would be the same in each of them and consequently, it would maintain the ratio of the spaces traversed. Whence it follows that, the media being different, and consequently, the resistance being different, one can no longer say that the resistance to the passage of the ray along the line *CB* is to the resistance to the passage of the ray along the line *BA* as the line *CB* is to the line *BA*; rather, in that case, the resistance along the line *CB* will be to the resistance along the line *BA* as *CB* is to some other line, the ratio of which to the line *BA* will express that of the two different resistances.

By speaking in this way, Fermat managed to mask a basic assumption he preferred not to make explicit. A comparison of both the passage just cited and of his later analysis with his 1646 letter to Gassendi reveals the underlying use of a theorem taken from Galileo's *Two New Sciences*: if two bodies move uniformly over different distances at different velocities, the times of motion are directly as the distances traversed and inversely as the velocities, i.e. $t_1:t_2 = (S_1:S_2)(v_2:v_1)$. If, then, the velocities are equal, the times are simply as the distances.

It was precisely the notion of the velocity of light that enabled Fermat to apply Cureau's principle to the problem of refraction. Relying perhaps on his Aristotelian training, Fermat made the simple assumption that the velocity of light is inversely proportional to the resistance, i.e. $v_1:v_2 = R_2:R_1$. In the case of equal media, the equality of the resistances reduced the ratio of the times to that of the paths traversed. But, in the case of a ray of light traversing path CB in a medium of resistance R_C and path BA in a medium of resistance R_A, the times of passage would be in the compound ratio $(CB:BA)(R_C:R_A)$. That Fermat was thinking in these terms is clear from the manner in which he moved from his preceding remarks to a concrete example in geometrical terms. He took the case in which R_A is either twice R_C or one half of R_C: "Given the two points C and A and the straight line DB, to find a point in the line DB such that, if you draw to it the lines CB and BA, the sum of CB and one half of BA constitutes the least of all sums similarly taken, or the sum of CB and twice BA constitutes the least of all sums similarly taken." That is, in the case that $R_A:R_C = 2:1$, the ratio of the times is $(CB:BA)(1:2)$, and the total time can therefore be represented as $CB + 2BA$. It is that total time that must be minimized. But it is the expression of that time as the sum of multiples of line lengths that hides the fact that time is being minimized and that masks the presence of a physical quantity in Fermat's mathematical derivation.[67]

Fermat had reason to mask its presence, for the attribution to light of a finite velocity placed him at odds with both Cureau and Descartes. Were the notion of velocity of light made explicit in the form of the derivative notion that light takes time to traverse a path, it would have led to further argument over the nature of light. Yet it is clear from both the letter under discussion and Fermat's subsequent papers on the subject that the main appeal of Cureau's approach lay in the manner in which it freed Fermat from making any overt commitment on the actual nature of light itself. Mathematics, he felt, could proceed without physics, overriding the difficult issues separating the various members of the scientific community. "Do not be surprised," Fermat told Cureau,

that I speak of "resistance" after you have decided that the motion of light takes place in an instant and that refraction is caused only by the natural antipathy between light and matter. For, whether you agree with me that the non-successive motion of light can be contested and that your proof is

---

[67]Fermat did make the factor of time explicit in his letter to Cureau of 1 January 1662, FO.II.460: ". . . il faut encore trouver le point qui fait la conduite en moins de temps que quelque autre que ce soit, pris des deux côtés . . ."

not completely demonstrative, or one must abide by your decision, i.e. that light avoids the abundance of matter which is alien to it, I find that, even in this last case, since light flees matter and one flees only that which causes one difficulty and which resists, I can, without deviating from your opinion, establish resistance where you establish flight or aversion.

Fermat could not agree with Cureau and Descartes that light is propagated instantaneously. He could not agree with Descartes that it is propagated more strongly in the denser medium. He could not accept the validity of Descartes' analogical demonstration. The beauty of Cureau's "principle of least action" lay precisely in the possibility of avoiding all these and similar issues.

Fermat ended his letter to Cureau of August 1657 with the promise that he would solve the mathematical problem to which Cureau's principle reduced the question of refraction.[68] The resumption in March 1658 of the controversy over the *Dioptrics*, however, together with the new dispute over number theory, interrupted his efforts, and for the next four years completely occupied his attention. The problem of refraction also presented difficulties he had not foreseen. As he reported to Cureau on 1 January 1662, he encountered two major obstacles: first, he could not be entirely sure of Cureau's principle nor of the manner in which he had applied it to refraction; second, the application of his method of maxima and minima to the problem led to formidable algebraic computations which he lacked the heart to tackle. He could see no simple result emerging, and "the fear of finding, after a long and difficult calculation, some irregular and fantastic proportion, and my natural inclination to laziness, left the matter in that state, until the most recent admonition that President Miremont just delivered to me on your behalf, which I take as a law stronger than either my fear or my laziness."

That admonition, one may gather from the opening of the letter of 1 January, took the form of a fervent plea to Fermat to bring the dispute with Clerselier to an end.[69] Cureau apparently begged Fermat to make good his earlier promise of a derivation of the law of refraction on the basis of Cureau's principle as a way of helping to decide the issue. There was, by that time, little question that the sine law itself was correct. Petit had carried out careful experiments that supported it. Indeed, Fermat's "fear" was precisely that he might arrive at some other law. The issue had become one of a convincing derivation of the sine law, and Cureau's plea finally pushed Fermat to carry out his calculations for better or worse. To Fermat's amazement, they resulted

---

[68]FO.II.359: "Je vous garantis par avance que j'en ferai la solution quand il vous plaira et que j'en tirerai même des conséquences qui établiront solidement la verité de notre opinion."

[69]Fermat to de la Chambre, 1.I.1662, FO.II.457: "Il est juste de vous obéir et de terminer enfin par votre entremise la vieux démêlé qui a été depuis si longtemps entre M. Descartes et moi sur le sujet de la réfraction, et peut-être serai-je assez heureux pour vous proposer une paix que vous trouverez avantageuse à tous les deux partis."

in the sine law. To the letter of 1 January, he proudly attached his analysis of the law of refraction.[70]

As one would expect, Fermat wasted no time in discussing the nature of light, but rather took up directly the mathematical problem: "Let *ACBI* be a circle of which the diameter *AFDB* separates two media of diverse nature, of which the rarer is on the side *ACB*, the denser on the side *AIB*. Let the center of the circle be the point *D*, on which falls the radius *CD* from a given point *C*. Sought is the diaclastic radius *DI*, that is, the point *I* toward which the refracted radius is directed." Since point *C* is given, so too is the foot *F* of the perpendicular from *C* to the diameter *AB*, and hence so too is the length *FD*.

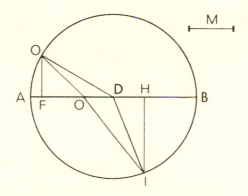

"Let the ratio of the media, i.e. the ratio of the resistance of the denser medium to the resistance of the rarer medium, be as the given length *DF* to a given length *M* outside, which moreover will be less than the line *DF*, since the resistance of the rarer medium is less than the resistance of the denser medium by a most natural axiom." Fermat then applied the rule of resistances discussed above: "Therefore we come to measure the motions along the lines *CD* and *DF* with the aid of the lines *M* and *DF*; that is, the motion along the two lines is comparatively represented by the sum of two rectangles, of which one is *CD · M* and the other *DI · DF*."[71] The problem was to minimize that sum; the method for doing so had been part of Fermat's mathematics for almost thirty years.

If $CD = DI = n, DF = b, DH = x$, and $M = m$, then one must minimize the expression $nm + nb$. Take, then, some interval $DO = y$, and draw *CO* and *OI*. By a theorem of Euclid, $CO^2 = n^2 + y^2 - 2by$ and $OI^2 = n^2 + y^2 + 2xy$. The motion along *CO* and *OI* is therefore $CO \cdot M + OI \cdot DF$, or

$$m\sqrt{n^2 + y^2 - 2by} + b\sqrt{n^2 + y^2 + 2xy} ,$$

[70] *Analysis ad refractiones*, FO.I.170–172.
[71] That is, time along CD:time along DI = (CD:DI) (M:DF), whence (time along CD + time along DI) can be expressed as CD·M + DI·DF.

and, in accordance with the method of maxima and minima, this sum is adequal to the sum $nm + nb$. The calculations from this point on, which Fermat does not supply, are distinguished only by their complexity; the procedures themselves follow strictly the patterns laid down by Fermat in his various papers on the method of maxima and minima.[72] Upon removal of the quadratic surds by repeated squaring, one eliminates the terms common to both sides, divides through by y, and then suppresses all terms still containing y. The result is simple: $x = m$. That is, $DF:DH = DF:M$, and inspection alone shows that this result is precisely the sine law of refraction.

Having based his derivation on the assumption that light travels more slowly in the denser medium, in direct contradiction to Descartes' principle, Fermat was taken aback at arriving at exactly the same law as Descartes.[73] One measure of his astonishment is the memoir that soon followed his analysis to Paris. For the first time in his career, Fermat felt the necessity of providing a synthetic demonstration of a result achieved by his analytic use of the method of maxima and minima.[74] It is worth presenting at least the first part of that demonstration if only as the sole representative of what Fermat had in mind when he so often remarked that "the regress from analysis to synthesis is easy."

In circle $AMBH$, of which the diameter AB divides two media, point M is given, let point H be determined by the given ratio $DN:NS$, where D and S are the feet of the perpendiculars to AB from M and H, respectively, and N is the center of the circle. In accordance with the result just derived analytically, $DN:NS$ is also the ratio of the resistances of the two media. Let, then, $DN:NS = MN:NI$. The result to be demonstrated is that the sum $IN + NH$ is the least of all similarly determined paths from M to H. For example, imagine the path MRH, in which $DN:NS = MR:RP$; one must show that $MR + RP > IN + NH$.

Let $MN:DN = RN:NO$ and $DN:NS = NO:NV$. Clearly, $NO < RN$ and $NV < NO$. By Euclid, $MR^2 = MN^2 + NR^2 + 2DN \cdot NR$, whence $MR^2 = MN^2 + NR^2 + 2MN \cdot NO$. Since, however, $NR > NO$, $MR^2 > MN^2 + NO^2 + 2MN \cdot NO =$

[72] Sabra, *Theories of Light*, pp. 146–147, carries them out in full. Fermat simply refers the reader to Herigone's account of the method in the 1644 *Supplément* to the *Cursus mathematicus*.

[73] Fermat to de la Chambre, 1.I.1662, FO.II.461–462: "Mais le prix de mon travail a été le plus extraordinaire, le plus imprévu et le plus heureux qui fut jamais. Car, après avoir couru par toutes les équations, multiplications, antithèses et autres opérations de ma méthode, et avoir enfin conclu le problème que vous verrez dans un feuillet séparé, j'ai trouvé que mon principe donnait justement et précisément la même proportion des réfractions que M. Descartes a établie.

J'ai été si surpris d'un événement si peu attendu, que j'ai peine à revenir de mon étonnement. J'ai réitéré mes opérations algébriques diverses fois et toujours le succés a été le même, quoique ma démonstration suppose que le passage de la lumière par les corps denses soit plus malaisé que par les rares, ce que je crois très vrai et indisputable, et que néanmoins M. Descartes suppose le contraire."

[74] *Synthesis ad refractiones*, FO.I.173–179; the proof is repeated in slightly different terms in Fermat to [?], 1664, FO.II.489–496.

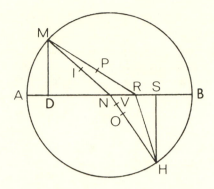

$(MN + NO)^2$, or $MR > MN + NO$. But $DN{:}NS = MN{:}NI = NO{:}NV$, whence $DN{:}NS = (MN + NO){:}(IN + NV)$. But from $DN{:}NS = MR{:}RP$, it then follows that $MR{:}RP = (MN + NO){:}(IN + NV)$, whence $RP > IN + NV$.

It remains only to show that $RH > HV$. By the same theorem of Euclid, $RH^2 = HN^2 + NR^2 - 2SN \cdot NR$. Since $MN = HN, MN{:}DN = HN{:}DN = NR{:}NO$, and $DN{:}NS = NO{:}NV$, it follows that $HN{:}NS = NR{:}NV$, or $HN \cdot NV = NS \cdot NR$. Therefore, $RH^2 = HN^2 + NR^2 - 2HN \cdot NV > HN^2 + NV^2 - 2HN \cdot NV = (HN - NV)^2 = HV^2$; i.e., $RH > HV$. Finally, then, $PR + HR > IN + NV + HV = IN + HN$. In a second part of his full demonstration, Fermat carried out a similar argument for a point R located on the other side of N from S.

The very fact that, despite the contradiction in the assumptions regarding the relation of the speed of light to the density of the medium, Fermat's and Descartes' analyses both led to the same conclusion made Fermat confident that his analysis and synthesis of the law of refraction would end the long debate.[75] The followers of Descartes, he hoped, would rejoice in this independent verification of Descartes' correct law, and Fermat in turn could claim the satisfaction of having provided a demonstration convincing to more people. His hope was ill placed. To a firm Cartesian such as Clerselier, the principle that nature always acts along the shortest path was a moral, rather than a physical principle, and he refused to countenance it or to accept the compromise Fermat offered.[76] To his dismay, Fermat found he had only stoked the

[75]Following the statement of surprise quoted in n.73, Fermat continued: "Que devons-nous conclure de tout ceci? Ne suffirait-il pas, Monsieur, aux amis de M. Descartes que je lui laisse la possession libre de son théorème? N'aura-t-il pas assez de gloire d'avoir connu les démarches de la nature dans la première vue et sans l'aide d'aucune démonstration? Je lui cède donc la victoire et le champ de bataille, et je me contente que M. Clerselier me laisse entrer du moins dans la société de la preuve de cette vérité si importante, et qui doit produire des conséquences si admirables."

[76]Clerselier to Fermat, 6.V.1662, FO.II.465: "Le principe que vous prenez pour fondement de votre démonstration, à savoir que la nature agit toujours par les voies les plus courtes et les plus simples, n'est qu'un principe moral et non point physique, qui n'est point et qui ne peut être la cause d'aucun effet de la nature."

fire of the controversy, as his own derivation now became the target of criticism.[77]

It is the final irony that the last extant scientific letter written by Fermat (to some unknown recipient) should contain his summary account of the dispute over the *Dioptrics* and a history and defense of his own derivation, including a repetition of the synthesis of the law of refraction. In his very first letter to Mersenne in 1636, Fermat had disclaimed expertise or interest in mathematical physics and had given evidence of his unwillingness to involve himself in the controversies of the day. And yet his last letter in 1664 found him deeply embroiled in just such a controversy, his mathematics almost forgotten. That letter is the sad note ending the career of a man who basically believed that mathematics and physics had little to do with one another.

III. PROBABILITY

Sometime during the spring of 1654 Blaise Pascal presumed on his late father's acquaintance with Pierre de Fermat to seek Fermat's advice and corroboration on a problem in probability.[78] Pascal's original letter has not survived, but Fermat's reply[79] has, and it indicates at least one of the questions Pascal put to his father's old friend. Suppose a player has wagered to cast a given number, say 6, with a single die in eight throws, and suppose that after three throws, the game is interrupted. How are the stakes to be divided?

Fermat's reply distinguishes two cases, both of which reduce to the basic principle that the probability of success is the same for each of the eight throws and in no way depends on the outcome of the previous throws. On each throw, the chances of success are 1 in 6. Suppose, then, on the one hand, that "after the stakes have been set, we agree that I do not make the first throw; then, by my principle, I should receive 1/6 of the stakes as fair compensation." Having made settlement, Fermat continues, and having left the remainder in the pot, suppose we agree that I will not make my second cast; then I receive again 1/6 of the stakes remaining, i.e. 1/6 of 5/6, or 5/36 of the original stakes. If we continue in this fashion, it is clear that my fourth cast will be worth 125/1296 of the original stakes. On the other hand, "you propose to me in the last example of your letter (I use your terms) that, if I undertake to make a 6 in eight throws, and I have made three without achieving it, and if my partner proposes that I do not make my fourth throw and

---

[77]See Clerselier's attack in the letter just cited.

[78]On the circumstances behind Pascal's first letter to Fermat, see Oystein Ore, "Pascal and the Invention of Probability Theory," *American Mathematical Monthly*, 67(1960), 409–419. The problem on which Pascal was seeking advice had been posed to him by a man named Méré, to whom much of the literature refers as a "gambler"; Ore questions that reputation. Although he mentions the resulting correspondence between Pascal and Fermat, Ore does not discuss it in any detail.

[79]Fermat to Pascal, 1654, FO.II.288–289. The letter is undated, but was probably written early in July.

wants to compensate me for the chance that I might succeed, he owes me 125/1296 of the total sum of the stakes." This situation, Fermat argues, represents a different case from the one just discussed. Here, after the first three casts, the stakes remain the same as at the outset. My chances of success on the fourth throw also remain the same as they were on each of the first three, i.e. 1 in 6. Hence, my fourth throw is worth 1/6 of the total stakes. "For, if the total stakes remain in the game, it not only follows from the principle but also from common sense that each throw should have an equal chance of success."

In this very first exchange of letters between Fermat and Pascal, then, the basic principle of the theory of probability already underlay the thinking of the two men, for Pascal agreed entirely with Fermat's analysis.[80] The chance of success in any situation is the ratio of favorable outcomes to all possible outcomes. Their agreement on this basic principle did not, however, preclude conflict over its application to specific problems, especially where the proper application was not clear. Pascal's response to Fermat's letter sets out such a problem.[81] Two (or more) players wager on a series of win-or-lose decisions (e.g. the flipping of a coin, or high-card draw), in which each player has an equal chance of success. The first player to win a given number of decisions, say three, wins the stakes (since each has an equal chance to win, the ante is the same for all players). In the event the game is prematurely ended, how should the stakes be divided? In a letter dated 29 July 1654, Pascal offered his solution of the problem. He began recursively.

Let the game involve just two players; the first to win three decisions wins the game. When player *A* has won two and player *B* one, the game is interrupted. Consider the possible outcomes of the next decision, had it been made. A win for *A* would have meant his winning the entire game, and the entire stakes would have been his. A win for *B* would have evened the score at two wins apiece; in that event, both players would have had an equal chance to win the game on the next throw, and each would have maintained his claim to his original ante. Hence, Pascal reasoned, whatever the outcome of the unplayed fourth decision, *A* would be entitled to his original ante. But there is also a 50–50 chance he could have won the entire game. In addition, then, to his original ante, *A* is also entitled to one-half of *B*'s; i.e. *A* should receive 3/4 of the total stakes.

Now let the game be interrupted at the point where *A* has won two decisions and *B* none, and consider again the possible outcomes of a third decision, had

---

[80]Pascal to Fermat, 29.VII.1654, FO.II.290: "... en un mot, vous avez trouvé les deux partis des dés et des parties dans la parfaite justesse: j'en suis tout satisfait, car je ne doute plus maintenant que je ne sois dans la vérité, après la rencontre admirable où je me trouve avec vous."

[81]*Ibid.* Pascal's letter suggests that he had already included the problem in his first letter to Fermat and that he had received Fermat's solution to it. If so, that solution has been lost; the first reply from Fermat contains only the problem of the dice game.

it been made. Had *A* won, he would have won the whole game and hence the total stakes. Had *B* won, the score would have been 2-1 in favor of *A*. The second possibility defines *A*'s minimum compensation; by the analysis just carried out, he is entitled to at least 3/4 of the stakes. But he has a 50-50 chance to win it all; hence, he should receive 7/8 of the total stakes. Finally, consider the situation in which *A* has won one decision and *B* none. A win for *A* on the unplayed second decision would entitle him to 7/8 of the stakes; a win for *B* would make the players even. Hence, *A* is entitled at least to his original ante and has a 50-50 chance to win 7/8 of the total stakes. To be fairly compensated, he should receive 11/16 of the total stakes.

In terms, then, of what *A* wins from *B*, the first decision is worth 3/8 of *B*'s ante; the second, another 3/8; and the last, 1/4. Operating inductively from this example, Pascal came to the general conclusion that, when the game in question requires *n* decisions to win, and the score stands at $n - 1$ to 0 when the game is interrupted, the value of the undecided last play is $1/2^{n-1}$ of the loser's ante.[82] But what if the game requires *n* decisions to win, and it is interrupted after the first decision? In Pascal's words, "Given any number of plays whatever, find the value of the first." In the example just analyzed, the first play was worth 3/8 of the loser's ante. Rather than challenge Fermat to find the answer for himself, Pascal offered his solution immediately:[83]

Let, for example, 8 be the given number of plays. Take the eight first even numbers and the eight first odd numbers, i.e. 2, 4, 6, 8, 10, 12, 14, 16, and 1, 3, 5, 7, 9, 11, 13, 15. Multiply the even numbers in this way: the first by the second, the product by the third, the product by the fourth, the product by the fifth, and so on. Multiply the odd numbers in the same way: the first by the second, the product by the third, and so on.

The final product of the even numbers is the *denominator*, and the final product of the odd numbers the *numerator*, of the fraction that expresses the value of the first play of eight. That is, if each man wagers the number of *pistoles* expressed by the product of the even numbers, the winner is due from the money of the other the amount expressed by the product of the odd numbers.

To generalize Pascal's solution for the arbitrary case of *n* plays, one should note first that the product of the first *n* even numbers is $2^n n!$ and, second,

[82]*Ibid.*, pp. 291-292: "Or, pour ne plus faire de mystère, puisque vous voyez aussi bien tout à découvert et que je n'en faisais que pour voir si je ne me trompais pas, la valeur (j'entends sa valeur sur l'argent de l'autre seulement) de la dernière partie de *deux* est double de la dernière partie de *trois* et quadruple de la dernière partie de *quatre* et octuple de la dernière partie de *cinq*, etc."

[83]As will become clear directly, Pascal here misstates his solution; he is computing the value of the first of *nine* plays, not the first of *eight*. The derivation to follow shows how easily such a mistake can be made.

that the product of the first n odd numbers is $(2n)!/2^n n!$. Setting the first product as denominator and the second as numerator yields the fraction $(2n)!/2^{2n} n! n!$. But $(2n)!/n! n! = \binom{2n}{n}$, or $_{2n}C_n$; therefore, the fraction reduces to $_{2n}C_n/2^{2n}$.

Pascal noted that he had been unable to derive this result by the recursive method used above, but rather had to rely on combinations. "Here are the propositions that lead to it, which are properly arithmetical propositions pertinent to combinations, of which I have found some rather pretty properties." The first theorem, which Pascal gave in both French and Latin ("because French is not suitable"),[84] is couched in terms of a specific instance, but is immediately generalizable: Given any set of $2n$ elements, the following combinatorial relation obtains: $\frac{1}{2}(_{2n}C_n) + {}_{2n}C_{n+1} + {}_{2n}C_{n+2} + \ldots + {}_{2n}C_{2n} = 2 \cdot 4^{n-1} = 2^{2n-1}$.[85] In his letter, Pascal offered no proof, but later letters show that Fermat was quick to recognize the underlying use of the famous "triangle of numbers," which displays in triangular array the successive expansions of $(1 + 1)^n$.[86] The $(k + 1)$th term of any such expansion is $_nC_k$, and Pascal's triangle shows clearly that the terms are symmetric about the middle term (or two equal middle terms in the case of an even number of terms), i.e. $_nC_k = {}_nC_{n-k}$. Since the expansion of $(1 + 1)^{2n}$ contains an odd number of terms, the theorem above does no more than split the middle term in half and add to the half all the terms on one side of the middle term. But that amounts to no more than taking half the sum of the terms of the expansion, and one half of $(1 + 1)^{2n}$ is 2^{2n-1}.

On the basis of this "purely arithmetical" theorem, Pascal then derived the theorem that represents his solution to the question of dividing the stakes after the first of n decisions:

> It is necessary to say first that, if one has made one out of five plays, and hence four are lacking, the game will be completely decided in eight plays, i.e. in twice four plays.
>
> The value of the first of five plays [in terms of the claim] on the oppo-

---

[84]*Ibid.*, p. 293: "Par exemple, et je vous le dirai en latin, car le français n'y vaut rien: . . ."

[85]*Ibid.*: "Si quotlibet litterarum, verbi gratia octo: A, B, C, D, E, F, G, H, sumantur omnes combinationes quaternarii, quinquenarii, senarii, etc., usque ad octonarium, dico, si jungas dimidium combinationis quaternarii, nempe 35 (dimidium 70), cum omnibus combinationibus quinquenarii, nempe 56, plus omnibus combinationibus senarii, nempe 28, plus omnibus combinationibus septenarii, nempe 8, plus omnibus combinationibus octonarii, nempe 1, factum esse quartum numerum progressionis quaternarii cujus origo est 2: dico quartum numerum, quia 4 octonarii dimidium est. Sunt enim numeri progressionis quaternarii, cujus origo est 2, isti: 2, 8, 32, 128, 512, etc., quorum 2 primus est, 8 secundus, 32 tertius et 128 quartus: cui 128 aequantur 35 dimidium combinationis 4 litterarum + 56 combinationis 5 litterarum + 28 combinationis 6 litterarum + 8 combinationis 7 litterarum + 1 combinationis 8 litterarum."

[86]Cf. Fermat to Pascal, 29.VIII.1654, FO.II.307-309.

nent's money is the fraction that has as its numerator one half of the combinations of 8 taken 4 at a time (I take 4 because it is equal to the number of plays lacking, and 8 because it is twice 4), and as its denominator this same numerator plus all superior combinations.

By the theorem just cited, the sum in the denominator is 2^7; hence, the value of the first of five plays is $\frac{1}{2}(_8C_4)/2^7$, or $_8C_4/2^8$. As shown above, one arrives at this value by dividing the product of the first four even numbers into the product of the first four odd numbers.

Only after reading Pascal's derivation of the solution by means of products does one catch the error in its original statement. The fraction consisting of the product of the first eight odd numbers as numerator and the product of the first eight even numbers as denominator does not yield the value of the first of eight plays, but that of the first of nine. From the derivation, it becomes clear that Pascal's procedure is the following: if a game requiring n plays to win is interrupted after the first play, to find what portion of the opponent's money is due the winner, divide the product of the first $n - 1$ odd numbers by the product of the first $n - 1$ even numbers.

Pascal provided no explanation for the particular combinations used in his derivation, apparently assuming that Fermat would understand the rationale behind them. Instead, he moved directly to a pair of tables showing the values of (1) each of n plays for $n = 1, 2, 3, 4, 5, 6$, and (2) the cumulative value up to and including each play. He concluded his remarks on probability with a new problem posed to him by Méré, who apparently refused to accept Pascal's mathematical results. Méré had noted that, on the one hand, the odds were in favor of making a six with a single die in four casts but against making a double six with two dice in twenty-four casts, while on the other hand $4 : 6 = 24 : 36$. "There you have the great scandal that leads him haughtily to proclaim that our propositions are not valid and that Arithmetic lies." Neither Pascal nor Fermat explicated the problem further, each confident that the other could handle it.

Fermat's reply to Pascal's letter is no longer extant, but its content can be reconstructed from Pascal's reaction to it.[87] Though Fermat agreed with Pascal's numerical results, he did not much like the method used to obtain them. His own reasoning was more direct, focusing on the portion of the entire stakes due the winner of the first play. Consider a game of n plays, which is interrupted after the first play. The winner of that play needs $n - 1$ more wins to gain the entire stakes. Were the game to continue, then, it would be decided in at most $2n - 2$ plays. Since the game consists of a series of independent win-or-lose decisions of equal probability, those $2n - 2$ plays have 2^{2n-2} possible outcomes. How many of those outcomes would be favorable to the winner of the first play, i.e. how many would award the whole game to him? The

[87]Pascal to Fermat, 24.VIII.1654, FO.II.300–307.

answer, clearly, is any outcome that involved $n - 1$ or more wins for him. To find the number of these favorable outcomes, compute then the number of combinations of $2n - 2$ things taken $n - 1$ at a time, the number of $2n - 2$ things taken n at a time, the number taken $n + 1$ at a time, and so on up to the number taken $2n - 2$ at a time. Pascal's arithmetical proposition provides immediate relief from the actual computation involved, for the sum just mentioned is clearly $2^{2n-3} + \frac{1}{2}(_{2n-2}C_{n-1})$. On the principle, then, that the winner of the first play is entitled to the portion of the entire stakes corresponding to the ratio of the number of outcomes favorable to him to the number of total possible outcomes, he should receive

$$\frac{2^{2n-3} + \dfrac{1}{2}\left(_{2n-2}C_{n-1}\right)}{2^{2n-2}}$$

of the entire stakes. To see that this result agrees completely with Pascal's, one need only rewrite it in the form

$$\frac{1}{2} + \frac{1}{2}\left[\frac{_{2n-2}C_{n-1}}{2^{2n-2}}\right]$$

and recall that Pascal's results spoke of the portion of the opponent's ante due the winner, i.e. the portion of 1/2 the entire stakes.[88]

Pascal reported Fermat's solution to the Parisian group of mathematicians, among whom it met with some resistance, particularly from Roberval.[89] He objected to the possibly counterfactual basis of the solution: though the game could end on any play from the $(n - 1)$th to the $(2n - 2)$th, Fermat was treating it as going on for the full $2n - 2$ plays. Since Pascal's own solution

---

[88] As given so far, Fermat's solution applies only to the first of n plays, i.e. to the situation in which the score is 1-0. It is, however, easily adapted to the more general case of a game requiring n decisions to win, which is interrupted when one of the players has won p decisions and the other q. In that case, the game would have been decided in at most $2n - p - q - 1$ further plays. Of the $2^{2n-p-q-1}$ possible outcomes, those involving $n - p$ or more wins for the player with p decisions are favorable to him. The number of favorable outcomes is:

$$_{2n-p-q-1}C_{n-p} + {}_{2n-p-q-1}C_{n-p+1} + \cdots + {}_{2n-p-q-1}C_{2n-p-q-1}.$$

Here, of course, Pascal's theorem for computing the sum of combinations offers limited help.

[89] Pascal to Fermat, 24.VIII.1654, FO.II.302: "Je communiquai votre méthode à nos Messieurs, sur quoi M. de Roberval me fit cette objection: Que c'est à tort que l'on prend l'art de faire le parti sur la supposition qu'on joue en *quatre* parties, vu que, quand il manque *deux* parties à l'un et *trois* à l'autre, il n'est pas de nécessité que l'on joue *quatre* parties, pouvant arriver qu'on n'en jouera que *deux* ou *trois*, ou à la vérité peut-être *quatre*. Et ainsi qu'il ne voyait pas pourquoi on prétendait de faire le parti juste sur une condition feinte qu'on jouera *quatre* parties, vu que la condition naturelle du jeu est qu'on ne jouera plus dès que l'un des joueurs aura gagné, et qu'au moins, si cela n'était faux, cela n'était pas démontré, de sorte qu'il avait quelque soupçon que nous avions fait un paralogisme."

rested on the same fiction, he took up the cudgel for both himself and Fermat. He saw, however, an implication of Roberval's position that could spell trouble for Fermat's approach to the matter.

For the case of a two-player game, Pascal argued, the assumption that the game continued for the full $2n - 2$ plays was no less valid for being possibly counterfactual. One need only suppose that the players agree to play out the full game regardless of an early decision; for example, treat a game requiring three wins as one decided by three out of five, where the players agree to go five rounds whatever the intermediate outcome. That the players would be willing to do so is clear; neither stands to lose or gain further.[90]

If the difference between the actual course of a two-player game and its assumed complete form is irrelevant to an analysis of the probabilities involved, the same is not true of a game involving three or more players. Consider, for example, a three-player game. Like the two-player game, it is carried out as a series of win-or-lose decisions in which each player has an equal chance of success but only one can win (e.g. high-card draw). The first to win n decisions gains the entire stakes of the game (the stakes, of course, consist of three equal antes). To see the difficulty inherent in a direct application of Fermat's method, consider the situation in which the game requires three decisions to win and is interrupted when player A has two and players B and C one each.

Clearly, three more decisions would suffice to determine the winner. Hence, by Fermat's method, one would first calculate the total possible outcomes of those three decisions; since each decision could go one of three ways, the three decisions have 27 possible outcomes. It would seem at first that, of those 27 outcomes, any that include one or more wins for A would be favorable to him; 12 include one win, 6 include two, and 1 three. Hence, 19 of the possible 27 outcomes are favorable to A, and it would seem that he should receive 19/27 of the total stakes if the game is not continued. As Pascal pointed out, however, that division would not in fact be equitable. Where three players are involved, it is fully possible for two of them to achieve winning totals in a

---

[90]*Ibid.*, p. 303: "N'est-il pas clair que les mêmes joueurs, n'étant pas astreints à jouer les quatre parties, mais voulant quitter le jeu dès que l'un aurait atteint son nombre, peuvent sans dommage ni avantage s'astreindre à jouer les quatre parties entières et que cette convention ne change en aucune manière leur condition? Car, si le premier gagne les deux premières parties de *quatre* et qu'ainsi il ait gagné, refusera-t-il de jouer encore deux parties, vu que, s'il les gagne, il n'a pas mieux gagné, et s'il les perd, il n'a moins gagné? Car ces deux que l'autre a gagné ne lui suffisent pas, puisqu'il lui en faut trois, et ainsi il n'y a pas assez de quatre parties pour faire qu'ils puissent tous deux atteindre le nombre qui leur manque.

Certainement il est aisé de considérer qu'il est absolument égal et indifférent à l'un et à l'autre de jouer en la condition naturelle à leur jeu, qui est de finir dès qu'un aura son compte, ou de jouer les quatre parties entières: donc, puisque ces deux conditions sont égales et indifférentes, le parti doit être tout pareil en l'une et l'autre. Or, il est juste quand ils sont obligés de jouer quatre parties, comme je l'ai montré: donc il est juste aussi en l'autre cas."

theoretically complete game; for example, in the three plays needed to complete the game in question, *A* might win one and *B* two, and both would then have the three wins necessary to claim the entire stakes. How does one account for this possibility?

In his letter of 24 August 1654, Pascal suggested splitting between two of the players any of the combinations favorable to both. In the example above, three combinations give one win to *A* and two to *B*, and another three give one win to *A* and two to *C*. Hence, Pascal reasoned, *A*'s portion of the 27 outcomes should consist of the 13 uniquely favorable to him plus one half of the six doubly favorable to him and one of the other players, for a total of 16. *B* and *C* each receive credit for 4 uniquely favorable outcomes and $1\frac{1}{2}$ doubly favorable outcomes. Pascal could see no other way to assign the probabilities involved, but he was unhappy at the result.[91] Fermat's method failed, he felt, because one had to take into account the real nature of the game, which stops when one of the players gains his three wins. To assume that the game continues beyond that point introduces into it an element unacceptable to the players, to wit, that two of them could claim victory and would have to share the stakes. The counterfactual game is not the same as the real one, and hence an analysis based on the former is not valid for the latter. For all his dissatisfaction with Fermat's solution, Pascal could offer no general approach of his own and left the problem open.[92]

Fermat's reply of 29 August required but one paragraph to settle Pascal's qualms.[93] The extension of his method to games involving more than two players did mean the introduction of a consideration irrelevant to the two-player situation, but that consideration was order. To win, *A* not only needed one or more wins, but he needed one win before either of the other two players gained two. Hence, among the six doubly favorable outcomes, two were in fact unfavorable to *A*: the cases in which *B* or *C* won both of the first two decisions and *A* the last. *A* had, therefore, no claim to those outcomes, and the correct division allotted 17 favorable outcomes to *A* and 5 each to *B* and *C*.

It was the rest of Fermat's letter of 29 August that attempted to change the

[91]*Ibid.*, p. 305: "Voilà, ce me semble, de quelle manière il faudrait faire les partis par les combinaisons suivant votre méthode, si ce n'est que vous ayez quelque autre chose sur ce sujet que je ne puis savoir. Mais, si je ne me trompe, ce parti est mal juste."

[92]*Ibid.*, p. 307: "Je ne laisse pas de vous ouvrir mes raisons pour en attendre le jugement de vous. Je crois vous avoir fait connaître par là que la méthode des combinaisons est bonne entre deux joueurs par accident, comme elle l'est aussi quelquefois entre trois joueurs, comme quand il manque *une* partie à l'un, *une* à l'autre et *deux* à l'autre, parce qu'en ce cas le nombre des parties dans lesquelles le jeu sera achevé ne suffit pas pour en faire gagner deux; mais elle n'est pas générale et n'est bonne généralement qu'au cas seulement qu'on soit astreint à jouer un certain nombre de parties exactement. De sorte que, comme vous n'aviez pas ma méthode quand vous m'avez proposé le parti de plusieurs joueurs, mais seulement celle des combinaisons, je crains que nous soyons de sentiments différents sur ce sujet."

[93]Fermat to Pascal, 29.VIII.1654, FO.II.309.

subject of his correspondence with Pascal to number theory, and it evoked the cold response that has been cited several times above.[94] Pascal accepted Fermat's final adjustment of the method for dividing stakes but then advised him to "look elsewhere for those who will follow you in your numerical researches." Their discussion of probability came to an abrupt end. Short-lived as it was, it helped to lay the foundations of the theory of probability. Pascal's investigation of the binomial expansion provided some of the mathematics necessary to translate into concrete results the basic principle that common sense dictated: if, of n possible outcomes, p are favorable to a player, then the player's chances of winning are p/n. Fermat's contribution to the discussion seems to have been limited to articulating the mathematics Pascal provided. Ingenious as it was, that mathematics would not itself take probability far. If Pascal and Fermat could resolve the two-player game without actually counting the combinations involved, they found themselves forced in the dispute over the three-player game actually to set out in tabular form all 27 possible outcomes and count those favorable to each player. They apparently lacked the mathematics to solve such an elementary problem as: of the 27 possible outcomes, how many contain one or more wins favorable to A? It is doubtful on the basis of the mathematics present in the letters exchanged that either man would have been able to compute without tables the number of outcomes unfavorable to a player because of the order of wins. Neither man went on to develop the tools probability would require, for neither man continued working on the subject. Pascal had finished with mathematics itself; Fermat had other matters on his mind, especially number theory. It was left to van Schooten and his pupil, Huygens, to pick up where Pascal and Fermat had left off and to develop further the classical foundations of probability.

[94] See above, Chap. VI, §IV.

I. THE PUBLICATION OF
FERMAT'S WORKS[1]

Bibliographical Essay and
Chronological Conspectus of
Fermat's Works

As a result of Fermat's own disregard for the publication and preser-
vation of his work, his death in 1665
left his memoirs and letters (many of
them unique copies) strewn all over
Europe. Although Mersenne, Hérigone,
Frenicle, Wallis, Lalouvère, and Clerselier had captured a small portion of
Fermat's mathematical achievement in
print,[2] the bulk of it remained in
manuscript among the papers of his
various correspondents. Much lay in
the hands of Pierre de Carcavi, who,
from 1636 on, had served as Fermat's
depositary in Paris. It was to Carcavi
that Huygens wrote in 1665 to express
his hope that, "one will not lose what
remains of [Fermat's] writings; since
you have always been one of his close
friends, I do not doubt that your intervention with his heirs will be of great
use in saving such excellent work from
obscurity."[3]

This prediction unfortunately proved
less accurate than Huygens' fear, voiced
some years earlier, that the rapid development of mathematics at the time
would obscure Fermat's seminal contributions. Immersed in the establishment of the new *Académie Royale des
Sciences*, Carcavi left to Fermat's heirs
themselves the task of preserving their
father's work for posterity. The eldest
son and executor, Clément-Samuel de
Fermat, began that task in 1670 by re-
publishing the 1621 Bachet edition of

---

[1] What follows is largely a *précis* of bibliographical information provided by the editors
of *FO*, most particularly in their "Advertissement" to FO.I.
[2] For details, see FO.I.ix-xiii.
[3] Huygens to Carcavi, 26.III.1665, FO.IV.137.

Diophantus' *Arithmetica*, to which he added his father's *Observations*, Billy's *Inventum novum*, and some isolated letters.[4] In the years following, he gathered together from various sources what he could of his father's extant papers, which he published in 1679 under the title *Varia opera mathematica*.[5] It is not clear what assistance, if any, he received from Carcavi in this endeavor. The *Varia* lacked several important memoirs and included only a small sample of Fermat's mathematical correspondence. At least some of the missing material certainly was at Carcavi's disposal; he had cited it in his *Eloge* of Fermat in the *Journal des Sçavans* in 1665. As the editors of the *Oeuvres* put it:[6]

> It is hard to believe that Carcavi, after what he had said in the *Eloge*, would have refused Fermat's son copies of the items he possessed, at least of those that were detailed in the aforesaid *Eloge*. It is no less certain that, if he did not refuse absolutely, he did not give over copies of all the works in his hands and did not wish to communicate any of the numerous letters Fermat had addressed to him personally.

Though Roberval proved more cooperative, a comparison of the *Varia* with material now available indicates that he did take the opportunity to select (and even reformulate) the material he forwarded, in order to enhance his own reputation.

For all its many faults, the *Varia* remained the only published collection of Fermat's papers until the late nineteenth century. The manuscripts themselves, both autographs and copies, gradually fell into the hands of collectors or, since several were untitled and anonymous, into manuscript collections attributed to other writers (e.g. the treatises uncovered by De Waard among the *Discepoli* of Galileo). A large number of those in private hands returned briefly to the public domain when, in the *Journal des Savants* for September 1839, Count Guillaume Libri, the well-known bibliophile and historian of mathematics, announced his purchase in Metz of a collection of manuscripts formerly belonging to Arbogast and containing many hitherto unedited papers of Fermat.[7] A subsequent article by Libri in 1841,[8] pointing out the defects

---

[4] *Diophanti Alexandrini Arithmeticorum libri sex et de numeris multangulis liber unus cum commentariis C. G. Bacheti V. C. et observationibus D. P. de Fermat Senatoris Tolosani. Accessit Doctrinae analyticae inventum novum, collectum ex variis eiusdem D. de Fermat epistolis.* Toulouse: Bernard Bosc, 1670.

[5] *Varia opera mathematica D. Petri de Fermat Senatoris Tolosani. Accesserunt selectae quaedam ejusdem Epistolae, vel ad ipsum à plerisque doctissimis viris Gallicè, Latinè, vel Italicè, de rebus ad Mathematicas disciplinas aut Physicam pertinentibus scriptae.* Toulouse: Joannis Pech, 1679; repr. Berlin, 1861; Brussels, 1969.

[6] FO.I.xvii.

[7] Louis-François-Antoine Arbogast (1759-1805) took active part in the reorganization of French education during the Revolution. A bibliophile and mathematician, he organized the library of the Committee of Public Instruction and taught mathematics at several institutions. See *Dictionnaire de biographie française*, Vol. VII (Paris, 1939), cols. 274-275; and *Dictionary of Scientific Biography*, Vol. I (New York, 1970), pp. 206-207 (Jean Itard).

[8] Libri, "Des manuscrits inédits de Fermat," *Journal des Savants*, 1839, 540-651; 1841, 267-279; 1845, 682-694.

in the 1679 *Varia* on the basis of the Arbogast material, induced the Minister of Public Education, Villemain, to institute in 1843 a project for a new edition of Fermat's works at state expense.[9] Designated head of the project, Libri was joined by an assistant, a young mathematician named Despeyrous. For reasons known only to himself, however, Libri refused to give his assistant access to the Arbogast manuscripts and instead sent him off to Vienna to search for more material. Other than some letters from Fermat to Clerselier, the search revealed nothing. Meanwhile, Libri himself was doing nothing to carry out his assignment. In 1848, rumors and some direct evidence led to a warrant for his arrest on charges of having appropriated for personal profit some 300,000 *livres* worth of books and manuscripts from French libraries.[10] In the turmoil of the political disorder at the time, Libri managed to escape to Italy with most of his personal library, including the Fermat manuscripts; a few fragments seized in his lodgings were turned over to the Bibliothèque Nationale. The editor having absconded with most of the material, the proposed edition was dropped.

Some thirty years later, Charles Henry renewed the search for Fermat's works and in 1879 published in Boncompagni's *Bolletino* an article entitled "Recherches sur les manuscrits de Pierre de Fermat, suivies de fragments inédits de Bachet et de Malebranche."[11] In 1881, Henry received word from Prince Boncompagni that he had acquired two manuscript volumes of Fermat's works, which he was ready to place at the disposal of the editors of any new edition. A comparison of those volumes with Libri's description of the Arbogast material revealed that they indeed constituted at least a major part of what had been in Libri's possession. Henry had reason to believe, however, that another, significant portion lay among the manuscripts Libri had sold to Lord Ashburnham following his flight from France. When successful negotiations brought the Ashburnham manuscripts back to their rightful place in the Bibliothèque Nationale, examination of them proved Henry's suspicion to be ill founded. With the exception of some material already published in the *Varia*, the Ashburnham collection contained nothing of Fermat's works, and it seemed likely that Boncompagni's two volumes in fact represented the whole of Libri's collection of Fermat's papers.

Armed, then, with the Boncompagni manuscripts, the material published during Fermat's lifetime, the 1670 Diophantus, the 1679 *Varia*, the results of Despeyrous's research in Vienna, and the "Recherches," Henry joined with Paul Tannery, later also engaged with Charles Adam in the publication of the *Oeuvres de Descartes*, to produce the *Oeuvres de Fermat*. The first volume, which appeared in 1891, contains all of Fermat's extant memoirs, the *Observations on Diophantus*, and Fermat's various philological efforts. The second

---

[9] Libri discussed the plans for the new edition in "Fermat," *Revue des Deux Mondes*, 1845, 679-707.

[10] Barbara McCrimmon, "The Libri Case," *Journal of Library History*, 1(1966), 7-32.

[11] The article, which appeared in Vol. XII of the *Bolletino*, pp. 477-568, 619-739, was published separately in Rome in 1880.

volume (1894) includes all of Fermat's correspondence extant at the time. The third volume (1896) provides French translations of the Latin material of the first two volumes, together with translations of Billy's *Inventum novum* and of those portions of Wallis' *Commercium epistolicum* that pertain to Fermat. The fourth and final volume (1912) of the original *Oeuvres* contains correspondence that had come to light since the preparation of Volume II, along with excerpts referring to Fermat from the correspondence of his contemporaries. The fourth volume concludes with a series of mathematical and historical notes concerning Fermat and his work.

Subsequent research done with the aid of the *Oeuvres* shows how completely Henry and Tannery carried out their task. In the years immediately preceding and following World War I, Cornelis De Waard, beginning the archival research that would culminate in his monumental *Correspondance de Mersenne*, uncovered in Groningen and Florence a few letters and memoirs of Fermat's—most importantly, the 1643 letter to Brûlart—and published them in 1922 as a *Supplément* to Volumes I-IV of the *Oeuvres*. He was also able, through excerpts from the correspondence of Ricci and Torricelli, to shed further light on Fermat's reputation in Italy. In 1943, Joseph E. Hofmann identified as stemming from Fermat two appendices to a special edition of Frenicle's *Solutio duorum problematum ...* ; one lends new insight into Fermat's number theory, but the other is only a *précis* of the *Tripartite Dissertation*.[12] De Waard's and Hofmann's discoveries represent to date the only additions to the *Oeuvres*.

There is little reason to suspect the existence of much more (if indeed any) original material. For example, a manuscript volume written by Michelangelo Ricci, which Henry and Tannery knew about but could not find, was recently (1966) offered for sale in Paris and purchased by the Municipal Library of Toulouse. According to the complete catalogue of contents prepared by the dealer with the assistance of Taton and Costabel, the volume contains copies of several of Fermat's works, apparently those that Mersenne took with him on his trip to Italy in 1644, but all of them have been published. More important, with the exception of an occasional letter, Fermat's extant works themselves give no indication of any major lacuna. There are, for example, no references in them to papers not contained in the *Oeuvres* or the *Supplément*. Fermat's work habits as discussed in Chapters I and II above offer little encouragement to the researcher seeking material Fermat did not cite. For example, Chapter VI suggests strongly that Fermat simply never committed to paper many of the number-theoretical proofs and derivations he claimed to possess.

One possibility for future developments remains open. If the *Observations*

---

[12] "Neues über Fermats zahlentheoretische Herausforderungen von 1657 (mit zwei bisher unbekannten Originalstücken Fermats), *Abhandlungen der Preussischen Akademie der Wissenschaften*, Jahrgang 1943, Math.-naturw. Klasse, Nr.9 (Berlin, 1944).

on Diophantus is any indication, Fermat had the habit of writing in the margins of his books. Unfortunately, his uncatalogued library was apparently dispersed following his death. Some volumes may still exist in various public and private collections. Where, for example, is his copy of Pappus' *Collection* which, in the edition by Commandino, also had generous margins? The hope that some of the missing volumes might still come to light induced Henry and Tannery to include in the introduction to Volume I of the *Oeuvres* a reproduction of a page from the autograph copy of the *Doctrine of Tangents*, as a sample of Fermat's hand. Their initiative has so far brought no response.

As the preceding biography has tried to show, what one now has of Fermat's work suffices to establish a picture of the man and his mathematics. In most cases, one may doubt whether Fermat ever recorded the answers to the questions that still remain open.

II. A CHRONOLOGICAL CONSPECTUS OF FERMAT'S WORKS

The following list is meant to serve several purposes. First, with rare exceptions, the chapters above have restricted bibliographical citations of Fermat's works to the standard edition in the *Oeuvres*, with no indication of the nature, number, and provenance of the sources used to establish it. In some cases, that information suggests the varying circulation of Fermat's memoirs. Second, again with some exceptions, the discussion above did not explicitly argue or defend the dating of the memoirs. It did so only in cases where subsequent research or analysis led to a result at variance with the findings of Henry and Tannery. It may help to specify in each case now the basis for dating the text. Third, by treating Fermat's career topically, and by considering within each topic the texts that offer the most pertinent insights, the main discussion has tended to blur the chronological development of Fermat's career as outlined in Chapter II and has also dropped from consideration some papers that are either pedestrian or repetitious in content. It may be of some use to return here again to the chronological order and to supply a *précis* of the texts not explicitly discussed.

In addition to the standard abbreviations used above, the conspectus employs the following:

A = MS Arbogast-Boncompagni, the first of the two manuscript volumes mentioned above;[13] for a detailed analysis of the MS, which contains copies made by Arbogast, see FO.I.xxii–xxvii. Present location: Bibliothèque Nationale, Paris, Fonds fr. nouv. acq. 6862.

C = MS Ashburnham 1848 I (= *BN* Fonds latin, nouv. acq., 2339 (cf. FO.I.xxii).

[13] The second of the Boncompagni manuscripts, formerly owned by Vicq-d'Azyr and then by Arbogast, contains only letters, most of them written in a poor seventeenth-century hand; for a detailed analysis, see FO.I.xxviii–xxx. Present location: Bibliothèque Nationale, Paris, Fonds fr. nouv. acq. 10556.

$F = $ MS Florence, Biblioteca Nazionale, MSS Galileiani, *Discepoli*, Vol. CIII (= Vol. XLV of MSS Viviani); analysis (based on research of Giovannozzi) in FO.*Suppl*.xvi-xvii.

$G = $ MS Library of the University of Groningen, 110, van Schooten collection; analysis, FO.*Suppl*.xi, xvi-xvii.

$R = $ MS Ricci, recently acquired by the Bibliothèque Municipale de Toulouse (information kindly supplied by J. E. Hofmann); description published in catalogue of the Libraire Alain Brieux, Paris, with the assistance of P. Costabel and R. Taton.

The chapter references in parentheses next to the titles of the works indicate where each is discussed in the main text of the biography.

1629-1636

Methodus ad disquirendam maximam et minimam et de tangentibus linearum curvarum (Chap. IV, § § II, III). FO.I.133-136, from *Varia*, 63-64. Copies in $C; F$, 83^r - 84^r; G, 6^v - 7^r; R, 37^r - 39^r. All evidence (cf. Chap. IV, n. 1) points to this treatise as the one Fermat reported having given to Etienne d'Espagnet in 1629. Despite De Waard's guess (FO.*Suppl*.xvi) of "Fin 1637" as the date of the memoir, Fermat's correspondence of 1636-1637 gives no hint of his having composed a new version; rather, he apparently asked d'Espagnet to turn his copy over to Mersenne and Roberval. This would be the same copy, then, that Mersenne sent on to Descartes early in January 1638.

⟨*Loci ad tres lineas demonstratio*⟩ (Chap. III, §III). FO.I.87-89, from *C*. Copies in F, 112^r - 113^r; G, 20^r. The untitled, anonymous copy in C contains the note, written in a contemporary hand: "Pour Monsr Carcavi rue Michel Leconte au milieu." Libri's attribution of the text to Fermat was initially corroborated by a reference to it in the letter of 20.IV.1637 from Fermat to Roberval (FO.II.105); its presence in F and G further confirms the attribution. Fermat reports in that letter having sent the paper to Beaugrand "il y a longtemps" but, as Chap. III argues, this could not have been much before 1635, when a breakthrough on Prop.II,5 of the *Plane Loci* gave Fermat the basis for the *Isagoge*. One may guess that the attached note stems from Beaugrand himself, whom Fermat had asked to forward the text to Roberval.

Apollonii Pergaei libri duo de locis planis restituti (Chap. III, §III). FO.I.3-51, from *Varia*, 12-43. No extant MS copies. According to letters to Mersenne (20.IV.1636, FO.II.5) and to Roberval (20.IV.1637, FO.II.105), Fermat began this work in Bordeaux and had restored all but Prop. II,5 before leaving for Toulouse. He announced its recent completion in the letter to Mersenne but, as of the letter to Roberval a year later, had only just then sent the final version to Carcavi, who was to forward it to Beaugrand. It was circulating in Paris by the end of 1637.

1636

⟨*Propositio D. de Fermat circa parabolen*⟩. FO.I.84-87, from *Varia*, 144-145. Copies in *F*, 110^v - 112^r; *G*, 19^r - 19^v. The paper contains the solution to the problem of constructing a parabola passing through four given points. Cavalieri had already heard of the solution in 1635 (cf. Chap. II, n. 58), but the *Varia* places it among letters of 1636.

⟨*De motu gravium descendentium*⟩ (Appendix I). FO.*Suppl*.36-43, from MS Florence, Biblioteca Nazionale, MSS Galileiani, Parte V, Tomo VII, 98^r - 100^v. The text is anonymous and untitled; De Waard (FO.*Suppl*.20-36) attributes it to Fermat and dates it 1636.

⟨*De quadratura helicis Galileiani*⟩ (Chap. V, §II). FO.*Suppl*.15-19, from MS Florence, Bibl. Naz., MSS Galileiani, Parte IV, Tomo IV, 34^r - 34^v; previously published in *Atti e Memorie della R. Accademia di scienze ecc.*, 296(1894-95), nuova ser., XI, 40-42. De Waard (FO.*Suppl*.1-15) gives a detailed justification of attributing this anonymous, untitled treatise to Fermat and comes to the conclusion (p. 15) that "le présent écrit forme un envoi à part destiné pour Carcavi seul. Sa date ne se peut préciser entre le 3 juin 1636 et le commencement de l'année 1637."

Ad locos planos et solidos isagoge (Chap. III, §II). FO.I.91-103, from *Varia*, 1-8 (controlled by *C*, which enables the restoration of the original Viètan notation, replaced in the *Varia* by the Cartesian system). Copies in MS Brit. Mus., Harleian 6083 (formerly owned by Chas. Cavendish); *F*, 75^r - 80^r; *G*, 1^r - 4^v; *R*, 28^r - 34^r. Though Fermat refers to this work from his earliest letters on, the concluding paragraph places its composition after that of the *Plane Loci*, hence after 1635; Fermat sent the text to Paris late in 1637. From all indications, however, he wrote it sometime early in 1636.

Appendix ad isagogen topicam continens solutionem problematum solidorum per locos (Chap. III, §V). FO.I.103-110, from *Varia*, 9-11. Copies in *F*, 80^v - 82^v; *G*, 5^r - 6^r; *R*, 34^r - 36^r. Despite the juxtaposition of this text with the *Isagoge* in the *Varia* and in *F*, *G*, and *R*, its absence from the other two MSS containing the *Isagoge* itself and various separate references in Fermat's correspondence suggest that the two works did not at first constitute a single treatise. As noted in Chap. III, §V, the *Appendix* represents an afterthought. It is difficult to determine just when it occurred to Fermat. Late 1636 seems likely, though it may have been even later. Fermat first mentions it in a letter to Mersenne of February 1638 (FO.II.134), but in a tone that suggests that it had been circulating in Paris for some time. Unlike the later *Tripartite Dissertation*, the *Appendix* does not refer to Descartes' treatment of the same material in the *Geometry*, a copy of which Fermat received in December 1637. Since it seems likely that, knowing about Descartes' work, Fermat would have mentioned it, the last date may perhaps be taken as a *terminus ante quem*.

⟨*Solutio problematis a Domino Pascal propositi*⟩. FO.I.70-74, from *Oeuvres de Pascal* (ed. Bossut, 1779), IV, 449-454. A geometric solution of the problem: given the vertex angle of a triangle and the ratio of the perpendicular to the difference of the sides, to find the species of the triangle (in essence, to construct the triangle). The text concludes with a counterproblem posed "tam Domino Pascal quam Domino Roberval" (to construct the tangent at any point of the Galilean spiral), which is referred to in the letter of 3.VI.1636 from Fermat to Mersenne (FO.II.12; cf. n. 2). Clearly, of course, the Pascal in question is not Blaise, but his father, Etienne.

1638

⟨*Ad eamdem methodum de maximis et minimis*⟩. *Inc:* "Volo meâ methodo . . ." FO.I.140-147, from *Varia*, 66-69. Copy in *R*, $47^v - 51^v$. *F*, $85^v - 88^v$, and *G*, $7^v - 9^v$, are copies of a contemporary French translation (possibly by Fermat himself, more probably by Mersenne) beginning, "Je veux par méthode . . ."; published in FO.*Suppl*.74-83. In the treatise, Fermat illustrates his method of maxima and minima by applying it to a series of examples. De Waard suggests (FO.*Suppl*.72-73) that the treatise was designed to explain the method to Mydorge or Desargues when, in the spring of 1638, they agreed to serve as arbiters in the dispute with Descartes. The conceptual and stylistic similarity between the text and the earlier *Method* would tend to support this conjecture, but unfortunately Fermat himself nowhere refers to the text in his correspondence, and hence no direct corroboration is possible.

⟨*Centrum gravitatis parabolici conoidis ex eadem methodo*⟩ (Chap. V, §III). FO.I.136-139, from *Varia*, 65-66. Copies in *F*, $84^r - 85^r$; *G*, $7^r - 7^v$. The treatise was written sometime during the spring of 1638; the editors of FO (FO.I.136, n. 3) suggest it was sent to Mersenne with the letter of 20.IV. 1638, for transmission to Roberval. By 15.VI.1638, Fermat was defending it against Roberval's objections.

1640

Analytica eiusdem methodi investigatio (Chap. IV, §II). FO.I.147–153, from *A*. Copies in *BN* Fonds fr., nouv. acq., 3280 (a fragment, published by Henry, "Recherches . . . ," 180-183); *F*, $93^v - 96^r$; *R*, $47^v - 51^v$. The textual tradition is discussed above, Chap. IV, n. 3; the basis for dating is argued in Chap. IV itself.

⟨*De tangentibus*⟩ (Chap. IV, §VII). *Inc:* "Doctrinam tangentium antecedit jamdudum tradita methodus de inventione maximae et minimae . . ." FO.I. 158-167, from *Varia*, 69-73. Autograph copy in BN Fonds fr., nouv. acq., 3280, 112-117 (untitled). Other copies in *A; F*, 7^v - 15^r and 89^r - 92^r; *G*, 10^r - 11^v; *R*, 51^v - 54^r. Judged from its content, this most sophisticated version of the method of tangents clearly postdates the "Méthode expliquée" of June 1638. Mersenne was circulating it by late October 1640 (FO.II.218, n. 2). Most likely, Fermat wrote it in conjunction with the *Analytic Investigation* when, around 1640, he had decided to answer Descartes' charges of being lucky by a full disclosure of the foundations of his methods.

1642

⟨*Problema missum ad Reverendum Patrem Mersennum 10^a die Novembris 1642*⟩. FO.I.167-169, from BN Fonds lat., 11197 (first publ. by Henry, "Recherches . . . ," 195-196). Copy in *G*, 20^v - 21^r. The text contains Fermat's solution of a problem he himself had posed in his first letter to Mersenne: to find the cylinder with maximum surface inscribable in a given sphere. It is solved by the techniques of the *Appendix to the Method of Maxima and Minima.*

1643

⟨*Letter to Brûlart*⟩ (Chap. IV, §VI). FO.*Suppl.*120-125, from *F*, 113^v - 115^v (first published by Giovannozzi, *Archivio di storia della scienza*, 1(1919), 137-140). The only question regarding the date of this important letter (the only letter included in this conspectus, because of its importance) is whether it was written in May 1643, as given in *F*, or in March. Since, as De Waard points out, Fermat refers to it in his letter to Mersenne of 7.IV. 1643 ("Vous aurez maintenant la réponse que je fais à M. de Brûlart . . ."), the *maii* of the copy in *F* is almost certainly a misreading of *martii* in the original.

De contactibus sphaericis. FO.I.52-69, from *Varia*, 70-88. Copy in *R*, 40^r - 47^r; the propositions are listed by Mersenne in his *Universae geometriae . . . synopsis* (Paris, 1644). Except for the obvious *terminus ante quem* of 1644, when Mersenne left for Italy with those copies of Fermat's works that are included in *F* and *R*, the dating is problematical. The text itself contains no clues permitting a firm fix. The style is purely geometrical in the mode of the *Plane Loci*, and the subject matter is an extension of Viète's restoration of Apollonius' *Contacts* (*Apollonius Gallus*, Viète, *Opera*, ed. Schooten, Leiden, 1646, pp. 325-346). Where Apollonius and Viète attacked the general problem of determining a circle tangent to (or passing through) each member of a triple of points, lines, or circles, Fermat here extends the problem to three dimensions by seeking a sphere tangent to (or passing through) each member of a given quadruple of points, planes, or spheres. By its subject, the treatise could well represent an exercise undertaken very early in Fermat's career, at a time when he was working his way into the style of Viète. However, the absence of any reference to it in his early letters to Paris, its first mention by Mersenne in 1644, and at least superficial relations with the *Introduction to Surface Loci* all combine to make 1643 the most likely date of composition.

Isagoge ad locos ad superficiem (Chap. III, §IV). FO.I.111-117, from *A*. Copy in *R*, 18^r - 21^r. Fermat refers to the subject matter of this treatise as early as 1636 (Fermat to Roberval, 16.XII.1636, FO.II.94; cf. also Fermat to Roberval, 20.IV.1637, FO.II.106, where Fermat states that he has not had the opportunity to put the material together in finished form). The version in *A*, dedicated to Roberval, bears the explicit date 6.I.1643.

De solutione problematum geometricorum per curvas simplicissimas et unicuique problematum generi proprie convenientes dissertatio tripartita (Chap. III, §V). FO.I.118-131, from *Varia*, 110-115. The dating is problematical; Chap. III discusses in detail the reasons for placing its composition sometime around 1643 or 1644.

⟨*Ad Bon. Cavalierii quaestiones responsa*⟩ (Chap. V, §IV). FO.I.195-199, from *A* and MS Vicq-d'Azyr-Boncompagni. According to the editors of FO (I.195, n. 1), the text is a fragment of a letter addressed to Cavalieri, via Mersenne, sometime before 1644; the latter date is fixed by Mersenne's inclusion in the *Cogitata* of 1644 of an almost verbatim copy. As shown in Chap. V above, the content represents Fermat's breakthrough on problems in quadrature that he could not resolve as late as 1638 and relates directly to the exchange of letters with Torricelli in 1646 (cf. Chap. V, n. 5). Its date of composition most likely lies close to the time it was dispatched to Cavalieri.

1644

Ad methodum de maxima et minima appendix (Chap. IV, §VII). FO.I.153-158, from *A* (Arbogast notes: "d'après le manuscrit de Fermat"). Copies in BM Harleian 6083; *F*, 115^v - 117^r. De Waard (FO.*Suppl*.xvi) dates it 21.IV.1644, but the *Problem Sent to Mersenne* (1642) shows that its contents at least date from some time earlier.

1650

Novus secundarum et ulterioris ordinis radicum in analyticis usus (Chap. IV, §VII). FO.I.181-188, from *Varia*, 58-62. Copies in *C*; BN Fonds latin 11196, 46-53. The contents of this treatise connect it closely with the 1644 *Appendix to the Method of Maxima and Minima*, for which it provides the algebraic foundations, and may well derive from that time. The text itself, however, was sent to Carcavi with a covering letter dated 20.VIII.1650.

1655

Porismata duo. FO.I.74-76, from *Oeuvres de Pascal* (ed. Bossut, 1779), IV, 449-454. This fragment contains Porisms *I* and *V* of the *Porismatum Euclideorum . . . , q.v.*

1655/56

Porismatum Euclideorum renovata doctrina et sub forma isagoges recentioribus geometris exhibita. FO.I.76-84, from *Varia*, 116-120. No extant copies. According to FO.I.77, n. 2, the date of this treatise (as well as that of the *Porismata duo*) is apparently fixed by the remark of Ismael Boulliau in the preface to his own essay on porisms in his *Exercitationes geometricae tres* (Paris, 1657). There he states that he was led to the subject by propositions (*Porismata duo?*) sent out by Fermat *ante biennium.* Fermat's attempts at restoring this most elusive of Euclid's lost works were not known in Paris before 1654; his own reference in the main treatise to Queen Christina of Sweden (cf. above, Chap. I, n. 71) definitely places it sometime after Bernard Medon's appeal to the Queen via Nicholas Heinsius in 1651. Indeed, Fermat's remark is the only evidence that Heinsius may actually have forwarded the request to the Queen and received an affirmative response. In that case, this treatise, otherwise unmotivated by the work in which Fermat was engaged at the time, may be explained as his attempt to satisfy the Queen while persisting in his refusal to publish.

1658/1659

De aequationum localium transmutatione et emendatione ad multimodam curvilineorum inter se vel cum rectilineis comparationem, cui annectitur proportionis geometricae in quadrandis infinitis parabolis et hyperbolis usus (*Treatise on Quadrature;* Chap. V, §IV). FO.I.255-285, from *Varia*, 44-57. As noted in Chap. V above, although much of the material in this treatise was developed in the early 1640s, Fermat did not conceive of presenting it in any formal manner until after the appearance of Wallis' *Arithmetica infinitorum* in 1656, or indeed until his criticism of Wallis' work became an issue in the correspondence with Digby in late 1657. A direct reference (FO.I.263) to the *Treatise on Rectification* written in 1659 suggests that the *Treatise on Quadrature* did not reach final form until that time. One may question whether Fermat ever sent his work out, in particular to Digby. It was clearly meant as a response to Wallis, and Digby was the intermediary; but no mention of it is made by the English, nor is any copy known. Huygens' reference to it in his letter to Leibniz of 1.IX.1691 (FO.IV.137) suggests that he first encountered it in the 1679 *Varia;*

Clément-Samuel may therefore have been working from the original copy when he included it there.

1659

De linearum curvarum cum lineis rectis comparatione dissertatio geometrica (*Treatise on Rectification;* Chap. V, §V). FO.I.211-254, from *Varia*, 89-109. Originally published in 1660 as Appendix I to Lalouvère's *Veterum geometria promota* . . . under the above title with the designation *Autore M.P.E.A.S.* There is every reason to accept the thesis of the editors of FO (I.211, n. 3) that the treatise does not represent old work brought forward, but rather a direct reply to Pascal's *Egalité des lignes spirale et parabolique* (*Lettres de A. de Dettonville*, Paris, 1659). Fermat discusses Pascal's results directly in letters to Carcavi dated IX.1659 and II.1660 (FO.II.441-446), which helps to confirm 1659 as the date of the completed treatise. Moreover, the treatise and its method are wholly dependent on the full-scale treatment of quadrature as presented in the *Treatise on Quadrature*.

⟨*Ad Laloveram propositiones*⟩ (Chap. V, §V). FO.I.199-209, from *Pars prior, Appendix secunda, Veterum geometria promota* . . . by Lalouvère (Toulouse, 1660). Though these propositions based directly on the *Treatise on Rectification* were published anonymously, Lalouvère's preamble to the second appendix (printed FO.I.199-200, n. 1) clearly identifies the text (*non res tantum, sed verba etiam ipsa sunt integerrimi Senatoris*) as being by Fermat, whom Lalouvère calls *alter seculi nostri Conon* (cf. in this regard above, Chap. VI, n. 119). The direct tie between these propositions and the *Treatise on Rectification* clearly make them simultaneous with the latter.

1659/1661

⟨*Ad Adriani Romani problema*⟩. FO.I.189-194, from MS Leiden, MSS Huygens, 30 (first published by Henry, "Recherches . . . ," 211-213). In 1594, the Dutch mathematician Adrien van Roomen challenged French mathematicians to solve a certain 45th-degree equation. Viète responded immediately by showing that the equation corresponded to the division of a given angle into forty-five equal parts; specifically, it represented the equation $2 \cos 45X = C$ expanded in terms of $x = 2 \cos X$). Viete systematized his solution in the treatise *Ad problema Adriani Romani responsum* (*Opera*, ed. van Schooten, 1646, pp. 305-324). Fermat here notes that Viète's technique of solution is limited not only to the class of higher-order equations comprising the expansions of $2 \cos nX$ for $n = 1, 2, 3, \ldots$, but also to the members of that class for which the constant term C is less than or equal to 2. He attempts to set out procedures for solving those equations in cases where the constant is greater than 2. The paper is dedicated to Huygens; indeed, it invites his participation in research aimed at the solution of higher-degree equations. Hence, it dates from sometime after 1656, when the two men opened correspondence. FO.I.191, n. 1, suggests that the memoir was sent to Huygens in 1661, but that date seems too late in light of the fact that they had largely ceased active correspondence by then; 1659-1661 seems the closest limit one can set on the date.

⟨*De cissoide fragmentum*⟩. FO.I.285-288, from *MS* Leiden, *MSS* Huygens, 30. According to FO.I.285, n. 1, the text in that *MS* follows a letter from Carcavi to Huygens dated 1.I.1662 and bears the note, "De M. Carcavi, qui l'avoit de M. de Fermat," along with Huygens' remark, "J'ai démontré cette proposition quatre ans auparavant." The fragment contains a quadrature of the cissoid that is essentially unconnected with the systematic procedures of the *Treatise on Quadrature*. Nonetheless, one may assume that it is roughly contemporaneous with the *Treatise*, and that Carcavi forwarded it to Huygens soon after receiving it himself.

1661

Analysis ad refractiones (Appendix I). FO.I.170-172, from *Lettres de Descartes* (ed. Clerselier, 1667). According to FO.I.170, n. 1, the text accompanied a letter to Cureau de la Chambre dated 1.I.1662. Since Fermat's correspondence with Clerselier during 1659 gives no indication of the new algebraic approach to the problem of refraction, it seems clear that Fermat sent it to Cureau shortly after conceiving it and setting it down.

1662

⟨*Synthesis ad refractiones*⟩ (Appendix I). FO.I.173-179, from *Lettres de Descartes* (ed. Clerselier, 1667). Promised to Cureau in the letter of 1.I.1662, it is referred to in a letter from Clerselier to Fermat dated 20.V.1662. According to Clerselier's copy, the text was sent to Cureau in February 1662.

INDEX

Académie des sciences, 53, 366, 411
Acta eruditorum, 366
adequality, 163–165, 186–191, 201, 210–
 213, 216, 241–242, 245–254, 256, 266,
 268, 271, 321; "Archimedean adequation," 250
Alberti, Leon Battista, 12, 73
Aleaume, Jacques, 27
Aleaume, Pierre, 27
algebra, and analysis, 12, 28–34, 41–45, 69
 (*see* algebraic analysis); Arabic, 4; Babylonian, 33, 156; Greek geometrical, 32,
 33, 99, 100; cossist, 4–8, 11, 32, 34–36,
 39, 156, 158; fundamental theorem of,
 150–152; symbolic, 28, 36–38, 149; and
 trigonometry, 38. *See also* algebraic analysis, application of areas, theory of
 equations
aliquot parts, 54, 286, 288–293, 314, 337,
 341, 344
analysis, and algebra, 28, 29, 31–34, 41,
 44, 45; algebraic, 12, 28, 38–41, 44, 47,
 69, 70, 75, 77, 78, 80, 92–94, 96, 98,
 152, 195, 243, 264, 278, 279, 283, 363–
 366; art of, 28, 34, 41, 51, 77, 78, 94,
 131, 215, 216, 340; definition of, 29; Diophantine, 332; exegetic, 28, 29, 34; field
 of, 30, 40, 47, 64, 75; Greek geometrical,
 28, 29, 31, 47, 96, 113, 123, 153; logic
 of, 29, 30; poristic, 28, 29, 34; problem-
 solving, 30, 33; reduction, 30, 37, 38, 59,
 83, 84, 87–90, 99, 101, 103, 122, 126,
 129, 131–140, 216, 245, 254–266, 273,
 287; restoration of, 56, 64, 74, 75, 94,
 96, 98, 115, 119, 285; rhetic, 28, 29;
 and synthesis, 31, 46, 85, 127, 268;
 theorem-proving, 30; zetetic, 28, 29, 34,
 37, 46
analytic program, 26, 93, 115, 119, 123,
 124, 125, 216, 339–340, 363, 366–367
anastrophe, 126, 133, 148
Anderson, Alexander, 27, 37, 38, 148
application of areas, 33, 113, 114
Apollonius of Perga, 27, 32, 49, 51, 124,
 152, 235, 268, 340; *Conics,* 8, 30, 33,

39, 41, 46, 73, 74, 85, 86, 91, 92, 113–
 116, 120; *Contacts,* 30, 40, 94, 419; *Determinate Section,* 30, 51, 152, 215; *Inclinations,* 30, 40; *Plane Loci,* 30, 40, 48,
 51, 53, 68, 77, 79, 91–93, 94–112, 161,
 174, 176, 214; *Section of Area,* 30, 40;
 Section of Ratio, 30, 40
Arbogast, Louis-François-Antoine, 412, 413,
 415
Archimedes, 3, 4, 29, 30, 32, 46, 100, 216,
 217, 243, 266, 267, 268, 335, 340, 346–
 347; *Conoids and Spheroids,* 120, 241;
 Equilibrium of Planes, 8, 360, 366; *Floating Bodies,* 3, 8, 372, 373; *Measurement
 of Circle,* 269; *Quadrature of Parabola,* 3,
 26, 220, 222, 235; *Sphere and Cylinder,*
 27, 33, 125, 127, 154, 209, 269; *Spirals,*
 26, 27, 52, 73, 218, 222–225, 228, 229,
 244, 267; statics, 110, 373, 374, 376–
 378, 380, 393
Archytas of Tarentum, 42, 125
Ariosto, Ludovico, *Orlando furioso,* 27
Aristaeus, 30, 40, 77, 96
Aristarchus, 3
Aristotle, 130, 268, 370, 383; *Posterior Analytics,* 46; paradox of wheel, 74
arithmetic, "Hindu-Arabic," 5; and geometry,
 9; of integers, 283, 284, 302, 338–339.
 See also number theory
Arsenal of Venice, 8
ars rei et census, 4; *see* algebra, cossist
arte della cosa, 4; *see* algebra, cossist
artists, 11, 12, 73
Ashburnham, Lord, 413–416
astrology, 11, 14
astronomy, 14
asymptote, 364
Augeard, 18, 19, 22, 23
Autolycus of Pitane, 3

Bachet, Claude Gaspar, sieur de Méziriac,
 163, 230, 307–310, 333, 328, 331, 351;
 ed. and comm. *Arithmetica* of Diophantus,
 3, 229, 286, 307, 315–318, 320, 321,
 411

Index